Citrus Production Manual

Louise Ferguson | Elizabeth E. Grafton-Cardwell

Technical Editors

University of California
Agriculture and Natural Resources

Oakland, California
Publication 3539

To order or obtain ANR publications and other products, visit the ANR Communication Services online catalog at http://anrcatalog.ucdavis.edu or phone 1-800-994-8849. You can also place orders by mail or FAX, or request a printed catalog of our products from

University of California
Agriculture and Natural Resources
Communication Services
1301 S. 46th Street
Building 478 - MC 3580
Richmond, CA 94804-4600
Telephone 1-800-994-8849 • 510-665-2195
FAX 510-665-3427
E-mail: anrcatalog@ucanr.edu

Publication 3539
ISBN-13: 978-1-60107-840-7

Library of Congress Cataloging-in-Publication Data

Citrus production manual / Louise Ferguson, Elizabeth E. Grafton-Cardwell, technical editors. -- First edition.

pages cm. -- (Publication ; 3539)

Includes bibliographical references and index.

ISBN 978-1-60107-840-7

1. Citrus--California. 2. Citrus--Diseases and pests--Integrated control--California. I. Ferguson, Louise. II. Grafton-Cardwell, Elizabeth Elliot. III. University of California (System). Division of Agriculture and Natural Resources. IV. Series: Publication (University of California (System). Division of Agriculture and Natural Resources) ; 3539.

SB369.2.C2C58 2014

634'.30409794--dc23

2014001433

Editing: Stephen Barnett. Design: Robin Walton. Illustrations: Will Suckow, except as noted in the captions. Proofreading: Hazel White. Indexing: Sherry L. Smith. Photo Research: Evett Kilmartin. Editorial assistance: Sueanne Johnson. Print coordination: Ann Senuta. Photo credits are given in the captions. Uncredited citrus photos courtesy the Citrus Variety Collection, University of California, Riverside. Cover photo: Elëna Zhukova.

UC PEER REVIEWED This publication has been anonymously peer reviewed for technical accuracy by University of California scientists and other qualified professionals. This review process was managed by ANR Associate Editors Larry Bettiga, Ben Faber, David Haviland, and Joe Nuñez.

Printed in Canada on recycled, acid-free, paper.

5m-pr-3/14-SB/RW

PRECAUTIONS FOR USING PESTICIDES

Pesticides are poisonous and must be used with caution. READ THE LABEL CAREFULLY BEFORE OPENING A PESTICIDE CONTAINER. Follow all label precautions and directions, including requirements for protective equipment. Use a pesticide only on crops specified on the label. Apply pesticides at the rates specified on the label or at lower rates if suggested in this publication. In California, all agricultural uses of pesticides must be reported. Contact your county agricultural commissioner for details. Laws, regulations, and information concerning pesticides change frequently, so be sure the publication you are using is up to date.

LEGAL RESPONSIBILITY. The user is legally responsible for any damage due to misuse of pesticides. Responsibility extends to effects caused by drift, runoff, or residues.

TRANSPORTATION. Do not ship or carry pesticides together with foods or feeds in a way that allows contamination of the edible items. Never transport pesticides in a closed passenger vehicle or in a closed cab.

STORAGE. Keep pesticides in original containers until used. Store them in a locked cabinet, building, or fenced area where they are not accessible to children, unauthorized persons, pets, or livestock. DO NOT store pesticides with foods, feeds, fertilizers, or other materials that may become contaminated by the pesticides.

CONTAINER DISPOSAL. Dispose of empty containers carefully. Never reuse them. Make sure empty containers are not accessible to children or animals. Never dispose of containers where they may contaminate water supplies or natural waterways. Consult your county agricultural commissioner for correct procedures for handling and disposal of large quantities of empty containers.

PROTECTION OF NONPEST ANIMALS AND PLANTS. Many pesticides are toxic to useful or desirable animals, including honey bees, natural enemies, fish, domestic animals, and birds. Crops and other plants may also be damaged by misapplied pesticides. Take precautions to protect nonpest species from direct exposure to pesticides and from contamination due to drift, runoff, or residues. Certain rodenticides may pose a special hazard to animals that eat poisoned rodents.

POSTING TREATED FIELDS. For some materials, reentry intervals are established to protect field workers. Keep workers out of the field for the required time after application and, when required by regulations, post the treated areas with signs indicating the safe reentry date.

HARVEST INTERVALS. Some materials or rates cannot be used in certain crops within a specific time before harvest. Follow pesticide label instructions and allow the required time between application and harvest.

PERMIT REQUIREMENTS. Many pesticides require a permit from the county agricultural commissioner before possession or use. When such materials are recommended in this publication, they are marked with an asterisk (*).

PROCESSED CROPS. Some processors will not accept a crop treated with certain chemicals. If your crop is going to a processor, be sure to check with the processor before applying a pesticide.

CROP INJURY. Certain chemicals may cause injury to crops (phytotoxicity) under certain conditions. Always consult the label for limitations. Before applying any pesticide, take into account the stage of plant development, the soil type and condition, the temperature, moisture, and wind direction. Injury may also result from the use of incompatible materials.

PERSONAL SAFETY. Follow label directions carefully. Avoid splashing, spilling, leaks, spray drift, and contamination of clothing. NEVER eat, smoke, drink, or chew while using pesticides. Provide for emergency medical care IN ADVANCE as required by regulation.

WARNING ON THE USE OF CHEMICALS

Pesticides are poisonous. Always read and carefully follow all precautions and safety recommendations given on the container label. Store all chemicals in their original labeled containers in a locked cabinet or shed, away from foods or feeds, and out of the reach of children, unauthorized persons, pets, and livestock.

Recommendations are based on the best information currently available, and treatments based on them should not leave residues exceeding the tolerance established for any particular chemical. Confine chemicals to the area being treated. THE GROWER IS LEGALLY RESPONSIBLE for residues on the grower's crops as well as for problems caused by drift from the grower's property to other properties or crops.

Consult your county agricultural commissioner for correct methods of disposing of leftover spray materials and empty containers. **Never burn pesticide containers.**

PHYTOTOXICITY: Certain chemicals may cause plant injury if used at the wrong stage of plant development or when temperatures are too high. Injury may also result from excessive amounts or the wrong formulation or from mixing incompatible materials. Inert ingredients, such as wetters, spreaders, emulsifiers, diluents, and solvents, can cause plant injury. Since formulations are often changed by manufacturers, it is possible that plant injury may occur, even though no injury was noted in previous seasons.

CONTAINER DISPOSAL. Dispose of empty containers carefully. Never reuse them. Make sure empty containers are not accessible to children or animals. Never dispose of containers where they may contaminate water supplies or natural waterways. Consult your county agricultural commissioner for correct procedures for handling and disposal of large quantities of empty containers.

PROTECTION OF NONPEST ANIMALS AND PLANTS. Many pesticides are toxic to useful or desirable animals, including honey bees, natural enemies, fish, domestic animals, and birds. Crops and other plants may also be damaged by misapplied pesticides. Take precautions to protect nonpest species from direct exposure to pesticides and from contamination due to drift, runoff, or residues. Certain rodenticides may pose a special hazard to animals that eat poisoned rodents.

Contents

Preface and Acknowledgments

This is the first comprehensive manual on citrus production produced by the University of California Agriculture and Natural Resources. The manual provides the most current information on all phases of fresh citrus production in arid climates. The text is largely nontechnical and is intended for field application; it includes a glossary of technical terminology and references for those interested in more detail. This manual is meant to be used in conjunction with *Integrated Pest Management for Citrus,* Third Edition (University of California Agriculture and Natural Resources Publication 3303), and the *UC IPM Citrus Pest Management Guidelines* (UC IPM website, http://www.ipm.ucdavis.edu), which provide more detailed information on pest and disease management.

Forty authors contributed to this publication. Information was drawn from the experts from the University of California, the U. S. Department of Agriculture (USDA), and the citrus industry. The authors include UC Cooperative Extension specialists and advisors, research assistants, UC faculty from Davis and Riverside, USDA scientists, emeritus advisors, and highly skilled experts working in the citrus industry. The efforts and dedication of these authors are greatly appreciated. Appreciation is also extended to the citrus growers who reviewed early versions of the chapters and the scientists who participated in the confidential peer review process.

Appreciation is also extended to the Agriculture and Natural Resources Communication Services staff, especially Ann Senuta, Stephen Barnett, and associate editors Ben Faber, Joe Nunez, and Mary Louise Flint.

We would particularly like to recognize one of our authors, Dr. David J. Gumpf, pathologist and major contributor, who passed away before the completion of this manual.

*~**Beth Grafton-Cardwell and Louise Ferguson***

Authors

Technical Editors

Louise Ferguson
Cooperative Extension Pomologist, Department of Plant Sciences, University of California, Davis

Elizabeth E. Grafton-Cardwell
Cooperative Extension IPM Specialist and Research Entomologist, Department of Entomology, University of California, Riverside, and Director, Lindcove Research and Extension Center

Contributing Authors

James E. Adaskaveg
Professor, Department of Plant Pathology and Microbiology, University of California, Riverside

Mary Lu Arpaia
Cooperative Extension Subtropical Horticulturalist, Department of Botany and Plant Sciences, University of California, Riverside

Richard Barker
Director, Citrus Roots Foundation

J. Ole Becker
Cooperative Extension Specialist and Nematologist, Department of Nematology, University of California, Riverside

Gary S. Bender
UC Cooperative Extension Farm Advisor, San Diego County

Patrick Brown
Cooperative Extension Pomologist and Professor, Department of Plant Sciences, University of California, Davis

Edwin L. Civerolo
USDA-ARS, San Joaquin Valley Agricultural Sciences Center, Parlier, Retired

Charles W. Coggins Jr.
Professor Emeritus of Plant Physiology, Department of Botany and Plant Sciences, University of California, Riverside

Donald A. Cooksey
Cooperative Extension Professor and Bacteriologist, Department of Plant Pathology, University of California, Riverside

Joseph M. DiTomaso
Cooperative Extension Weed Specialist, Department of Plant Sciences, University of California, Davis

Reza Ehsani
Associate Professor, Citrus Research and Education Center, University of Florida

Robert Elliott
Director, Food Safety, Sunkist Growers, Fontana, CA

Ben Faber
UC Cooperative Extension Farm Advisor, Ventura County

Louise Ferguson
Cooperative Extension Pomologist, Department of Plant Sciences, University of California, Davis

Helga Förster
Project Scientist, Department of Plant Pathology and Microbiology, University of California, Riverside

Allan Fulton
UC Cooperative Extension Irrigation and Water Resources Advisor, Tehama County

David A. Goldhamer
Cooperative Extension Water Management Specialist, Department of Land, Air, and Water Resources, University of California, Davis, Retired

Elizabeth E. Grafton-Cardwell
Cooperative Extension IPM Specialist and Research Entomologist, Department of Entomology, University of California, Riverside, and Director, Lindcove Research and Extension Center

David A. Grantz
Cooperative Extension Specialist and Plant Physiologist, Department of Botany and Plant Sciences, University of California, Riverside

David J. Gumpf
Former Professor, Department of Plant Pathology, University of California, Riverside

David H. Headrick
Horticulture and Crop Science Department, California Polytechnic State University, San Luis Obispo

Tracy L. Kahn
Cooperative Extension Principal Museum Scientist, Department of Botany and Plant Sciences, University of California, Riverside

Craig E. Kallsen
UC Cooperative Extension Farm Advisor, Kern County

Robert R. Krueger
Horticulturist, USDA-ARS National Clonal Germplasm Repository for Citrus and Dates, University of California, Riverside

Carol J. Lovatt
Professor of Plant Physiology and Plant Physiologist, Department of Botany and Plant Sciences, University of California, Riverside

Peggy A. Mauk
Cooperative Extension Subtropical Horticulture Specialist, Department of Botany and Plant Sciences, University of California, Riverside

Joseph G. Morse
Professor of Entomology, Department of Entomology, University of California, Riverside

Neil V. O'Connell
UC Cooperative Extension Farm Advisor, Tulare County

MaryLou Polek
Plant Pathologist, California Citrus Research Board

Mikeal L. Roose
Professor, Department of Botany and Plant Sciences, University of California, Riverside

Blake Sanden
UC Cooperative Extension Irrigation and Agronomy Farm Advisor, Kern County

Lawrence J. Schwankl
UC Cooperative Extension Irrigation Specialist, Kearney Agricultural Research and Extension Center

Joseph Smilanick
Research Plant Pathologist, USDA ARS, San Joaquin Valley Agricultural Sciences Center, Parlier, California

Richard L. Snyder
Cooperative Extension Biometerology Specialist, Department of Land, Air, and Water Resources, University of California, Davis

Timothy M. Spann
Assistant Professor, Department of Agricultural and Horticultural Sciences, University of Florida

Scott Steinmaus
Professor of Biological Sciences Department, California Polytechnic State University, San Luis Obispo

Etaferahu Takele
UC Cooperative Extension Area Advisor, Farm Management/Agricultural Economics, Southern California

Georgios Vidalakis
Cooperative Extension Specialist and Plant Pathologist, Department of Plant Pathology and Microbiology, University of California, Riverside

Herbert John Webber
Former Director, University of California Citrus Experiment Station

Laosheng Wu
Cooperative Extension Water Management Specialist and Professor, Department of Environmental Sciences, University of California, Riverside

Part 1 California Citrus Industry Overview

1 History and Development of the California Citrus Industry

Herbert John Webber, Richard Barker, and Louise Ferguson

Introduction of Citrus into California

No definite records exist indicating when the first citrus seed or trees were planted in California. However, it appears that disputes among the Jesuit, Franciscan, and Dominican orders led to the beginning of California's citrus industry. In 1767, the Jesuits were expelled from their missions in Baja, or Lower, California, and their holdings were given to the Franciscans. A subsequent dispute between the Franciscans and Dominicans resulted in a segment of the Franciscans departing to develop missions in Alta, or Upper, California. In 1769, under the leadership of Father Junípero Serra, they entered what is now California and founded the first of their missions at San Diego. Twenty-one missions were ultimately established in a chain along the California coast as far north as San Rafael. Because the missions had to be self-sufficient, most established gardens and orchards. The Franciscans of Alta California got their initial supplies of seed and plants from the Baja California missions. Thus, their first gardens and orchards contained the same trees cultivated in the earlier settlements.

Dated records established that oranges and lemons were cultivated in the mission gardens of Baja California prior to 1739. By the time the Franciscan expedition, which established the first California mission at San Diego in 1769, departed from the missions of Baja California, oranges and lemons must have been fruiting there, and it is very probable that orange and lemon seed or plants were taken along with grapes, olives, and other important products for propagation in the prospective new settlements. Thus 1769, the date of the establishment of the first mission at San Diego, is probably also when citrus was introduced into California.

The first orange orchard of any size in California was planted in the garden of San Gabriel Mission in 1804. By the time Los Angeles was settled in 1831, the trees had been bearing fruit for 25 years. Interestingly, while there are accounts of pears, peaches, nectarines, apples, and pomegranates being cultivated near Los Angeles, most citrus orchards apparently did not become established outside the missions until after secularization in 1833. Apparently, the mission fathers prized citrus highly. In 1877, Jose del Carmen Lugo of the San Bernardino rancho recalled that owners of fields could not obtain seed of oranges and lemons from the missions because the fathers "refused to allow these fruit to be raised elsewhere than at their missions." Jean Louis Vignes, a Frenchman, planted the second orange orchard in California in 1834. He procured 35 large seedling sweet orange trees from Mission San Gabriel, which he transplanted in an orchard on Aliso Street in Los Angeles for his personal use. By this time, planting citrus trees in gardens and courts for home use had become a common practice, but no orchards had been planted to provide fruit for sale.

California's First Commercial Citrus Grower: William Wolfskill

William Wolfskill (fig. 1.1), a Kentucky trapper of German descent who came overland to Los Angeles in 1831, planted the first truly commercial citrus orchard in California. His first trees, seedling sweet oranges, were obtained from Mission San Gabriel and were set out in 1841 on a small tract on the site occupied later by the Southern Pacific Railroad's Arcade Station, currently bordered by South Central Avenue and South Alameda Street between East 4th and East 6th Streets. Wolfskill's neighbors ridiculed him and

Figure 1.1 The only known photograph of William Wolfskill (1802–1866), taken by Henri Penelon, Los Angeles, September 27, 1866.

his idea of growing oranges for sale, but with tenacity and patience he maintained his trees and gradually extended his plantings. The orchard increased to approximately 70 acres. The last crop to be disposed of in Wolfskill's lifetime, from about 28 acres of his orchard, sold on the trees for $25,000.

Influence of the Gold Rush on Citrus Industry Development

After the planting of the Wolfskill's Mission San Gabriel orchard, the next great stimulus to the citrus industry came with the ceding of California to the United States. This transaction followed the Mexican War in 1846, the conquest of California by the United States, and was ratified by treaty in 1848. Almost immediately after came the discovery of gold in the new territory, followed by the great gold rush of 1849, which vastly increased the population of California and created a nearby lucrative market for all the fruit the existing orchards could produce. This was the real birth of the commercial citrus industry in California. The fruit could be shipped by ocean freight to San Francisco and then by riverboat up the Sacramento, American, and Feather Rivers to points near the mines. San Francisco became the great market for the industry and remained so for three decades.

Apparently, however, there was no great haste to take advantage of this opportunity. It required too long to produce a bearing orchard, and the people then populating California were seeking wealth through gold metal, not golden fruit. The reaction to these new markets was slow, but by providing a potential market, the stimulation that was needed to expand the industry was created and a few new orchards were planted. In 1852, Don Benito Wilson purchased the small orchard that thereafter was known as the Don Benito Wilson Grove and began to extend the plantings. In 1853, Mathew Keller began planting an orchard opposite the Wolfskill orchard with seedlings grown from fruit obtained from Central America and Hawaii. In 1857, L. Van Luven planted a small orchard in Old San Bernardino with seedlings grown by him and others from Los Angeles. In 1869, L. F. Cram set out a small orchard of some 200 seedling trees near Highland. These last two orchards were important as they were the first in the interior valleys, though some claim that orange trees were grown at Old San Bernardino in the mission period. By 1865, when Myron H. Crafts set out a small orchard of some 200 trees at Crafton, the citrus industry in the interior valleys had become reasonably well established. Four years later, F. A. Kimball planted a small orchard of orange, lemon, and lime trees at National City in San Diego County, the first orchard planting in that section of the state. In 1867, according to the U.S. Department of Agriculture, there were 17,000 orange trees and 3,700 lemon trees in California, of which 15,000 orange trees and 2,300 lemon trees were in the Los Angeles region.

Extension of Citrus into Northern California

Simultaneously, citrus growing had been developing in the northern part of the state. In 1856, Judge Joseph Lewis purchased three sweet orange seedlings from Jesse Morrill in Sacramento. These were planted in the vicinity of Oroville in Butte County. One of these trees was planted at the west end of the famous suspension bridge at Bidwell's Bar, and it is still living and bearing large annual crops. It has come to be known as the Mother Orange Tree and is now the oldest and largest living orange tree in California. In 1965, this tree was moved to a new site overlooking the Oroville Dam. Measurements made by an official committee on November 27, 1926, gave the size of the tree as follows: height, 33 feet 6 inches; spread of top, 31 feet 5 inches; circumference of trunk 1 foot from the ground, 66 inches. The success of these trees soon led to the planting of other trees in the vicinity.

The visiting committee of the State Agricultural Society stated in 1858 that they had found "oranges" and "citrons" in the ornamental garden of the venerable pioneer General John A. Sutter on the bank of the Feather River, and they also found orange trees at Marysville that had endured the winters for 2 years. In the *Transactions* of the Society for 1872, the variety committee reported that they had found bearing orange and lemon trees in the garden of Judge Sexton at Oroville; a 12-year-old orange tree in the garden of Mr. Glaucauf, also in Oroville, which had produced a crop of 400 fine oranges during the year; and the 16-year-old Mother Orange Tree at Bidwell's Bar, which had borne 1,500 oranges that year. The following statement by Isaac N. Hoag, associate editor of the *Sacramento Record-Union*, in 1879 shows the extent of citrus growing in northern California in the late 1870s: "Marysville, Sacramento, and many other cities and towns from San Diego to Red Bluff have large numbers of orange trees now in bearing. Contrary to general expectation the orange ripens from two weeks to one and one-half months earlier in nearly every locality north of San Francisco than in Los Angeles." In 1862, H. M. White planted two orange trees in Frasier Valley east of Porterville, Tulare County, which later formed the nucleus of a 40-acre orchard; the first orchard in this area was the A. R. Henry orchard at Porterville, which was set out in 1883.

By 1870 many individual trees and small orchards had been planted in all the principal citrus-growing regions of California, and the commercial industry, one of the most important horticultural industries in the state at that time, could be considered well established. Before this time citrus had been grown for home use only. Fresh fruit were scarce and expensive; even a large share of the oranges consumed in the state were being imported. In 1866, San Francisco, then the largest city in California, imported some 3 million oranges from Mexico and the islands of the Pacific, whereas only 250,000 were received from the vicinity of Los Angeles. The rapid extension of fruit production in the late 1860s, with many new orchards coming into bearing, soon resulted in competition that led to lowered prices. In 1862, there were about 25,000 orange trees in the state, but by 1882 there were over half a million. However, due to competition for markets within the state the citrus industry began to decline sharply.

Introduction of the Washington Navel Orange

In 1870, Judge J. W. North founded Riverside, the southern California community that became famous in the early history of citrus culture: the first citrus seed and trees were planted there in 1871. The development of orange, lemon, and lime orchards in Riverside was very rapid between 1871 and 1880. However, it was the 1873 introduction of the Washington navel orange to Riverside that made it a famous citrus-producing region. The U.S. Department of Agriculture had obtained cuttings of this navel orange, a mutation from an orange tree in a monastery garden in Bahia, Brazil. Mr. William Saunders, superintendent of gardens for the USDA, sent two or three starter trees to spiritualist and suffragette Eliza Tibbets in Riverside in 1873. The newly introduced navel orange was far superior to other varieties at the time because it was seedless, sweet, and ripened in winter under California's Mediterranean climate. The navel orange also changed how farmers produce citrus and other fruit trees. Prior to the introduction of the Washington navel, citrus was grown mostly from seed, which meant the trees retained their individual biological diversity, bearing markedly different fruit that lacked standardization.

This tree produced the fruit that became the basis of the commercial California citrus industry, established Riverside as California's first leader in citrus production, and propelled the growth of California citrus, which in turn fueled the economic and social development of California. One of the original three trees planted in Riverside (fig. 1.2) is still bearing fruit and has been designated a California historical landmark. The tree produced its first fruit in 1878. Within a few years the variety had become so famous that citrus culture in California was firmly established. A census of the citrus industry of Riverside in 1880 counted 17,038 orange trees, 3,199 lime trees, and 2,480 lemon trees. It was thought at the time that the lime was likely to be more successful as a commercial fruit than the lemon, and limes were planted in all the citrus-growing regions of the state. During this period Riverside contained the largest citrus plantings in the state, other than those immediately around Los Angeles.

The citrus trees planted during this period were largely seedlings grown from seed taken from local trees, mainly from the orchard of Mission San Gabriel, or from special fruit shipped in from Mexico, Hawaii, and Tahiti. Considerable effort was expended to obtain good varieties. Thomas A. Garey, who established a nursery at Los Angeles in 1865 and who was one of the most outstanding of the pioneer nurserymen of California, introduced a large number of varieties from 1868 to 1875. He is said to have received shipments of varieties and seed from Mexico, Central America, Australia, southern Europe, and Florida from such famous nurserymen as Elwanger and Barry of Rochester, New York, and Thomas Rivers of Sawbridgeworth, England. His most outstanding orange importations were the Mediterranean Sweet and the Paperrind St. Michael, both extensively planted in the early orchards. He

Figure 1.2 California's original navel orange tree, started from cuttings imported from Brazil and planted by Eliza Tibbets of Riverside. This tree was the source of the navel oranges that spurred the development of the California citrus industry and established Riverside as California's first major commercial citrus production area. *Photo:* UC ANR.

also named and introduced the Eureka lemon, which was produced from an exceptionally good seedling grown by C. R. Workman, a Los Angeles pioneer orange grower. These early cultivars, and later hundreds more cultivars and rootstocks, are now preserved at the University of California Riverside Citrus Collection, http://www.citrusvariety.ucr.edu/. A history of this collection, including information about the founders, H. J. Webber and W. P. Bitters, and scientists C. N. Roistacher, J. W. Wallace, and E. C. Calavan, can be found at the Citrus Variety Collection website. Roistacher later used the collection to develop the practices and regulations that now ensure the quality of California's citrus nurseries.

The Completion of Transcontinental Railways

The third great impetus to the extension of the citrus industry in California came in the late 1870s and early 1880s with the completion of three transcontinental railways: the Southern Pacific Valley Line in 1876, making connections with the Central Pacific and Union Pacific to the East; the southern line of the Southern Pacific to New Orleans in 1881; and the Santa Fe in 1885. These railways provided competing carriers for eastern shipments and had an immediate effect on the industry. The first carload of fruit to be shipped east from California was sent from the famous Wolfskill orchard in Los Angeles to St. Louis in 1877. The fruit arrived in good condition, though it was a month in transit. The freight charge on this car of 300 boxes is said to have been $500. In 1881, as competition among emerging rail companies increased, the Southern Pacific Railroad cut the rate on carload lots from Los Angeles to Chicago from $650 to $350, with similar reductions to intermediate points. The first carload shipment from any point in the state other than Los Angeles was probably when G. W. Garcelon and A. J. Twogood of Riverside sent a mixed car of oranges and lemons to Denver from San Bernardino in 1882. According to W. A. Spalding, the first special train loaded exclusively with oranges left Los Angeles February 14, 1866, on the Southern Pacific and Union Pacific railways. By the following year, ventilated boxcars were carrying oranges across the continent to New York. The refrigerated boxcar made its appearance in 1889. A. T. Hatch, a pioneer in refrigeration, reported that merchants at first viewed the fruit dubiously and commented that it would likely decay upon being removed from refrigeration. In 1892, five carloads of fruit were shipped by the California Fruit Transportation Co. to New York and transferred to steamers for a 14-day voyage to Liverpool and London. Great crowds attended the fruit sale in London, and out of the first shipment some samples were sent to Queen Victoria, who found them to be acceptable.

The Citrus Industry Organizes

As acreage expanded and citrus became the new economic basis of California, individual growers began to realize that they would have more influence over their net returns if they organized themselves into a marketing force. Individually they were at the mercy of the wholesale agents and the railroads. In 1893, to better control their marketing, the citrus growers formed the Southern California Fruit Exchange, now known as Sunkist. In 1904 Tulare County joined, and the name, to reflect the entire state, became the California Fruit Growers Exchange. The Exchange was the first agricultural industry in the United States to market a fresh commodity with a major advertising campaign. They earmarked their annual budget as not to exceed $250 annually for advertising in England and Europe.

However, it was the Exchange's early cooperative domestic marketing program with the Southern Pacific Railroad that indelibly linked California and its oranges with their respective reputations for health and wealth. In the beginning of the twentieth century, as the number of orchards and rail transportation increased, supplies of oranges and lemons quickly saturated the undeveloped domestic markets. Price undercutting was rampant, and growers began abandoning or removing orchards. E. O. McCormick, president of the Southern Pacific Railway, needing stable long-haul traffic to sustain his railroad, approached his friend Francis Q. Story, president of the Exchange, and committed to match an advertising campaign up to $10,000. The objective was increasing consumption of fresh oranges in the eastern states. Interestingly, the offer almost died within the Exchange's board when some members argued that the massive advertising would also benefit the 50% of the industry who were not members of the Exchange, an argument that persists within California's commodity groups to this day. The advertising campaign kicked off in March 1908 in Des Moines, Iowa. The $7,000 campaign increased domestic consumption in 1907 by 50% in Iowa and 17.7% nationwide. In response, the Exchange committed to, and the Southern Pacific company matched, a $25,000 advertising budget in 1908. The resulting campaign was not only the first perishable food advertising campaign in the United States but also the model for future market saturation campaigns. The California orange quickly became a symbol of California, the land of sunshine and opportunity. Orange sales soared, as did migration to California. This migration was greatly enhanced by the Southern Pacific Company's "colonist" fare to California. From 1900 to 1916 over 795,000 people, heeding the admonition "Oranges for Health—California for Wealth," paid $25 for a one-way ticket to California.

The advertising agency suggested the Exchange needed a trademark for advertising. They suggested Sunkissed, which became Sunkist in 1909. The Exchange, still sensitive competitors, further distinguished their best product with individual fruit tissue wrappers, Sunkist for premium oranges and Red Ball for next best. To ensure that customers purchased Sunkist fruit, these wrappers were exchangeable for Sunkist Orange Blossom silverware. A 1915 *Ladies Home Journal* advertisement for the silverware exchange is shown in figure 1.3. From 1910 though 1917 the Exchange was the largest silverware purchaser in the world, and the silverware became a constant reminder of California oranges. In 1911, at the conclusion of the 3-year advertising campaign, the Exchange and Southern Pacific Railroad were each spending $100,000 annually. The price of advertising was never again thought to be excessive.

This advertising campaign, while precipitating great changes in the citrus and railroad industries, also generated other industries. By 1926 the Exchange had 12,000 growers in 230 local packinghouse associations that packed from 190,000 acres, 75% of California's citrus. Marketing this volume required the valued tissue wrappers and wooden crates. To meet these needs the Exchange added another division in 1907, Fruit Growers Supply. The most immediate need was wood for the crates, called shooks. The Exchange purchased lumber mills, box factories, logging railroads, and timberlands through Fruit Growers Supply to supply the wood for the crates. The crates from the different citrus regions and packinghouses begged for distinctive labels. The resulting citrus crate labels, which combine the history of California and the citrus and printing industries, are a California story of its own. These labels are now collected for printing technique, historical, and esthetic value (see the Citrus Label Society website, http://www.citruslabelsociety.com/).

Figure 1.3 An advertisement in the *Ladies Home Journal* in 1915, offering Orange Blossom silverware in exchange for the tissue paper wrappers on Sunkist oranges. Courtesy Sunkist Growers, Inc.

A second set of industries precipitated by the California's fresh citrus industry were industries that used fruit that could not be sold as fresh. In 1915 the Corona lemon growers established the Exchange By-Products Company to produce citric acid, lemon oil, and citrus pectin. That same year, the Exchange introduced orange juice with a "Drink an Orange" campaign; rapidly expanding glass companies were soon producing millions of glass reamers. By 1922 a newly established orange by-product organization was shipping orange oil to Canada and England. In 1922, Sunkist, in another advertising first, promoted vitamin C in fresh and processed citrus. Shortly thereafter the Sunkist Electric Fruit Juice Extractor for oranges and lemons was introduced in 60,000 soda fountains. In the 1930s the Exchange shipped its first substantial volumes of canned orange juice. Orange juice became a diet staple rather than a luxury product. The grape, tomato, apple, prune, and other industries followed the citrus industry's lead in producing juices from fruit not acceptable for fresh sales.

As the science of citrus production and processing progressed, it became apparent that the fruit qualities that produce premier marketable fresh oranges or lemons are different from the qualities that produce economically profitable fruit for the juice industry. The science of maintaining fresh citrus fruit quality has markedly improved, allowing a higher percentage of fruit to be marketed as fresh. Because California, with its arid, sunny climate and day-to-night temperature changes, can produce a large, brightly colored fruit with high internal and external quality, it has remained primarily a fresh citrus producer. In contrast, the steadily hot and humid climates of Florida and Brazil allow for the production of excellent orange juice.

By the 1950s, 80% of California's citrus growers belonged to Sunkist Growers. Today, however, there are multiple cooperative citrus grower organizations in the state. All are dedicated to the same mission defined by the original California Fruit Growers Exchange: cooperation among the members for the benefit of the industry.

The Citrus Experiment Station

On July 2, 1862, President Abraham Lincoln signed into law the Morrill Land Grant Act, which granted federal lands to each state to establish colleges for the teaching of agriculture and allied arts. In that year he also signed the law that created the U.S. Department of Agriculture; the Homestead Act, which granted western lands to those who would farm; and the Pacific Railway Act, which cleared the way for the transcontinental railway. These federal laws signed in 1862 were crucial to the development of the California citrus industry because they led to the rail transportation of fruit and the migration of people to California, and they also led to the founding of the University of California and the USDA.

Just as the Exchange partnered with the Southern Pacific Railroad Company early in the twentieth century, the University of California also combined efforts with the railway in the common goal of developing agricultural commodities in the state. From 1875 to 1904, Eugene W. Hilgard was professor of agricultural chemistry at the University of California (in Berkeley) and director of the state Agricultural Experimental Station. To produce a soil map of California, Hilgard enlisted the aid of the Southern Pacific Company. The railroad provided Hilgard with data on soils, elevation, rainfall, temperature, water, and other features of the state, and the railroad used Hilgard's chemical analyses of soil and the maps he created to market their land. This cooperation led to local railroad depot agents distributing agricultural research bulletins published by the University. Just as the California Fruit Express advertised the California citrus industry, the railway now advertised the University of California's agricultural education programs. Southern Pacific also offered drastically reduced fares to farmers attending University of California programs.

In 1908 the University and the railway launched their most widespread cooperative effort, funded by the railway. Bringing farmers to the College of Agriculture at Berkeley was not practical in this era before automobiles. The University was convinced that it had to deliver "scientific agriculture" to growers. The University's first demonstration train (fig. 1.4), called the University on Wheels and renamed An Evangel Train, brought data-based scientific information translated into production practices to the state's farmers. When the University of California demonstration train visited citrus districts, G. Harold Powell, a USDA pomologist, gave presentations on citrus pests, spray methods, predatory insects, and new varieties of citrus. Encouraged by the overwhelming public response to the first year's efforts, the Southern Pacific continued at their own expense dispatching the train from 1909 through 1912.

Each year the touring University grew larger and more elaborate, added more rail miles, and attracted larger crowds. Richard J. Orsi noted, "By 1911–12 they traveled 4,000–5,000 miles (double the beginning year), made 238 stops and attracted 102,000 visitors. Criticism of the University of California, College of Agriculture diminished paving the way for more funding, and more buildings." Also,

Figure 1.4 The Evangel Train, so named because it delivered the gospel of agriculture to the farmers in cooperation with the Southern Pacific Railroad. Courtesy The Bancroft Library, University of California.

collaborating in this University on Wheels enabled both the University of California and Southern Pacific Railroad to shed their undeserved reputations as elitist, a reputation undeservedly accrued due to their size and the threat of change and industrialization they embodied.

In December 1912 the $175-million Southern California citrus industry experienced the worst freeze on record: 39% of the crop was destroyed. The Southern Pacific Railroad and University of California dispatched the "Frost Education Special" through Los Angeles, Riverside, and San Bernardino Counties to advise growers how to prune, irrigate, and fertilize to minimize long-term tree damage and also how to prepare for a future freeze. The schedule was publicized in local newspapers of the 24 cities within the citrus districts. At each stop University officials solicited the assembled growers and local leaders to sign a petition supporting pending legislation appropriating $385,000 for establishment of a University experimental station for citrus research. The California legislature responded with $185,000 for the Citrus Experiment Station. On December 14, 1914, the UC Regents approved Riverside as the new location. The community of Riverside celebrated by ringing the Mission Inn bells and blowing the electrical plant steam whistle for 15 minutes. The Riverside Daily Press opined that it was "the most important day that has occurred in all the history of Riverside." Figure 1.5 shows the University of California Citrus Experiment Station in 1930. This building is now the nucleus of the University of California, Riverside, where citrus research is still done.

The research, education of new scientists, and extension effort to the citrus industry started in the Citrus Experiment Station continue on multiple University of California campuses and in cooperation with citrus scientists worldwide. The California citrus industry continues to be closely involved with its academic partner, identifying problems and approaches to research, supporting research with industry funding, and cooperating in developing educational programs and guiding industry regulation. The many benefits to both the California citrus industry and the University of California are too numerous and specific to recount in this chapter. Every chapter in this University of California Cooperative Extension *Citrus Production Manual* is built on more than a century of research and testing by the University in cooperation with the California citrus industry.

Figure 1.5 The original Citrus Experiment Station in Riverside, founded with the support of California citrus growers. This building still stands as the nucleus of the campus that became the University of California, Riverside. Courtesy Citrus Roots Society.

The California Citrus Industry Today

As of 2011, California is the largest producer of fresh market citrus in the United States. California leads in navel production and also produces significant percentages of the nation's Valencia oranges, lemons, grapefruit, and tangerines. The following seven tables detail the development of the modern California commercial citrus industry. The USDA historical tables (tables 1.1 and 1.2) give the bearing acres, yield, production, unit value, and total value of navel and Valencia oranges from 1920 through 2008. Table 1.3 gives the estimated statewide bearing acreage as of 2010. Table 1.4 is a more detailed 2010 picture of citrus, dividing acreage by type and bearing status. Table 1.5 gives 2010 citrus acreage by type, county, and year planted. Table 1.6 gives 2010 citrus acreage by type, variety, and year planted. Tables 1.7 and 1.8 give the top 15 markets for California's fresh orange and lemon exports, which accounted for over 30% of the state's annual production in 2010.

Over 75% of the oranges produced in California are navel oranges. Valencia orange acreage continues to decline due to competition with summer stone fruit, later-ripening navels, and earlier and later mandarin cultivars. The volume of the oranges sold fresh ranges from 70 to 90% of total production. Total statewide average annual crop values have remained near $600 million for the past decade, with 89% from sales of fresh fruit. Per capita U.S. domestic consumption of fresh oranges declined from 15 pounds per person in the 1970s to 12 pounds per person in 2006, a decline of about 20%. However, from 2000 to 2010, international demand increased by 6% to represent over 30% of total production. Demand from Canada, Korea, and Japan, the major export markets, has decreased, while demand from Southeast Asia has increased.

California produces 90% of the nation's lemon crop. In the last decade, production has remained over 750 million tons, of which 66% is marketed fresh. Sales of fresh lemons account for over 99% of the average annual crop value of $300 million. Although the lemon harvest begins in the desert in August and moves up along the coast and into interior valleys later in the year, demand for lemons is strongest in the summer. There has been a steady increase in fresh lemon and lemon juice consumption in the last decade, from 2.7 to 3.3 pounds of fresh lemons and 0.14 to 0.16 gallons of juice per person. The increase in juice is a result of using lemon juice in soft drinks and juices. Imports of lemons account for only 8% of U.S. fresh lemon consumption, but juice imports account for over 50% of juice consumption.

California's grapefruit production is only 10% of Florida's and is also less than that of Texas, though cold weather and hurricanes in the past decade have decreased grapefruit production in both states. However, as with oranges and lemons, California grapefruit have a larger percentage of the fresh market, 17%, because they are marketed fresh year round, both before and after the Texas and Florida grapefruit. Over 88% of California's grapefruit are sold fresh; if processed as juice, they would produce a net negative return.

The biggest change in California's citrus industry in the past decade has been the increasing production of mandarins. While California has long had a small Satsuma mandarin industry, particularly in the Sacramento Valley, these newer plantings are primarily Clementine cultivars. Over the past decade, California Clementine acreage has more than doubled, from 16,000 to over 40,000 acres, with an average annual value of over $60 million. Most mandarins are marketed fresh and fill the harvest niche between Valencia oranges and other summer fruit and navels. As with all fresh citrus, researchers are seeking earlier and later varieties to expand the current marketing window.

References

Anonymous. 1858. Report of the visiting committee. Transactions of the California State Agricultural Society 1858:167–168.

———. 1872. Report of corresponding secretary. Transactions of the California State Agricultural Society 1872:178–204.

Brown, J., Jr., and J. Boyd. 1922. History of San Bernardino and Riverside Counties. Madison, WI: The Western History Association.

Butterfield, H. M. 1963. History of subtropical fruit and nuts in California. University of California Division of Agricultural Science.

California State Board of Horticulture. 1892. Annual report of the State Board of Horticulture of the state of California for 1892. Sacramento: State Board of Horticulture.

Clavijero, F. J. 1852. Historia de la antigua ó baja California. Mexico: J. R. Navarro.

Cobo, B. 1890–95. Historia del nuevo mundo … Escrito en el año 1653. 4 vols. Seville: E. Rasco.

Coit, J. E. 1915. Citrus fruit. New York: Macmillan.

Diaz del Castillo, B. 1908–16. The true history of the conquest of New Spain. Edited and published in Mexico by Genario Garcia. Translated into English with introduction and notes by Alfred Percival Maudslay. 5 vols. London: Hakluyt Society.

Englehardt, Z. 1927. San Gabriel Mission and the beginnings of Los Angeles. San Gabriel: San Gabriel Mission.

Evans, T. 1874. Orange culture in California. Overland Monthly 12:235–244.

Fawcett, H. S. 1936. Citrus diseases and their control. New York: McGraw-Hill.

———. 1941. Adventures in the plant-disease world. Berkeley: University of California Press.

Garey, T. A. 1882. Orange culture in California. San Francisco: Pacific Rural Press.

Gómara, F. L. de. 1554. Historia de México. Anvers: Iuan Bellero.

Hoag, I. N. 1879. Orange culture in California. Transactions of the California State Agricultural Society 1879:132–138.

Hume, H. H. 1926. The cultivation of citrus fruit. New York: Macmillan.

Kimball, F. A. 1897. Lemon culture. Proceedings of the American Pomological Society 25:95–99.

Kino, E. F. 1919. Kino's historical memoir of Pimería Alta; a contemporary account of the beginnings of California, Sonora, and Arizona, 1683–1711. Translated into English, edited, and annotated by Herbert Eugene Bolton. 2 vols. Cleveland, OH: Arthur H. Clark.

Lelong, B. M. 1902. Culture of the citrus in California. Revised by the State Board of Horticulture. Sacramento: State Printing.

Lugo, J. del Carmen. 1950. Vida de un ranchero. Dictated to Thomas Savage for H. H. Bancroft in October, 1877. Translated by Helen Pruitt Beattie. Manuscript. Berkeley: University of California Bancroft Library.

NASS (USDA National Agriculturaal Statistics Service, California Field Office). 2012. California historic commodity data. NASS website, www.nass.usda.gov.

Navarro de Andrade, E. 1933. Manual de citricultura. Parte I. Cultura e estatistica. Sao Paulo: Edicao de Charcaras e Quintaes.

Rasmussen, W. D. 1960. Readings in the history of American agriculture. Urbana: University of Illinois Press.

Robinson, A. 1846. Life in California. New York: Wiley and Putnam.

Roistacher, C. N. n.d. History of the parent Washington Navel orange tree. Slide show. Ecoport website, http://ecoport.org/ep?SearchType=slideshowView&slideshowId=79.

Spalding, W. A. 1885. The orange: Its culture in California with a brief discussion of the lemon, lime, and other citrus fruit. Riverside Press and Horticulturist.

———. 1922. Early chapters in the history of California citrus culture. California Citrograph 7:66, 103, 150.

Swingle, W. T. 1932. A monograph of the Satsuma orange, with special reference to the occurrence of new varieties through bud variation. Memoirs of the Faculty of Science and Agriculture, Taihoku University 4:1–626.

———. 1943. The botany of citrus and its wild relatives of the orange family. In H. J. Webber and L. D. Batchelor, eds., The citrus industry. Vol. 1. Berkeley: University of California Press. 129–474.

U.S. Census Bureau. 2012. Foreign trade statistics. Census bureau website, http://www.census.gov/foreign-trade/.

Vancouver, G. 1798. A voyage of discovery to the North Pacific Ocean and round the world; performed in the years 1790-1795. 3 vols. London: C. G. & J. Robinson.

Venegas, M. 1757. Noticia de la California, y de su conquista temporal y espiritual hasta el tiempo presente. 3 vols. Madrid: Viuda de M. Fernandez.

Webber, H. J. 1896. The two freezes of 1894–95 in Florida, and what they teach. U.S. Department of Agriculture Yearbook 1894:193–202.

———. 1925. A comparative study of the citrus industry of South Africa. South Africa Department of Agriculture Bulletin 6:106.

———. 1927a. The mother orange tree. California Citrograph 12:186–187.

———. 1927b. California's oldest orange tree. Sunset 59:21, 79–80.

———. 1928. Notes on the history of citrus varieties. California Citrograph 14:36–39.

———. 1943. Historic romance in modern horticulture. California Citrograph 28:58, 80–81.

Wilson, I. H. 1965. William Wolfskill: 1798-1866. Frontier trapper to California ranchero. Glendale, CA: Arthur H. Clark. 268 p.

Table 1.1. California bearing acreage, yield per acre, production, and value of navel and other oranges, 1920–2008*

Year	Bearing acres	Yield per acre*		Production*		Value per unit*†		Total value ($1,000)
		Boxes	Cartons	1,000 boxes	1,000 cartons	$/box	$/carton	
1919–20	92,000	106	213	9,794	19,588	3.00	1.50	29,382
1920–21	93,700	139	278	13,021	26,042	1.56	0.78	20,313
1921–22	94,500	86	172	8,126	16,252	2.95	1.48	23,972
1922–23	95,000	123	246	11,695	23,390	1.87	0.94	21,870
1923–24	95,400	148	296	14,106	28,212	1.28	0.64	18,056
1924–25	95,600	122	244	11,672	23,344	2.74	1.37	31,981
1925–26	95,400	123	247	11,762	23,524	2.74	1.37	32,228
1926–27	95,300	148	297	14,141	28,282	2.56	1.28	36,201
1927–28	94,900	133	266	12,634	25,268	3.45	1.72	43,587
1928–29	94,500	198	395	18,680	37,360	1.91	0.96	35,680
1929–30	94,200	113	225	10,605	21,210	3.64	1.82	38,602
1930–31	93,700	180	359	16,834	33,668	1.43	0.72	24,073
1931–32	93,400	165	330	15,416	30,832	1.17	0.59	18,037
1932–33	92,300	162	324	14,941	29,882	0.79	0.39	11,803
1933–34	89,500	134	268	11,974	23,948	1.11	0.56	13,304
1934–35	90,000	211	422	18,990	37,988	1.10	0.55	20,876
1935–36	90,400	160	320	14,469	28,938	1.30	0.65	18,756
1936–37	88,600	149	299	13,234	26,468	1.71	0.86	22,677
1937–38	87,600	190	381	16,680	33,360	0.81	0.41	13,582
1938–39	88,500	203	406	17,970	35,940	0.63	0.32	11,342
1939–40	88,700	198	395	17,521	35,042	0.97	0.48	16,936
1940–41	88,600	221	441	19,555	39,110	1.06	0.53	20,693
1941–42	87,600	251	502	21,974	43,948	1.05	0.53	23,132
1942–43	88,000	162	324	14,241	28,482	2.53	1.27	36,045
1943–44	88,300	239	477	21,071	42,142	2.71	1.36	57,159
1944–45	88,000	251	502	22,100	44,200	2.91	1.45	64,270
1945–46	87,200	203	406	17,680	35,360	2.87	1.43	50,656
1946–47	86,000	229	457	19,670	39,340	2.24	1.12	44,126
1947–48	82,700	229	457	18,900	37,800	1.66	0.83	31,428
1948–49	81,500	146	292	11,910	23,820	2.34	1.17	27,865
1949–50	80,200	195	390	15,630	31,260	2.16	1.08	33,786
1950–51	79,200	184	369	14,610	29,220	2.58	1.29	37,627
1951–52	77,900	162	323	12,600	25,200	2.67	1.33	33,627
1952–53	74,200	224	448	16,630	33,260	1.97	0.99	32,819
1953–54	72,200	200	401	14,460	28,920	2.44	1.22	35,326
1954–55	70,300	218	436	15,330	30,660	2.46	1.23	37,767
1955–56	67,600	224	449	15,170	30,340	2.96	1.48	44,971
1956–57	62,300	247	494	15,400	30,800	2.96	1.48	45,602
1957–58	62,000	147	294	9,100	18,200	4.75	2.37	43,187
1958–59	60,800	278	556	16,900	33,800	3.19	1.60	53,973
1959–60	60,000	225	450	13,500	27,000	3.68	1.84	49,742
1960–61	60,100	150	300	9,000	18,000	5.41	2.71	48,730
1961–62	57,700	128	256	7,400	14,800	5.92	2.96	43,778
1962–63	57,800	215	429	12,400	24,800	4.52	2.26	56,104
1963–64	58,200	258	515	15,000	30,000	3.78	1.89	56,640

Year	Bearing acres	Yield per acre*		Production*		Value per unit*†		Total value ($1,000)
		Boxes	Cartons	1,000 boxes	1,000 cartons	$/box	$/carton	
1964-65	60,200	248	495	14,900	29,800	3.93	1.97	56,557
1965-66	62,500	293	586	18,300	36,600	2.90	1.45	53,070
1966-67	65,600	260	519	17,025	34,050	3.20	1.60	54,480
1967-68	69,200	132	264	9,150	18,300	4.34	2.17	39,711
1968-69	73,700	252	505	18,600	37,200	3.01	1.51	55,986
1969-70	78,800	269	538	21,200	42,400	2.75	1.38	58,300
1970-71	85,400	204	407	17,400	34,800	3.75	1.88	65,250
1971-72	93,800	238	475	22,300	44,600	3.11	1.56	69,353
1972-73	101,100	185	370	18,700	34,400	3.52	1.76	65,824
1973-74	107,300	204	408	21,900	43,800	4.03	2.02	88,257
1974-75	112,000	249	498	27,900	55,800	3.53	1.77	98,487
1975-76	114,900	246	493	28,300	56,600	3.12	1.56	88,296
1976-77	115,100	222	445	25,600	51,200	3.66	1.83	93,696
1977-78	112,300	178	356	20,000	40,000	5.99	3.00	119,800
1978-79	111,700	186	372	20,800	41,600	7.05	3.53	146,718
1979–80	111,800	292	583	32,600	65,200	3.93	1.97	128,189
1980–81	110,200	352	703	38,750	77,500	4.01	2.00	155,341
1981–82	108,000	245	491	26,500	53,000	7.22	3.61	191,436
1982–83	109,000	369	738	40,200	80,400	4.61	2.31	185,328
1983–84	109,000	309	618	33,700	67,400	5.57	2.79	187,721
1984–85	108,000	243	485	26,200	52,400	11.36	5.68	297,610
1985–86	107,000	308	617	33,000	66,000	7.15	3.58	235,980
1986–87	106,000	325	651	34,500	69,000	7.04	3.52	242,840
1987–88	106,000	297	594	31,500	63,000	7.82	3.91	246,462
1988–89	108,000	315	630	34,000	68,000	7.57	3.78	257,328
1989–90	106,000	418	836	44,300	88,600	8.03	4.01	355,661
1990–91	109,000	145	290	15,800	31,600	10.08	5.04	159,330
1991–92	112,000	313	627	35,100	70,200	8.71	4.36	305,811
1992–93	114,000	384	768	43,800	87,600	6.32	3.16	277,008
1993–94	114,000	321	642	36,600	73,200	7.97	3.99	291,807
1994–95	118,000	297	594	35,000	70,000	7.82	3.91	273,750
1995–96	121,000	314	628	38,000	76,000	7.35	3.68	279,300
1996–97	124,000	323	646	40,000	80,000	9.35	4.68	373,890
1997–98	129,000	341	682	44,000	88,000	8.84	4.42	388,870
1998–99	130,000	162	324	21,000	42,000	10.46	5.23	219,649
1999–00	128,000	313	626	40,000	80,000	6.55	3.28	262,160
2000–01	130,000	273	546	35,500	71,000	9.98	4.99	354,320
2001–02	130,000	246	492	32,000	64,000	12.88	6.44	412,305
2002–03	134,000	313	627	42,000	84,000	7.98	3.99	335,040
2003–04	136,000	290	581	39,500	79,000	10.60	5.30	418,773
2004–05	138,000	319	638	44,000	88,000	9.82	4.91	431,904
2005–06	140,000	336	671	47,000	94,000	9.63	4.81	452,375
2006–07	141,000	245	489	34,500	69,000	11.35	5.67	391,565
2007–08	141,000	319	638	45,000	90,000	10.46	5.23	470,591

Source: NASS 2012.

Notes: *Net weight per carton = 37.5 pounds; net weight per box = 75.0 pounds. †Equivalent packinghouse door returns.

Table 1.2. California bearing acreage, yield per acre, production, and value of Valencia oranges, 1920–2008

Year	Bearing acres	Yield per acre*		Production*		Value per unit*†		Total value
		Boxes	Cartons	1,000 boxes	1,000 cartons	$/box	$/carton	
1919–20	63,800	107	214	6,838	13,676	3.70	1.85	1.85
1920–21	66,500	162	323	10,750	21,500	2.41	1.21	1.21
1921–22	68,900	86	171	5,895	11,790	3.48	1.74	1.74
1922–23	71,700	134	267	9,588	19,176	2.00	1.00	1.00
1923–24	74,400	135	270	10,047	20,094	2.40	1.20	1.20
1924–25	77,400	88	177	6,834	13,668	4.21	2.10	2.10
1925–26	80,400	155	309	12,438	24,876	2.71	1.36	1.36
1926–27	85,500	165	330	14,111	28,222	3.17	1.59	1.59
1927–28	88,000	115	230	10,103	20,206	4.72	2.36	2.36
1928–29	91,400	224	448	20,479	40,958	1.88	0.94	0.94
1929–30	95,900	110	221	10,590	21,180	5.06	2.53	2.53
1930–31	98,800	186	371	18,345	36,690	1.45	0.72	0.72
1931–32	104,000	185	370	19,242	38,484	1.11	0.56	0.56
1932–33	108,500	178	356	19,324	38,648	1.03	0.52	0.52
1933–34	112,700	146	292	16,465	32,930	1.92	0.96	0.96
1934–35	117,300	222	444	26,057	52,114	1.24	0.62	0.62
1935–36	122,600	150	299	18,340	36,680	1.86	0.93	0.93
1936–37	129,100	129	257	16,593	33,186	2.14	1.07	1.07
1937–38	134,500	217	435	29,234	58,468	0.81	0.41	0.41
1938–39	138,100	170	340	23,450	46,900	0.88	0.44	0.44
1939–40	139,900	192	385	26,904	53,808	1.14	0.57	0.57
1940–41	141,900	220	440	31,223	62,446	1.50	0.75	0.75
1941–42	142,800	211	423	30,181	60,362	2.18	1.09	1.09
1942–43	144,200	209	417	30,088	60,176	3.11	1.55	1.55
1943–44	146,000	212	423	30,890	61,780	3.40	1.70	1.70
1944–45	146,600	262	524	38,400	76,800	2.67	1.34	1.34
1945–46	148,000	178	356	26,330	52,660	3.45	1.72	1.72
1946–47	148,100	229	457	33,860	67,720	1.45	0.73	0.73
1947–48	146,000	184	369	26,930	53,860	1.84	0.92	0.92
1948–49	141,200	178	356	25,100	50,200	1.68	0.84	0.84
1949–50	134,800	195	389	26,230	52,460	1.87	0.93	0.93
1950–51	132,500	231	462	30,600	61,200	1.77	0.89	0.89
1951–52	129,900	199	397	25,810	51,620	1.81	0.90	0.90
1952–53	126,500	232	465	29,400	58,800	1.66	0.83	0.83
1953–54	120,000	150	299	17,940	35,880	2.97	1.49	1.49
1954–55	112,900	213	427	24,090	48,180	2.29	1.14	1.14
1955–56	102,600	226	452	23,200	46,400	2.68	1.34	1.34
1956–57	89,600	229	458	20,500	41,000	2.76	1.38	1.38
1957–58	86,600	163	326	14,100	28,200	4.60	2.30	2.30
1958–59	84,300	276	553	23,300	46,600	2.94	1.47	1.47
1959–60	78,700	220	440	17,300	34,600	3.78	1.89	1.89
1960–61	76,300	210	419	16,000	32,000	3.63	1.82	1.82
1961–62	72,100	182	363	13,100	26,200	3.61	1.81	1.81
1962–63	68,100	238	476	16,200	32,400	3.86	1.93	1.93
1963-64	66,800	250	500	16,700	33,400	4.35	2.18	2.18

Year	Bearing acres	Yield per acre*		Production*		Value per unit*†		Total value
		Boxes	Cartons	1,000 boxes	1,000 cartons	$/box	$/carton	
1964–65	62,200	254	508	15,800	31,600	3.25	1.63	1.63
1965-66	63,200	278	557	17,600	35,200	3.26	1.63	1.63
1966–67	64,900	305	609	19,775	39,550	2.56	1.28	1.28
1967–68	70,700	141	283	10,000	20,000	4.45	2.23	2.23
1968–69	76,800	335	669	25,700	51,400	1.90	0.95	0.95
1969–70	81,300	219	438	17,800	35,600	2.74	1.37	1.37
1970–71	82,500	238	475	19,600	39,200	2.83	1.42	1.42
1971–72	86,500	244	488	21,100	42,200	2.55	1.28	1.28
1972–73	87,700	267	534	23,400	46,800	2.83	1.42	1.42
1973–74	88,700	209	417	18,500	37,000	3.71	1.86	1.86
1974–75	85,200	318	636	27,100	54,200	2.93	1.47	1.47
1975–76	82,800	296	592	24,500	49,000	2.80	1.40	1.40
1976–77	76,800	257	513	19,700	39,400	3.94	1.97	1.97
1977–78	76,300	296	592	22,600	45,200	5.76	2.88	2.88
1978–79	75,400	219	438	16,500	33,000	6.67	3.34	3.34
1979–80	74,000	362	724	26,800	53,600	3.75	1.88	1.88
1980–81	73,100	363	725	26,500	53,000	5.15	2.57	2.57
1981–82	72,200	213	427	15,400	30,800	11.51	5.76	5.76
1982–83	68,400	525	1,050	35,900	71,800	3.66	1.83	1.83
1983–84	68,100	217	435	14,800	29,600	14.22	7.11	7.11
1984–85	67,300	389	779	26,200	52,400	6.17	3.09	3.09
1985–86	67,600	309	618	20,900	41,800	6.17	3.09	3.09
1986–87	66,900	350	700	23,400	46,800	8.17	4.09	4.09
1987–88	66,600	413	826	27,500	55,000	7.70	3.85	3.85
1988–89	69,600	358	716	24,900	49,800	8.23	4.12	4.12
1989–90	60,100	392	784	27,100	54,200	7.36	3.68	3.68
1990–91	69,400	141	282	9,800	19,600	21.7	10.90	10.90
1991–92	69,800	463	926	32,320	64,640	3.97	1.98	1.98
1992–93	70,000	329	657	23,000	46,000	8.48	4.24	4.24
1993–94	71,000	380	760	27,000	54,000	6.32	3.16	3.16
1994–95	73,000	288	576	21,000	42,000	9.48	4.74	4.74
1995–96	73,000	267	533	20,000	40,000	10.51	5.26	5.26
1996–97	75,000	316	632	24,000	48,000	8.77	4.39	4.39
1997–98	76,000	342	684	25,000	50,000	9.03	4.52	4.52
1998–99	73,200	204	408	15,000	30,000	12.28	6.14	6.14
1999–00	71,000	338	676	24,000	48,000	3.48	1.74	1.74
2000–01	68,000	279	558	19,000	38,000	8.43	4.22	4.22
2001–02	65,000	300	600	19,500	39,000	7.51	3.76	3.76
2002–03	64,000	313	625	20,000	40,000	6.53	3.27	3.27
2003–04	57,000	193	386	11,000	22,000	12.48	6.24	6.24
2004–05	53,000	387	774	20,500	41,000	8.38	4.19	4.19
2005–06	50,000	280	560	14,000	28,000	12.93	6.46	6.46
2006–07	49,000	235	469	11,500	23,000	11.04	5.52	5.52
2007–08	47,000	362	723	17,000	34,000	8.12	4.06	4.06

Source: NASS 2012.

Notes: *Net weight per carton = 37.5 pounds; net weight per box = 75.0 pounds. †Equivalent packinghouse door returns.

Table 1.3. Estimated statewide bearing citrus acreage as of 2010*

Type	2008–09	2009–10
grapefruit[†]	9,600	9,600
lemons	47,000	47,000
oranges, navel	141,000	141,000
oranges, Valencia	45,000	43,000
mandarins and mandarin hybrids	27,000	30,000

Source: NASS 2012.

Notes: *Reported data with an allowance for incompleteness. [†]Includes pummelos and hybrids.

Table 1.4. Bearing, nonbearing, and total California citrus acreage by type, through 2010

Type	Acres standing in 2008			Acres standing in 2010		
	Bearing	Nonbearing	Total	Bearing	Nonbearing	Total
grapefruit*	7,832	1,276	9,108	7,826	1,103	8,929
lemons	43,811	2,759	46,570	41,535	2,942	44,477
limes	384	114	498	462	11	473
oranges, navel	130,260	7,185	137,446	130,469	4,437	134,906
oranges, Valencia	43,978	492	44,471	41,862	678	42,540
pummelos and hybrids	1,757	40	1,796	1,559	20	1,579
mandarins and mandarin hybrids	25,685	5,707	31,392	32,734	6,092	38,826

Source: NASS 2012.

Note: *Excludes pummelos and hybrids.

Table 1.5. Bearing, nonbearing, and total California citrus acreage by type, county, and year planted, through 2010*

Type and county	2002 and earlier	2003	2004	2005	2006	2007	2008	2009	2010	Acres standing in 2010		
										Bearing	Nonbearing	Total
	Acres											
Grapefruit[†]												
Fresno	16	0	32	48	27	22	65	0	0	123	87	210
Imperial	454	0	0	122	82	95	0	70	0	658	165	823
Kern	353	0	128	10	26	105	0	0	0	491	131	622
Riverside	3,814	43	214	23	8	40	75	0	41	4,102	156	4,258
San Bernardino	407	0	0	0	0	0	0	0	0	407	0	407
San Diego	1,069	27	91	40	52	75	47	0	0	1,279	122	1,401
Tulare	303	23	53	100	123	157	61	63	23	479	427	906
Ventura	206	0	0	17	6	5	10	0	0	229	15	244
All other counties	58	0	0	0	0	0	0	0	0	58	0	58
STATE TOTAL	6,680	93	518	360	324	499	258	133	64	7,826	1,103	8,929
Lemons												
Fresno	844	52	25	60	70	66	156	35	0	981	327	1,308
Imperial	3,837	89	83	29	0	0	200	103	0	4,038	313	4,341
Kern	3,182	23	74	51	49	20	8	15	0	3,330	92	3,422
Monterey	923	0	0	0	0	0	0	0	0	923	0	923
Orange	370	0	0	4	0	0	34	0	0	374	34	408
Riverside	5,002	103	76	48	33	39	27	0	0	5,229	99	5,328
San Bernardino	257	0	0	0	0	0	0	0	0	257	0	257
San Diego	2,209	207	12	2	5	34	14	10	50	2,430	113	2,543
San Luis Obispo	1,449	24	10	0	0	54	22	0	0	1,483	76	1,559
Santa Barbara	1,338	45	10	14	35	18	4	6	0	1,407	63	1,470
Tulare	3,391	149	72	212	276	257	206	286	53	3,824	1,078	4,902
Ventura	16,673	194	205	128	128	288	167	98	66	17,200	747	17,947
All other counties	59	0	0	0	0	0	0	0	10	59	10	69
STATE TOTAL	39,534	886	567	548	596	776	838	553	179	41,535	2,942	44,477

Table 1.5. *Continued*

Type and county	2002 and earlier	2003	2004	2005	2006	2007	2008	2009	2010	Acres standing in 2010		
										Bearing	Nonbearing	Total
	Acres											
Limes												
Riverside	34	0	14	4	1	0	5	0	0	52	6	58
San Diego	258	2	0	1	0	0	0	0	3	261	3	264
All other counties	59	0	0	0	0	2	0	0	0	59	2	61
STATE TOTAL	317	2	0	95	1	2	5	0	3	462	11	473
Oranges, navel, and other												
Butte	113	1	0	0	0	1	0	0	0	114	1	115
Fresno	19,586	437	756	433	321	249	266	181	49	21,533	745	22,278
Glenn	324	4	0	0	0	0	0	0	0	328	0	328
Kern	28,209	213	746	1,076	417	396	168	150	167	30,661	881	31,542
Kings	62	0	0	80	0	0	0	0	0	142	0	142
Madera	2,654	0	0	0	38	0	0	0	0	2,692	0	2,692
Riverside	2,200	1	1	1	8	1	6	3	0	2,211	10	2,221
San Bernardino	1,460	8	6	29	15	6	12	4	0	1,518	22	1,540
San Diego	541	0	8	7	5	0	0	0	7	561	7	568
Tulare	63,356	1,465	2,083	1,590	998	1,147	623	697	238	69,492	2,705	72,197
Ventura	733	104	27	9	43	18	4	0	29	916	51	967
Yolo	100	4	2	0	0	0	9	0	0	106	9	115
All other counties	142	21	0	2	30	0	6	0	0	195	6	201
STATE TOTAL	119,480	2,258	3,629	3,227	1,875	1,818	1,094	1,035	490	130,469	4,437	134,906
Oranges, Valencia												
Fresno	2,770	25	0	0	20	1	0	0	0	2,815	1	2,816
Imperial	233	0	0	0	0	0	0	0	300	233	300	533
Kern	7,268	0	0	88	0	0	0	0	50	7,356	50	7,406
Madera	908	0	0	0	0	0	0	0	0	908	0	908
Orange	67	0	0	0	0	0	0	0	0	67	0	67
Riverside	2,955	20	0	0	2	37	37	0	0	2,977	74	3,051
San Bernardino	1,182	0	0	7	0	0	0	0	0	1,189	0	1,189
San Diego	4,399	7	2	3	7	8	0	0	2	4,418	10	4,428
San Luis Obispo	219	0	1	0	0	0	0	0	0	220	0	220
Tulare	16,728	6	9	82	71	35	37	14	79	16,896	165	17,061
Ventura	4,701	0	0	21	0	44	24	5	5	4,722	78	4,800
All other counties	61	0	0	0	0	0	0	0	0	61	0	61
STATE TOTAL	41,491	58	12	201	100	125	98	19	436	41,862	678	42,540
Pummelos and hybrids												
Fresno	75	0	0	9	0	0	2	0	0	84	2	86
Imperial	90	0	0	0	0	0	0	0	0	90	0	90
Kern	190	19	0	0	0	0	0	0	0	209	0	209
Riverside	244	0	0	0	0	0	0	0	0	244	0	244
San Diego	55		1	0	0	0	0	0	0	56	0	56
Tulare	824	9	20	6	2	1	2	16	0	862	18	880
All other counties	14	0	0	0	0	0	0	0	0	14	0	14
STATE TOTAL	1,492	29	20	15	2	1	4	16	0	1,559	20	1,579
Mandarins and mandarin hybrids												
Butte	80	3	0	2	0	1	0	0	0	86	0	86
Fresno	1,604	635	871	567	441	142	368	1,028	137	4,260	1,533	5,793
Glenn	83	0	0	0	0	0	0	0	0	83	0	83
Imperial	552	154	190	55	0	20	50	30	42	971	122	1,093
Kern	7,173	1,759	1,262	2,786	143	50	224	1,362	3	13,173	1,589	14,762
Madera	630	388	272	250	571	436	0	29	10	2,547	39	2,586
Placer	114	5	1	2	4	5	3	0	0	131	3	134
Riverside	1,623	251	111	69	109	0	1	0	0	2,163	1	2,164
San Diego	429	82	3	53	5	5	3	3	6	577	12	589
Stanislaus	226	0	0	0	0	0	0	0	3	226	3	229
Tulare	3,005	389	1,456	676	1,294	420	1,067	1,381	125	7,240	2,573	9,813
Ventura	27	46	76	304	162	227	130	44	3	1,092	177	1,269
All other counties	164	4	5	10	2	0	17	3	20	185	40	225
STATE TOTAL	15,710	3,716	4,247	4,774	2,731	1,306	1,863	3,880	349	32,734	6,092	38,826

Source: NASS 2012.

Notes: *"All other counties" include counties with less than 50 acres. Totals may not add between county and variety tables due to rounding. †Excludes pummelos and hybrids.

Table 1.6. California bearing, nonbearing, and total citrus acreage by type, county, and year planted, through 2010*

Type and Variety	2002 and earlier	2003	2004	2005	2006	2007	2008	2009	2010	Acres standing in 2010		
										Bearing	Nonbearing	Total
	Acres											
Oranges, navel and other												
Blood	480	0	14	9	0	22	47	39	25	503	133	636
Cara Cara	1,467	144	328	347	542	500	302	140	48	2,828	990	3,818
Late varieties†	13,482	611	1,431	730	443	239	427	323	20	16,697	1,009	17,706
All other varieties	104,049	1,504	1,856	2,141	889	1,057	317	533	397	110,439	2,304	112,743
STATE TOTAL	119,478	2,259	3,629	3,227	1,874	1,818	1,093	1,035	490	130,467	4,436	134,903
Grapefruit												
Rio Red	1,115	11	19	139	133	128	0	70	0	1,417	198	1,615
Ruby	1,104	0	5	3	0	5	2	0	0	1,112	7	1,119
Ruby Red	1,399	0	0	0	0	10	2	0	0	1,399	12	1,411
Star Ruby	1,180	8	239	117	162	279	131	11	30	1,544	613	2,157
All other varieties	1,881	74	255	100	28	76	122	53	33	2,310	312	2,622
STATE TOTAL	6,679	93	518	359	323	498	257	134	63	7,782	1,142	8,924
Pummelos and hybrids												
Chandler	752	0	12	11	0	1	4	16	0	775	21	796
Melogold	93	28	9	4	1	0	0	0	0	134	1	135
Oroblanco	540	1	0	0	0	0	0	0	0	541	0	541
All other varieties	108	0	0	0	0	0	0	0	0	108	0	108
STATE TOTAL	1,493	29	21	15	1	1	4	16	0	1,558	22	1,580
Mandarins and mandarin hybrids												
Mandarins/tangerines	11,224	3,394	4,023	4,470	2,605	1,211	1,814	3,615	285	26,927	5,714	32,641
Clementines	5,127	1,756	1,196	2,409	307	25	195	1,608	85	10,820	1,888	12,708
Algerian	233	0	189	58	0	0	0	60	0	480	60	540
Caffin	255	16	25	0	174	15	30	179	0	485	209	694
Clemenules (Nules)	4,273	1,597	979	2,342	118	10	165	1,221	85	9,319	1,471	10,790
Fina Sodea	316	0	0	0	0	0	0	0	0	316	0	316
Marisol	50	0	3	0	0	0	0	0	0	53	0	53
Oro Grande	0	143	0	9	15	0	0	148	0	167	148	315
Gold Nugget	140	47	32	108	52	56	88	65	7	435	160	595
Fairchild	990	0	0	48	13	0	0	0	0	1,051	0	1,051
Pixie	115	14	4	14	16	6	14	9	0	169	23	192
Satsuma	2,643	22	29	11	39	32	119	39	0	2,776	158	2,934
Shasta Gold	0	2	27	45	23	7	4	0	0	104	4	108
Tango	12	0	20	0	0	70	1,112	1,322	145	102	2,579	2,681
W. Murcott Afourer	1,433	955	2,488	1,552	1,704	851	85	179	0	8,983	264	9,247
All other varieties	764	598	227	283	451	164	197	393	48	2,487	638	3,125
Tangelos	4,566	315	218	301	126	95	36	242	64	5,629	343	5,972
Minneola	4,473	315	218	301	126	95	36	118	61	5,528	215	5,743
All other varieties	93	0	8	0	0	0	1	124	3	101	128	229
Tangelos/temples	167	6	0	0	0	0	15	23	0	173	38	211
Royal Mandarin	146	6	0	0	0	0	10	0	0	152	10	162
All other varieties	21	0	0	0	0	0	5	23	0	21	28	49
STATE TOTAL	37,041	9,186	9,686	11,951	5,769	2,637	3,926	9,368	783	76,278	14,078	43,773

Source: NASS 2012.

Notes: **Shaded/bold* area indicates nonbearing years. Totals may not add between county and variety tables due to rounding. †Includes Autumn Gold, Barnfield, Chsiat, Lane Late, and Powell varieties.

Table 1.7. Top 15 export markets for U.S. oranges, 2005–2006

	Cumulative to date quantities and values, in thousands of dollars					
	2005		2006		2007	
	Value	Qty	Value	Qty	Value	Qty
Canada	106,596	171,380	107,856	164,260	84,420	105,873
Hong Kong	41,595	65,442	53,576	77,332	34,853	41,209
South Korea	82,655	101,299	79,953	92,568	75,794	77,126
Japan	52,246	73,681	61,072	83,409	40,563	42,947
China	15,678	33,141	19,587	33,026	11,972	16,742
Malaysia	12,741	22,026	14,415	25,087	7,132	9,520
Singapore	8,865	15,974	9,140	15,958	3,816	4,800
United Arab Emirates	353	1,088	595	1,210	50	76
Australia	8,796	8,540	10,762	9,401	8,456	11,141
Taiwan	4,739	9,435	3,631	6,428	2,577	3,481
Mexico	5,266	18,936	4,344	16,401	4,287	11,579
Philippines	708	1,509	1,219	2,559	400	857
Indonesia	682	1,254	845	1,550	490	555
New Zealand	2,867	3,311	1,769	2,844	3,818	4,568
Bangladesh	111	256	27	43	0	0
TOTAL	343,898	527,272	368,791	532,076	278,628	330,474
All other countries	5,572	7,142	4,749	6,267	2,973	3,558
Combined total*	349,470	534,414	373,540	538,343	281,601	334,032
	Cumulative to date quantities and values, in thousands of dollars					
	2008		2009		Jan–Apr 2010	
	Value	Qty	Value	Qty	Value	Qty
Canada	105,758	152,981	125,204	148,825	76,942	93,215
Hong Kong	42,419	55,141	47,757	64,146	31,301	39,948
South Korea	74,950	83,788	57,829	59,701	35,146	36,460
Japan	59,055	68,123	41,053	49,105	20,566	28,182
China	21,145	33,846	20,258	33,546	9,903	15,550
Malaysia	12,734	19,485	11,122	17,278	8,680	12,730
Singapore	9,236	14,265	8,593	13,042	5,772	8,171
United Arab Emirates	9,335	16,874	6,829	11,407	4,273	5,934
Australia	14,004	16,123	8,265	9,120	5,960	5,723
Taiwan	4,893	7,537	2,624	4,507	1,843	2,720
Mexico	4,138	13,836	1,092	4,269	812	3,561
Philippines	1,721	2,787	2,159	4,000	2,041	3,128
Indonesia	1,515	2,430	1,576	2,543	1,582	2,202
New Zealand	5,806	7,087	2,072	2,302	1,934	1,662
Bangladesh	1,707	2,547	1,361	1,711	1,197	1,314
TOTAL	368,416	496,850	337,794	425,502	207,952	260,500
All other countries	9,363	12,432	8,529	11,207	6,577	8,429
Combined total*	377,779	509,282	346,323	436,709	214,529	268,929

Source: U.S. Census Bureau 2012. Users should be cautious when interpreting quantity reports that use mixed units of measure. Quantity line items will include only statistics on the units of measure that are equal to, or are able to be converted to, the assigned unit of measure of the grouped commodities.

Note: *Numbers may not add due to rounding.

Table 1.8. Top 15 export markets for U.S. lemons, 2005–2006

	Cumulative to date quantities and values, in thousands of dollars					
	2005		2006		2007	
	Value	Qty	Value	Qty	Value	Qty
Japan	34,973	53,093	36,747	51,570	81,021	83,549
Canada	24,276	30,366	27,254	30,164	31,515	27,072
Hong Kong	6,158	6,501	6,094	7,008	2,121	1,837
Australia	4,414	3,097	4,710	4,269	7,748	8,795
South Korea	3,367	3,573	5,118	5,000	7,711	7,397
China	757	1,072	1,531	1,776	2,971	2,923
Russia	0	0	10	6	4	2
United Arab Emirates	0	0	0	0	0	0
Philippines	169	279	179	223	34	44
New Zealand	141	151	157	144	965	977
Taiwan	386	521	479	685	106	113
Mexico	148	168	235	256	368	394
Singapore	113	238	0	0	0	0
Colombia	0	0	0	0	0	0
Malaysia	111	73	74	25	10	12
TOTAL	75,013	99,132	82,588	101,126	134,574	133,115
All other countries	977	853	569	420	773	799
Combined total*	75,990	99,985	83,157	101,546	135,347	133,914
	Cumulative to date quantities and values, in thousands of dollars					
	2008		2009		Jan–Apr 2010	
	Value	Qty	Value	Qty	Value	Qty
Japan	66,646	66,205	47,261	35,810	18,813	13,346
Canada	46,499	40,734	34,129	29,641	16,336	12,572
Hong Kong	6,167	6,182	7,806	8,907	1,373	1,401
Australia	6,819	6,022	9,120	6,589	982	657
South Korea	7,825	8,000	5,570	4,103	2,218	1,578
China	2,284	2,759	2,775	3,229	1,429	1,646
Russia	449	344	1,159	1,036	1,649	1,539
United Arab Emirates	175	162	224	351	0	0
Philippines	309	286	328	342	195	181
New Zealand	896	868	421	272	214	152
Taiwan	91	77	115	129	69	74
Mexico	137	118	118	129	245	221
Singapore	38	44	91	95	103	137
Colombia	23	27	100	91	33	23
Malaysia	100	102	101	91	218	194
TOTAL	138,458	131,930	109,318	90,815	43,877	33,721
All other countries	2,737	2,167	528	354	1,008	12,242
Combined total*	141,195	134,097	109,846	91,169	44,885	45,963

Source: U.S. Census Bureau 2012. Users should be cautious when interrelating quantity reports that use mixed units of measure. Quantity line items will include only statistics on the units of measure that are equal to, or are able to be converted to, the assigned unit of measure of the grouped commodities.

Note: *Numbers may not add due to rounding.

2 Costs of Establishment and Production

Etaferahu Takele

Citrus is one of California's major crops. In 2012, there were 269,000 acres in citrus production throughout the state. In gross value, citrus production amounts to about $1.27 billion (CASS 2012). The major varieties of citrus grown in California are oranges, lemons, and grapefruit; other varieties include mandarins, tangerines, and tangelos (see table 2.1). Seventy-three percent of citrus production is in the San Joaquin Valley (Fresno, Kern, and Tulare Counties), with the largest share in Tulare County. Most of the navel orange and mandarin production is also in this area; most of the lemon production is in southern California, particularly in Ventura County. Valencia orange production is located mainly in the Central Valley, with a sizable but declining industry in southern California.

Costs of Citrus Establishment and Production Practices

It is not possible to generalize the costs of production and enterprise budgets for citrus for California as a whole, as the cost of inputs and production practices vary by region and variety. Costs are generally higher in rapidly urbanizing areas where land and water uses are shifting away from agriculture. This chapter presents a general outline of production activities

Table 2.1. Citrus acreage in California counties by variety, 2012

County	Lemons	Valencia oranges	Navel oranges	Grapefruit	Mandarins and mandarin hybrids*	Other†	Total
Fresno	1,334	2,611	21,701	287	9,506	85	35,524
Imperial	3,854	130	0	943	1,048	180	6,155
Kern	3,173	6,665	29,341	679	15,013	181	55,052
Riverside	5,112	2,735	1,928	4,155	2,192	302	16,424
San Bernardino	270	919	1,253	327	0	0	2,769
San Diego	2,509	4,184	543	1,448	580	324	9,588
Tulare	5,257	15,742	71,455	953	12,358	853	106,618
Ventura	17,469	4,344	862	182	1,615	0	24,472
Others‡	4,491	1,249	3,003	61	3,608	76	12,488
Total	**43,469**	**38,579**	**130,086**	**9,035**	**45,920**	**2,001**	**269,090**

Source: CASS 2012.

Notes:

*Includes mandarins, tangerines, clementines, tangelos, tangors, and temples.

†Includes limes and pummelos.

‡Includes the following counties: Lemon: Monterey, Orange, San Luis Obispo, Santa Barbara, and others. Valencia oranges: Madera, Orange, San Luis Obispo, and others. Navel oranges: Butte, Glenn, Kings, Madera, Yolo, and others. Mandarins and mandarin hybrids: Butte, Glenn, Madera, Placer, Stanislaus, and others.

and their influences on costs of orchard establishment and production of citrus in California. For detailed information on production activities, see the appropriate chapters in this manual. For detailed information on the cost of lemons, mandarins, and oranges production in California, see the UC Davis Cost Studies website, http://coststudies.ucdavis.edu.

Land Preparation

The cost of land preparation before orchard establishment includes in some cases clearing the land, layout, marking, and planting, but it can also include ripping and leveling if planting is on open ground. If planting is on land previously used for an orchard, the cost must also include orchard removal operations such as removing trees and roots using a bulldozer or tractor and burning and shredding old trees. These operations are commonly done by contract operators. Land preparation charges also depend on the job type (see O'Connell et al. 2009, 2010) and the topography of the orchard.

Fumigation before Orchard Establishment

Fumigation for weed and disease control was previously one of the major costs during land preparation, especially in orchards being planted back to citrus. Fumigation using methyl bromide is currently banned on many crops (see DPR 2009; for information on fumigation alternatives to methyl bromide, see the UC Postharvest website, http://ucanr.org/sites/phmba/).

Planting

The major factor causing planting costs to vary in citrus orchard production in California is tree spacing (orchard density). In Riverside County, most plantings use a narrow (closer) spacing (152 trees per acre, or 12 ft by 24 ft regardless of variety), while wider spacing is used in San Diego County and the Central Valley (108 to 110 trees per acre), with Ventura County falling in between (121 to 136 trees per acre). Many new orchards, especially in the Central Valley, are planted at closer spacings to get heavier early production, but historical data show that the trees begin to crowd at age 8 to 9, requiring that some trees be removed during their productive years. This loss must be accounted for when comparing this practice with planting at wider spacings. A small variation may exist between the cost of plants by variety and also by region (see O'Connell et al. 2009). Labor costs can be estimated using the amount of time required to plant a tree and wage rates. The labor time to plant a tree, about 15 to 20 minutes, is about the same regardless of variety and region (approximated from growers' data collected for cost studies).

Irrigation

The cost of water varies widely among citrus production regions. Although the desert regions of Riverside County require more irrigation than do other regions, the price of water there is much lower than in other regions. Most water in Riverside County is purchased from water districts and municipalities. Prices in 2010 in the Palo Verde Valley include a flat fee of $50 per acre per year, regardless of the amount used. The calculation in the Coachella Valley, according to the water district there, uses three components: water charges of about $25 per acre foot of water, plus flat gate and availability charges. The impact of the gate and availability charges is not significant for farms where the costs will be distributed over many acres. Prices of water in these areas have remained stable over the years.

In other areas of southern California, the price of water is significantly higher than in the desert areas of Riverside County. The price varies among regions, reaching to up to $1,000 per acre-foot in western Riverside County and up to $1,200 per acre-foot in San Diego County in 2011. In Ventura County, water costs are lower; current cost studies for a moderately steeply sloped orchard were estimated using water costs at $350 per acre-foot (Takele et al. 2012). The higher price in San Diego County is due to the fact that most of the water is supplied by districts and municipalities, whereas in Ventura County, many growers draw most of their water from wells. In San Diego County, some growers have attempted to invest in wells to mitigate the high cost of purchased water. However, they have encountered several problems with well water, including short supply (many wells produce only a few gallons per minute) and salinity. Furthermore, the price of diesel and electricity has made well development economically infeasible in San Diego County.

Fertilization

The cost of fertilization in citrus production is generally more for citrus grown in the desert areas, where trees require more nitrogen, than in other areas. Because nitrogen is generally applied in irrigation water, the cost of fertilizer application is insignificant. Also, citrus grown in the desert region of California can often be deficient in micronutrients and can require one or two micronutrient fertilizer applications before the fruit are fully expanded.

Pruning and Hedging

Pruning and hedging includes hand labor and mechanical topping and hedging. Pruning costs for oranges and grapefruit include annual light operations mainly consisting of skirt pruning low-hanging branches and removing deadwood; extensive pruning is required every 4 to 5 years. Occasional mechanical topping and hedging of mature orchards is done to keep picking costs down. Lemons require more frequent pruning: annually after the first year of establishment in Ventura and San Diego Counties and in the Central Valley, and about every 3 years in the Coachella Valley. The time required to prune and hedge an orchard depends on the number of trees and their age; however, the cost of pruning mature trees can be calculated by assuming that it would take 10 to 15 minutes to prune and hedge each tree, regardless of variety and region (see the UC Davis Cost Studies website, http://coststudies.ucdavis.edu).

Pest and Disease Control

Because citrus pest and diseases vary from region to region, the cost of pest control varies among regions. For more information, see the UC Davis Cost Studies website, http://coststudies.ucdavis.edu, and the UC IPM Citrus Pest Management Guidelines, http://www.ipm.ucdavis.edu/PMG.

Labor

Hourly wage rates differ among regions and also because of availability. In general, the agricultural wage rates published by the National Agricultural Statistics Service (NASS) can serve as a guide for estimating labor wages. Machine and skilled labor wages should be estimated at 10 to 20% higher than manual labor wages. Table 2.2 provides agricultural wage rates (no benefits included) in California from 1995 to 2012. The table is useful in understanding how the wage rate changes from year to year and how to make projections for the future. On average, wages increase approximately 3.25% per year. Benefits would include federal and state payroll tax, workers' compensation, and other miscellaneous benefits as determined by law. These benefits, according to most current cost studies, increase wages by as much as 38%.

The accounting of labor hours could be misrepresented when occasional help is involved, especially when the owner or family members may be involved in production and also when one person performs both management and labor. In all cases, a good record-keeping system along with prompt recording of activities must be in place so that charges accurately measure the hours of labor.

Harvesting and Marketing

Harvesting costs, including picking and hauling, are estimated on a piece-rate basis, so they depend on

Table 2.2. California average agricultural labor wages and percentage changes, 2000–2012

Year	All hired ($/hr)	Fieldworkers ($/hr)	Field- and livestock workers ($/hr)	% Change from previous year for all hired workers	% Change from previous year for fieldworkers	% Change from previous year for field- and livestock workers
2000	8.21	7.48	7.56	4.19	4.18	3.99
2001	8.67	7.89	8.02	5.60	5.48	6.08
2002	9.14	8.34	8.46	5.42	5.70	5.49
2003	9.25	8.34	8.50	1.20	0.00	0.47
2004	9.33	8.40	8.58	0.86	0.72	0.94
2005	9.76	8.81	9.00	4.61	4.88	4.90
2006	10.09	9.00	9.20	3.38	2.16	2.22
2007	10.51	9.56	9.72	4.16	6.22	5.65
2008	11.03	9.98	10.17	4.95	4.39	4.63
2009	11.14	10.05	10.25	1.00	0.70	0.79
2010	11.24	10.14	10.31	0.90	0.90	0.59
2011	11.04	10.10	10.24	–1.78	-0.39	-0.68
2012	11.64	10.56	10.74	5.43	4.55	4.88
Average	9.32	8.46	8.61	3.21	3.25	3.08

Source: NASS 2000–2012.

Note: Wages do not include benefits.

the yield of the orchard. Picking and hauling piece rates also differ by variety (see O'Connell et al. 2009, 2010). Most citrus fruit are hand-picked, which is labor intensive and involves hundreds of labor hours per acre (e.g., 242 hours per acre for oranges; see O'Connell et. al 2009). Picking and hauling are generally done by contract harvesters. The marketing costs of packing are charged on a piece-rate basis by packinghouses; there may also be other marketing costs, such as assessment charges. Overall, harvesting and marking costs account for a significant part of the cost of production.

Costs of Business Transactions and Cash Overhead

Calculating the costs of production is relatively straightforward when using financial accounting and direct cash costs methods. However, economic accounting of costs is needed to evaluate the direct and indirect costs (noncash and opportunity) and the profitability or viability of enterprises. Determining many of the indirect costs can be complex. For more information, see the UC Davis Cost Studies website, http://coststudies.ucdavis.edu.

Management

Determining management costs is especially difficult since there is no clear, established information on management charges in agricultural practices. Small-scale orchard operations do not normally hire managers, mainly because the owner usually works as both manager and laborer. Even in cases of absentee owners who hire management companies, management and labor charges are rarely separate. Hence, most cost studies consider the net returns to be split as a reward for management and profit.

Interest on Operating Capital

The financial charge for borrowed money or the opportunity cost for using one's own money is significant, particularly for capital outlay during the establishment period. Short-term lending rates are often used as the basis for estimating interest on operating capital; they are calculated from the period of expenses to harvesting (the period considered when payment will begin on the loan or money invested).

Office Expenses

Costs for office leases, utilities, supplies, office equipment, bookkeeping, tax accounting, and office staff may not usually be accounted for by individual enterprises of a farm. There are several ways to allocate costs to individual enterprises or activities. One way is to allocate costs according to the proportion of gross returns received by the enterprise; another is to make the allocation proportional to production expenses (as production expenses are reflections of the amount of work needed). A very common way is to allocate the expenses equally to all enterprises.

Property Taxes

Most commonly, counties charge 1% tax on the assessed value of the property, including land, equipment, buildings, and improvements. Special assessment districts in some counties may charge additional taxes on property.

Property and Liability Insurance

Growers carry insurance for property protection and liability to cover accidents on the farm. Property insurance is based on valuation of assets and was reported to be $7.75 per $1,000 of valuation in 2012 (California Department of Insurance Rate Regulation Branch). Liability insurance is usually based on the size of a farm or enterprise. Approximate liability rates by farm size as estimated by the California Department of Insurance Rate Regulation Branch are given in table 2.3. The coverage in table 2.3 is for $1 million, but growers can purchase insurance for more coverage. Liability insurance premiums are approximated to increase by a little over 1% per year.

Field Sanitation

The costs of sanitation required by state and federal regulations include the availability of one toilet and hand-washing facility for every 20 employees of each sex, located within a quarter-mile walk of the work site, or if not feasible, at the closest point of vehicular access. As an alternative to providing the required toilet and hand-washing facilities, employers may transport employees to toilet and hand-washing facilities (see Cal/OSHA 1995). Field toilets can be built or rented on a monthly basis.

Long-Term Assets or Investment Costs

Citrus production requires a substantial investment in land, an irrigation system, and machines and equipment that serve the farm for multiple years. The cost of long-term assets or investments includes annual repairs, maintenance or improvements, insurances, fuel and lubrication, as well as ownership costs of depreciation and interest on investment or capital

Table 2.3. Liability insurance premiums by farm size ($1 million coverage), 2008

Farm size (ac)	Liability insurance premium, 2008 ($)	Liability insurance premium, 2012 ($)
> 25	453	477
26–100	559	587
101–150	693	728
151–200	774	813
201–250	831	873
251–350	884	928
351–500	970	1,018
501–750	1,073	1,127
751–1,000	1,216	1,278
1,001–2,000	1,350	1,419
2,001–3,000	1,438	1,511
3,000+	1,533	1,611

Source: K. Klonsky, UC ANR, pers. comm.

recovery, which is a combined charge of interest on investment and depreciation. Formulas for calculating these costs are given in the UC Davis Cost Studies website, http://coststudies.ucdavis.edu. Most farm management resources and computer software on cost studies have formulas for calculating these costs. Repairs and maintenance costs in most cases are also kept by growers. For capital recovery calculation, the AJDesigner Capital Recovery Factor calculator website also provides a formula (see http://www.ajdesigner.com/phpdiscountfactors/capital_recovery_equation.php). The interest on investment for long-term assets can be approximated using the long-run rate of return to agricultural assets from current income. In California these rates ranged from 4 to 6% over the past 20 years.

Land

The charge for land use—lease or the opportunity cost—is a significant cost in citrus production, especially in urban areas where the land is being considered for alternative uses such as residential or commercial development. Land values are generally higher in southern California than in the Central Valley. Before the economic downturn in 2008, the rise in land values was one of the causes for the decline in the planting of citrus and other crops in southern California. The California Chapter of the American Society of Farm Management and Rural Appraisers provides up-to-date land values for citrus production at their website, http://www.calasfmra.com/. Land rental or leases are estimated as the value of land times the long-run rate of return for agricultural assets to current income.

Other Investments

Costs of other investments, including the irrigation system, orchard heaters or wind machines, buildings, and farm equipment and implements are for the most part calculated using their prices, usage (hours of use), and life. Consult local companies and dealers for prices and management resources such as Boehlje et al. (1984) and O'Connell et al. (2009, 2010) for more information.

Irrigation System

Citrus irrigation for the most part uses microsprinklers around the trees with main and lateral lines to distribute water. The life of the irrigation system, with proper maintenance, is expected to match the life of the orchard—at least 35 years.

Orchard Heaters and Wind Machines

Not all citrus production regions require frost protection. In San Diego and Ventura Counties, wind machines are used for Valencia orange orchards. Wind machines are also used in the Central Valley for frost protection for both navel and Valencia oranges, but their use is usually delayed until significant fruit is produced, which may be during the 6th or 7th year. One wind machine may be sufficient to serve 10 acres and is expected to last the life of the orchard.

Machines, Equipment, and Buildings

The machine and equipment complement of citrus production includes tractors, sprayers, shop tools, fuel tanks, and pumps, as well as buildings to store farm machinery. Their number and size depend on the size of the farm. The lives of these machines vary. Information on these items can be found in resources such as Boehlje et al. (1984) and factory machine specifications.

Trees (Establishment Cost)

Most citrus varieties begin to bear fruit in year 3, which is considered to be the end of the establishment period for tax purposes. Growers should consult the IRS publication *Farmer's Tax Guide* (IRS 2011) or a tax accountant for tax updates and regulations. In most cost studies, however, the establishment period is considered to end after 5 or 6 years, when the trees reach reasonable yielding age. The cumulative net cash costs (income minus cash costs) during the period of establishment is the asset value for trees (establishment cost).

Gross Returns

Growers calculate their gross returns by multiplying their yield by the price per box or carton they receive from their packinghouse. Packinghouses determine the prices after they record the grades and qualities of the fruit received and deduct their costs for handling. Counties publish crop values (total citrus crop in the county times prices) on the web (see http://ucanr.org/sites/Farm_Management/County_Agricultural_Reports100/). These reports provide the average yield (total citrus crop divided by total citrus acreage) and prices in the counties. The average county yield and price may be lower than that of the typical grower because the average county yield includes yield of all acreage (orchards), including very young and newly bearing ones. Average prices also include all prices received by growers in the county and may not necessarily reflect the typical grower price for good-quality mature orchard production.

Profitability Analysis

The viability and sustainability of producing citrus can be evaluated in several ways.

Gross Margin and Net Margin

In many cases, growers estimate only their operating costs without including their investment depreciation and interest on investment, in which case their interpretation of profit would be gross margin instead of net margin (economic profit). This is often done when the grower has no investment debt. Gross margin is total returns minus operating and overhead cash costs; it is useful when estimating taxable income (gross margin minus depreciation). The long-term viability of the enterprise, however, can be determined only if the net, or economic, margin is greater or equal to zero. The net margin is obtained by subtracting all costs (including investment and management) from the gross income.

Break-Even Analysis

Another method of profitability analysis is determining break-even costs, which are unit costs such as dollars per box or per carton of production that can be compared with the price of the product. Break-even costs are calculated as the cost of production divided by the yield. The difference between the price and the unit cost is the profit margin. The gross margin break-even cost reflects the price needed to cover all cash costs; the net margin break-even cost reflects the price needed to cover all costs, including investment and management. Sometimes growers would like to know the yield level that would obtain the break-even cost for a given price ($/unit). This break-even yield level can be calculated as the total cost divided by the price.

Range Analysis

Growers can also calculate the impacts of yield and price variations on profitability. A range analysis allows growers to determine the gross and economic margins based on a combination of expected yield and prices.

How to Determine the Cost of Production and Returns

Several templates can be used to develop costs of citrus orchard establishment and production. Most cost studies done at the University of California use the Budget Planner software, version 2.0, by Karen Klonsky. Also, a new Excel-based template has been developed by Eta Takele and can be requested directly from the author. For detailed cost studies on California lemons, mandarins, and oranges, see the UC Davis Cost Studies website, http://coststudies.ucdavis.edu. These cost studies are based on assumptions of practices and costs that are applicable to the areas for which they are made.

References

Agricultural County Reports, 2008-2012. UC Agricultural Economics website, http://ucanr.org/sites/Farm_Management/County_Agricultural_Reports100/.

Boehlje, M. D., and V. R. Eidman. 1984. Farm management. New York: Wiley.

California Department of Insurance, Rate Regulation Branch. 2012. Website http://www.insurance.ca.gov/0500-about-us/0500-organization/0400-rate-regulation/.

Cal/OSHA (California Department of Industrial Relations, Division of Occupational Safety and Health). 1995. Cal/OSHA field sanitation fact sheet. Cal/OSHA website, www.cal-osha.com/articles/COR04-20060606-032.pdf.aspx.

CASS (California Agricultural Statistics Service). 2012. California citrus acreage report, August 23, 2012.

DPR (California Department of Pesticide Regulation). 2009. California restricted materials requirements. DPR website, http://www.cdpr.ca.gov/docs/enforce/pr-enf-013a.pdf.

IRS (U.S. Department of the Treasury, Internal Revenue Service). 2011. Farmer's tax guide. IRS Publication 225. IRS website, http://www.irs.gov/pub/irs-pdf/p225.pdf.

NASS (USDA National Agricultural Statistics Service). 1995–2012. Farm labor. NASS website, http://usda.mannlib.cornell.edu/MannUsda/viewDocumentInfo.do?documentID=1063.

O'Connell, N. V., C. E. Kallsen, K. M. Klonsky, and R. L. De Moura. 2009. Sample costs to establish a grove and produce oranges. UC Davis Department of Agricultural and Resource Economics website, http://coststudies.ucdavis.edu/files/orangevs2009.pdf.

———. 2010. Sample costs to establish a grove and produce lemons. UC Davis Department of Agricultural and Resource Economics website, http://coststudies.ucdavis.edu/files/lemonvs10.pdf.

Takele, E., B. Faber, and M. Vue. 2012. Avocado sample establishment and production costs and profitability analysis for Ventura, Santa Barbara, and San Luis Obispo Counties, 2011. UC Davis Department of Agricultural and Resource Economics website, http://coststudies.ucdavis.edu/files/2011/AvocadoConventionalVSBSLO2011.pdf/.

University of California, Davis, Department of Agricultural Resource Economics. Cost studies website, http://coststudies.ucdavis.edu.

University of California Postharvest Alternatives to Methyl Bromide Use website, http://ucanr.org/sites/phmba/.

University of California Statewide Integrated Pest Management Program. UC IPM citrus pest management guidelines. UC IPM website, http://www.ipm.ucdavis.edu/PMG.

Part 2 Citrus Botany and Physiology

3 Physiology and Phenology

Carol J. Lovatt

Commercial citrus cultivars are evergreen, woody perennial fruit trees. Botanically, they are classified as Anthophyta, flowering plants that produce seed enclosed in a fruit, and eudicots, dicots that produce seed with two cotyledons. Unlike annual vegetable crops in this group, for which the focus is on a single season's harvest with the opportunity to begin anew each year, perennial fruit trees require growers to support the current season's crop and also to maintain tree health and vigor for future years. Citrus, like many woody perennials, is polycarpic and not genetically programmed to die. Thus, theoretically, citrus species could produce fruit annually in perpetuity if the trees are kept healthy. In essence, woody perennial fruit trees have "memories": factors influencing the tree one year can have repercussions for several years thereafter. The evergreen character of citrus means that they drop their leaves a few at a time throughout the year, not all at once in the fall as is typical of deciduous fruit trees. The greatest number of leaves is shed at the time of flowering and early fruit set, with more leaves shed during a heavy bloom than during a light bloom. Note that there are deciduous citrus relatives, such as members of the genus *Poncirus* (e.g., *Poncirus trifoliata* L. Raf. cv. trifloliate orange); in the citrus belt, most *Poncirus* species are used as rootstocks. Thus, their deciduous growth habit is not readily apparent.

Phases in the Life Cycle of a Citrus Tree

All plants, even woody perennials like citrus, have a basic life cycle that begins with germination of a seed and growth of the seedling through a juvenile phase, a transitional immature adult phase, and an adult phase. Citrus trees grown from seed have a long juvenile phase, lasting 5 to 13 years, depending on the cultivar, in comparison with 20 to 30 years for other tree fruit crops and forest tree species and less than 1 to 2 years for annual and biennial herbaceous species, respectively. In the juvenile phase, flowering plants are sexually incompetent and unable to flower and reproduce sexually even in the presence of the proper environmental cues; thus, in the juvenile phase they are restricted to vegetative growth. Subsequently, plants undergo phase transition and enter the adult phase, in which they are sexually competent and can flower and sexually reproduce in response to developmental and environmental cues. Sexual reproduction occurs within the carpel of the flower and typically leads to the development of a zygotic embryo within a zygotic seed. The seed is within the fruit. The relationship between an ovule and a seed, an ovary and a fruit, pollination, self-incompatibility, out-crossing, fertilization, and embryo development are discussed later in this chapter under "Flowering." Among commercial citrus cultivars there are several variations in the life cycle, including parthenocarpy, stimulated parthenocarpy, and nucellar embryony; these are also discussed under "Flowering." Woody perennial plants undergo additional secondary growth that results in the formation of the wood (xylem tissue) and the outer layer of bark of the trunk, scaffold branches, and main roots of the tree.

To shorten the time to flowering (and for other reasons discussed later in this chapter and in other chapters), scion cultivars in the adult phase are grafted onto rootstocks for planting in a commercial citrus grove. Despite the grafting of adult scion tissue onto the selected rootstock, flowering is still delayed 3 to 6 years after grafting, as the scion material passes through an immature adult stage to a mature adult stage. This delay in flowering results in a significant economic loss to growers. "Precocity" refers to the shortness of the time to flowering. Thus, scion-rootstock combinations and cultural practices that promote earlier flowering (precocity) and, hence, earlier commercial bearing, have significant economic value. For example, if huanglongbing (citrus greening disease), which typically shortens tree life, reaches California, it will be imperative to develop cultivars and cultural practices that optimize precocious production.

Rootstocks grown from seed or from adventitious buds from the roots or trunk base of adult trees are juvenile. Passage of progressively newer growth through the transitional phase to the adult phase does not alter the juvenile tissues at the base of a tree. The base of the tree remains juvenile, and the tissues and buds at the base of the tree are juvenile. Plants in the juvenile phase have differences in gene expression, physiology, and hormone composition and subtle to extreme differences in morphology compared with adult plants. Among citrus cultivars, the juvenile and adult phases are characterized by the following differences: It is easier to promote root formation on juvenile stems than on adult stems, juvenile stems (shoots and branches) tend to be more thorny than adult stems, and it is difficult to promote flowering on juvenile plants, whereas flowering is easily induced on adult plants.

Grove Location

Successful commercial citrus production requires wise selection of the grove location; choice of a scion cultivar and a rootstock that are appropriate to the site; implementation of cultural management practices that ensure tree health and optimize yield, fruit size, and fruit quality; and production strategies that provide the greatest economic advantage in the market. Important characteristics of grove location that influence yield parameters are climate, soil, and quality of the irrigation water supply. For more information, see chapter 9, "Establishing the Citrus Orchard."

Climate

In general, the lowest temperature at which a citrus tree can survive and reproduce determines the areas in which citrus can be grown. The citrus belt lies between latitude 35°N and latitude 35°S. This encompasses the area where average winter and summer temperatures are 60°F or greater. The maximum northern latitude at which citrus is grown is in Spain at 44°N, due to the warming effects of the Canary Current. The maximum northern latitude in California is 39°N due to the temperature-modifying effect of the Japanese Current. Citrus fruit with the best quality are grown at the outer limits of the citrus belt, where there is a potential for frost; cold damage to the trees or loss of the crop can be expected to occur in some years. Tropical regions of the citrus belt produce fruit of poor quality—that is, low acidity, low sugar content, and pale color. Fruit quality is further compromised by the fact that flowering in the tropics is weak and continuous, so that harvests are unavoidably comprised of fruit at various degrees of maturity. In addition, high temperature and humidity lead to fungal and insect pest damage to fruit and trees. Thus, temperature not only limits where citrus can be grown but also impacts the uniformity of annual bearing and fruit quality.

California has a subtropical climate characterized by low humidity in the inland areas and rainfall of less than 10 inches per year, typically concentrated in the winter months. In contrast, Florida has a semitropical climate characterized by high humidity and significant annual rainfall (50 in/yr); summers are hot and winters are generally warm, with occasional cold spells and devastating freezes.

Microclimate

Citrus trees grow best between 70° and 90°F. When temperatures reach 100°F or drop to 50°F, the growth rate of citrus trees is markedly reduced. Reduced growth in response to cold winter temperatures approaches dormancy (ecodormancy, quiescence), but citrus trees do not go dormant (endodormancy, rest) the way deciduous fruit trees do. During this period of reduced growth, trees maintain a low rate of water transport, hydrolysis of starch to glucose, and glucose metabolism via cellular respiration, which produces energy and maintains tree physiology. When choosing a grove location, microclimate must be considered even if the grove is within the citrus belt. Nearby bodies of water moderate the climate, making it cooler in summer and warmer in winter.

Air drainage

A grove should have good air drainage. At night, cold air moves down a slope and settles at the base of the slope or at the lowest part of a valley. For each 100 feet of rise in elevation out of a valley, the nighttime minimum temperature increases about 5° to 11°F. In California, south-facing slopes are warmer than north-facing slopes in a grove. A slight wind can keep cold air from settling. Strong, hot, desiccating winds increase water-deficit stress, break limbs bearing fruit, and shred leaves or defoliate the tree, whereas cold winter winds can cause freeze damage. Local obstructions should be considered, as obstructions below a grove can block air drainage and create frost pockets. Solid obstructions aggravate wind problems by increasing turbulence on both sides of the obstruction. Properly placed windbreaks can provide an area of complete wind protection five to six times their height, and they can provide some protection against advective freezes caused by horizontally moving cold air. On the other hand, improperly placed windbreaks can trap cold air moving down a slope, creating a frost pocket. Other disadvantages of windbreaks include shading, possible harboring of insect pests, and reducing the area that can be planted in citrus (see Platt 1973).

Soil

The chemical and physical properties of soil are important to citrus. Soil pH can limit nutrient uptake and productivity when it is below 6.0 or above 7.5. Citrus can survive in soils from pH 5.5 to 8.5, but nutritional problems may occur as the pH at the ends of this range are approached (see chapter 8, "Soil and Water Analysis and Amendment," and chapter 10, "Nutrient Deficiency and Correction"). Salt and toxic ions can negatively impact citrus production. Low levels of calcium (Ca) and magnesium (Mg) relative to high levels of sodium (Na) can also affect the physical properties of soil. The ratio of sodium to calcium plus magnesium in a saturated paste extract of the soil is the sodium adsorption ratio (SAR). For more information, see chapter 8, "Soil and Water Analysis and Amendment," and chapter 10, "Nutrient Deficiency and Correction."

Irrigation Water

The quality of the irrigation water influences the chemical properties of the soil. Irrigation water must be compatible with both the citrus cultivar to be grown and the soil to which it will be applied. The quality of irrigation water is determined by its salt, boron, and bicarbonate content. An analysis of water for irrigation should include the cations calcium, magnesium, and sodium, and the anions bicarbonate, carbonate, sulfate, borate, and chloride. For additional information on irrigation water quality and water quality analysis, see chapter 8, "Soil and Water Analysis and Amendment."

Roots

Since citrus scion cultivars of commercial importance are grafted (usually as buds) onto a rootstock that is typically a different cultivar or even a different species, the roots of the trees in a commercial citrus grove have a different genetic makeup than the canopy and fruit. Rootstocks are selected first for their compatibility with the scion cultivar to be grown and second for a wide range of other characteristics, which include nutrient uptake, pH sensitivity or tolerance, performance under the soil conditions in the grove, salinity tolerance, fungal and nematode resistance, yield, fruit size, time of fruit maturity, fruit quality, precocity, vigor, tree size and shape, cold hardiness, and drought resistance. The choice of rootstock also influences hormone production, source-sink relations within the tree, and efficiency of nutrient uptake, water use, and yield. In fact, it is nearly impossible to think of an aspect of tree physiology and phenology or yield that is not affected by the choice of rootstock. When you change the rootstock, you change the roots. Whereas roots are underground and frequently overlooked, their impact on the tree is best summarized by the words of the citrus researcher Frank Turrell: "As the roots grow, so grows the tree" (Turrell et al. 1969).

Plant Water Relations

Plant water relations begin at the roots (fig. 3.1). Water, which carries dissolved nutrients and gases and other compounds from the soil, enters the plant through the cells of the epidermis, the outer layer of cells covering the root. Root hairs, specialized epidermal cells, are very active in the uptake of water, dissolved nutrients, and gases. From the epidermis, water flows laterally through the root cortex toward the center of the root, or stele. The stele contains the xylem and phloem transport tissues of the plant, which run from the root up the stem, out into the branches and young shoots, and finally into the leaves and fruit. In the root, there is a ring of cells around the stele called the endodermis.

Cells of the endodermis are encircled with the Casparian strip, a band of suberin, a waxy material. However, the front and back surfaces of the endodermal cells have no suberin. The flow of water through the cortex occurs in the walls of the cells of the cortex and in the spaces between cells, both are apoplastic (nonliving) parts of the plant and part of the plant's apoplastic transport system. The Casparian strip of the endodermis blocks the lateral flow of water in the apoplast and forces the water to enter the living part (symplast) of a cell. This forced symplastic movement results in the selection of dissolved substances that are allowed to move beyond the endodermis into the stele in contrast to substances that are prevented from moving any farther. The Casparian strip is responsible for saving the life of the plant by forcing material in the apoplast to the symplast, where, to varying degrees, toxins can be

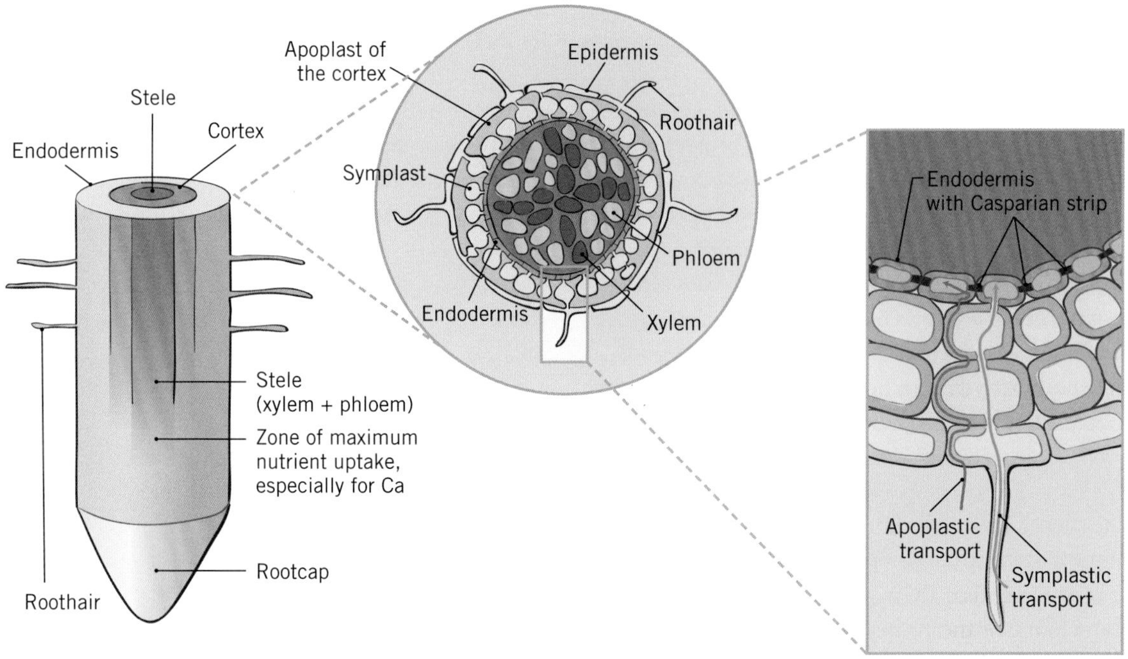

Figure 3.1 Anatomy of a dicot root. (A) Longitudinal section of a root; (B) cross section of a root; and (C) close-up of the cortex, illustrating symplastic and apoplastic transport.

detected and prevented from entering into the stele. Some citrus rootstocks are able to sequester sodium or chloride in the cortex of the root and thus protect the canopy.

Once in the stele, water and dissolved substances can again move apoplastically in the xylem transport tissue. (The actual conducting cells of the xylem, tracheids and vessel elements, are not living when they are functioning in transport.) Water, and substances dissolved in the water, moves in the tracheids and vessels of the xylem transport tissue up the tree, out into the branches and young shoots, into the leaves, and eventually out through the open stomates (pores in the leaf formed and opened and closed by two guard cells each) of the leaves into the air. This continuous column of water from the soil through the plant to the air is known as the soil-plant-air continuum (SPAC). In fact, water is so prevalent in the nonliving parts of the plant that make up the apoplast and in the living cells that make up the symplast that water accounts for 80 to 90% of the fresh weight of a plant.

Nutrient Uptake and Assimilation

Plants require 17 essential nutrient elements: carbon (C), oxygen (O), hydrogen (H), nitrogen (N), phosphorus (P), potassium (K), calcium (Ca), magnesium (Mg), sulfur (S), iron (Fe), chlorine (Cl), manganese (Mn), boron (B), zinc (Zn), copper (Cu), molybdenum (Mo), and nickel (Ni). An element is considered essential if it meets the following three criteria (Arnon and Stout 1939):

- The organism cannot complete its life cycle if this element is lacking.
- The element cannot be replaced by another element.
- The element must have a distinct function.

For more information on these essential nutrients, see chapter 10, "Nutrient Deficiency and Correction."

All nutrients must be in solution to be taken up by the cells of any plant. Ion channels in plant cell membranes facilitate the uptake of some essential elements. Plants contain highly selective channels for the cations H^+, K^+, and Ca^{+2} and the anion Cl^-. Carrier proteins are essential for the uptake of some other nutrients such as nitrate (NO_3^-). Nutrients taken up by the roots are moved in solution to other parts of the plant via the conducting cells (tracheids and vessels) of the xylem transport tissue by the transpiration-cohesion-tension mechanism. Transpiration is the evaporation of water from the cells within the leaf through the open stomates of the leaf. Transpiration creates tension on the column of water in the SPAC. Because water molecules are attracted to each other by cohesive forces, the whole column of water moves from the soil through the roots, up the stem, out into the branches, and out of the leaves into the air via the open stomates. In this manner nutrients and other substances dissolved in the water in the apoplast of the plant are also moved throughout the plant.

Transpiration occurs only when stomates are open. Thus, the movement of dissolved nutrients and other substances occurs only when the stomates are open. Low soil temperatures ($\leq$ 55°F) from January to April in citrus-growing areas of California reduce solubility of nutrients in the soil solution, root metabolic activity, and nutrient uptake and nutrient transport in the transpiration stream (Hamid et al. 1988). Transpiration and nutrient uptake and movement are further reduced when air temperatures are also low or when humidity or precipitation is high. Thus, the ability of trees to use nutrients applied to the soil depends on many factors unrelated to nutrient demand or the presence of adequate amounts of each nutrient in the soil, including soil temperature; soil moisture; soil pH; soil sand, clay, or organic matter content; salinity; air temperature; and humidity. It is not unusual for leaves of citrus trees to be pale or yellow in a cold, wet spring due to an inability to take up sufficient nutrients to meet the demand of the developing flush. Citrus flowering and fruit set, periods of high nutrient demand, frequently occur when soil temperatures are still low and soils are still wet despite the onset of sunny days and moderate air temperatures. Although typically transient, foliar fertilizers should be used to overcome these deficiencies to reduce potential negative impacts on fruit set, fruit growth rate, and fruit quality.

Once in the plant, nutrients can also move from one part of the plant to another by translocation in the conducting cells of the phloem transport tissue. Unlike the transport cells of the xylem, the conducting cells of the phloem tissue are living when they are functioning in transport. The living transport cells of the phloem are sieve cells, which are aligned end-to-end to form continuous sieve tubes from the leaves out into the shoots and branches and down the trunk to the roots. The rate of phloem transport is nutrient specific. For more information on the phloem mobility of the 17 essential nutrient elements, see chapter 10, "Nutrient Deficiency and Correction."

Ultimately, nutrients dissolved in the water in the apoplast enter the living cells of the root, xylem and phloem tissues of the stem, branches, leaves, flowers, and fruit, where they are used in the metabolism, physiology, and structure of the plant (table 3.1). For a more detailed discussion of the role of each of the 17 essential nutrients in citrus, see chapter 10, "Nutrient Deficiency and Correction."

Root Growth

The first year that a citrus tree is planted in the grove is the most important for root development. Stresses caused by diseases, nematodes, weed competition, insufficient watering, or overwatering can hinder root and scion development and are particularly devastating in the first year. Citrus roots tend to grow where the soil is moist and can extend up to twice as far as the drip line, that imaginary circle below the canopy edge. These facts must be considered when fertilizing your grove other than through the irrigation system. Roots grow in flushes that tend to alternate with vegetative shoot flushes.

In general, in a commercially bearing navel orange grove in California, the first root flush occurs about 4 to 6 weeks following flowering and fruit set (approximately mid-June to July) (fig. 3.2). The first root flush coincides with rising soil temperatures and the initiation of irrigation. This is likely coincidental and not cause-and-effect because soil temperatures at this time are above the threshold temperature (about 45°F) that limits root growth. The second root flush occurs from late summer into fall about 4 to 6 weeks following the first root flush when soil temperatures have reached their peak of about 73°F. Irrigation is scheduled most frequently at this time. A third root flush occurs in late fall, about 6 weeks following the second root flush.

New roots are the most active in terms of water and nutrient uptake, assimilation, and metabolism. Thus, new roots of each flush must be protected from mechanical, fungal, nematode, and insect damage.

Root Metabolism

Roots play an important role in the synthesis, catabolism, and storage of metabolites, including hormones. Roots are sinks that import sucrose (the transported form of carbohydrate in citrus) from the leaves. Roots either catabolize carbohydrate imported as sucrose through the process of cellular respiration, which provides the energy and building materials to support further anabolism (synthesis of large, complex molecules) and root growth (spring through fall), or they store the imported carbohydrate as starch during the winter. In cellular respiration, oxygen (O_2) is used to break down the glucose molecules that make up starch. Carbon dioxide (CO_2), water (H_2O), and energy are released from the catabolism of glucose. The energy produced is stored in ATP molecules. ATP is the "energy currency" of living organisms (including plants), supplying energy for growth, maintenance, repair, reproduction, and responding to the environment. Nonphotosynthetic tissues such as roots

Table 3.1. The role of essential nutrients in plant metabolism, physiology, and structure

Element	Abbr.	Form used	Major function
			Macronutrients
carbon	C	CO_2	
oxygen	O_2	O_2 CO_2 H_2O	sugars; amino acids; fats; nucleotides; hormones
hydrogen	H_2	H_2O	
nitrogen	N_2	NO_3^- NH_4^+	amino acids; nucleotides; hormones
potassium	K	K^+	role in ionic balance of cells; role in stomate opening and closing; cofactor in protein synthesis
calcium	Ca	Ca^{+2}	component of the middle lamella; influences permeability of membranes; possible role in gravitropism
magnesium	Mg	Mg^{+2}	part of chlorophyll molecule; ATP-Mg complexes in energy metabolism; stabilizes ribosomes in protein synthesis
phosphorus	P	$H_2PO_4^-$ HPO_4^{-2}	ATP; nucleotides; phospholipids; sugar phosphates
sulfur	S	SO_4^{-2}	part of two amino acids: cysteine and methionine; therefore, essential to protein synthesis
			Micronutrients
chlorine	Cl_2	Cl^-	required for splitting $H_2O \rightarrow 2H + O_2$ in photosynthesis
iron	Fe	$Fe^{+2,+3}$	molecular component in enzymes in electron transport chains; necessary for the synthesis of chlorophyll
manganese	Mn	Mn^{+2}	enzyme cofactor; ATP-Mn complex in energy metabolism; required for splitting $H_2O \rightarrow 2H + O_2$ in photosynthesis
boron	B	H_3BO_3 $H_2BO_3^-$	unknown, but most likely required for some aspect of cell division and cell wall structure
zinc	Zn	Zn^{+2}	enzyme cofactor; necessary for synthesis of chlorophyll
copper	Cu	$Cu^{+1,+2}$	enzyme cofactor in electron transport; lignin synthesis
molybdenum	Mo	Mo^{+2}	enzyme cofactor in nitrate reduction
nickel	Ni	Ni^{+2}	urease cofactor (urea catabolism)

Note: The leaf takes up CO_2 and O_2. The roots take up H_2O and dissolved O_2, HCO_3^- and all other nutrients.

rely heavily on respiration to acquire the energy and building materials needed for the synthesis of new molecules and growth. Thus, it is essential to have well-aerated soil in your grove to provide the roots with the oxygen needed to carry out respiration. The energy and carbon skeletons provided by respiration are used to synthesize nitrogen-containing compounds. Citrus roots synthesize predominantly glutamate, glutamine, proline, and arginine, which are transported in the xylem to leaves, flowers, and developing fruit. However, citrus roots take up a significant amount of nitrate nitrogen that is not reduced in the roots but rather transported to the leaves, where reduction and assimilation occur. Roots also play an important role in hormone metabolism, which is discussed below in the section on hormones.

Canopy

Shoot Growth

Citrus shoot development is sympodial, a type of plant development in which the terminal bud of the shoot stops growing—due either to its abortion or its development into a flower or an inflorescence—and the uppermost lateral bud takes over the further growth of the shoot. In California, citrus trees generally produce three distinct vegetative shoot flushes: spring, summer, fall (see fig. 3.2). The latter two flushes are diminished when trees are setting a heavy crop. The relationship between the phenology of shoot and root growth in citrus is not well understood. The many studies reported in the literature present conflicting results, which are likely due to differences in climate or tree age or differences associated with the scion or rootstock cultivar; the influence of air and soil temperature and soil moisture on shoot and root growth have been well studied. However, Bevington and Castle (1985) found that when environmental factors were non-limiting, shoot growth was the major factor limiting root growth in young trees, 16-month-old Valencia orange trees on Rough Lemon *(Citrus jhambiri)* and Carrizo Citrange *(Poncirus trifoliata* × *C. sinensis)* rootstocks. Alternation of shoot and root growth, with shoot growth preceding root growth, was most often reported with seedlings, recently budded nursery trees, and young trees up to 6 years of age (Crider 1927; Marloth 1950; Reed and McDougal 1937). To date, there is no scientific basis to justify ascribing this phenology to older trees in commercial production. Even with seedlings and recently budded nursery trees, Marloth (1950) observed sufficiently frequent variations from the general trend of alternate shoot and root growth that he proposed that alternation was not a physiological characteristic of citrus, but one imposed by climate. In the only study of mature trees in commercial production, Schneider (1952) reported that after the initial spring shoot flush, shoot and root growth were concurrent for 30-year-old Valencia orange trees *(C. sinensis)* on Sweet Orange rootstock *(C. sinensis)*. Concurrent shoot and root growth was also observed for bearing trees of Hamlin sweet orange *(C. sinensis)*, Nagpur Santra *(C. reticulata)*, and sweet lime *(C. aurantifolia)* despite the fact that they were only 7 years old (Krishnamurthi et al. 1960).

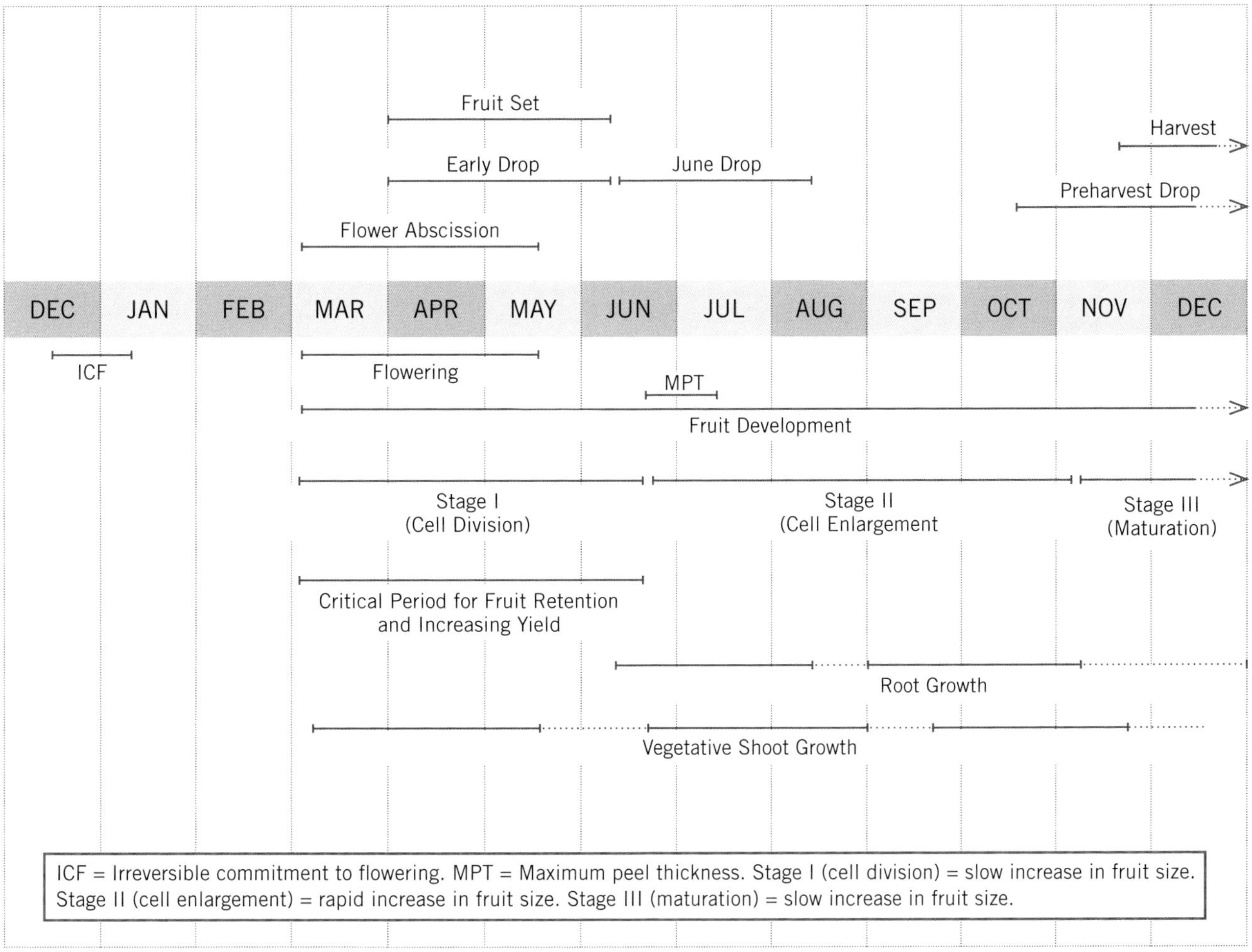

Figure 3.2 Phenology of the Washington navel orange in California. *Source:* Adapted from Lovatt 1999.

Hamid, Van Gundy, and Lovatt (1985, 1988) documented that flushes of shoot and root growth comprised of both root initiation and elongation occurred concurrently on healthy 20-year-old Washington navel orange trees on Troyer Citrange rootstock during off-crop and on-crop years of an alternate-bearing grove. For comparison, carbohydrate depletion was induced by infection of the roots by the citrus nematode (*Tylenchulus semipenetrans* Cobb) at a rate of greater than 2,000 female nematodes per gram of roots each year. Hamid et al. (1985) previously demonstrated that this level of root infection by the citrus nematode reduced the carbohydrate status of Washington navel orange trees on Troyer Citrange rootstock. Starch levels, which are a good indicator of carbohydrate reserves, were 70% lower in the canopy and 96% lower in the roots of nematode-infected trees than in trees in the same block recovering from prior nematode infection and protected from subsequent infection by application of oxamyl (Hamid et al. 1985). For nematode-parasitized trees, shoot and root flushes alternated. Concurrent shoot and root growth increased when parasitized trees were treated with oxamyl. Shoot and root growth are coordinated by root-produced cytokinins that move in the xylem to the leaves and stimulate shoot growth and leaf-produced cytokinins translocated in the phloem that stimulate root growth. Cytokinin biosynthesis in each organ is directly controlled by nitrogen concentration and indirectly controlled by carbohydrate availability (Sakakibara 2006).

Mature tree size depends primarily on the scion and rootstock cultivars used, but it is also affected by growing conditions and cultural practices. Mature, standard-sized orange and grapefruit trees are 20 and 30 feet tall, respectively. The only true genetic dwarf rootstock currently available is Flying Dragon trifoliate orange. On Flying Dragon rootstock, mature trees grow to less than half of standard size. Other rootstocks may have a dwarfing effect on mature tree size of certain scions; such trees may be labeled semidwarfs and grow to about two-thirds the standard size.

Trunk, Branches, and Shoots

The role of the trunk, branches, and shoots of a citrus tree is three-fold. The first is supporting the leaves and reproductive structures of the tree. In this regard, the goal of the grower should be to minimize the length and mass of the trunk, branches, and shoots in favor of complexity with regard to shoot age; that is, there should be shoots that will mature and flower the next spring and others that will mature and flower the following spring. This complexity generates more wood on which to bear inflorescences and more fruit per unit of canopy volume for maximum yield; provides an adequate number of leaves to produce sufficient photosynthate to support the growth of the developing fruit to achieve optimal size, as well as the growth of new shoots and roots. In addition, the greater complexity of the branching system improves the distribution of the fruit relative to the source leaves. The second role of the trunk, branches, and shoots is in transport. A good balance between the number of leaves and fruit per shoot ensures that water, dissolved nutrients, gases, and root-produced hormones and metabolites transported from the roots to shoots in the tracheids and vessels of the xylem are distributed uniformly to the leaves and developing flowers and fruit. This balance also ensures that adequate carbohydrate and shoot-produced hormones and metabolites are translocated in the phloem and evenly distributed to flowers and fruit and the root system. There is evidence in citrus that transport in both the xylem and phloem is predominantly up and down with limited lateral movement. Therefore, a better distribution of branches around the trunk is needed to ensure that the supply of photosynthate is uniformly distributed to roots. This, in turn, helps ensure that root growth is uniform around the trunk of the tree so that the canopy is uniformly supplied with water and nutrients. Gaps in the canopy or roots lead to unproductive areas of the tree. Remember also that some essential mineral nutrient elements are redistributed from the leaves via the sieve tubes of the phloem in leaves, shoots, and branches to developing flowers, fruit, and new shoots and leaves. The third role of the trunk, branches, and shoots, along with the leaves and roots, is to store nutrient and carbohydrate reserves. In addition, waste products are sequestered in the oldest (centermost) nonconducting xylem (heartwood) of the trunk.

Xylem and phloem are complex tissues made up of four cell types that develop from the vascular cambium, a meristem located between the xylem and phloem tissues in the roots, trunk, branches, shoots, and veins of leaves and fruit. The vascular cambium produces new cells by cell division from approximately March through August. These cells differentiate into phloem starting in March. Sieve tubes of the phloem begin active transport in late March and continue active transport through October. New xylem tissue forms from May through August, but xylem transport continues all year as long as transpiration is occurring. Transpiration can occur from any surface with stomates, such as leaves, fruit peels, and buds. Because transpiration is essential to transport in the xylem, transport is always from roots to shoots. Phloem transport is always from source to sink. A source, typically a leaf, is a site of active (energy-requiring) or passive (by plasmodesmata) loading of photosynthate, typically sucrose, into the sieve cells of a sieve tube. The increased concentration of photosynthate in the sieve tube results in the influx of water from the tracheids and vessels of the xylem. This creates water pressure that pushes the contents of the sieve tube to a sink. Being a site of unloading, sieve cells in the sink have a low photosynthate concentration, low water content, and low water pressure. Typically a sink is a site of active growth, such as a young developing tissue or organ (e.g., a shoot, young leaf, or fruit) that cannot produce enough photosynthate to support its own needs. A sink can also be a site of storage, such as the roots of citrus, which store starch from fall through winter. However, in the spring, the roots become a source and the flowers become a sink: the roots are the site of starch catabolism and are the site of loading of sucrose into the sieve tubes for translocation to and use in flowering. Movement in the phloem can be up or down the tree, depending on the position of a source and its sink. Parallel sieve tubes can simultaneously transport photosynthate in opposite directions. Source and sink activity is also controlled by hormones: loading at the source is stimulated by indole-3-acetic acid (IAA) and inhibited by the accumulation of abscisic acid (ABA), whereas sink activity is stimulated by ABA and inhibited by the accumulation of IAA. Sink strength, the capacity of a tissue or organ to import metabolites needed for growth, is maintained by the export of IAA from the sink organ and by cytokinin activity within the sink.

Leaves

Citrus trees have thick, leathery leaves with the classic anatomy of a dicot leaf (fig. 3.3). The outer layer of upper and lower epidermal cells (epidermis) has a waxy cuticle that protects the mesophyll cells that perform photosynthesis from mechanical, pathogen, or insect damage, as well as from desiccation. There are two types of mesophyll cells in dicot leaves: the upper, elongate, palisade cells and the lower, round, loosely packed, spongy cells. Both types contain chloroplasts, the organelles in which photosynthesis occurs. In plants, the leaf is the major organ for photosynthesis. In citrus, the evergreen leaves, along

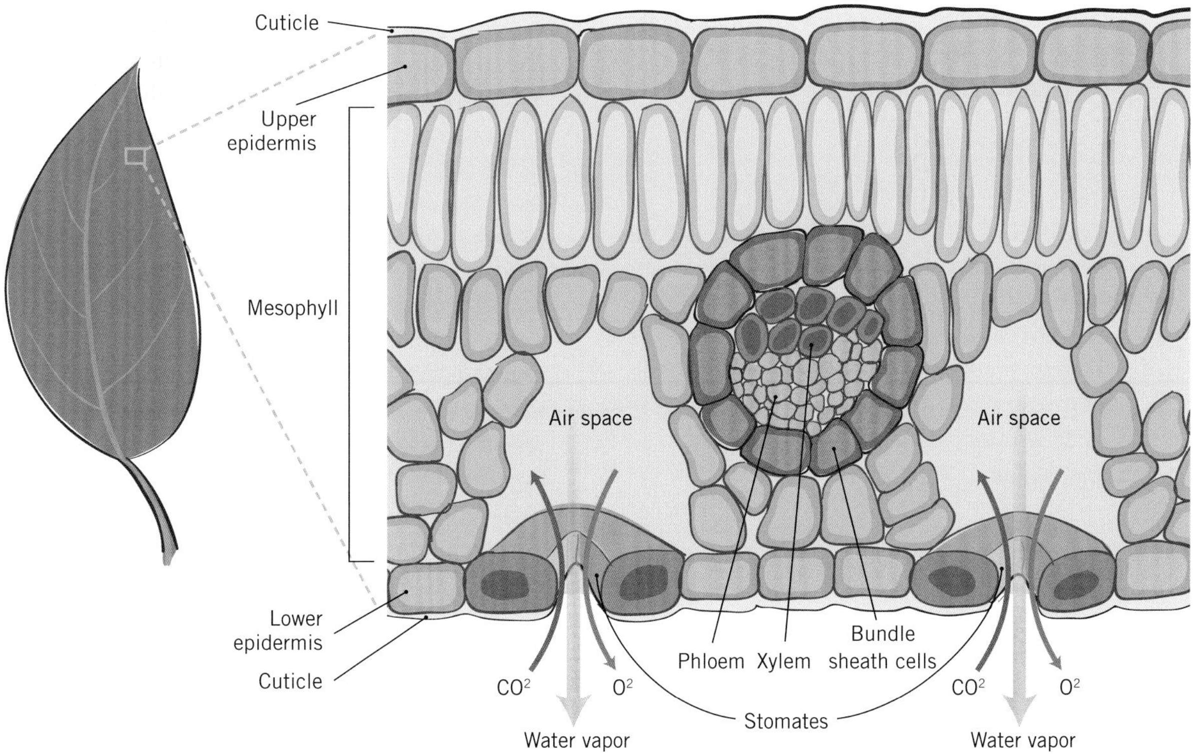

Figure 3.3 The anatomy of a dicot leaf.

with twigs, branches, and roots, are also involved in storing excess photosynthate as starch through the winter. In citrus leaves, the greatest amount of starch accumulates by late February, just before the spring bloom and vegetative shoot flush (Hamid et al. 1985).

Leaves are arranged spirally on the stem and typically remain on the tree for 1 to 2 years. Environmental factors such as high temperatures, wind, low soil moisture, low relative humidity, nutrient deficiencies, and high soil salinity can cause premature leaf drop. Other causes are disease or pest problems and scion-rootstock incompatibility. An axillary (lateral) bud occurs in the axil of each citrus leaf. The meristem in each bud can produce a vegetative shoot, or it can transition to a floral meristem and produce a floral shoot. During the spring flowering period of most citrus types, leaf drop can be pronounced and new leaf growth is most vigorous.

The primary functions of citrus leaves are photosynthesis, transpiration, nitrate reduction, and primary and secondary metabolism. The leaves of some citrus species contain secondary metabolites that are important sources of pharmaceutical products, perfumes, and condiments. For photosynthesis and transpiration to take place, the stomates must be open. Two specialized epidermal cells called guard cells regulate the opening and closing of each stomate (for more information, see "Potassium" in chapter 10, "Nutrient Deficiency and Correction"). In most dicots, stomates open in the morning in response to sunrise and remain open until they close for a couple of hours during the hottest part of the day, the so-called midday depression in photosynthesis, and then reopen for a period of time before closing for the night. Citrus leaves are unique in that the guard cells open and close in an unsynchronized cyclic manner throughout the day and thus have a poorly defined midday depression. Intake of carbon dioxide through open stomates for photosynthesis, the export of oxygen generated as a waste product of photosynthesis, and the loss of water vapor during transpiration are illustrated in fig. 3.3.

Photosynthesis

Once a new grove is planted, the goal is to bring the grove into bearing and to generate grower income. Growth of new shoots, leaves and roots; increased diameter of branches, trunk and roots; and production of flowers and development of the ovaries into mature fruit require building materials and energy. Plants are unique among most organisms because they are autotrophic, able to produce their own food and energy through photosynthesis. This process uses the simple inorganic molecules of carbon dioxide and water along with light energy to synthesize energy-rich organic food molecules. Plants must also break down the energy-rich food molecules that result from photosynthesis by the process of cellular respiration to obtain the energy and building materials stored in these molecules.

Transpiration

Transpiration is the evaporation of water from the surface of cells inside the leaf. Evaporation and transpiration can occur only when the stomates of the leaf are open (see fig. 3.3). Transpiration exerts a pull that contributes to the uptake and movement of water and dissolved nutrients and dissolved gases (CO_2 and O_2) from the soil through the roots and up the xylem to all parts of the tree. Thus, water and dissolved nutrients and gases move efficiently throughout the plant via the xylem only when the stomates are open. The opening and closing of the stomates is controlled by two guard cells that form each opening in the leaf. Guard cells of the stomates have the tough job of managing the photosynthesis-transpiration compromise. Stomates have to be open for sufficient photosynthesis and transpiration to take place to meet the tree's needs for energy and building materials (products of photosynthesis and water and nutrient uptake) for growth and fruit development, but should not be open so long that the tree loses too much water and becomes stressed.

Nitrate reduction and primary and secondary metabolism

To be used by the plant, nitrate (NO_3^-) must be reduced to ammonium (NH_4^+), which is incorporated into glutamate; additional ammonium is incorporated into glutamate to form glutamine. These two amino acids are the predominant sources of amino nitrogen for the synthesis of other nitrogen-containing compounds. Nitrate reduction to ammonium is a multistep process carried out in roots and leaves. Research (Lovatt and Cheng 1990) has shown that leaves of numerous citrus scion cultivars have much greater rates of nitrate reduction and glutamine and protein synthesis than the citrus rootstock cultivars commonly used in California. Glutamine, along with the amino acid asparagine, link carbon and nitrogen metabolism in plants. Primary, or basal, metabolism produces the building materials for plant growth, development, reproduction, and secondary metabolism. Primary metabolites include sugars, amino acids, organic acids, fatty acids, and nucleotides. Secondary metabolites include a broad range of compounds in three major groups: nitrogen-containing secondary products (alkaloids), propenoids, and terpenoids.

Secondary metabolites within these three groups include toxins and feeding deterrents to insects and herbivores, antimicrobial compounds (phytoalexins), chemical attractants of pollinators and beneficial insects, pigments of flowers and fruit, UV radiation protectants, and structural compounds like lignin found predominantly in wood.

Hormones

Hormones are chemical signals found in all multicellular organisms. In plants, hormones coordinate the plant's growth, development, physiology, and response to its environment. Hormones are produced by one part of the tree and transported to other parts, where they elicit a specific response. There are five classes of plant hormones: auxins, gibberellins, cytokinins, ABA, and ethylene; other compounds are being evaluated as potential plant hormones, including salicylic acid and jasmonic acid, signal molecules that induce defense reactions and wound repair in response to pathogen, insect, herbivore, or mechanical damage to the plant. Many growth, developmental, and physiological processes are regulated by the relative amounts of two or more hormones. For a more detailed discussion of the role of hormones in citrus physiology and production, see chapter 14, "Plant Growth Regulators."

Indole-3-acetic acid (IAA)

Indole-3-acetic acid (IAA) is the major auxin produced by plants. Recent research indicates that plants also produce indolebutyric acid (IBA). IAA controls growth by cell enlargement, for example, cell elongation or cell expansion, with a supporting role in cell division. Thus, it plays a role in fruit size. IAA causes positive phototropism, the growth or bending of the plant toward the light. It also controls the plant's response to gravity (gravitropism), with roots exhibiting positive gravitropism and shoots negative gravitropism, regardless of how the seed or plant is initially oriented. IAA, along with cytokinin, controls apical dominance and correlative inhibition. In apical dominance, growth of the apical bud inhibits growth of the lateral (axillary) buds below the apical bud. Apical dominance results from the relative amounts of IAA to cytokinin in the lateral buds. The apical bud produces IAA, which moves by polar transport down the shoot to the lateral buds and inhibits the synthesis of cytokinins in the lateral buds, thus inhibiting their growth. Removing the source of IAA (the apical bud) decreases the IAA concentration moving past the lateral buds, allowing cytokinin biosynthesis to proceed and lateral buds to undergo budbreak and grow out into new shoots, inflorescences, or flowers. Correlative inhibition is basically the same as apical dominance, except that the apical bud is not the source of IAA. Instead, some other part of the plant is the source (e.g., the fruit), and this part controls the growth or inhibition of both the lateral and apical buds. IAA, along with ethylene, also controls leaf, flower, and fruit abscission. The auxin 2,4-dichlorophenoxyacetic acid (2,4-D) is used worldwide to reduce preharvest fruit drop of citrus. It should be noted that in California, when 2,4-D is used to reduce preharvest drop, it is applied at different times, depending on the cultivar, and it is typically used with the gibberellin GA_3 to delay rind senescence so that treated fruit remain physiologically young for a longer period of time. The combined treatment extends on-tree storage and lengthens the period for harvest. For more information, see chapter 14, "Plant Growth Regulators," or the UC IPM Citrus Pest Management Guidelines, http://www.ipm.UCDavis.edu/PMG.

IAA generally moves from shoots to roots, a process known as polar transport. In young, actively growing tissues, IAA is transported in nonvascular tissues. Auxin carrier proteins ensure that once IAA enters a cell, it can exit the cell only at its base and move toward the roots. IAA would accumulate in roots if it weren't for key metabolic reactions that catabolize it to keep it from reaching concentrations that would inhibit root growth. In mature shoots and stems, petioles of leaves, and pedicels of fruit, IAA is transported in the phloem.

Gibberellins (GAs)

More than 120 different gibberellins are known, but the function of many remains unresolved. The functionally important gibberellin is GA_1. Its primary role is to promote cell elongation or enlargement, with a supporting role in cell division. High concentrations of gibberellins are associated with the juvenile phase of citrus development; several, including gibberellic acid (GA_3), have been shown to stimulate budbreak and vegetative shoot growth and to inhibit flowering in citrus if applied prior to irreversible commitment to flowering (discussed in detail under "Flowering" below). GA_3 prevents senescence in citrus and can maintain chloroplast integrity and keep citrus fruit green longer. Gibberellins, along with ABA, control seed dormancy. When the gibberellin concentration is lower than the ABA concentration, a tissue, organ, or plant becomes dormant; when the gibberellin concentration is greater than the ABA concentration, dormancy is overcome and growth resumes. Gibberellins are produced by young tissues, including roots, and are transported within the canopy and from the canopy to the roots in the phloem. Roots are thought to play a role in converting gibberellins from the shoots, including those applied to the foliage, into different gibberellins that are then transported back to the canopy in the xylem. In some plants, gibberellins promote flowering and increase fruit set and fruit size. For commercial uses of gibberellins in citrus production, see chapter 14, "Plant Growth Regulators" and the UC IPM Citrus Pest Management Guidelines, http://www.ipm.ucdavis.edu/PMG.

Cytokinins

The major cytokinins in plants are isopentenyladenine, zeatin, and dihydrozeatin. Each is also found bonded to a five-carbon ribose sugar (i.e., isopentenyladenosine, zeatinriboside, and dihydrozeatinriboside) that improves transport within the plant and uptake by cells. Cytokinins promote cell division and play a role in flowering, fruit set, and fruit growth. In conjunction with IAA, they control shoot and root production as well as apical dominance and correlative inhibition as discussed above. In conjunction with ABA, cytokinins play a role in bud dormancy and spring budbreak. When ABA is greater than the cytokinins in the bud, it is dormant. An increase in cytokinins stimulates budbreak. Cytokinins prevent senescence and keep tissues young. Roots are the major sites of cytokinin synthesis, but leaves, buds, and fruit carry out cytokinin biosynthesis. Root-produced cytokinins are transported in the transpiration stream through the conducting cells of the xylem, predominantly as zeatinriboside and dihydrozeatinriboside. Cytokinins synthesized in the leaf are translocated in the conducting cells of the phloem to the roots predominantly as isopentenyladenosine. Because the synthesis of cytokinins is regulated directly by nitrate (or ammonium) concentration and indirectly by the availability of carbohydrate in these organs, shoot and root growth are coordinated (Sakakibara 2006). Sinks such as fruit import or synthesize significant amounts of cytokinins to maintain the strength of the sink—that is, its ability to import materials required for growth.

Abscisic Acid (ABA)

Leaves and roots produce abscisic acid (ABA) in response to many abiotic stresses that cause osmotic stress, including water deficit (drought), high temperature, desiccating wind, low temperature, and salinity. Leaves and roots can detect a lack of water and respond by synthesizing ABA, which moves in the apoplast, including the transpiration stream in the transport cells of the xylem if root produced, to the guard cells of the leaf to close the stomates, reducing the amount of water lost by the plant through transpiration (see fig. 3.3) and saving the tree from desiccation, loss of leaves and fruit, shoot or root dieback, and possible death. ABA regulates dormancy in combination with cytokinins and gibberellins, as discussed above. Though not its major function, ABA also plays a role in leaf, flower, and fruit abscission (hence the name "abscisic acid"). The role of ABA and other hormones in abiotic stress is discussed in greater detail in the section "Abiotic Stress," below.

Ethylene

Ethylene, the only hormone that is a gas, promotes flowering in some species. Plants produce ethylene when wounded or damaged. Ethylene causes senescence (aging) of tissues and controls fruit maturation and ripening. Citrus fruit are nonclimacteric: they do not undergo true botanical ripening as do climacteric fruit such as apples or bananas, which produce a burst of ethylene prior to, during, or just after a rise in respiration. Citrus fruit also lack the autocatalytic ethylene biosynthesis pathway and cannot sustain ethylene synthesis after being exposed to ethylene. Ripening is associated with fruit softening, conversion of starch to sugars, and an increase in the synthesis and accumulation of anthocyanins (red to purple propeniod pigments). Rather than ripen, citrus fruit simply mature and senesce (age), a process associated with a lower level of ethylene, leading to an increase in sugars at the expense of tricarboxylic acids, including citric acid, and the loss of chlorophyll to reveal the carotenoids (orange and yellow terpenoid pigments) that accumulate in the flavedo and juice vesicles of citrus fruit. The maturation and senescence of citrus fruit can be hastened by exposure to ethylene or slowed by inhibiting endogenous ethylene synthesis. For additional information, see chapter 14, "Plant Growth Regulators." The role of the hormones in citrus fruit development is discussed in the section "Fruit Development," below, and illustrated in figure 3.5.

Flowering

Citrus flowers, which are white to purplish, are known for their alluring fragrance. Oils from the flowers of the sour orange *(Citrus aurantium)*, Bouquet de Fleurs, are used in perfumes, as are the oils of Bergamot orange *(C. bergamia)*, which is grown in southern Europe. Nectar from citrus flowers is a favorite of bees and is the raw ingredient of citrus-based honey.

In California's subtropical climate, citrus trees usually have three vegetative growth flushes: spring, summer, and fall. In the adult phase, each growth flush can produce flowers at spring bloom the following year and set fruit, but in most cultivars, the previous year's spring flush is the most productive, producing approximately 60% of the bloom (Verreynne and Lovatt 2009). In California, citrus trees typically yield a single crop of fruit as early as fall or as late as the following summer, depending on the cultivar. Citrus types that are ever-bearing in California, such as lemons, limes, and citrons, can bloom and set fruit throughout the year, but they tend to bloom most abundantly in the spring.

Citrus flowering is recurrent under tropical and semitropical conditions, with a low number of inflorescences and flowers being produced throughout the year. As a result, fruit at different stages of development and maturation can be present on a tree. This complicates the harvest of properly mature fruit, resulting in a preponderance of juice-quality fruit typical of the citrus industries of Florida and Brazil. Thus, yields of fresh market fruit are typically low. Relatively low temperatures (night: 50° to 55°F; day: 59° to 65°F) promote citrus flowering (Lovatt, Zheng, and Hake 1988a, 1988b; Monselise 1985; Monselise and Goren 1969). Increasing the exposure to chilling temperature from 2 to 8 weeks increases both the number of floral shoots and flowers per shoot (Lovatt, Zheng, and Hake 1988a, 1988b; Southwick and Davenport 1986), resulting in a concentrated spring bloom, synchronized fruit development, and a concentrated period of fruit maturation and harvest. In Florida and other semitropical and tropical citrus-growing areas, winter water-deficit stress is imposed on citrus trees of all cultivars to compensate for inadequate exposure to low temperature during mild winters (Davies and Albrigo 1994; Spiegel-Roy and Goldschmidt 1996). Citrus is day neutral, meaning that day length does not induce flowering as it does in many annual and biennial plants. Flowering in citrus is autonomous; the vegetative shoot apical meristem (SAM) within the shoot terminal bud produces a specific number of nodes (points of leaf attachment) and leaves along the shoot and then transitions to a floral meristem. If environmental conditions support flowering, an inflorescence will develop; if not, the SAM will continue the vegetative growth of the shoot. Low temperature and water deficit are the primary environmental factors promoting annual flowering in adult citrus (by favoring the transition of the vegetative shoot apical meristem in a bud to transition to a floral meristem). Floral intensity increases with the severity or duration of the stress as long as the stress imposed does not reach a lethal level (Davies and Albrigo 1994; Lovatt, Zheng, and Hake 1988a, 1988b; Southwick and Davenport 1986). When winters are warm or accompanied by rain, flowering and yield are reduced.

Transition of the apical meristem in a bud to a floral meristem initiates the development of the apical flower in cultivars that produce a single flower or a cymose inflorescence (Lord and Eckard 1985). Development of the terminal flower is followed by near-simultaneous development of additional flowers in the axils of six or seven potential leaves in an inflorescence. The terminal flower is always the most advanced in development. The flower in the position just below the apical flower is always the least advanced due to the apical dominance exerted by the terminal flower. Basipetally, flowers are less affected by apical dominance and show advancing stages of development, flower opening, petal fall, and fruit set. Citrus flowers are perfect, complete with male (stamen) and female (carpel) structures. The outer layer of sepals, green leaf-like structures that protect the bud, is initiated first, and with their initiation, the bud is irreversibly committed to flowering, meaning that the bud cannot revert to a vegetative bud even when treated with GA_3 (Lord and Eckard 1987).

In seeded fruit, pollen carrying the sperm must land on the stigma of the carpel and germinate to produce a pollen tube that grows through the style of the carpel to deliver the sperm to the egg within the ovule. Fertilization, the fusion of the egg and sperm, results in a zygote that grows by cell division into the embryo, the young citrus tree. Fertilization triggers the development of the ovule containing the embryo into a seed and the ovary containing one or more ovules into a fruit. Seed are a source of hormones that contribute to maintaining the sink strength of the developing fruit. Thus, the number of seed within a fruit can influence both fruit set and fruit size.

Most citrus types are self-compatible, which means that they can be fertilized by their own pollen (self-pollinated) and can produce a crop even if a single tree is planted in isolation from other citrus trees. Unlike most citrus, however, Clementine mandarins, some grapefruit hybrids, and pummelos are self-incompatible. To set a large crop of fruit, some of these cultivars will benefit from pollination by an appropriate cultivar planted nearby. This cross-pollination or out-crossing introduces new genetic material, resulting in hybrid offspring with new characteristics. In many plants, species within a genus do not readily interbreed, but if they do, the offspring (a hybrid cross) is sterile. But in citrus, interbreeding is usual and common. Thus sweet oranges *(Citrus sinensis)* have crossed successfully with mandarin oranges *(C. reticulata)* to yield tangors, and mandarin oranges have crossed with grapefruit *(C. paradisi)* to yield tangelos. Some botanists believe that the grapefruit is a hybrid cross between a pummelo *(C. grandis)* and a sweet orange. There is also evidence the sweet orange originated from a cross between a mandarin orange and a pummelo.

Nucellar Embryony

Many citrus cultivars produce two types of embryos in the same seed, a condition known as polyembryony: a sexual (zygotic) embryo develops after fertilization, and one or more asexual embryos arise from female nucellar tissue within the same ovule. The asexual embryos that derive from cells of the mother only are known as nucellar embryos. Nucellar embryos are common in cultivars of *Citrus* (oranges, grapefruit, lemons, mandarins), *Fortunella* (kumquat), and *Poncirus* (trifoliate orange), but their occurrence is not universal. The stimulation by pollination is required for the nucellar embryos to develop. Nucellar embryos often crowd out the sexual embryo. Because nucellar embryos develop asexually from the nucellar tissue of the female parent without any input from a male parent, they are genetically identical to (are clones of) the female parent, which is of practical importance in citrus breeding, particularly in producing clonal propagating material such as clonal rootstocks. The traits of clonal rootstocks are those of the female parent rootstock, which was selected by breeders for predictable effects on fruit quality, yield, and tolerance to diseases and pests. To increase the probability that a seedling rootstock is a clone, young seedlings are evaluated for trueness to type, and rogue seedlings, which are typically zygotic, are discarded.

Polyembryony can be observed by germinating citrus seed; if more than one seedling develops from one seed, one of the embryos (usually the weakest one) is of sexual origin and the others are nucellar. Reproduction primarily by nucellar embryony is an important trait of clonal citrus rootstocks in use today, such as *Poncirus trifoliata*: about 90% of its seed are nucellar. The highly nucellar polyembryony of citrus rootstocks allows propagators in the citrus industry to rely on seedling clonal rootstock production. For more information on clonal rootstocks, see chapter 7, "The California Citrus Clonal Protection Program."

Seedlessness and Parthenocarpy

Navel oranges and Satsuma mandarins do not produce a high percentage of viable pollen, eggs, or ovules. The ovary (base of the carpel) of a navel orange or a Satsuma develops into a fruit without the stigma of the carpel in the flower ever being pollinated, without the pollen germinating to produce a pollen tube to deliver the sperm to the egg, or without fertilization. Thus, no embryo is formed and the ovules abort instead of developing into seed. However, the ovary does develop into a fruit, resulting in a seedless, or parthenocarpic, fruit. Undeveloped ovules can be seen in parthenocarpic fruit and, occasionally, a fully developed seed with a viable embryo can be found. For the Washington navel orange, one viable seed can be found in approximately every 100 fruit. In some cases, pollination, but not fertilization, is required to stimulate the development of the ovary into a fruit (stimulated parthenocarpy). Research with the Washington navel orange demonstrated that pollination with the pollen from certain scion cultivars increased the number of ovaries that developed into mature fruit without increasing seed number. In the absence of pollen from another source, self-incompatible Clementine mandarins produce parthenocarpic fruit, but, in general, yield is lower and fruit size smaller than when Clementine mandarin fruit produce viable seed.

Fruit Set

Fruit set is determined by dividing the number of fruit harvested by the number of flowers produced. Citrus trees usually flower abundantly, but most flowers are shed without setting fruit that survive to maturity. Thus, fruit set values for most citrus cultivars are typically low, approximately 1% for Valencia and 0.2% for the Washington navel orange (Erickson and Brannaman 1960). Early-opening inflorescences tend to be totally floral due to the failure of the leaf primordia to develop into leaves. There is a trend during bloom for the number of leaves per floral shoot to increase, with a concomitant decrease in the number of flowers per floral shoot. Flowers opening late in the bloom period tend to be on leafy floral shoots and produce more fruit that survive to harvest. This is likely due to the benefits of having leaves in the inflorescence: increased availability of photosynthate, greater influx of root-derived resources due to leaf transpiration, increased flower and fruit transpiration, and the milder climate during fruit set for later-opening flowers. Any weather conditions, pests, or diseases that lead to blossom injury reduce the season's fruit production. In addition, endogenous factors within the fruit influence fruit set. Faster-growing fruit have a greater potential to set and survive to harvest than do slower-growing fruit, which are more likely to abscise at an early stage of development. As discussed above, an increased number of seed in a given cultivar is related to increased growth rate, greater fruit size, and a higher fruit set percentage.

Fruit Development

The citrus fruit is a type of berry with a thick, leathery, removable rind (peel) and a soft, segmented pulp. Scientists identify the parts of the fruit using the terminology shown in figure 3.4. Citrus fruit usually attain edible quality 8 to 16 months after bloom and have a variable harvest season of about 2 to 6 months, depending on the climate. They exhibit a single sigmoid growth curve, with little increase in fruit size during Stage I of fruit development, an exponential increase in growth in Stage II, and a slow increase in Stage III.

Stage I

During Stage I of fruit development, growth is predominantly by cell division. During this stage, all the tissues that will make up the mature fruit are formed. Indeed, all the cells that will comprise the mature fruit are produced during this stage, with the exception of the cells in the flavedo, the only tissue that continues cell division during Stages II and III. The number of cell divisions that occur before and after anthesis is not known for citrus fruit, but it is clear that cell number has a greater impact on final fruit size than does cell size. Stage I begins with the development of the carpel in the flower. Maximum peel thickness, which marks the end of Stage I of fruit development, was determined experimentally over 5 years to occur from June 11 to July 26 for navel and Valencia oranges grown from 33° to 37°N latitude in California (Ali et al. 2000).

Stage II

In Stage II of fruit development, growth is predominantly by cell enlargement (expansion). This stage is characterized by an exponential increase in fruit growth and fruit size that results from the uptake of water by and expansion of the juice vesicles. The cells of the albedo expand to their maximum cuboid size, then stretch and become progressively thinner and more elongated, which accommodates the increased size of the endocarp; the albedo cells remain connected to each other only by fingerlike projections. Thus, at the end of Stage I, the peel is at its maximum thickness because during Stage II the albedo tissue becomes progressively thinner as albedo cells stretch and the juice vesicles and segments enlarge. The flavedo continues to grow by cell division to accommodate the increase in fruit size.

Stage III

Stage III is the maturation stage of citrus fruit development. The flavedo changes from green to orange or yellow, depending on the cultivar, as chlorophyll degrades and reveals the carotenoids accumulating in the plastids in the flavedo. Fruit respiration slows, decreasing the rate at which sugar is converted to acid, resulting in sugar accumulation and reduced acid production. Moreover, citric acid and other tricarboxylic acids produced during respiration are used in the synthesis of other compounds found in citrus juice and peel tissues. It is interesting to note that only about one-fourth of the ascorbic acid (vitamin C) in citrus fruit is found in the juice; the remainder is in the peel, concentrated primarily in the flavedo of the fruit (Silva et al. 2002).

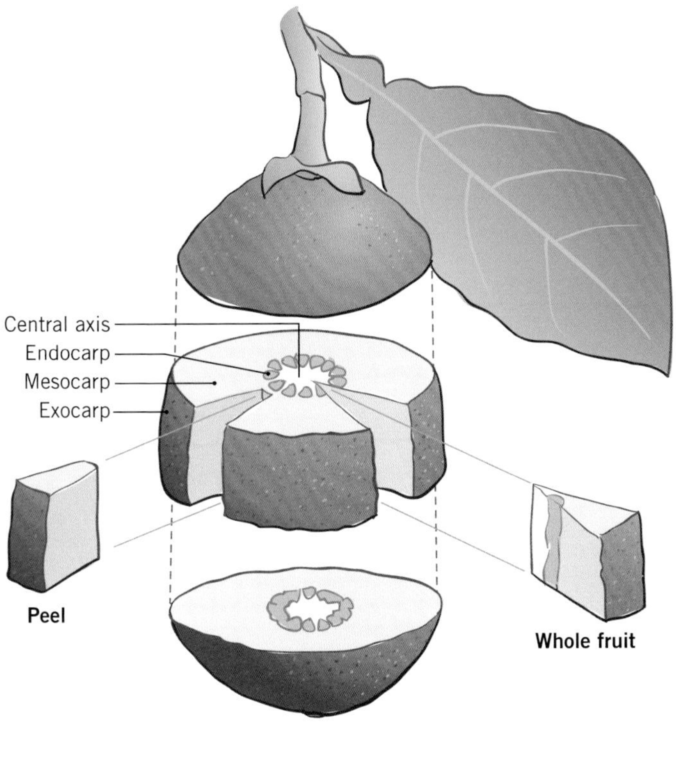

Figure 3.4 Anatomy of a young Washington navel orange fruit, about 0.8 inch in diameter. The outer colored layer of the peel is the flavedo (exocarp). Inside the peel is the whitish albedo (mesocarp), which surrounds the segments, or locules (endocarp), containing the juice vesicles (juice sacs, pulp). In the center of the fruit is the central axis. *Source:* Redrawn from Tomlinson and Lovatt (1987).

Early Drop, June Drop, and Preharvest Drop

Fruit tree (and even forest tree) species shed flowers and fruit in three distinct periods: early drop, June drop, and preharvest drop. For Valencia and Washington navel orange trees, the number of flowers and young fruit that abscise during the early drop period represents 87% and 97% of the current year's crop for the two cultivars, respectively (Erickson and Brannaman 1960). Early drop occurs during the initial period of fruit set, coincident with Stage I of fruit development (see figs. 3.2 and 3.5). In early drop, the flower or fruit abscises by the separation of the pedicel from the shoot. Early drop occurs from the period of initial flower opening (late March to early April) to maximum peel thickness in early June to late July. Fruit set (early fruit drop) is the most critical stage of fruit development from the grower's point of view. It is during this period that the greatest gains in fruit retention (influencing final yield) can be made. Events during this period also impact fruit size and quality.

A considerable number of young, immature, very small citrus fruit, less than 1 inch in diameter, abscise from trees during the period of June drop. These fruit separate from the floral disk, not the pedicel as in early drop. June drop is a natural fruit thinning process related to the simultaneous demand for resources by the fruit, which have entered Stage II and are undergoing an exponential increase in fruit size, initiation of the first root flush, and initiation of the summer flush of vegetative shoot growth. Whereas a lower number of reproductive structures abscise during June drop than in early drop, each structure is of greater biomass. June drop is exacerbated by deficiencies in nitrogen or other nutrients, sudden high temperatures, lack of water, heavy pruning, infestations of thrips or mites, or the occurrence of strong winds, such as the Santa Ana winds in southern California. Keeping trees in good health minimizes fruit drop to expected levels. When citrus fruit are greater than 1 inch in diameter, they tend to stay on the tree. June drop occurs from approximately maximum peel thickness through the end of August in California.

Preharvest drop is the abscission of fruit in Stage III; abscission at this time occurs at the floral disk. Again, whereas the number of fruit abscising is less than in early and June drop, the fruit are of significant size and are ready or nearly ready for harvest.

Fruit development provides an excellent example of how physiology, developmental processes, and growth are coordinated by the five classes of hormones (see fig. 3.5). IAA is essential for the formation of the ovary at the base of the carpel during flower development. IAA up-regulates the synthesis of GA_1, which plays a critical role in fruit set and Stage I of fruit development, especially in parthenocarpic and stimulated parthenocarpic fruit development. Stage I of fruit development, in which fruit growth is dominated by cell division, is under the control of IAA and cytokinins, which together control both cell division and sink strength. In Stage II, cell expansion, under the control of gibberellin, becomes the dominant mode of growth supported by IAA and cytokinins, which maintain sink strength. Stage III fruit maturation and eventually senescence are characterized by an increase in ABA and ethylene and loss of IAA, cytokinins, and gibberellin. Note that it is important to prevent abiotic stress due to water deficit during Stages I and II of fruit development. The lack of water will prevent the plant from transpiring sufficiently to cool the leaves and protect photosynthesis, and it will cause ABA and ethylene to accumulate, reducing fruit growth, causing leaf senescence, and increasing leaf and fruit drop.

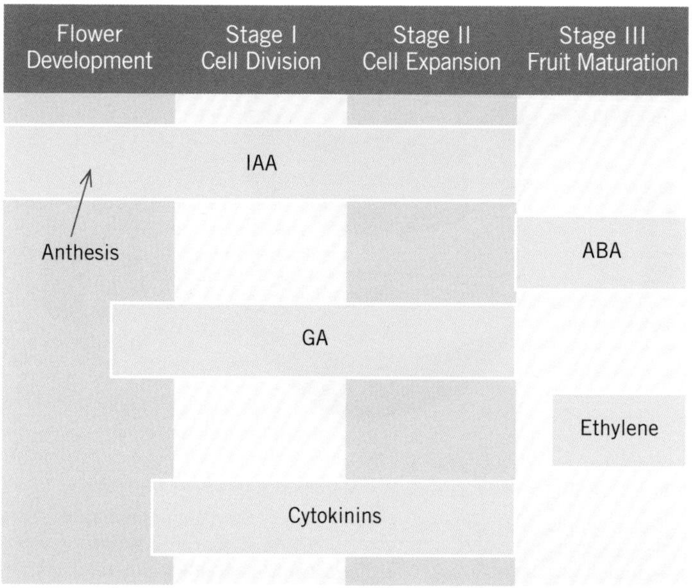

Figure 3.5 Hormones associated with the three stages of citrus fruit development.

Measures to Increase Yield

Foliar fertilizers

Winter prebloom foliar applications of low-biuret urea (46-0-0; ≤ 0.25% biuret) or potassium phosphite (0-28-26) have been shown to increase flower number and yield (Albrigo 1999; Ali and Lovatt 1992, 1994; Lovatt 2013; Lovatt, Zheng, and Hake 1988a, 1988b). Proper timing is important to achieve the desired outcome. The winter prebloom spray is designed to increase flower number and yield without reducing fruit size. The most effective application time is around the time of irreversible commitment to flowering; in California, the period from December 15 to February 15 seems to be appropriate in most years. However, it is important to note that this window was extrapolated from research conducted in southern California. Later applications (March and April) of fertilizer increase the retention of abscising reproductive organs. However, as time progresses from flowering through early fruit set to June drop, more flowers and fruit will have abscised. Hence, yield potential decreases and the contribution that any treatment can make to increase yield diminishes. When used as a winter prebloom foliar spray, low-biuret urea (46-0-0, ≤ 0.25% biuret) is applied at the rate of 50 pounds in 200 to 250 gallons of water per acre. Lower volumes can be used as long as tree coverage is good, but higher volumes, 500 to 700 gallons per acre, show greater incidence of tip burn. The potassium phosphite formulation that has been used in all research trials reported in the literature thus far is Nutri-Phite (0-28-26). Nutri-Phite is applied as a winter prebloom spray at the rate of 2.6 quarts (0.64 gal) in 200 to 250 gallons of water per acre (Albrigo 1999; Lovatt 1999; Lovatt and Mikkelson 2006).

Plant growth regulators

Plant growth regulators hold much promise for reducing early and June drop of citrus. Gibberellins were shown to increase fruit set (Hield et al. 1958; Krezdorn 1969). GA_3 has been shown to enhance early fruit growth, resulting in a delay or inhibition of early fruit drop and an increase in final fruit set (El-Otmani et al. 1992; Van Rensburg et al. 1996). However, GA_3 treatments used in South Africa, Morocco, and Spain to increase yield of Clementine mandarins have not proven to be effective or reliable when implemented in California.

As a result, a comprehensive experiment to determine GA_3 concentrations and application times required to maximize the yield of commercially valuable fruit (transverse diameter 57.2 to 76.2 mm) of Nules Clementine mandarin was conducted (Chao et al. 2011). The GA_3 concentrations tested were 5, 10, 15, and 25 mg/l. Application times at 60% bloom, 90% bloom, 75% petal fall, 10 days after 75% petal fall, and July were tested in numerous combinations from two to four applications. Some treatments were initiated early at 60% bloom; others were delayed until 90% bloom. Some treatments were completed by 10 days after 75% petal fall, whereas others continued until July. Based on 3 years of yield data, the use of GA_3 gives variable results, due in part to differences in crop load. In the year of the light bloom (550 fruit per untreated control tree), it was necessary to apply 15 or 25 mg/l GA_3 at 60% bloom, 90% bloom, 75% petal fall and 10 days after 75% petal fall to significantly increase the number of fruit per tree and yield of commercially valuable fruit (pounds and number of fruit per tree) ($P < 0.0001$) above that of control trees, with no reduction in total pounds per tree. In the following year with an intense bloom (approximately 1,200 fruit per tree), it was better not to apply GA_3: no treatment increased total yield or fruit size and five of seven GA_3 treatments tested reduced total yield (as pounds and number of fruit per tree ($P = 0.0003$). They also significantly reduced the yield of large fruit as both pounds and number of fruit per tree. The results also demonstrate that a high amount of GA_3 (four applications at 25 mg/l) applied one year had no negative effect on yield the next year. The results provided strong evidence that for mandarins GA_3 efficacy is crop load dependent, a factor that should be considered when using GA_3 to increase fruit set or fruit size of mandarins. For more information, see chapter 14, "Plant Growth Regulators," and the UC IPM Citrus Pest Management Guidelines, http://www.ipm.ucdavis.edu/PMG.

Cytokinins and aminoethoxyvinylglycine (AVG), an ethylene biosynthesis inhibitor, are showing promise in reducing fruit abscission during early drop and June drop, respectively. The auxin 2,4-D is used worldwide to reduce preharvest drop of citrus. It should be noted that in California, when 2,4-D is used to reduce preharvest drop, the timing of application depends on the cultivar, and it is typically applied with GA_3 to delay rind senescence so that treated fruit remain physiologically young for a longer period of time, improving on-tree storage and extending harvest. For more information, see chapter 14, "Plant Growth Regulators," and the UC IPM Citrus Pest Management Guidelines, http://www.ipm.ucdavis.edu/PMG.

Fruit Size

The amount of building materials, energy, and hormones that are available to the developing fruit controls fruit size. When the tree sets a heavy crop, fruit size will be reduced if there is an insufficient amount of any one of these factors. Typically, when a tree sets a low number of fruit, there is no problem with fruit size. In fact, the problem can be that fruit become too large, with thick, rough peels.

The auxin 2,4-D is registered for use in increasing fruit size of some citrus cultivars. For more information, see chapter 14, "Plant Growth Regulators," and the UC IPM Citrus Pest Management Guidelines, http://www.ipm.ucdavis.edu/PMG.

Summer applications of low-biuret urea (46-0-0, ≤ 0.25% biuret) or potassium phosphite made after the completion of early drop and after most of June drop increase fruit size without affecting total yield (Lovatt 1999), although they occasionally increase total yield. Timing of the application is important. The treatments increase fruit size by extending the cell division stage. The end of cell division is characterized by maximum peel thickness; this was experimentally determined to be between about June 11 to July 26 in California. Low-biuret urea is most effective when applied between July 1 and July 26. Applications of low-biuret urea that are too early (May and June) increase fruit retention and are less effective in increasing fruit size. Potassium phosphite was more effective in increasing fruit size when applied two times, May 15 ± 7 days and July 15 ± 7 days. How long after the end of July low-biuret urea or potassium phosphite applications would still effectively increase fruit size is not known. When applied in the summer at maximum peel thickness, low-biuret urea should be applied as a single spray on July 15 ± 7 days at the rate of 50 pounds in 200 gallons of water per acre. To increase fruit size, potassium phosphite (Nutri-Phite, 0-28-26) should be applied in two sprays at the rate of 2 quarts (0.49 gal) in 200 gallons of water per acre for each application. The first application should be on May 15 ± 7 days and the second on July 15 ± 7 days (Lovatt 1999).

All applications should be made to give good canopy coverage, much like applying a pesticide or plant growth regulator. The pH of solutions should be between 5.5 and 6.5. Applications of low-biuret urea or potassium phosphite should be made when the air temperature is below 80°F. Taking into consideration the potential maximum daytime temperature, applications should be made in the early morning or late afternoon to early evening. Information to help guide you in deciding whether to use foliar-applied low-biuret urea or potassium phosphite as part of your production management strategy to increase fruit size is provided in chapter 10, "Nutrient Deficiency and Correction."

Fruit Quality

Fruit color

Fruit coloration in citrus (a change from green to orange or yellow) is associated with a loss of chlorophyll (green pigment) and an increase in carotenoids (yellow, orange, and orange-red terpenoid pigments) and occurs naturally as fruit mature. Chlorophyll and carotenoid pigments are found together in a pigment-protein complex within the chloroplast (Mackinney 1961). When chlorophyll disappears during maturation, the chloroplast becomes less organized. If there is little alteration in the shape of the chloroplast, it becomes a chromoplast. However, loss of green color and an increase in yellow or orange color in the fruit flavedo may not be correlated with internal fruit quality. For example, in the fall, when temperatures are decreasing, chlorophyll loss occurs and carotenoids increase in the fruit peel. According to Mackinney (1961), with the exception of the blood orange, "it is a safe generalization that the color, both of the external flavedo and internal juice sacs, will depend on the nature and quantity of the carotenoids present, as well as on the completeness of the disappearance of the original chlorophyll." Color change, which is related to temperature, is not a good indicator of eating quality. Despite the mature-looking color of the flavedo, chilling may be required to increase sugar content or to decrease acidity in the juice vesicles before eating quality becomes acceptable. Conversely, Clementine mandarins may reach eating quality before developing good color.

Regreening

The flavedo of maturing Valencia oranges that remain on the tree may revert from orange to green as chlorophyll is resynthesized, a process called regreening. The only accepted explanation for regreening is that new chloroplasts are organized (Mackinney 1961). Regreening does not change internal fruit quality or fruit flavor. The degree of regreening is a function of temperature distribution during the 21 months required for the maturation of the Valencia orange in California. Years with a high degree of regreening versus years with good color were characterized by lower than normal temperatures from February prior to bloom through September and warmer than normal temperatures from November through harvest the following year (as late as August) (Jones 1961). Conversely, warmer than normal temperatures from February (prebloom) to September followed by cooler temperatures from November to August (harvest) resulted in fruit with good color. In addition to high heat, high nitrogen fertilization may also contribute to regreening. The green color can be removed by preharvest or postharvest treatment with ethylene.

Crease

Crease (also known as albedo breakdown) is a rind blemish that occurs in the flavedo when underlying albedo cells separate (albedo breakdown), causing the flavedo cells to sink and form a crease (Jones et al. 1967; Storey and Treeby 1994; Ali et al. 2000) (fig. 3.6). When fruit have multiple creases, the areas between creases have the appearance of the disorder known as puff. Growers in California frequently refer to the disorder as puff and crease. Whereas it is likely that both puff and crease are influenced by several common factors, it is believed that the underlying physiology of crease is different from that of puff (see the section "Puff," below). Once a disorder limited in California to the Valencia orange, crease has become a serious economic problem for California navel orange growers. In California, crease is historically most common in the southern coastal citrus-growing areas, where fruit tend to be smaller and have thinner peels, followed by inland southern California. The occurrence of crease in the Central Valley of California is typically lower.

Peel thickness (flavedo plus albedo) has been identified as a factor consistently related to the incidence and severity of crease in both navel and Valencia oranges in California. Cultural practices that cause thin peels, including irrigation and fertilization management, rootstock, crop load, and climatic conditions, have been reported to increase the occurrence of crease (Jones et al. 1967; Holtzhausen 1981; Nagy et al. 1982). Crease is worse in the thinner-peeled Valencia orange than in the thicker-peeled navel orange. For both navel and Valencia oranges, crease is worse in an on-crop year due to the greater number of small, thin-peeled fruit than in an off-crop year (Miller 1945).

In all years, crease increases as the fruit age. Nitrogen, phosphorus, and potassium fertilization rates are known to affect peel thickness and have been reported to influence the occurrence of crease. High nitrogen fertilization rates that make the peel coarser and thicker reduce crease (Reitz and Koo 1960). In contrast, high nitrogen fertilization accompanied by an increase in yield increases the severity of crease (Embleton et al. 1978; Jones et al. 1967). In most cases, increased potassium fertilization reduces the severity of crease (Embleton and Jones 1973; Fourie and Joubert 1957; Koo and Reese 1977). The effects of high nitrogen plus high potassium fertilization on crease are not additive (Jones et al. 1967). High rates of phosphorus fertilization, especially in the absence of potassium, result in thinner peels and more crease (Chapman and Rayner 1951; Embleton et al. 1978; Fourie and Joubert 1957). Research conducted in Australia linked low concentrations of calcium in the albedo tissue during early fruit development to the separation of the albedo cells but not to peel thickness (Treeby and Coote 1997). Despite the documented effects of fertilizer applications, peels of creased fruit at harvest, independent of fertilizer treatments, were characterized by significantly higher concentrations of nitrogen, phosphorus, and potassium and lower concentrations of calcium, magnesium, and sodium than the peels of fruit without crease (Jones et al. 1967). It is likely that peel nutrient concentrations at harvest are symptoms that develop considerably later than the initial events that cause crease.

Whether foliar gibberellic acid (GA_3) applied in South Africa when fruit are 1.2 to 1.6 inches in diameter reduces crease by increasing peel thickness or influencing peel nutrient status is unknown (Gilfillan et al. 1981). It is worth noting that this treatment has no effect on crease in California (C. Coggins, unpublished data). Preharvest GA_3 application reduces crease in California, presumably because the treatment delays rind senescence (Coggins 1981; El-Otmani et al. 2000).

Effect of peel thickness and peel nutrient content

Ali et al. (2000) quantified the relationships among crease, leaf and peel nutrient concentrations, peel thickness, and temperature for commercial navel and Valencia orange groves located in coastal and inland southern California and in California's Central Valley. The objective was to develop a model to predict crease early in the season that would be applicable to both navel and Valencia orange groves, with different rootstocks and cultural practices, located in the various climate zones in which citrus is grown in California. It was anticipated that the research and resulting model would provide insight into the cause of crease and suggest treatments to mitigate it and the stage of fruit development at which they should be made.

Crease appeared significantly less often when peels were thicker at maximum peel thickness and in October for both navel and Valencia orange.

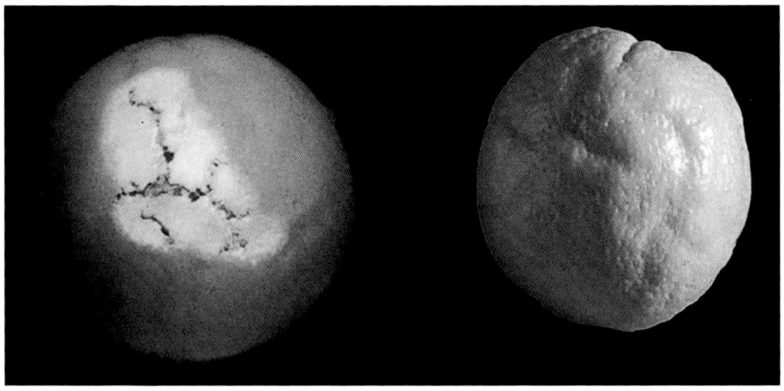

Figure 3.6 Navel orange fruit with crease. The left photo shows creasing of the peel (sunken areas in the flavedo) with associated areas of puff. The right photo shows separation of the cells of the albedo underlying the crease in the flavedo. *Photo:* C. Lovatt.

Maximum peel thickness, which marks the end of the cell division stage of fruit development, was experimentally determined in this study to be approximately July 7 ± 20 days. The data were the first to quantify the relationship between peel thickness at an early stage of fruit development and the incidence of crease at harvest, and they were consistent with the many reports that thin peels at harvest are associated with the occurrence of crease. In contrast, several reported relationships between crease and tree nutrient status that were based either on fertilization rates or leaf nutrient analysis were not true for peel nutrient concentrations determined at maximum peel thickness.

When navel orange peels had lower concentrations of nitrogen and calcium at maximum peel thickness, peels were thicker in October and the incidence of crease was low. The results obtained for Valencia oranges were the same. Whether the inverse relationship between peel nitrogen concentration and crease was due to an increase in yield, resulting in many thin-peeled fruit as noted by Embleton et al. (1978), was not determined. The results Ali et al. (2000) obtained for calcium were opposite to findings in Australia, where low calcium concentration in the albedo was associated with a greater incidence of crease (Treeby and Coote 1997). Embleton et al. (1978) reported that increasing tree potassium nutrition made peels thicker and reduced crease, whereas increasing tree phosphorus nutrition made peels thinner and increased crease. For navel orange in California, higher peel potassium and phosphorus concentrations at maximum peel thickness resulted in significantly thicker peels by October and less crease. The same trends were obtained for Valencia orange. Ali et al. (2000) were the first to report the relationship between peel nutrient concentrations at maximum peel thickness and crease. It is important to point out that in every case, higher peel nutrient concentrations at maximum peel thickness increased peel thickness in October and reduced the incidence of crease, providing additional evidence of the cause and effect relationship between peel thickness and crease in California.

Effect of temperature

Ali et al. (2000) also used linear regression analysis to analyze the effect of monthly average maximum and minimum temperatures on crease, as well as the effect of differences between the average maximum and minimum temperatures for each month. We also analyzed the effect of critical temperature ranges during two periods identified by Jones et al. on the incidence of crease at harvest (1967): the first was the range between the mean minimum temperature in June and the mean maximum temperature in July, and the second was the range between the mean minimum temperature from December 15 to January 15 and the mean maximum temperature from January 15 to February 15. For the 2 years of our study, high average maximum and minimum temperatures in February were associated with significantly lower potassium concentration in peels, thinner peels, and higher levels of crease. These results suggest that shoot growth was promoted but low soil temperature prevented the tree's demand for potassium from being met.

Management strategies for reducing the incidence of crease

In 2001, D. Holden tested whether crease on Valencia oranges in Ventura County could be reduced by foliar application of calcium and potassium just after petal fall. He treated two groves with 50 pounds of KNO_3 and 1 quart per acre of Metalosate calcium. All treatments were applied in 200 gallons of water containing 0.05% of the adjuvant Silwett L-77. In June of the year after application, D. Holden and J. Voelcker compared 50 fruit from 10 treated trees with the same number of fruit from two untreated groves. The results showed an average 65% reduction in crease in groves that received the application of potassium and calcium after petal fall. He noted that "this initial report is not meant to give an unequivocal endorsement of this practice" (Holden 2001). Based on Holden's results and those of Ali et al. (2000), properly timed foliar-applied potassium and calcium warrant further investigation as a mitigation strategy for crease.

As citrus fruit age, the capacity of the peel to produce ethylene increases. Ethylene is known to promote senescence in many plant tissues, suggesting that inhibition of peel ethylene biosynthesis might slow peel senescence and reduce crease. Aminoethoxyvinylglycine (AVG) inhibits ethylene biosynthesis and is now commercially sold for horticultural use. Foliar-applied AVG at 38 mg/l at 75% petal fall (May 26) or during June drop (June 15) significantly reduced the incidence and severity of crease of Washington navel oranges harvested late in the season (March) by 30% and 42%, respectively, compared with untreated control fruit (Gonzalez and Lovatt 2007). The June application was more effective than the May application in reducing the severity of crease, but a preharvest (November) application of AVG (38 mg/l) was without effect. The results with AVG and preharvest GA_3 confirm that delaying rind senescence to reduce crease is effective in California. The petal fall application of AVG had the additional benefit of significantly increasing fruit retention through early January (the midharvest period for Washington navel oranges in California). However, AVG applied at either stage of fruit development reduced the fruit diameter by approximately 0.08 inch; the reduction was statistically significant for early-, mid-, and late-harvested fruit treated with AVG at petal fall. Fruit treated with AVG during June drop were significantly smaller only at early harvest in November. This was the first test of AVG on citrus and the first to demonstrate the positive effect of AVG on reducing crease. Based on the results of this preliminary study, using AVG to reduce crease warrants further testing as a possible management tool to assist the California citrus industry in its continued efforts to provide picture-perfect navel oranges for the fresh-fruit market.

Puff

The separation of the innermost layer of albedo cells from the outermost wall of the locules creates an air space between the two tissues that makes the fruit puffy. Puffy fruit are fragile and do not handle, pack, or ship well. Puff is common in Satsuma mandarins and is frequently visible between crease lines on sweet orange fruit. The underlying physiological causes for crease and puff are suspected to be distinctly different. It is tempting to speculate that puff is related to increased activity of enzymes that hydrolyze pectins, especially pectinmethylesterase, hemicellulose, or cellulose. Nutrition and climate might be factors common to both puff and crease. It is clear that both disorders are a function of aging, suggesting that they share changes in metabolism or hormones that are associated with the aging processes of fruit. Although no plant growth regulator is used to control puff directly, since it is a disorder observed in fruit past maturity, GA_3 treatments used to delay peel senescence reduce puff significantly (Agustí et al. 1986; García-Luís et al. 1985; Kuraoka et al. 1977). Less puff was found on Satsuma mandarin when the auxin 2,4-DP was used to improve fruit size (Agustí et al. 1995). This was the result of improved peel firmness of the treated fruit.

Split

Splitting, which is most common in sweet orange and mandarin cultivars, occurs at the stylar end of the fruit, appearing as one or more fissures, although it may occasionally appear as a single fissure midway between the stem and stylar ends of the fruit. The problem is caused by the pressure exerted by expanding juice vesicles on the peel. Splits seem to result from the failure of the peel to expand as fast as the interior locules of the fruit. It is likely related to fluctuations in temperature, humidity, soil moisture, and cultural practices including irrigation and nutrition. Calcium nitrate applied during flowering (Barry and Bower 1997) and 2,4-D applied after June drop (Theron et al. in press) reduced splitting of mandarin cultivars. Splitting of navel oranges has been observed in response to winter rain.

Granulation

Granulation is characterized by enlarged, hardened, nearly colorless juice vesicles that develop first at the stem end of the fruit (Erickson 1968). In fruit with severe granulation, one-third to one-half of the juice vesicles may be granulated. Cell walls of granulated juice vesicles are thicker. The sugar, organic acid, and carotenoid concentrations of granulated juice vesicles decrease with gelling and hardening of the juice vesicles, whereas mineral nutrient and pectin concentrations increase. An imbalance in the ratio of demethoxylated pectin to methoxylated pectin in the presence of high concentrations of divalent cations, particularly calcium and magnesium, occurs in the juice vesicles (Chanana and Nijjar 1984; Shomer et al. 1988). This imbalance results in gelling of the pectin through the formation of calcium pectate, creating hardened, dry juice sacs. Low concentrations of calcium and magnesium reduce gelling.

Gelling of pectin is part of the natural senescence of citrus fruit. However, gelling and granulation can occur early in the harvest season in some years, significantly reducing the quality of the crop. Environmental, cultural, or internal conditions that induce a quicker rate of fruit development such that fruit are larger or more advanced physiologically increase the occurrence of granulation in terms of number of affected fruit and how early the disorder appears. In a given grove, granulation occurs in fruit that are larger and have lower acidity than fruit in which granulation does not develop, which is consistent with granulated fruit being at a more advanced physiological stage. Rootstock cultivar likely influences the occurrence of granulation by influencing the rate of fruit growth and development, how early nutrient uptake begins each year, and the total uptake of calcium and magnesium. Irrigation, fertilization, and soil amendments (lime, gypsum) would be expected to influence the severity of granulation, but the critical periods are not known. Heavy summer rain has been reported to increase granulation (Erickson 1968).

Consistent with this effect, Goldhamer (2003) found that an early-summer regulated deficit irrigation (RDI) treatment reduced granulation of navel oranges without reducing fruit size or yield or other quality parameters. Significant reductions in the mean weight per fruit resulted in significant reductions in total harvest in tons per acre for all but the early-summer RDI treatment compared with the fully irrigated control. This RDI treatment is being tested further for its efficacy in reducing granulation of large fruit in groves that have this problem. Additional research is needed to identify the key factors that cause granulation in California in order to further reduce the economic loss caused by this disorder.

Alternate Bearing

Alternate, biennial, or uneven bearing is the production of a heavy (on) crop followed by a lighter (off) crop. Alternate bearing is a significant economic problem for the California and worldwide citrus industries. In an on-crop year, there are too many small fruit of low cash value. In an off-crop year, fruit are generally of good size, although frequently some of the fruit are too large and of poor quality due to having rough, thick peels; net return to the grower is low because there are too few high-value fruit. Thus, alternate bearing results in price instability and erratic annual returns to growers. The limited supply of fruit in the off-crop year can lead to loss of market share and can hamper the development of value-added products.

Alternate bearing is initiated when climatic conditions (freeze, low or high temperatures, drought) during flowering and fruit set cause flowers or fruit to abscise, resulting in an off-crop year that is followed in 1 to 2 years by an on-crop year, depending on how long it takes for the trees to recover. Conversely, climatic conditions that are optimal for flowering and fruit set such that natural crop thinning fails to take place result in an on-crop year that is followed by an off-crop year. Once initiated, alternate bearing becomes entrained through the effect of crop load on endogenous tree factors that ultimately impact floral intensity the following year.

The low floral intensity of the return bloom in an alternate-bearing grove has been attributed to the low carbohydrate reserves in the trees after they have produced a heavy crop. Similarly, after an off-crop year, trees have a large carbohydrate reserve that supports a large number of inflorescences and the setting of an on-crop. Tree nitrogen status and reserves of other essential mineral nutrients follow the same alternating pattern. However, a role for hormones in alternate bearing cannot be ruled out. Fruit thinning during the on-crop year has shown potential for increasing the size of fruit in that on-crop and also increasing yield the following year (Hield and Hilgeman 1969). Alternate bearing can occur at the level of a shoot or branch, or it can occur for entire trees, entire groves, or for a whole industry, with the caveat that in each case some small subset might be out of synchrony with the majority.

Mechanism and Underlying Physiology

The Pixie mandarin (*Citrus reticulata* Blanco) was used as a model system to determine the physiological basis by which crop load (fruit number) one year influences the number of inflorescences the next year to perpetuate the alternating light and heavy yields. The results of this research provided evidence that fruit inhibit budbreak at two key stages in the phenology of the Pixie mandarin tree (Verreynne 2005; Verreynne and Lovatt 2009). First, young, developing fruit inhibit budbreak in the summer and early fall, reducing the number of summer and fall shoots that develop and thus reducing the number of nodes on which to produce floral shoots the following spring. Second, the mature fruit inhibit spring budbreak and subsequent development of the predominantly floral shoots, as well as the relatively small number of vegetative shoots that characterize the spring bloom, thereby further reducing the floral intensity of the return bloom the following year. For Pixie mandarin trees in California, about 60% of the floral shoots in the return bloom are produced by the previous year's spring vegetative shoots, about 32% by the summer vegetative shoots that developed the previous year, and about 8% by the previous year's fall vegetative shoots. Thus, 40% of the return bloom is contributed by summer and fall shoots.

To confirm the importance of the summer and fall shoots to the floral intensity of the return bloom, all summer and fall shoots were removed from a set of off-crop trees. This removal reduced the number of floral shoots and flowers produced by these off-crop trees to a value equaling on-crop trees. This research is the first to characterize and quantify the contribution of summer and fall shoots to the return bloom of alternate-bearing citrus trees and to demonstrate that summer vegetative shoot growth is key to a good return bloom; the more summer vegetative shoots, the greater the return bloom and yield. The results of this research also demonstrated that return bloom was affected more by a localized effect of fruit on the shoot that set it (bearing shoot) than by the total crop of fruit on the tree (whole tree effect). Thus, shoots that did not set fruit (nonbearing shoots) were the ones that produced the inflorescences and crop the following year; the more nonbearing shoots, the greater the return bloom. These results apply to other mandarins and sweet oranges, taking into consideration that some cultivars are not harvested as late as Pixie.

Inhibition of buds that produce the summer vegetative shoots was determined to be due to correlative inhibition—high concentrations of IAA and low concentrations of the cytokinin isopentenyladenosine in the buds (Verreynne 2005). Fruit were documented to export IAA, which accumulated in the buds for the summer vegetative shoots and inhibited cytokinin biosynthesis in the buds in a manner analogous to apical dominance. Removal of the on-crop in July eliminated the source of IAA, and restored cytokinin accumulation in the buds and summer vegetative shoot growth. For Pixie mandarin, the presence in spring of nearly mature fruit causes inhibition of spring budbreak that results from the accumulation of fruit-produced ABA and a low cytokinin concentration in the buds (Arbona and Lovatt, unpublished). In addition, bud carbohydrate (starch) concentration is low. Inflorescence number during return bloom was related positively to the starch concentration and negatively to the ABA to IPA ratio of buds just before bloom. A cause and effect relationship was confirmed by removing fruit, which increased inflorescence number and bud concentrations of starch and cytokinins, and reduced bud ABA concentration. The results of this research are the first to provide evidence that correlative inhibition is one mechanism by which fruit perpetuate alternate bearing in citrus and the first to clearly demonstrate a role for endogenous hormones in regulating alternate bearing in citrus.

Others have proposed that fruit exert their influence during the winter by inhibiting the transition of vegetative buds to reproductive buds. Thus, in an on-crop year, the number of floral shoots would be low and the number of vegetative shoots would be high. Winter fruit removal should result in a shift toward more floral and fewer vegetative shoots. In all cases during the research, the spring vegetative shoot number was very low, rarely influenced by the fruit removal treatments, and never inversely related to floral shoot number (Verreynne 2005; Verreynne and Lovatt 2009). Thus, the results of this research provided no evidence to support the contention that fruit inhibit phase transition. However, the possibility that fruit inhibit floral development at a later stage cannot be ruled out. Progressively later fruit removal from January through bloom results in fewer inflorescences and more dormant (inactive) buds, with no effect on vegetative shoot number.

Practical Measures for Reducing Alternate Bearing

Fruit thinning

The incremental decrease in inflorescence number due to the presence of the on-crop past December is a problem for Pixie mandarin, Valencia, and other late-maturing cultivars. For these cultivars, it is important that on-crop trees be harvested as soon as possible after the fruit reach legal maturity. Similarly, alternate bearing is exacerbated by the cultural practice of holding the on-crop on the tree to extend the commercial harvest period to increase fruit size or obtain a better price for the fruit. Having two crops on the trees into early summer should be avoided to reduce the risk of inhibiting summer vegetative shoot growth and creating a second off-crop and back-to-back low crop years.

Thinning (fruit removal by hand, by chemical, or by pruning) the on-crop prior to summer vegetative shoot growth is the most effective time for increasing return bloom and yield; spring, summer, and fall flush shoots will contribute inflorescences. Summer fruit thinning is also optimal for increasing the size of the young, developing fruit. Thinning the on-crop prior to fall shoot growth will increase return bloom and yield to a lesser degree than summer thinning because fall vegetative shoots contribute fewer inflorescences to return bloom than summer vegetative shoots. Note that December of the on-crop year is the latest time to thin and still obtain a positive effect on return bloom and yield. In this case, only the contribution of the spring shoots to return bloom will be significantly increased. The goal is to reduce the number of fruit in the on-crop uniformly over the tree sufficiently early to promote summer vegetative shoot growth. Research in a commercial Nules Clementine mandarin orchard in Grapevine provided evidence that fruit drop of the young fruit is minimal by the third week of August in both the on- and off-crop years (Chao et al. 2011). It is possible to wait until this time to thin the crop, which will reduce the risk associated with high temperatures that can occur during the June drop period (mid-June through the end of July) but still increase return bloom to a greater degree than fall fruit thinning would accomplish. If the crop is to be thinned by pruning, prune with caution. The goal is a light pruning to promote summer vegetative shoot growth. It must be done sufficiently early (spring or early summer) so the new vegetative shoots have time to mature and develop floral buds. Shoots that develop in response to fall pruning will contribute marginally to return bloom the following spring, but will provide shoots that will flower 1 year later. Pruning off summer shoots from spring shoots that did not set fruit (nonbearing shoots) defeats the purpose. Avoid overpruning, which will interfere with floral bud development. When deciding how many fruit to remove or how much to prune from on-crop trees, keep in mind that next year's crop will be produced predominantly on (current season) spring and summer shoots that did not set fruit (nonbearing shoots) or from which fruit have been removed by hand or pruning early in the season. In addition to pruning to reduce crop load, pruning should be done at other times of the year to increase shoot number (fruiting wood) and tree complexity with regard to the age structure of the shoots so that some contribute to increased floral shoot number and yield the following year and some the year after.

Fertilization

Growers should fertilizer their trees to meet the total demands of the tree throughout the development of the fruit—from inflorescences to setting fruit, young fruit during exponential growth, and mature fruit, especially when this crop remains on the tree during bloom and fruit set of the next crop—the growth of the first major root flush in approximately June, and also the growth of the summer vegetative shoots, which contribute significantly to bloom. Not only is matching the amount of fertilizer applied to tree nutrient demand cost effective, it also protects the environment, especially our water resources. The amount of fertilizer to be applied is dependent on yield, canopy size, and tree and soil nutrient reserves based on leaf and soil nutrient analyses. In addition, irrigation water can contain significant amounts of essential nutrients, which should be factored into a grower's fertilization program. Note that the fertilization goal for the heavy crop year is to provide adequate fertilizer to support the developing crop and support summer vegetative shoot growth. Research shows that a winter prebloom foliar application of low-biuret urea to Nules Clementine mandarin trees in Grapevine, California, increased 2-year cumulative yield of commercially valuable size fruit (transverse diameter 63.5 to 69.9 mm) as both pounds and number of fruit per tree with no reduction in total yield or fruit quality (Gonzalez et al. 2010). In a subsequent experiment with Nules Clementine mandarin trees in Clovis, California, a winter prebloom foliar application of low-biuret urea also significantly increased the 2-year cumulative yield of commercially valuable size fruit with no reduction in total yield or fruit quality compared to untreated control trees (Zheng et al. 2013). In both cases, a greater effect was achieved in the heavy crop year. The interpretation is that the extra nitrogen is necessary when the tree is setting a heavy crop to support the early growth of fruit to maintain size.

Abiotic Stress

Hormones mediate the effects of abiotic stresses. Stresses, such as low or high temperature, salinity, and water deficit (drought), cause changes in a plant's capacity to make energy and to take up water and essential nutrients. These stresses also cause changes in the relative amounts of hormones in tissues, which in turn influence plant physiology and development. This can have negative effects such as excessive leaf, flower, and fruit drop, small fruit size, or poor fruit quality. Some plants respond to stress with changes in hormone concentrations that protect the plant from the negative effects of stress on the plant's ability to make energy and take up water and nutrients, such that homeostasis (normal equilibrium) is maintained and no loss in fruit number or fruit size results.

Roles played by ABA in protecting plants from stress are well studied. As discussed above, ABA is produced by roots in response to a lack of water in the root zone and is transported in the xylem to the leaves, where it affects guard cell physiology and causes the closing of the stomates, preventing further loss of water by transpiration from the plant. Similarly, when hot, dry winds cause severe loss of water by transpiration from leaves, ABA is exported from the chloroplasts of the leaf to its guard cells, resulting in stomatal closure. ABA has also been shown to stimulate the synthesis of other plant metabolites that serve to protect the plant from stress.

Other hormones may also play positive roles in mitigating the effects of stress. However, it should be noted that increased ethylene biosynthesis is a negative consequence of stress that causes senescence and can contribute to abscission of leaves, flowers, and fruit. IAA and cytokinins have been suggested to play a role in mitigating the negative effects of ethylene; cytokinins are known to prevent ethylene-induced senescence. It is also possible that plants that synthesize polyamines in response to stress produce less ethylene. Because the ethylene and polyamine biosynthetic pathways compete for the same precursor, increased polyamine synthesis would reduce the availability of this precursor for ethylene synthesis.

Polyamines are nitrogen compounds that protect plants from the negative effects of stress. Other nitrogen compounds reported to be stress protectants are proline and members of the large family of late embryogenesis-abundant proteins, which include dehydrins, osmotins, heat shock proteins, osmotic shock proteins, and antifreeze proteins. Additionally, a plant's capacity to remove ammonia ammonium generated during stress and prevent its accumulation to toxic levels also mitigates the negative effects of stress (Lovatt 1990). Detoxification of ammonia ammonium proceeds via the arginine biosynthetic pathway and is related to an increase in arginine, proline, and polyamine metabolism and a likely reduction in ethylene biosynthesis (Lovatt 1990). Moreover, this sequence of events is physiologically linked to low-temperature and water-deficit stress-induced flowering in citrus (Ali and Lovatt 1995; Lovatt, Zheng, and Hake 1988a, 1988b). Positive relationships between polyamines and flowering, fruit set, and fruit development were also demonstrated (Ali and Lovatt 1995; Lovatt, Sagee, and Ali 1992; Lovatt, Sagee, Ali, Zheng, and Protacio 1992; Sagee and Lovatt 1991). This led to the use of winter prebloom foliar applications of low-biuret urea to increase flowering, fruit set, yield, and fruit size discussed under "Measures to Increae Yield" and "Fruit Size," above (Ali and Lovatt 1992, 1994; Lovatt 1999, 2013). For further details on the use of urea to increase yield or fruit size and factors to be considered when selecting this treatment, see chapter 10, "Nutrient Deficiency and Correction." For more information on abiotic stress, see chapter 25, "Environmental, Physiological, and Cultural Injuries and Genetic Disorders."

Chimeras and Sports

Chimeras arise when a cell undergoes mutation. This mutation may be spontaneous, or it may be induced by irradiation or treatment with chemical mutagens. If the cell that mutates is located in a meristem, all other cells that are produced by cell division from the mutated cell will also be mutated. The result will be cells of different genotypes growing adjacent to each other in a plant tissue. Thus, a plant (or organ such as a fruit or leaf) is said to be a chimera when cells of more than one genotype are found growing adjacent to each other in the tissues of the plant. Variegated plants are perhaps the most common types of chimeras: "The cells in a variegated leaf all originated in the apical meristem of the shoot, but some cells are characterized by the inability to synthesize chlorophyll … These appear white rather than green even though they are components of the same tissue system … A mutation that results in colorless rather than green cells (variegation) is easily detectable, whereas a mutation that results in greater sugar accumulation in the cells would not be observed" (Lineberger n.d.). Stable chimeras have resulted in plant sports that are propagated by budding to ensure trueness to type. Pink Lemonade, a citrus cultivar with pink flesh and variegated foliage, is an example of such a sport. Not only are the leaves variegated, the flavedo of the fruit is also variegated and the locules contain pink juice vesicles (Silva et al. 2002).

Sports are genetic mutations (chimeras) that affect only part of the tree, perhaps a branch or two. Typically, these mutations are not desirable, but if they are favorable, citrus breeders may choose to develop a sport into a new cultivar. The orange cultivars Skaggs Bonanza and Robertson navel, among others, began as sports of the Washington navel. Washington navel was likely a sport of the Bahia navel from Brazil. Sports have foliage and fruit that are different from the rest of the tree. Pink Lemonade began as a sport of Eureka lemon (fig. 3.7). Development of sports is common in navel orange and grapefruit varieties (Silva et al. 2002).

Figure 3.7 Variegated pink Eureka lemon. *Photo:* D. Karp.

References

Agustí, M., V. Almela, and J. L. Guardiola. 1986. Recolección tardía del fruto en el mandarino Satsuma. Efecto del ácido giberéllico. Actas II Congreso, SECH. 1:352–361.

Agustí, M., V. A. Almela, M. Juan, M. Aznar, and M. El-Otmani. 1995. Quality control of citrus and stone fruit. In A. AitOubahou and M. El-Otmani, eds., Postharvest physiology, pathology and technologies for horticultural commodities: Recent advances. Agadir, Morocco: Institut Agronomique et Veterinaire Hassan II. 26–33.

Albrigo, L. G. 1999. Effects of foliar applications of urea or Nutriphite [sic] on flowering and yields of Valencia orange trees. Proceedings of the Florida State Horticultural Society 112:1–4.

Ali, A. G., and C. J. Lovatt. 1992. Winter application of foliar urea. Citrograph 78:7–9.

———. 1994. Winter application of low-biuret urea to the foliage of 'Washington' navel orange increased yield. Journal of the American Society for Horticultural Science 119:1144–1150.

———. 1995. Relationship of polyamines to low-temperature stress-induced flowering of the `Washington' navel orange (*Citrus sinensis* L. Osbeck). Journal of Horticultural Science and Biotechnology 70:491–498.

Ali, A. G., L. L. Summers, G. J. Klein, and C. J. Lovatt. 2000. Albedo breakdown in California. Proceedings of the 9th Congress of the International Society of Citriculture 2:1090–1093.

Arnon, D. I., and P. R. Stout. 1939. The essentiality of certain elements in minute quantity for plants with special reference to copper. Plant Physiology 14:371–375.

Barry, G. H. and J. P. Bower. 1997. Manipulation of fruit set and stylar-end fruit split in 'Nova' mandarin hybrid. Scientia Horticulturae 70:243–250.

Bevington, K. B., and W. S. Castle. 1985. Annual root growth pattern of young citrus trees in relation to shoot growth, soil temperature, and soil water content. Journal of the American Society of Horticultural Science 110:840–845.

Chanana, Y. R., and G. S. Nijjar. 1984. Note on foliar mineral status of trees in relation to granulation in sweet orange. Indian Journal of Horticulture 41(3/4): 240–243.

Chao, C.-C. T., T. Khuong, Y. Zheng, and C. J. Lovatt. 2011. Response of evergreen perennial tree crops to gibberellic acid is crop load-dependent. I: GA_3 increases the yield of commercially valuable 'Nules' Clementine mandarin fruit only in the off-crop year of an alternate bearing orchard. Scientia Horticulturae 130:743–752.

Chapman, H. D., and R. S. Rayner. 1951. Effect of various maintained levels of phosphate on the growth, yield, composition and quality of 'Washington' navel oranges. Hilgardia 20:325–358.

Coggins, C. W., Jr. 1981. The influence of exogenous growth regulators on rind quality and internal quality of citrus fruit. Proceedings of the 4th International Citrus Congress 1:214–216.

Crider, F. J. 1927. Root studies of citrus trees with practical applications. Citrus Leaves 7:1–3, 27–30.

Davies, F. S., and L. G. Albrigo. 1994. Citrus. Vol. 2. CAB International, Wallingford, UK.

El-Otmani, M., M. C. Ben Ismail, A. AitOubahou, and M. Achouri. 1992. Growth regulators use on clementine mandarin to improve fruit set. Proceedings of the 7th Congress of the International Society of Citriculture 1:500–502.

El-Otmani, M., C. W. Coggins Jr., M. Augustí, and C. J. Lovatt. 2000. Plant growth regulators in citriculture: World current uses. Critical Review of Plant Science 19:395–447.

Embleton, T. W., and W. W. Jones. 1973. 'Valencia' orange crease, fruit color and other factors affecting crop value as influenced by N, P, K and Mg and their interactions. International Citrus Congress 1:93–101.

Embleton, T. W., W. W. Jones, C. Pallares, and R. G. Platt. 1978. Effect of fertilization on citrus fruit quality and groundwater nitrate pollution potential. Proceedings of the 3rd Congress of the International Society of Citriculture 1:280–285.

Erickson, L. C. 1968. Physiology of citrus. In W. J. Reuther et al., eds., The citrus industry.

Vol. 2. Berkeley: University of California Division of Agricultural Sciences Press. 86–126.

Erickson, L. C., and B. L. Brannaman. 1960. Abscission of reproductive structures and leaves of orange trees. Proceedings of the American Society for Horticultural Science 75:222–229.

Fourie, W. P., and G. F. Joubert. 1957. The effect of potash and phosphate on yield and "crease" of navel oranges in the Citrusdal area. Citrus Growers 277:1–3.

García-Luís, A., M. Agustí, V. Almela, E. Romero, and G. L. Guardiola. 1985. Effect of gibberellic acid on ripening and peel puffing in 'Satsuma' mandarin. Scientia Horticulturae 27:75–86.

Gilfillan, I. M., J. A. Stevenson, and J. P. Wahl. 1981. Control of creasing in navels with gibberellic acid. Proceedings of the 4th Congress of the International Society of Citriculture 1:224–226.

Goldhamer, D. 2003. Using regulated deficit irrigation to optimize size in late harvest navels. Citrus Research Board 2003 annual report. Visalia, CA: Citrus Research Board 9–10.

Gonzalez, C. M., and C. J. Lovatt. 2007. Foliar-applied aminoethoxyvinylglycine (AVG) reduces albedo breakdown of late-harvested navel orange fruit: Preliminary results. Proceedings of the 10th Congress of the International Society of Citriculture 3:1062–1065.

Gonzales, C., Y. Zheng, and C. J. Lovatt. 2010. Properly timed foliar fertilization can and should result in a yield benefit and net increase in grower income. Acta Horticulturae 868:273–286.

Hamid, G. A., S. D. Van Gundy, and C. J. Lovatt. 1985. Citrus nematode alters carbohydrate partitioning in the 'Washington' navel orange. Journal of the American Society for Horticultural Science 110:642–646.

———. 1988. Phenologies of the citrus nematode and citrus roots treated with oxamyl. Proceedings of the 6th International Citrus Congress 2:993–1004.

Hield, H. Z., and R. H. Hilgeman. 1969. Alternate bearing and chemical fruit thinning of certain citrus varieties. Proceedings of the 1st International Citrus Symposium 3:1145–1153.

Hield, H. Z., C. W. Coggins Jr., and M. J. Garber. 1958. Gibberellin tested on Citrus. California Agriculture 12(5): 9–11.

Holden, D. 2001. Holden news and views©, "Providing a prescription for premium plant productivity" (electronic newsletter), June 19, 2001, issue.

Holtzhausen, L. C. 1981. Crease: Formulating a hypothesis. Proceedings of the 4th Congress of the International Society of Citriculture 1:201–204.

Jones, W. W. 1961. Environmental and cultural factors influencing the chemical composition and physical characters. In W. B. Sinclair, ed., The orange. Berkeley: University of California Division of Agricultural Sciences. 25–55.

Jones, W. W., T. W. Embleton, M. J. Garber, and C. B. Cree. 1967. Crease of orange fruit. Hilgardia 38:231–244.

Koo, R. C. J., and R. L. Reese. 1977. Influence of nitrogen, potassium, and irrigation on citrus fruit quality. Proceedings of the 2nd Congress of the International Society of Citriculture 1:34–38.

Krezdorn, A. H. 1969. The use of growth regulators to improve fruit set in citrus. Proceedings of the 1st International Citrus Symposium 3:1113–1119.

Krishnamurthi, S., G. S. Randhawa, and P. C. Sivaraman Nair. 1960. Growth studies in some citrus species under subtropical conditions. Indian Journal of Horticulture 17:171–184.

Kuraoka, T., D. Iwasaki, and T. Ishii. 1977. Effects of GA_3 on puffing and levels of GA-like substances and ABA in the peel of Satsuma mandarin (*Citrus unshui* Marc.). Journal of the American Society for Horticultural Science 102:651–654.

Lineberger, R. D. n.d. Origin, development and propagation of chimeras. Texas Cooperative Extension Aggie Horticulture website, http://aggie-horticulture.tamu.edu/tisscult/chimeras/chimeralec/chimeras.html. Accessed Oct. 9, 2012.

Lord, E. M., and K. J. Eckard. 1985. Shoot development in *Citrus sinensis* L. (Washington navel orange). I. Floral and inflorescence ontogeny. Botanical Gazette 146:320–326.

———. 1987. Shoot development in *Citrus sinensis* L. (Washington navel orange). II. Alteration of developmental fate of flowering shoots after GA_3 treatment. Botanical Gazette 148:17–22.

Lovatt, C. J. 1990. Stress alters ammonia and arginine metabolism. In H. E. Flores, R. N. Arteca, and J. C. Shannon, eds., Polyamines and ethylene: Biochemistry, physiology, and interactions. Rockville, MD: American Society of Plant Physiology 166–179.

———. 1999. Timing citrus and avocado foliar nutrient applications to increase fruit set and size. HortTechnology 9:607–612.

———. 2013. Properly timing foliar-applied fertilizers increases efficacy: A review and update on timing foliar nutrient applications to citrus and avocado. HortTechnology 23:536–541.

Lovatt, C. J., and A. H. Cheng. 1990. Comparison of some aspects of nitrogen metabolism of avocado with citrus. Acta Horticulturae 2:489–495.

Lovatt, C. J. and R. L. Mikkelsen. 2006. Phosphite fertilizers: What are they? Can you use them? What can they do? Better Crops 90:11–13.

Lovatt, C. J., O. Sagee, and A. G. Ali. 1992. Ammonia and/or its metabolites influence flowering, fruit set, and yield of the 'Washington' navel orange. Proceedings of the 7th Congress of the International Society of Citriculture 1:412–416.

Lovatt, C. J., O. Sagee, A. G. Ali, Y. Zheng, and C. M. Protacio. 1992. Influence of nitrogen, carbohydrate, and plant growth regulators on flowering, fruit set, and yield of citrus with special emphasis on the role of ammonia and/or its metabolites in flowering, fruit set, and yield of the 'Washington' navel orange. In Proceedings, Second International Seminar on Citrus, Sao Paulo, Brazil. 31–54.

Lovatt, C. J., Y. Zheng, and K. D. Hake. 1988a. A new look at the Kraus Kraybill hypothesis and flowering in citrus. Proceedings of the 6th International Citrus Congress 1:475–483.

———. 1988b. Demonstration of a change in nitrogen metabolism influencing flower initiation in citrus. Israel Journal of Botany 37:181–188.

Mackinney, G. 1961. Color matters. In W. B. Sinclair, ed., The orange. Berkeley: University of California Division of Agricutural Sciences. 302–333.

Marloth, R. H. 1950. Citrus growth studies. I. Periodicity of root-growth and top-growth in nursery seedlings and budlings. Journal of Horticultural Science 25:50–59.

Miller, E. V. 1945. Some observations on crease in Florida oranges. Proceedings of the Florida State Horticultural Society 58:159–160.

Monselise, S. P. 1985. *Citrus* and related genera. In A. H. Halevy, ed., CRC Handbook of flowering. Vol. 2. Boca Raton: CRC Press. 257–294.

Monselise, S. P., and R. Goren. 1969. Flowering and fruiting: Interactions of exogenous and internal factors. Proceedings of the 1st International Citrus Symposium 3:1105–1112.

Nagy, S., W. F. Wardowski, and R. L. Rouseff. 1982. Postharvest crease of 'Robinson' tangerines. Proceedings of the Florida State Horticultural Society 95:237–239.

Platt, R. G. 1973. Planning and planting the orchard. In W. Reuther, ed., The citrus industry. Vol. 3. Rev. ed. Berkeley: University of California Press. 48–81.

Reed, M. S., and D. T. McDougal. 1937. Periodicity in the growth of the orange tree. Growth 1:371–373.

Reitz, H. J., and R. C. J. Koo. 1960. Effects of nitrogen and potassium fertilization on yield and quality of Valencia and Hamlin oranges. Proceedings of the American Society for Horticultural Science 75:244–252.

Sagee, O., and C. J. Lovatt. 1991. Putrescine content parallels ammonia and arginine metabolism in the 'Washington' navel orange. Journal of the American Society for Horticultural Science 116:280–285.

Sakakibara, H. 2006. Cytokinins: Activity, biosynthesis, and translocation. Annual Review of Plant Biology 57:431–449.

Schneider, H. 1952. The phloem of the sweet orange tree trunk and the seasonal production of xylem and phloem. Hilgardia 21:331–366.

Shomer, I., E. Chalutz, E. Lomaniec, M. Berman, and R. Vasiliver. 1988. Granulation in pummelo [*Citrus grandis* (L.) Osbeck] juice sacs as related to sclerification. Proceedings of the 6th Congress of the International Society of Citriculture 3:1407–1415.

Silva, D., C. J. Lovatt, and M. L. Arpaia. 2002. Citrus. In D. R. Pittenger, ed., California master gardener handbook. Oakland: University of California Division of Agriculture and Natural Resources Publication 3382. 531–580.

Southwick, S. M., and T. L. Davenport. 1986. Modification of the water stress–induced floral response in Tahiti lime. Journal of the American Society for Horticultural Science 112:231–236.

Spiegel-Roy, P., and E. E. Goldschmidt. 1996. Biology of citrus. Cambridge, UK: Cambridge University Press.

Storey, R., and M. T. Treeby. 1994. The morphology of epicuticular wax and albedo cells of orange fruit in relation to albedo breakdown. Journal of Horticultural Science 69:329–338.

Theron, K., J. Stander, and P. Conje. In press. Foliar 2, 4-D spray reduces fruit splitting in mandarin. Acta Horticulturae.

Tomlinson P. and C. J. Lovatt. 1987. Nucleotide metabolism in 'Washington' navel orange fruit. I. Pathways of synthesis and catabolism. Journal of the American Society for Horticultural Science 112:529–535.

Treeby, M. T., and M. J. Coote. 1997. Calcium sprays for albedo breakdown protection in navel oranges. In S. Falivene, ed., National citrus field day. Victoria, Australia: The Sunnyland Press. 61–63.

Turrell, F. M., M. J. Garber, W. W. Jones, W. C. Cooper, and R. H. Young. 1969. Growth equations and curves for citrus trees. Higardia 39:429–445.

Van Rensburg, P. J. J., S. Peng, A. García-Luís, F. Fornes, and J. L. Guardiola. 1996. Increasing crop value in 'Fino' Clementine mandarin with plant growth regulators. Proceedings of the 8th Congress of the International Society of Citriculture 2:970–974.

Verreynne, J. S. 2005. The mechanism and underlying physiology perpetuating alternate bearing in 'Pixie' mandarin (*Citrus reticulata* Blanco). PhD diss., University of California, Riverside.

Verreynne, J. S., and C. J. Lovatt. 2009. Effect of crop load on bud break influences alternate bearing in 'Pixie' mandarin. Journal of the American Society for Horticultural Science 134:1–9.

Zheng, Y, T. Khuong, C. J. Lovatt, and B. A. Faber. 2013. Comparison of different foliar-fertilization strategies on yield, fruit size and quality of 'Nules' Clementine mandarin. Acta Horticulturae 984:247–255.

Part 3 Citrus Orchard Establishment

ORCHARDS
and CITRUS IN SAN DIEGO
NET WT.
ORCHARDS
IN SAN DIEGO

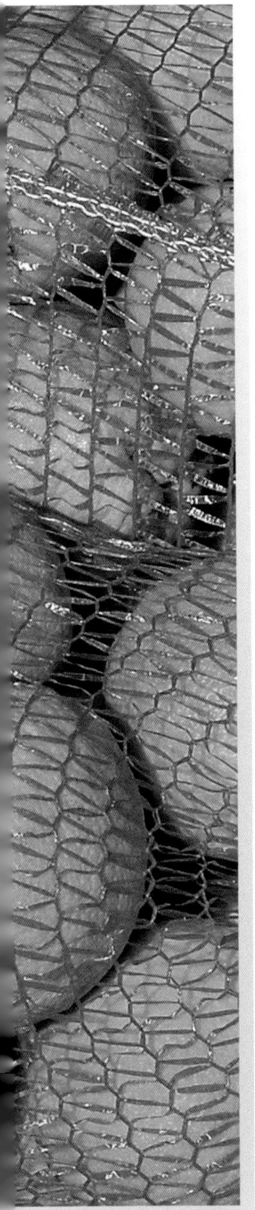

4 Scion Cultivars

Tracy L. Kahn and Georgios Vidalakis

Attractive and flavorful citrus fruit are an important component of the domestic and international fresh-fruit market. Citrus fruit such as oranges, mandarins, grapefruit, and lemons provide a significant source of folate and vitamin C for a healthy diet. As the population of California and the United States becomes more ethnically diverse, there is an increasing interest in an assortment of citrus cultivars. From late-maturing to pink-fleshed navel oranges to dozens of varieties of Satsuma mandarins, these new citrus types allow the U.S. market to meet the demands of the changing domestic and world markets. Once only available from countries where traditionally consumed or used, citrus cultivars such as Yuzu and Kaffir lime *(Citrus hystrix)* are now part of the growing interest for specialty fruit in the United States.

Citrus trees are generally grafted trees with two components: the scion, which includes the tree limbs, shoots, or canopy of the desired fruiting type, variety, or cultivar; and the rootstock, another citrus variety onto which the scion is grafted, which serves as a portion of the trunk and roots of the tree. For information on rootstock varieties, see chapter 5.

This chapter details the citrus scion types and cultivars grown in California, provides criteria to consider when deciding what to grow, and describes major existing and new scion varieties for large-scale and specialty markets. The Citrus Variety Collection (CVC) website, http://www.citrusvariety.ucr.edu, has further description of varieties, photographs of fruit and trees for varieties discussed in this chapter, data evaluating fruit quality for selected varieties, and additional information for selections mentioned briefly in this chapter as well as information about new varieties as they become available to the citrus industry. The CVC website also has links to the Citrus Clonal Protection Program (CCPP) website, http://www.ccpp.ucr.edu, which provides more fruit quality data and a list of registered citrus varieties in California. The USDA National Clonal Germplasm Repository for Citrus and Dates website, http://www.ars-grin.gov/npgs/acc/acc_queries.html has additional information about varieties. For information about varieties available prior to 1967 and the botany of citrus and related genera, see *The Citrus Industry,* Volume I (Reuther et al. 1967). Publications such as *Citrograph* and *Topics in Subtropics* can also be sources of information about new varieties. Familiarity with these resources and websites in the references at the end of the chapter provides a way to stay aware of new and existing commercial varieties available in California.

Climatic Requirements and Acreage

California's citrus industry primarily focuses on the production of fresh citrus for the domestic and international markets. California's fresh citrus fruit have a brightly colored rind and a balanced flavor due to high levels of acid and sugar. The subtle differences in climate between the production regions in California help define which types of citrus are most adapted to a region.

California has a subtropical climate characterized by long, dry summers with little rain and cool winters during which the majority of the rain occurs. The humidity is relatively low, and the warm daytime temperatures contrast with relatively cool nighttime temperatures. Within the state, the citrus production regions have climatic differences:

- San Joaquin Valley (Kern, Tulare, and Fresno Counties). The San Joaquin Valley has hot, dry summers and cold, wet winters, with substantial differences between daytime and nighttime temperatures. Over 80% of the precipitation occurs during the winter, when dense fogs can last a week or more at a time. Although the valley is relatively flat, most of the citrus production is in an area that includes rolling hills east of California State Highway 99.

- Coastal-intermediate region (San Diego, Orange, Ventura, Santa Barbara, and San Luis Obispo Counties). Due to the influence of the Pacific Ocean, the coastal-intermediate region has a milder climate than the San Joaquin Valley, with cool summers and relatively warm winters, and smaller differences in daytime and nighttime temperatures. The area is influenced by hot, dry Santa Ana winds in the fall. This region's low yearly mean temperatures cause citrus trees to grow slowly and bear and mature late, with a very long shipping season.
- Interior region (Riverside and San Bernardino Counties). The interior region is separated from the coastal-intermediate region by mountain ridges. As a result, the climate is warmer and drier in the summer and colder in the winter than the coastal-intermediate region, but not as hot or cold as the San Joaquin Valley. The low humidity of this area is exacerbated by the hot, dry Santa Ana winds.
- Desert valleys of southern California (Coachella and Imperial Valleys). The Coachella and Imperial Valleys have an extreme climate marked by very hot summers, large differences between daytime and nighttime temperatures, and very low humidity. The high yearly mean temperatures cause trees to grow rapidly and bear and mature early, with a fairly long shipping season.
- Northern Sacramento Valley (near the cities of Marysville, Oroville, Chico, and Redding). Like the San Joaquin Valley, the northern Sacramento Valley has a hot, dry summer, but it has a colder and wetter winter, with dense fogs.

Even within these production regions there are smaller localized climates, or mesoclimates, that differ from each other. Cold air movement, sunlight exposure, and large bodies of water contribute to differences that help determine these microclimates. Regions where cold air settles in low spots will be colder than nearby hillsides. Exposures that face south or west are generally warmer than those that face east or north. The storage of heat in large bodies of water during the day acts as a buffer against cold weather. The climate in a growing region and microclimates within those regions determine to a large extent what types and varieties are most suitable to grow in that area.

Cold hardiness or frost hardiness, along with heat requirements and tolerance, must be considered when selecting citrus for a given region. Commercial citrus types differ in cold hardiness. Mandarin trees are most hardy, followed by (in order from most to least) sour orange, sweet orange, grapefruit, pummelo (shaddock), Lisbon lemon, Eureka lemon, Tahitian lime, Mexican lime, and citron. Trees of citron, lime, and lemon are least hardy because they tend to grow continuously and flower and fruit in cool weather. On the other hand, grapefruit, sweet oranges (navel, Valencia, and blood oranges), sour oranges, mandarins, and kumquats are more frost hardy because their growth stops in cold weather. Also, a tree's tolerance to frost is not the same as the fruit's tolerance to frost. For example, although Satsuma mandarin trees are very cold hardy, the fruit are small and thin skinned, which makes them more susceptible to frost damage than grapefruit, which are large and have a firm rind. Another factor responsible for differences in frost tolerance is the maturity of the fruit. Fruit that is closer to maturity at the time of a freeze can withstand more cold than immature fruit. For example, although Valencia and navel trees are equally cold hardy, navel orange fruit are generally more cold tolerant than Valencia fruit because navel fruit mature earlier and have a higher sugar content during the coldest months.

Heat plays a role in determining the time of fruit maturity and the level of fruit quality. The heat requirement is determined by the number of hours that the temperature is above 55°F from the time of bloom to maturity. In general, grapefruit have the highest heat requirement, about 11,000 hours, in contrast to that of Valencia oranges, about 10,000 hours, and Washington navel, about 8,000 hours. Varieties that have lower heat requirements mature earlier than do those with high heat requirements. Also, in general, varieties that mature early in the season (late fall in California) require less time between bloom and fruit maturity than do varieties that mature later (winter or spring).

Citrus types also differ in their adaptation to temperature ranges. Grapefruit is the most resistant to high temperatures, which is why most grapefruit grown in California are in the desert valleys of southern California. In addition to grapefruit, Valencia oranges, lemons, and a number of mandarin and tangerine varieties produce well in the desert. On the other hand, Satsuma, Clementine, and navel orange varieties are likely to produce poorly when high temperatures occur during bloom and fruit set. Navel orange varieties require relatively high mean temperatures for best fruit quality, which is why they are best adapted to the warm interior region of California. Valencia orange varieties and certain varieties of mandarins such as Daisy and Fairchild are adapted to a wide range of temperatures and can be grown in multiple regions of the state.

Since the early 1900s, the types of citrus grown in California have been sweet oranges, lemons, and grapefruit, along with a small acreage of mandarins, mandarin hybrids, pummelos, and limes. In the past decade, and especially in the past 5 years, however, there has been a tremendous increase both in the diversity and in the acreage of citrus production, especially that of mandarins (see table 1.6 in chapter 1). The majority of the acreage of mandarins in California is located in the San Joaquin Valley, but a significant acreage is grown in all of the principal growing regions of California. This includes not only mandarins but mandarin hybrids such as tangors (mandarin × sweet orange hybrids), tangelos (mandarin × grapefruit hybrids), and other complex hybrids. This increase is mostly due to plantings of Clementine mandarin selections, and the varieties W. Murcott Afourer and Tango mandarin, released in 2006 from the UC Riverside (UCR) citrus breeding program. The acreage of Clementine, W. Murcott Afourer, Tango, and Minneola accounts for approximately 25%, 23%, 14%, and 13% of the mandarin and mandarin hybrid acreage respectively (NASS 2012). Clementine and Satsuma varieties, Minneola, W. Murcott Afourer, and Tango are grown primarily for the large-scale commercial tangerine market. In California, several Satsuma selections, Gold Nugget, Fairchild, Pixie, Shasta Gold, and numerous other mandarin and mandarin hybrid varieties are grown and marketed for specialty and farmers' markets.

Sweet oranges, including navel, Valencia, blood, or pigmented, oranges, and other common oranges account for the largest proportion (63.4%) of California's citrus acreage; navel and blood oranges contribute 48.9%, and Valencia oranges contribute 14.5% of the entire acreage (NASS 2012). Navel oranges are a main focus of the Californian citrus industry, with differentiation into early, midseason, and late-maturing varieties, and Cara Cara, a pink-fleshed navel orange. The acreage of Cara Cara has increased significantly since 2006 and currently accounts for 3.5% of the navel and blood orange acreage. Yet there has been a reduction in the acreage of navel (–1.8 %) and Valencia oranges (–4.3%) in California since 2010 (NASS 2012). The importation and planting of new late-season navel varieties into California, the increased competition of California's early-season Valencia oranges with imported navels from South Africa and Australia, and the fact that navel orange fruit has a brighter rind color and is easier to peel than Valencia have impacted the marketability of Valencia oranges and significantly reduced their production acreage. The majority of the navel and Valencia oranges are grown in the San Joaquin Valley region. A small number of acres of other sweet oranges, principally blood oranges, is grown in California, which accounts for slightly less than 0.5% of the statewide orange acreage. These niche market varieties are available for specialty displays at grocery store chains and for restaurants, farmers' markets, and other specialty markets. Blood oranges are also grown in the San Joaquin Valley, where significant differences in daytime and nighttime temperatures are present. Blood oranges require cold nighttime and warm daytime temperatures for good internal and external coloration and pigmentation. These conditions are not found in the coastal-intermediate region.

Lemons are one of the major types of citrus grown predominantly in the coastal and desert regions of California. Smaller numbers of acres of lemons are grown in the San Joaquin Valley, taking advantage of the range of climatic regions in the state and providing the market with year-round production. Lemon trees tend to produce flowers and marketable fruit throughout the year in coastal California due to the cool summers and mild winters. Most of the lemon production in the coastal region occurs in the late spring and early summer, when prices for lemons are highest. Lemon trees grown in the desert and interior regions of California have a shorter season, with a higher percentage of the crop produced during the fall and winter. Most of the lemons grown in California are selections of Lisbon, and to a lesser degree selections of Eureka lemon. A number of other lemon varieties are grown for the specialty markets, such as Meyer lemon, Femminello selections, and sweet lemons or limettas such as Millsweet limetta.

Along with sweet oranges, mandarins, and lemons, grapefruit are another major type of citrus grown in California. Grapefruit are grown primarily in the desert valleys of southern California, representing approximately 60% of the state's total grapefruit production in 2012 (NASS 2012). From the early through mid-1900s, most of the grapefruit varieties were white fleshed, but now the majority of the grapefruit acreage in California consists of pink- or red-fleshed varieties such as Star Ruby, Flame, and Rio Red. Grapefruit × pummelo hybrids such as Oroblanco and Melogold (products of the UCR citrus breeding program) and white-fleshed varieties such as Marsh are grown on a small scale for the specialty markets.

All citrus types, including pummelos, limes, sour oranges, citrons, and papeda hybrids such as Yuzu, Sudachi, and *Citrus hystrix* (Kaffir lime), are grown in the state due to California's diversity of climates. Kumquats, although not included in the genus *Citrus* in the most widely accepted taxonomic system, are closely related to commercial citrus types and are also grown on a small scale in California. Popular kumquats and kumquat hybrids include the Nagami and Meiwa kumquats and Eustis and Tavares limequats. Small to very small numbers of acres of these types are grown in California for specialty markets. Of all of the citrus types and relatives grown in California, pummelo and pummelo hybrid varieties such as Cocktail accounted for 0.58% of the citrus acreage in 2012 (NASS 2012).

Deciding Which Citrus Cultivars to Plant

The sheer number of cultivars of citrus available at wholesale citrus nurseries or at local nurseries makes the choice of which citrus varieties to plant a difficult task. If you are growing citrus for fresh-fruit production, a number of criteria must be considered when deciding what cultivars to plant.

One major consideration is that the type and cultivar you select must be compatible with the climatic conditions of the growing region where your orchard is located to ensure good production and high fruit quality. The previous section outlined the climatic differences between the citrus-growing regions of California. However, the presence of microclimates within a region can make it worthwhile to grow a type and variety that is not typically grown in a given region. For example, if you live in the San Joaquin Valley and have protected property that is warmer than other properties in your region, you might be able to grow lemons that can yield higher a price per acre when your fruit would arrive at maturity. Yet climatic conditions are just one consideration when determining which citrus variety to plant. Overall, you should determine which variety or varieties would likely generate the most income per acre based on your knowledge of your orchard (location, soil, climate, etc.), characteristics of the cultivars being considered, and current market data.

Seedless citrus fruit usually generate higher income than do seedy fruit. If you are interested in cultivars that produce seedless or low-seeded fruit, you must know whether the cultivar produces seedless fruit only when secluded from other citrus types that can cross with it (i.e., whether pollen from another citrus type or variety can fertilize it) or whether the cultivar produces seedless fruit no matter what is grown surrounding it. Cultivars that have little or no pollen or few or no functioning ovules will produce seedless fruit no matter what citrus is growing around them (for more information on citrus biology, see chapter 3, "Physiology and Phenology"). This is the case with navel orange selections and the pummelo hybrid varieties Melogold and Oroblanco. This is also the case with some (but not all) mandarin varieties, which makes the choice of a mandarin cultivar a bit confusing. Some mandarin varieties produce seedless or low-seeded fruit whether isolated or not; these include Gold Nugget, all Satsuma varieties, Pixie, Seedless Kishu, Yosemite Gold, Tahoe Gold, Shasta Gold, Tango, USDA 88-2, and the new UCR citrus breeding program releases Daisy SL, Fairchild LS, and Kinnow LS.

On the other hand, other mandarin varieties produce seedless fruit only if isolated from cross-pollination by other varieties that produce functional pollen. Cultivars that produce abundant functional pollen and ovules but are self-incompatible (meaning that self-pollination does not yield seed) will be seedless if grown in isolation and seedy if grown near types that can cross-pollinate them. Examples of these mandarin varieties are Clementine selections W. Murcott Afourer, Page, and Minneola. Pummelo varieties are also self-incompatible; they will be seedless only if isolated from other varieties that could cross-pollinate them. The distances necessary for isolation are quite large. For example, Chao et. al have shown that W. Murcott Afourer and Clementine trees separated by as many as 100 rows of Washington navel orange trees, which do not produce pollen, can still produce fruit with seed (Chao et al. 2005). This research has also shown that pollen from Clementine mandarins can pollinate flowers of W. Murcott Afourer (and vice versa), resulting in seedy fruit of both varieties. Because of the necessity of isolation between self-incompatible varieties and possible pollinating varieties and since most citrus growers do not have control of what varieties are grown on adjacent property, ensuring seedless fruit of self-incompatible varieties may be difficult. This is compounded by the fact that honey bees, which are pollinators of citrus, are prevalent in California due to their role as pollinators for the almond industry and other California crops. Currently, growers of W. Murcott trees are choosing to net the trees to prevent cross-pollination. An estimated 75% of the W. Murcott acreage in

California is netted to ensure that the fruit will be seedless, since fruit from unnetted W. Murcott trees lose about half of their value due to the presence of seed or suspected seediness (Parsons 2010).

Another major consideration is how you plan to market your fruit. In most cases, growers decide to take their fruit to a packinghouse to pack, ship, and sell their fruit. (For more information, see chapter 26, "Postharvest Handling.") If you are planning to sell your fruit directly to markets that specialize in local or organic or specialty produce or directly to restaurants, you must investigate which varieties are most needed and desired. This would also be the case for community supported agriculture (CSA), in which a farmer or farmers offer a certain number of shares as a membership or subscription for a range of produce. These markets may be interested in the some of the same varieties that packinghouses would purchase, or they may be interested in specialty cultivars that packinghouses would not be interested in. If a particular cultivar you would like to sell is unknown or if there is not an established market for it, you must develop a market. If you plan to market your fruit yourself at farmers' markets, specialty markets, restaurants, or to a CSA, you will have greater choices of the possible varieties available to plant. However, you may want to plant smaller numbers of trees of more than one variety since the volume needed will be lower. In addition, there may be an advantage to having a diversity of varieties available to meet seasonal demands and needs. (For more information, see the USDA Agricultural Marketing Service Fruit and Vegetable Market News website, http://www.marketnews.usda.gov).

Classification of Commercial Citrus Types

Almost all commercial citrus varieties are classified botanically in the genus *Citrus*. In the taxonomic classification system described by Walter T. Swingle and revised by Philip C. Reece in *The Citrus Industry* (1967), the genus *Citrus* and other related genera were classified within the subfamily Aurantiodeae of the Rutaceae plant family. In this system, the genus *Citrus* is divided into two subgenera, the subgenus *Papeda* with 6 species and the subgenus *Citrus* with 10 species. Most of the species in the *Papeda* subgenus are not used commercially except for the species *C. hystrix,* which is commonly called Mauritius papeda or Kaffir lime (although it is not a lime). The subgenus *Citrus* encompasses most of the commercial citrus varieties grown in California and includes the species commonly referred to as citron *(C. medica)*, lemon *(C. limon)*, lime *(C. aurantifolia)*, sour orange *(C. aurantium)*, sweet orange *(C. sinensis)*, mandarin *(C. reticulata)*, pummelo or shaddock *(C. maxima)*, and grapefruit *(C. paradisi)*. The common names of the above citrus species (i.e., lemon, pummelo, etc.) are very often used, as in this chapter, to group fruit together, but they are not used in a strict taxonomic manner based on the Swingle system of taxonomy.

A few commercial types such as the kumquats (genus *Fortunella*) and the trifoliate oranges (genus *Poncirus*) were not included in the *Citrus* genus by Swingle, but they have been included in that genus by other taxonomists. These genera and several others in the citrus subfamily Aurantiodeae have recently become of commercial interest. One is the genus *Microcitrus*. The flesh or juice vesicles of the fruit of *Microcitrus australasica*, commonly called Australian finger lime, are being used as a garnish for gourmet foods and drinks.

In the mid-1970s, results of research by several taxonomists suggested that there are really only three basic biological citrus species in the *Citrus* subgenus: *C. maxima* (pummelo), *C. reticulata* (mandarin), and *C. medica* (citron) (Scora 1988; Barrett and Rhodes 1976). The three-species taxonomic system is supported by biochemical and molecular data showing actual genetic relationships, and it is also being used to understand the evolutionary ancestry of current commercial types. All other species of citrus recognized in the Swingle system arose from single or sequential crossing events, which produced hybrids between these three species or their offspring. The lemon, lime, sour orange, sweet orange, and grapefruit types are now thought to have arisen directly or indirectly as first- or second-generation hybrids of the three basic biological citrus species and an additional citrus species, *C. micrantha* in the *Papeda* subgenus of *Citrus.*

Small genetic changes or mutations can result in a new variety, and these changes are believed to be responsible for generating varieties within the hybrids of the three basic biological species. Nearly all cultivars in the orange, grapefruit, and lemon groups are believed to have originated by either selection of bud sports or nucellar seedling sports (Soost and Roose 1996). This means that, despite having many named cultivars, there is relatively little genetic diversity within oranges, grapefruit, lemons, and limes. In contrast, in pummelo, mandarin, and, to a lesser extent, citron, many cultivars have arisen by sexual hybridization, and the levels of genetic diversity within these basic species are much higher (Roose et al. 1995).

In addition, some cultivars have also arisen as sexual hybrids between types but have often been classified within one of the parental types or a third type based on the characteristics of the new hybrid. For example, Meyer lemon is a low-acid natural hybrid of a lemon and sweet orange, but it is usually marketed as a lemon because of the lemonlike appearance of the fruit. Cocktail grapefruit is a hybrid of a low-acid pummelo (Siamese Sweet pummelo) and Frua mandarin, but the fruit has a size and shape similar to many grapefruit. A number of terms were developed by Webber and Swingle in the early 1900s to classify hybrids between citrus types. The two most commonly used terms are "tangelo" and "tangor." Hybrids of mandarin and grapefruit or pummelo were called tangelos; Minneola is an example of a tangelo variety, and a number of varieties that resemble mandarins such as Fairchild and Nova are hybrids between a mandarin and a tangelo. Hybrids of mandarin and orange were designated tangors; most of these resemble mandarins and include varieties such as Temple, Ellendale, and Ortanique. Three triploid mandarin hybrids developed by the UCR citrus breeding program, Tahoe Gold, Shasta Gold, and Yosemite Gold, have the tangor Temple as one of their three parents; the other two parents are mandarins. The presence of orange in the heritage of this complex hybrid makes it difficult to determine how to classify these varieties that resemble mandarins. Since a number of varieties have a small proportion of their heritage that is not mandarin, and since oranges and grapefruit have been shown to be first- or second-generation hybrids of the three primordial species, we now commonly refer to all varieties that produced fruit that resemble mandarins as mandarins and hybrids.

A question that sometimes arises is whether different citrus cultivars can be identified. At present, only cultivars that originated by hybridization between types or hybridization within the three primordial species can be distinguished with DNA fingerprinting. Usually questions arise about cultivars that originated as mutations of existing types, such as, "Which navel orange or Clementine is this?" This question cannot be reliably answered at present, although improved technology may eventually provide an answer.

Origin of New California Citrus Varieties

The majority of new citrus varieties originate from a mutation or genetic change in a vegetative bud, commonly called a bud sport or limb sport. The selection of a limb sport that has different fruit or tree characteristics from the mother tree, such as a change in fruit maturity or the color of the fruit flesh, is the most common method for "developing" or identifying new citrus varieties. For example, a bud sport on a Washington navel orange in Venezuela generated a new navel orange variety called Cara Cara, which has pink internal flesh instead of the normal orange. Another way new varieties originate is the selection of nucellar seedlings that have undergone a mutation or genetic change. Nucellar seedlings develop from cells of the nucellus, a tissue in the ovule of the flower but outside the embryo sac. In some types of citrus, the nucellus tissue can develop into embryos and then seedlings. Since these nucellar seedlings develop asexually, with no male cells contributing to their formation, they are usually genetically identical to the seed parent, but a mutation in the nucellus can generate changes in characteristics of the fruit or tree and result in the selection of a new variety.

Some of California's new varieties are products of the UCR citrus breeding program, currently directed by Mikeal Roose, professor of genetics in the department of botany and plant science. Citrus breeding programs use a variety of methods to develop new citrus varieties. The traditional method involves crossing selected parent varieties by moving pollen from one cultivar to the female portion of a flower of another cultivar for the production of hybrid seed, which will grow to new potentially promising varieties with recombined genes during this controlled sexual reproduction. This method was used by UCR citrus breeders to develop varieties such as Oroblanco and Melogold, which are pummelo and grapefruit hybrids, and the mandarin hybrids Tahoe Gold, Shasta Gold, and Yosemite Gold.

Another method used by the breeding programs to develop seedless to low-seeded varieties involves inducing mutations or genetic change by irradiating budwood of existing seedy varieties. After irradiation, the buds are propagated on the appropriate citrus rootstocks and grown to mature fruiting trees. Trees that produce fruit with similar characteristics to the original variety but with no or very few seed are selected for consideration as a new variety. After further propagation from the selected bud and evaluations of this new variety in replicated trials, if a variety is found to consistently produce fruit with characteristics similar to the parent variety but with few or no seed, with good yield efficiency on common rootstocks, then the variety will be released for commercial production. Varieties chosen for mutation breeding are chosen based upon having other important commercially advantageous characteristics. Recently the UCR citrus breeding program used this method to release a number of new low-seeded mandarin varieties such as Tango and Daisy SL, Fairchild LS, and Kinnow LS.

New citrus varieties are also introduced into California from other citrus-growing regions around the world. The Citrus Clonal Protection Program (CCPP) at UCR provides a safe mechanism for the introduction of citrus varieties under federal and state quarantine protocols, ensuring that citrus disease agents (i.e., graft-transmissible pathogens) are not introduced into the state. After disease testing and therapy during introduction, the CCPP maintains registered screenhouse and foundation field blocks of trees under a program of periodic disease retesting regulated by the California Department of Food and Agriculture (CDFA). These registered tree blocks serve as the primary source of disease-tested and true-to-type citrus budwood. The CCPP distributes limited quantities of registered budwood primarily to California nurserymen and growers, as well as growers in other states or countries, three times per year (January, June, and September). (For more information about distribution of registered budwood, see the CCPP website, http://ccpp.ucr.edu/budwood/budwood.php.) Recipients of budwood from the CCPP normally use the foundation block budwood to produce their own nursery- or grower-owned registered budwood source trees or budwood increase blocks that are also regulated by CDFA. Citrus varieties available through the CCPP are assigned a variety index (VI) number. The VI numbers are included in the descriptions of varieties in this chapter. Privately owned or patented cultivars and selected other cultivars are available from citrus nurseries with registered trees that have been produced by budwood tested by the CCPP and comply with the disease-testing requirements of the CDFA.

Commercial Citrus Scion Cultivars

The sections that follow provide general information on commercial scion cultivars available in California. This is not an exhaustive list of all commercially available varieties in California, but rather a list of the more commonly grown cultivars. Most include the variety index (VI) number that can be used by citrus nurseries or growers to order budwood from the CCPP. Some also include an alternate name in parentheses. The scion cultivars are organized based on the Swingle taxonomic system but using the common names for these species (sweet orange, sour orange, mandarin, pummelo, grapefruit, lemon, lime, citron, papeda, and kumquat). Due to the growing interest in specialty citrus, we have chosen to discuss cultivars commonly grown in large acreages, such as navel orange selections, as well as those grown in smaller acreage common to specialty markets. Varieties are listed alphabetically in the sections.

Sweet Oranges

The sweet orange appears to have arisen as a natural hybrid of pummelo and mandarin, although the exact proportion of genes derived from each of these parental species is not yet known. Four major types exist within the sweet oranges: navel oranges, common oranges (including Valencia oranges), blood oranges, and acidless oranges.

Navel oranges have a small secondary fruit, called a navel, in the stylar end of the fruit. Navel orange trees are generally not very vigorous. They have a round, somewhat drooping canopy and grow to a moderate size at maturity. The flowers lack viable pollen, so navel orange trees will not pollinate other citrus trees. Navel oranges such as the Washington cultivar produce seedless fruit because the flowers lack functional pollen and viable ovules. Washington navel orange, which is often also called the parent navel orange, is the best-known navel orange and is very often used as a standard for the industry. Many other navel orange cultivars that are now important to the citrus industry because they mature at slightly different times (extending the season from fall through summer) were developed as bud sport selections of Washington navel. The thicker rinds of navel oranges facilitate peeling, and the deep orange flesh and rind color makes navel oranges attractive as fresh fruit.

The common oranges, called blond oranges in the Mediterranean region, include Valencia orange selections and others such as Shamouti. Valencia orange selections mature in the late spring or summer, whereas most other common sweet oranges mature from November to February. The thinner adherent rind makes the Valencia oranges more difficult to peel and more suitable for juicing.

Blood, or pigmented, oranges are sweet oranges that have the pigment anthocyanin, which is responsible for the red pigmentation of the flesh and the reddish color or blush of the rind in some blood oranges. Blood oranges have a distinctive flavor compared with other sweet oranges and generally mature in winter and early spring. The acidless oranges are a small group of sweet oranges with the same level of sugars as other sweet oranges but very low acid, giving these oranges an insipid flavor.

Navel orange cultivars

Early-season selections

Atwood (Atwood Early) (VI 363) navel orange originated as a bud sport of Washington navel on the property of Frank Atwood in Lemon Cove, California, around 1935. The fruit characteristics are very similar to Washington navel, but the rind develops color slightly earlier and is slightly smoother in texture. In addition, the fruit store especially well without a significant loss of quality, but the fruit quality is not quite as good during the early part of the season as that of Washington navel. The harvest period for Atwood is from November to early May in the San Joaquin Valley.

Beck-Earli (Beck) originated as a Washington navel selection found to produce early-maturing fruit on the Beck property in Delano, California, about 1956 and was patented as Beck-Earli in 1991. Beck reaches color break slightly before Fisher and Fukumoto, early-season navels, and Washington navel; it reaches legal maturity in late September to early October, about 2 weeks before Washington navel. Beck fruit has good flavor, rind quality, color, and size and ships well. The fruit is slightly elongate but not as elongated as Newhall. Beck also does have alternate bearing to the same degree as does Newhall. The harvest period for this variety is from mid-October to early May in the San Joaquin Valley.

Bonanza (Skaggs Bonanza) (VI 515) navel orange is a bud sport from the Skaggs orchard in Lindsay, California. Bonanza reaches color break about 2 weeks earlier than Washington navel, and it reaches an 8:1 solids to acid ratio about a week earlier, on average, than Washington navel (Nauer et al. 1985). Bonanza also reaches color break ahead of Fisher navel, but it reaches an 8:1 solids to acid ratio after Fisher during most seasons. The Bonanza fruit has good size and a much thinner rind than Washington, Newhall, Fisher, and Lane Late navel oranges. The fruit of Bonanza tend to drop earlier in the season than the other selections listed above, so they must be harvested early in the season. The harvest period for Skaggs Bonanza is November or December through March in the San Joaquin Valley. Bonanza trees exhibit numerous bumps on the trunk; these have not been associated with any pathogens or production problems.

Fisher (Fischer) (VI 106) navel orange also arose as a bud sport of Washington navel orange; it was introduced by Armstrong Nurseries, who had received it from a Mrs. Fisher. Fisher is an early-season navel orange cultivar that reaches an 8:1 solids to acid ratio ahead of Atwood, Bonanza, Newhall, and Washington navel in Lindcove, California, but it does not reach color break before midseason cultivars such as Atwood and Washington navel (Nauer et al. 1985). Based on another study, Fisher reaches an 8:1 solids to acid ratio about the same time as Beck-Earli and Fukumoto but ahead of Washington navel. The harvest period for Fisher is November or December through March in the San Joaquin Valley.

Fukumoto (VI 430) navel orange selection (fig. 4.1) was introduced into California from Japan through the USDA Plant Introduction Station at Glenn Dale, Maryland, at the request of the late Professor W. P. Bitters of UCR in 1983. S. Fukumoto, from Kokawa-cho, Japan, was the first to grow this selection, in Wakayama Prefecture. This early-maturing navel orange selection reaches legal maturity 3 to 4 weeks before Washington navel and is one of the earliest to reach color break. One of the most outstanding characteristics of this selection is the deep reddish rind. The fruit are harvested from mid October to late December in California. Fukumoto trees tend to be smaller than other navel selections and appear to produce more chimeras than most navel orange selections. One additional major concern is reports in 2002 of decline of Fukumoto trees grown on C35 and Carrizo Citrange rootstocks (Marais and O'Connell 2002). The trees propagated on Carrizo exhibited symptoms of foamy bark rot syndrome, which includes lanky, zigzag shoot growth with abnormally large leaves and oozing of gum at the base of shoots and through splits in the bark at the crotches of branches (Marais and O'Connell 2002). These symptoms have also been observed when Fukumoto is grown on Swingle Citrumelo rootstock but not when grown on Sour Orange or Volkameriana rootstocks. The Fukumoto trees propagated on C35 have symptoms that may indicate a genetic rootstock-scion incompatibility. A series of research

Figure 4.1 Fukumoto (VI 430) navel orange. *Photo:* T. Siebert and D. Karp.

projects are currently in progress to identify biotic or abiotic factors responsible for this phenomenon. Due to concerns regarding these bud union problems, Fukumoto was recently reintroduced from Spain as VI 778. The Spanish bud source was introduced into California in 1992.

Navelina (Dalmau) (VI 532) orange was exported to Spain in 1933 from the Citrus Research Center in Riverside. This cultivar was originally called Smith's Early at the Citrus Research Center when it was selected as a bud sport from the Rubidoux Tract variety block about 1910. This cultivar was reintroduced to California from Spain in 1990 under the name Navelina. Navelina received extensive selection in Spain and differs from Smith's Early as the latter exists in California. Navelina trees are not very vigorous, being semidwarf. The fruit is medium to medium large and slightly pear shaped with a small navel. It has a very smooth rind that is reddish orange at maturity. The fruit mature very early and are juicy with a sweet flavor that is less sprightly than Washington navel. The harvest period for Navelina begins in mid-October in the San Joaquin Valley.

Newhall (VI 387) navel orange is a nucellar seedling selection that originated from the old budline Newhall, which originated as a limb sport of a Washington navel orange in the Duarte area and was selected by Paul Hackney of Newhall Land and Water Company of Piru. The fruit of Newhall mature slightly earlier and are slightly smaller with deeper rind and flesh color than Washington navel. Newhall trees are somewhat less vigorous and have slightly darker leaves than Washington navel trees. The harvest period for Newhall is mid-October through January in the San Joaquin Valley.

Rocky Hill (VI 180) navel orange was selected by the Rocky Hill farming company in Lindcove, California, in 1962. Although planted in the early 1980s, this selection has gone out of favor because its characteristics were not found to be superior to Washington navel and newer selections.

Thomson and T.I. (Thomson Improved) (VI 424 and VI 425) are navel orange selections. Thomson was one of the bud sport selections from Washington navel, on the property of A. C. Thomson in Duarte about 1891. Thomson was planted extensively for years because the fruit matured approximately 2 weeks earlier than Washington navel. The fruit is a bit more elongated and has a thinner, smoother rind than Washington navel. The juice vesicles have a ricelike texture and are less distinct than Washington navel. Although Thomson fruit mature earlier than Washington navels, they do not hang well on the tree and are inferior to Washington navel in all other respects. Two Thomson selections are available through the CCPP: Thomson Zimmer (VI 424) and Thomson Sheldon (VI 425). The harvest period for these selections is November through February in the San Joaquin Valley.

Midseason selections

Cara Cara (VI 471) navel orange (fig. 4.2) originated as a mutation that occurred on a Washington navel orange tree in 1976 at Hacienda Cara Cara in Venezuela. It was brought to Florida and introduced into California, where it has become well adapted. Most tree and fruit characteristics reflect its Washington navel orange ancestry, but the flesh is deep pink, similar to the darkest of the red grapefruit varieties. This rich color is due to the presence of lycopene, a carotenoid in the same family as beta carotene. The harvest period for Cara Cara begins in mid-November in the San Joaquin Valley.

Palmer (VI 507 and VI 526) is a navel selection imported from South Africa in 1985. Palmer trees appear to be more vigorous and more erect than Washington navel trees. The trees bear very well, and the internal quality and timing of maturity of the fruit are very similar to Washington navel. The two VI selections were obtained from the same institution but were apparently different budlines.

Summer Gold navel orange originated in the early 1980s on the Marrow property near Mourquong in the Sunraysia region of New South Wales, Australia. The variety was evaluated as a late-season navel orange along with Autumn Gold, Barnfield, Chislett, and Lane Late, but based on solids to acid ratio data in California, Summer Gold fruit reaches legal maturity in late December to early January,

Figure 4.2 Cara Cara (VI 471) navel orange. *Photo:* D. Karp.

approximately the same time as Washington navel. When individual locations were evaluated, Summer Gold still has higher solids to acid ratios at all locations, but in the coastal locations and other locations during certain years, the differences were not statistically significant. The percentage of titratable acids, or acid level, was considerably lower in Summer Gold than in Autumn Gold, Barnfield, Chislett, and Lane Late. The overall higher solids to acid ratios of Summer Gold relative to the other late selections are due more to the lower percentage acid of Summer Gold fruit than the total soluble solids or sugars in the fruit. During years when granulation was prevalent, Summer Gold, Autumn Gold, Barnfield, and Powell had a lower percentage of granulated fruit than did Lane Late. The fruit of Summer Gold were similar in size to Autumn Gold, Chislett, Powell, and Lane Late, but smaller on average than Barnfield. Data pooled over locations and years for the ratio of fruit length to width confirmed observations that Summer Gold fruit were slightly rounder than the others. (For more information about this variety and Autumn Gold, Barnfield, Chislett, Powell, and Lane Late, see Kahn et al. 2007).

Washington (VI 376) navel orange, or Parent navel orange (fig. 4.3), is also known as Bahia for the Brazilian city from which it was imported into the United States. The Washington Navel is reported to have originated from a bud sport of a Selecta orange tree in the early 1800s in Bahia (Hodgson 1967). More recent historical searches have indicated that the first written report of the navel orange in Italy was by a monk of the Society of Jesus, Giovanni Baptista Ferrari, in his work *Hesperides, sive, De malorum aureorum cultura et usu libri quatuor* (from *Hesperidoeidē,* the Greek botanical name for citrus) in 1646. There is also evidence that the navel orange was imported from Europe into Bahia, Brazil, at the end of the eighteenth century. In 1871, cuttings of this variety collected in Bahia by researchers for the USDA were successfully imported into the United States. Trees of this seedless variety of sweet orange were sent to California and Florida. Although the Florida trees did not flourish, those sent to Eliza Tibbets in Riverside found an ideal climate for their culture. These exceptionally delicious, seedless, easy-peeling fruit quickly attracted the attention of citrus growers, and within a decade the Washington navel orange, as it came to be known, was the most widely planted variety in the area. (For more information on the history of the introduction of the Washington navel orange to Riverside, see Roistacher 2007, 2009, and 2010). Washington navel orange trees are not very vigorous. They have a round, somewhat drooping canopy and grow to a moderate size at maturity. The large, round fruit have a slightly pebbled orange rind that is easily peeled, and the navel sometimes protrudes from the apex of the fruit. Washington navel orange is at its best in the late fall to winter months but will hold on the tree for several months beyond maturity and stores well. The harvest period for this variety is from mid-November through mid-May in the San Joaquin Valley.

Figure 4.3 Washington navel (VI 376) orange. *Photo:* T. Siebert and D. Karp.

Late-season selections

Autumn Gold navel orange originated on the Pollock property in New South Wales, Australia, in the mid-1980s as a bud selection of Washington navel. Based on data collected in California, the fruit of Autumn Gold are similar in size to Chislett, Powell, Summer Gold, and Lane Late but smaller on average than Barnfield. Based on the solids to acid ratio averaged over all locations, Autumn Gold, Barnfield, Chislett, and Lane Late fruit reached legal maturity in mid-January, up to about 4 weeks later than Washington navel. The percentage of titratable acids was very similar on average from late January to mid-July for Autumn Gold, Chislett, and Lane Late. Granulation varies from year to year, but during 2 years when granulation was prevalent during the evaluations, Autumn Gold had a lower percentage of granulated fruit than Barnfield, Chislett, Powell, and Lane Late.

Barnfield navel orange was a bud sport from an unknown navel selected from the Barnfield property in New South Wales, Australia, in the mid-1980s. Based on data collected in California, Barnfield fruit had the largest average weight, length, and width among the late-season Australian selections evaluated, including Autumn Gold, Chislett, Powell, and Lane Late. Barnfield fruit also had slightly thicker rinds than these other varieties, but rind thickness is related to fruit size because larger fruit tend to have thicker rinds. Based on solids to acid ratio data used as a characteristic of maturity and with data averaged over multiple locations, Barnfield fruit, as well as Autumn Gold, Chislett, and Lane Late, reached legal maturity around mid-January, up to about 4 weeks later than Washington navel. The percentage of titratable acids was slightly higher in Barnfield and Powell than in Chislett, Autumn Gold, and Lane Late. On average, based on puncture resistance data, Barnfield and Chislett had slightly firmer rinds than Autumn Gold, Powell, and Lane Late. Granulation varies from year to year; during years when Barnfield fruit had slightly lower granulation percentages than Lane Late, Barnfield fruit had been evaluated at fewer locations, which may account for the difference.

Chislett navel orange is a Washington navel orange selection first observed in 1988 on the property of Norm and Greg Chislett at Kenley in Victoria, Australia. Chislett and Powell fruit were slightly heavier on average than Lane Late and Autumn Gold fruit. Based on solids to acid ratio data averaged over multiple locations, Chislett fruit, as well as Autumn Gold, Barnfield, and Lane Late, reached legal maturity around mid-January, up to about 4 weeks later than Washington navel. The percentage of titratable acids was very similar on average over all locations from late January to mid-July for Chislett, Autumn Gold, and Lane Late. On average, based on puncture resistance data, Chislett and Barnfield fruit had a slightly firmer rind than Autumn Gold, Lane Late, and Powell. During years when granulation was prevalent, Chislett, Autumn Gold, Barnfield, and Powell had a lower percentage of granulated fruit than Lane Late.

Lane Late (VI 352) navel orange is the first of the late-maturing bud sports of Washington navel. It was discovered in Australia in 1950 and is named for the tree's owner. Based on solids to acid ratio data, Lane Late and Autumn Gold, Barnfield, and Chislett fruit reach legal maturity around mid-January, up to about 4 weeks later than Washington navel. The percentage of titratable acids was very similar on average over all locations from late January to mid-July for Lane Late, Autumn Gold, and Chislett. On average, based on puncture resistance data, Lane Late, Autumn Gold, and Powell had intermediate rind firmness. Granulation varies from year to year, but during years when granulation was prevalent during evaluations, Lane Late had the highest percentage of granulated fruit than all others, with only slightly higher percentages than Barnfield fruit, but Barnfield fruit had been evaluated at fewer locations, which may account for the difference.

Navelate (VI 548) navel orange originated as a limb sport from a Washington navel orange tree in 1930 and was selected in 1948 from an orchard in the Castellón province of Spain as a limb sport on a Washington navel orange tree. This selection was released for propagation in Spain in 1957 and was imported into California in 1991. Navelate trees are reported to be vigorous and slightly larger than Washington navel orange trees. The fruit are somewhat smaller than Washington navel fruit, and the navel structure is smaller and often concealed. The rind has a similar texture to that of Washington navel, but it is thinner and more difficult to peel. Navelate fruit are reported to hang on the tree for 4 months or more without appreciable loss of quality. The harvest period begins in January in the San Joaquin Valley.

Powell is one of the late-season navel orange selections that was imported into California. Neil Powell of New South Wales, Australia, made the selection in 1986. Powell (and Chislett) fruit are slightly heavier on average than Lane Late and Autumn Gold. Based on pooled solids to acid ratio data and percentage of titratable acidity, Powell fruit matured slightly later and had slightly higher acidity than other late-season oranges. On average, based on puncture resistance data, the rinds of Powell, Autumn Gold, and Lane Late were slightly less firm than Chislett and Barnfield. During years when granulation was prevalent, Powell, Chislett, Autumn Gold, and Barnfield had a lower percentage of granulated fruit than Lane Late.

Common orange cultivars

Campbell (VI 28) is thought to have been borne on a parent seedling tree, since budded trees were not frequently used in 1871, when the tree was planted in the Campbell orchard near Santa Ana. This vigorous cultivar was discovered in 1942. Campbell fruit are indistinguishable from Valencia, but it is not thought to be identical to the Valencia because it was discovered 5 years before the introduction of the Valencia into California, and Campbell trees are broader on the top and slower to come to bear than Valencia and its selections. The fruit also have a slightly lower juice content and a greater tendency to regreen than Valencia. A nucellar selection of Campbell was developed sometime prior to 1942 by Professor H. S. Fawcett at the University of California Citrus Experiment Station. The nucellar selection is more vigorous and thorny than the original budline. The harvest period for Campbell is from mid-March through August in the San Joaquin Valley.

Cutter Valencia (VI 30) was a nucellar seedling selected about 1935 by H. S. Fawcett of the UC Citrus Experiment Station from an outstanding old Valencia tree on the property of J. C. Cutter in Riverside. This nucellar selection was released in 1957 and was popular during the late 1960s. Cutter is known to be slow to come to bear, very vigorous, thorny, and more productive than the standard Valencia. The harvest period for Cutter is from March through August in the San Joaquin Valley.

Delta (Delta Seedless) (VI 474) is a Valencia selection discovered by P. Niven near Rustenburg, Western Transvaal, South Africa, in 1952. The trees are similar in vigor to Valencia, but slightly more erect. The fruit are virtually seedless, but occasionally may have one or two seed. The total soluble solids and acidity are reported to be slightly lower than other Valencia selections. The harvest period for Delta is late January through July in the San Joaquin Valley.

Figure 4.4 Olinda Valencia (VI 242) orange. *Photo:* T. Siebert.

Figure 4.5 Moro (VI 362) blood orange. *Photo:* D. Karp.

Frost Valencia (VI 240) was a nucellar seedling of Valencia orange. H. B. Frost, the first citrus breeder at the Citrus Research Center, selected the nucellar seedling in 1915, the first nucellar budline selected in California. This older selection was exceptionally vigorous and productive but is not now commonly grown in California.

Midknight (VI 460) orange was first observed as a slightly earlier-maturing Valencia orange tree in Addo, Cape Province, South Africa, by A. P. Knight, who made the selection about 1927. Midknight trees are reported to be slower growing and more frost susceptible than Valencia. In Lindcove, the fruit reach legal maturity in late January or early February. The solids to acid ratio of Midknight fruit is higher in April than Delta and Olinda. Midknight fruit in Lindcove also have a firmer and thicker rind than Delta or Olinda Valencia but lower juice content than Delta or Olinda, possibly due to thickness of the rind.

Olinda Valencia (VI 242) (fig. 4.4) originated as a chance seedling of Valencia found by H. J. Webber and L. D. Batchelor of the Citrus Research Center in the yard of O. Smith at Olinda in 1939; it was released in 1957. Since the fruit and tree characteristics of Olinda Valencia are considered to be indistinguishable from Frost Valencia, the chance seedling is now thought to have a nucellar origin and thus is genetically the same as Frost Valencia. The harvest period for Olinda is from March through August in the San Joaquin Valley.

Rocky Hill Oldline Valencia (VI 181) orange was once popular in California because it matured later than most Valencia selections, but it is no longer grown extensively in California.

Blood orange cultivars

Moro (VI 362) (fig. 4.5) blood orange is the most common of the pigmented oranges marketed in the United States. Of relatively recent origin, it is believed to be from Sicily. The tree is of moderate vigor and size, with a round and somewhat spreading growth habit. The fruit is round, of moderate size, and frequently borne in clusters. The orange-colored peel is commonly blushed with burgundy. Moro is the most highly colored of the blood oranges, owing its distinctive flesh color and rind blush to the presence of anthocyanin, the same pigment that colors purple grapes. The fruit is usually low seeded, with a flesh that can become very dark red late in the season. The fruit holds well on the tree but develops an off-flavor if held too long. The color and flavor develop best in the warm inland valleys of California. The harvest period for Moro is from December through April in the San Joaquin Valley.

Sanguinelli (Spanish Sanguinelli) (VI 413) pigmented orange originated as a bud sport of Doblefina blood orange in 1929 at Almenara in Castellón Province, Spain. Sanguinelli trees are vigorous, thornless, productive, and medium sized. The fruit are oval, with an extremely smooth, shiny rind that is pigmented over much of the surface. The fruit are small to medium sized. Internally the fruit are not as highly pigmented as Moro, and most of the red pigmentation is concentrated adjacent to the segment walls. The harvest period for Sanguinelli is late February through May in the San Joaquin Valley.

Smith's Red (VI 647) is a pigmented sport of a sweet orange tree in a home garden in Ventura County that was observed to be producing fruit with a pronounced red rind blush and strong red internal color. The tree is a vigorous grower and carries good crops of fruit. The fruit is of good size and flavor and is very low seeded. The rind frequently carries a heavy red blush, and the flesh is heavily pigmented by anthocyanin. The fruit shape is somewhat variable at present, globose to ovoid with a depressed base, possibly due to the juvenility of the subject trees. Although the fruit matures in late winter, it holds well on the tree into late spring, past the season for Moro, the most common commercial blood orange.

Tarocco (VI 384) (Tarocco #7) (fig. 4.6) blood orange appeared at the turn of the twentieth century, and is the favorite blood orange in Sicily. The trees are medium sized, moderately vigorous, and thorny. The fruit resembles a toy top because it often has a neck, like a Minneola tangelo. The fruit of Tarocco is medium to large, with few seed, firm but juicy flesh, and anthocyanin pigmentation giving it the red color internally and externally. The flavor is rich with raspberry overtones. The rind is medium in thickness and moderately tightly adherent. The selections vary in the amount of internal pigmentation. Two other Tarocco selections have been recently released to the industry: Bream Tarocco (VI 576) and Thermal Tarocco (VI 544). Bream Tarocco has levels of internal red coloration similar to Tarocco #7, and both have greater internal coloration than Thermal Tarocco. Tarocco fruit matures in the winter, slightly later than Moro blood orange. The harvest period for Tarocco is December through April in the San Joaquin Valley.

Acidless orange cultivars

Vaniglia Sanguigno (VI 442) is an acidless sweet orange with a pink flesh pigmented by lycopene, a carotenoid. The tree is small to medium sized at maturity, with a somewhat spreading form. The round fruit is medium sized, seedy, with a smooth orange rind of medium thickness. The flesh adjacent to the albedo is pink, which often extends to the interior of the fruit, especially near the section membranes. Because of its lack of acidity, the fruit can be eaten as early as late fall or early winter. The fruit is very juicy. A more recent introduction from Texas, Rio Farms Vaniglia (VI 676), seems to produce slightly more pigmentation than VI 442 based on observations in Riverside.

Sour Oranges

Sour oranges appear to have originated as a natural hybrid of pummelo and mandarin, although from a different cross than the one that gave rise to the sweet orange. The specific parents are not known. Sour orange and sweet orange trees resemble each other, but sour orange trees are more upright, thorny, and more resistant to cold than sweet orange trees. Sour orange leaves are darker, more pointed, and have longer petioles and larger wings on the petiole than

Figure 4.6 Tarocco (VI 384) (Tarocco #7) blood orange. *Photo:* T. Siebert and D. Karp.

sweet oranges. The fruit is flatter on each end than sweet oranges and more deeply colored. Because the fruit is too sour and bitter to eat fresh, they are commonly used to make marmalade, liqueurs, and perfume. Sour orange was the most widely used rootstock because of its compatibility with virtually every type of citrus, the excellent quality of the fruit on the scion, and its resistance to soil fungi, especially *Phytophthora,* which causes the serious diseases gummosis and root rot. However, the use of sour orange rootstock has greatly declined in California since the 1940s and 1950s, when the disease tristeza killed over 3 million citrus trees propagated on sour orange rootstock.

Bergamot (VI 420) is believed to be of hybrid ancestry, with sour orange and citron as the parents. The tree is small to medium at maturity, thornless, and somewhat spreading in habit. The fruit is medium sized and variable in shape, with obovoid the most commonly occurring form. The yellow rind has a slightly rough texture and a distinctive oil that is used commercially to flavor Earl Grey tea and as a component in perfumes, with Italy producing the majority of the world's supply. The fruit contain few seed and are monoembryonic. The flesh is pale yellow, acidic, and has moderate juice content. The fruit mature in late winter.

Bouquet de Fleurs (VI 589) is a sour orange planted primarily as an ornamental. The tree has a low, spreading, umbrella shape. The mature foliage is dense and bright green and the new growth is lighter green, giving the tree an attractive appearance. Because of its low growth habit, it is sometimes used as a hedge. The fruit is small to medium sized, somewhat oblate, and has a rough orange rind. The low-seeded fruit mature in winter and hold very well on the tree. Bouquet de Fleurs oranges are bitter and acidic and typically are not consumed.

Chinotto (VI 427) is sometimes referred to as the Myrtle-leaf orange. The tree grows very slowly and has a dwarf, compact habit. The leaves are small (usually under 2 in long), dark green, and lance shaped, and they are carried densely on the thornless branches and twigs. The tree flowers well and sets very good crops of small orange fruit of moderate seediness. These fruit mature in winter and are moderately tart. They hang on the tree for most of the year, making the tree highly ornamental. The primary use of Chinotto is as a decorative element in the landscape.

Seville (VI 95) orange is the sour orange traditionally used to make orange marmalade. The tree is attractive, large, vigorous, and cold tolerant. The fruit is medium large, round, with a depressed apex, and has pebbled, dark orange rind. Seville fruit mature in winter and are seedy, bitter, and acidic. Commercially, the fruit are valued for their oil content and juice. The fragrant flowers are used in China to flavor tea, and in Europe the flowers are the source of oil of neroli, used in perfumes.

Mandarins and Hybrids

The words "mandarin" and "tangerine" are often used interchangeably in the United States to refer to the mandarin group. "Mandarin" is the older and more widely used name. The word "tangerine" was coined for Dancy mandarins, which were possibly introduced from Tangiers, and it became associated with mandarins with orange-red rinds. A large and varied group, mandarins are now considered one of the three fundamental citrus species. Mandarin trees are small to medium sized with slender twigs and small branches. The trees are known to be the cold hardiest of the common commercial citrus species, but the mandarin fruit are not particularly cold hardy. The leaves are dark green and lanceolate, with smooth or entire leaf margins. The leaves have long petioles with very small wings. The fruit are almost always flattened and depressed at the stem and stylar ends. Mandarins have a greater tendency to alternate bearing than other types of citrus. The peel of mandarin fruit is loose at maturity, which explains why they are sometimes referred to as zipper skinned. They have also been called kid glove oranges because the pickers needed to use gloves so as not to damage these fruit. Mandarin fruit have a hollow core at maturity, and the segments tend to pull away easily from each other. The seediness of mandarins is variable: some may have no seed, a few seed, or as many as 20 to 25 seed per fruit when grown in mixed- or single-variety block plantings. Yet others produce no seed when grown as the only variety due to self-incompatibility but produce many seed if grown around trees that can cross-pollinate them. The mandarin and hybrid varieties discussed in this section are organized into categories based on their seediness.

Mandarins capable of producing seedless fruit

These mandarin and mandarin hybrids produce fruit with fewer than an average of three seed per fruit whether cross-pollinated or not by other citrus varieties.

Daisy SL (VI 763) is a mandarin selection developed by the UCR citrus breeding program from irradiation of the midseason-maturing variety Daisy. Daisy SL fruit have very low numbers of seed (average 2.2 seed per fruit), even when planted in the presence of pollinizer varieties. Fruit of Daisy SL are shaped like an upside-down cone, with an extremely smooth, thin, and very deep reddish orange rind color, making the fruit only moderately easy to peel. The fruit is moderately large (classed as Mammoth by State of California standards and size 21 for industry packing standards). The flesh of the fruit is deeply colored and finely textured. The tree growth habit is spreading, and mature trees are medium sized. The harvest period is from mid-December through February in Riverside. For more information, see the CVC (2013) and CCPP (2013) websites.

Fairchild LS (VI 762) is a mandarin selection developed by the UCR citrus breeding program from irradiation of the midseason variety Fairchild. Fairchild LS fruit have very low numbers of seed (average 2.4 per fruit), even in situations where cross-pollination occurs. Fruit of Fairchild LS are deeply oblate with no neck, large (classed as Jumbo by State of California standards and size 24 for industry packing standards), and extremely smooth skinned with a thin but pale orange rind. The peel is moderately thin and adherent at maturity, making the fruit less peelable than Satsuma mandarins. Fairchild LS trees are medium sized at maturity with a rounded growth habit and dense foliage. The harvest period is from mid-December through February in Riverside. For more information, see the CVC (2013) and CCPP (2013) websites.

Gold Nugget (VI 422) (fig. 4.7) is a mandarin variety developed by the UCR citrus breeding program as a cross between Kincy and Wilking (where Wilking = Willowleaf × King) (Roose et al. 2000). The tree grows vigorously and is upright in form. It can grow to be moderately large at maturity. Gold Nugget fruit are usually medium sized and oblate with a somewhat bumpy orange rind. The aromatic rind is moderately easy to peel. The flesh is bright orange, finely textured, and seedless, with less than 1 seed per fruit when grown in a mixed-variety planting. The flavor is rich and sweet. Depending upon location, the harvest period for Gold Nugget begins between mid-December and early March, but fruit hold exceptionally well on the tree, with summer-harvested fruit in June still being of good quality in some locations. This variety has a tendency to severe alternate bearing if no pruning is done to remove vigorous vertical branches. Fruit sometimes have a small blossom end opening that can lead to decay. For more information, see Roose et al. 2000 or the CVC (2013) and CCPP (2013) websites.

Pixie (VI 10) (fig. 4.8) is a mandarin variety developed by H. B. Frost at the UC Citrus Experiment Station in Riverside. Pixie is the result of an open pollination of Kincy (King × Dancy) that took place in 1927 and was eventually released in 1965. The tree is a vigorous grower with an upright growth habit. The fruit is usually globose to slightly oblate and sometimes has a neck. The rind is yellow-orange with a slightly pebbled texture and is easy to peel. The flesh is seedless, orange, and juicy. The flavor is mild and sweet. Pixie matures in late winter and holds exceptionally late on the tree; in certain mild locations, the fruit is known to hold well into summer. Fruit are small unless trees are carefully managed to prevent alternate bearing.

Satsuma selections

Satsumas represent a distinctive group of mandarins thought to have been imported from Wenzhou, China, to Nagashima, Japan, where an original chance seedling became established sometime before

Figure 4.7 Gold Nugget (VI 422) mandarin. *Photo:* T. Siebert.

Figure 4.8 Pixie (VI 10) mandarin. *Photo:* D. Karp.

AD 1600. The trees are cold tolerant, and the distinctive, seedless fruit have low heat unit requirements for ripening. Some varieties are therefore among the earliest citrus fruit that can be harvested in the fall. Numerous selections of Satsuma mandarins are currently being evaluated in California. Fruit quality data, including fruit size, shape, rind color and texture, and indicators of maturity are posted at the CVC and CCPP (2013) websites. Additional early evaluations of a number of the new selections were also published by T. Chao (2005). Early maturity is an important commercial characteristic of Satsuma selections, since they yield the highest prices. For information about evaluations of recently introduced selections, see Kahn et al. 2013 or the CVC (2013) and CCPP (2013) websites.

Okitsu Wase Satsuma (Wase Satsuma) (VI 389) is an early-maturing nucellar seedling selection of Miyagawa made by M. Kajiuera and T. Iwasaki at the Horticultural Research Station in Okitsu, Japan, in 1940. This selection generally matures in October in California, much earlier than the more common Owari Satsuma. Of the early selections, Okitsu Wase is probably currently the most widely grown in California.

Owari Satsuma (VI 33) (fig. 4.9) is the most common of the many Satsuma mandarin selections grown in the United States, although other selections have been introduced and additional introductions are being made. Because of its many fine qualities, Owari Satsuma has been the focus of extensive efforts to extend its season of availability. Owari Satsuma trees are cold hardy, productive, and vigorous, but they are small when mature and have a spreading and somewhat drooping character. The shape of the fruit depends on the conditions where they are grown, but they are most commonly oblate, with a smooth, thin, orange rind that is easily peeled. The flesh is bright orange, tender and juicy, seedless, and mildly flavored. Owari Satsuma is considered to be a midseason fruit, with its season of harvest being December to January in most areas of California. The fruit does not hold well on the tree, but it stores well after harvest.

Figure 4.9 Owari Satsuma (VI 33) mandarin. *Photo:* D. Karp.

Seedless Kishu (VI 433) is also known as Mukaku-kishu in Japan, where this mandarin of probable Chinese origin became popular. The tree grows well but is small and round at maturity. Indications are that it would make a good container specimen for ornamental plantings. The fruit is small, no more than 2 inches in diameter, but trees set reliable crops of seedless fruit. Seedless Kishu fruit have a thin, orange rind that is easily peeled. The flesh is bright orange, seedless, mild flavored, sweet, and juicy. It matures early, beginning in mid-December in Riverside, and the fruit holds well on the tree, although the peel puffs and becomes less attractive with time.

Shasta Gold (VI 490) mandarin hybrid was developed by the UCR citrus breeding program and released in 2002 at the same time as Tahoe Gold and Yosemite Gold. The tree grows vigorously and is somewhat spreading in form, with a tendency to alternate bearing, and can be moderately large at maturity. Trees are often thorny but may outgrow this as they age. Shasta Gold is a hybrid with Temple tangor, Dancy, and Encore mandarins in its parentage. The large, oblate fruit have an attractive, dark orange, relative thin rind. The flesh is seedless, bright orange, and juicy. The flavor is rich and sweet when mature. The harvest season for Shasta Gold is mid-February to mid-March in Riverside, and the fruit hold well on the tree into April or May.

Tahoe Gold (VI 506) mandarin hybrid was developed by the UCR citrus breeding program. The tree grows vigorously and is somewhat spreading, with a tendency to alternate bearing, and can achieve a moderately large size at maturity. Trees are often thorny but may outgrow this as they age. Tahoe Gold is a hybrid with Temple tangor, Dancy, and Encore mandarins in its parentage. The large, oblate fruit have an attractive dark orange rind that is relatively thin and very few seed. The flesh is seedless, bright orange, finely textured, and juicy. The flavor is rich and sweet when mature. The harvest season for Tahoe Gold is mid-January to mid-February in Riverside. The fruit do not hold well on the tree.

Tango (VI 765) (fig. 4.10) is an irradiated selection of W. Murcott Afourer mandarin developed by the UCR citrus breeding program. Fruit of this variety and fruit of W. Murcott are very similar in quality characteristics, with the exception of seed number. Tango fruit has a fine-textured rind and very few seed per fruit (average of about 0.2 seed per fruit in a mixed-block planting) in comparison with W. Murcott Afourer (average of 11 seed per fruit in a mixed-block planting). Other slight differences between Tango and W. Murcott fruit may be due to the seediness of W. Murcott. Tango trees have good yield and other fruit quality characteristics. The selection has very low pollen viability, which means that it does not cause other mandarins such as Clementine to be seedy if planted adjacent to it.

USDA 88-2 (VI 501) is a recently released (2008) hybrid mandarin that is a cross between Lee and Nova mandarins by Jack Hearn, USDA, Orlando, Florida. The fruit have very few to no seed under all conditions. The fruit are rounded and small to medium sized (mean fruit size 24), with a tendency to have a blossom end nipple. The rind is medium orange, thin, and fairly easy to peel. The trees have a moderate level of production with a tendency to alternate bearing. The fruit mature in November and hold on the tree through February.

Yosemite Gold (VI 509) mandarin hybrid was developed by the UCR citrus breeding program, with Temple tangor, Dancy, and Encore mandarins in its parentage. The tree grows vigorously to a moderately large size at maturity and is somewhat spreading in form, with a tendency to alternate bearing. The large, oblate fruit have an attractive, smooth, dark orange rind that is relatively thin. The flesh is seedless, bright orange, finely textured, and juicy. The flavor is rich and sweet. The harvest season for Yosemite Gold is January to mid-March, and the fruit hold well on the tree into April.

Mandarins capable of producing seedless or seedy fruit

These varieties can produce seedless, low-seeded, or seedy fruit, depending on the location or conditions under which they are grown. The fruit will be seedless or have low numbers of seed if grown in isolated blocks or under conditions in which cross-pollination is prevented. However, the fruit will be seedy if these varieties are grown near or adjacent to varieties that have pollen that can be transferred by honey bees.

The Clementine mandarin is currently one of the most popular mandarins in the world. Clementines represent a distinctive group of small, sometimes seedless mandarins originating in Algeria. The tree is medium sized, with a fine-textured appearance. The leaves are long and thin but not as narrow as the leaves of Willowleaf mandarin. The fruit is globose to oblate and small to medium sized. The dark orange rind has a pebbled texture due to the presence of prominent oil glands. The flesh is bright orange, finely textured, and juicy. The flavor is sweet and very rich. Clementine requires cross-pollination to ensure good fruit set. Unfortunately, cross-pollination also usually ensures a fruit containing many seed. The season of maturity for Clementine mandarins is approximately from early November to late December, but this varies with the individual selection and growing region. Algerian Clementine (VI 9) is the oldest selection in California, but it is not now commonly grown. The most commonly grown Clementine selection in California is Clemenules, or Nules, (VI 630); there is also significant acreage of Fina Sodea (VI 498) in production. A number of other available selections are available in California, including Caffin (VI 491), Corsica #1 (VI 604), Fina (VI 518), Marisol (VI 629), Monreal (VI 360), Nour (VI 632), Oroval (VI 517), and Sidi Aissa (VI 508). These selections differ slightly in fruit quality characteristics such as maturity and whether they are susceptible to granulation, a physiological disorder that causes fruit to become dry. More information and evaluation data for these selections are posted at the CVC (2013) and CCPP (2013) websites.

Figure 4.10 Tango (VI 765) mandarin. *Photo:* D. Karp.

Clemenules (Nules) (VI 630) (fig. 4.11) is reported to be a sport of Fina Clementine discovered in Nules Castellon de la Plana, Spain, in 1953. It is also known as Nules, Nulesina, Clementina Reina, Clementina Victoria, and Reina y Gorda de Nules and is the most widely grown Clementine selection in Spain and in California. The fruit are larger and mature slightly earlier than Fina, seedless in isolated blocks, and hang moderately well on the tree, but the fruit tend to puff if held too long after late December.

Minneola (VI 20, VI 174, and VI 208) is a tangelo, a hybrid of Duncan grapefruit and Dancy mandarin, sometimes marketed under the name

Figure 4.11 Clemenules (VI 630) mandarin. *Photo:* D. Karp and T. Siebert.

Honeybell. It was developed by the USDA and released in 1931. The tree grows vigorously to a large size. The fruit is round with a pronounced neck and a smooth red-orange rind that can be easily peeled. The flavor is rich and juicy, with a touch of grapefruit tartness. Minneola blossoms are self-incompatible and must be cross-pollinated by a suitable pollinator to ensure good fruit set. Most mandarin types are suitable pollinators, with the exception of Satsuma and the sibling cultivars of Minneola, Orlando, and Seminole. Unfortunately, when cross-pollinated, Minneola fruit tend to be seedy. Three different introductions of Minneola are available through the CCPP, but there are very little differences among them. Two (VI 20 and VI 174) are early nucellar seedling selections. The third, VI 208, was selected by Chase and Bailey. This budline produces fruit that are more rounded, with less of a neck during some years but this is not consistent. The three introductions are being kept available by the CCPP since some nurseries have a definite preference of one over another. Significant acreage of Minneola is currently in production in California due to the attractive and juicy characteristics of this late-maturing mandarin hybrid. Minneola should be harvested late in the season (late December) to ensure that the fruit reaches a desirable sugar to acid ratio.

Nour (VI 632) is reported to be a selection of Cadoux that was discovered in Morocco in 1962, and it is extensively planted there. It is also called Sunerine in the United States and Morocco. The fruit are medium sized with a slightly coarse rind and mature slightly earlier than Clemenules. This selection has shown very little evidence of granulation over several years of evaluation. The season of maturity for this selection is from October through December in Riverside.

Orlando (VI 19) is a sibling of Minneola tangelo, with Duncan grapefruit and Dancy mandarin as its parents. The tree is moderately vigorous and slightly more cold resistant than Minneola. The leaves of Orlando tend to be cupped, and the fruit is almost round or slightly flattened. The rind is orange, thin, slightly textured, and not easily peeled. The flesh is orange, juicy, and sweet. As with Minneola, Orlando's blossoms are self-incompatible and must be pollinated by a suitable pollinator (Dancy, Clementine, Kinnow, Temple) to ensure satisfactory fruit set for commercial production. As with Minneola, cross-pollinated fruit are seedy, but unlike Minneola, Orlando is not currently commonly grown in California. The season of maturity is early (October to November).

Ortanique (VI 343) is a natural tangor that was discovered in Jamaica. The name is a combination of "or" for orange, "tan" for tangerine, and "ique" for unique. The medium-sized tree is dense and round. The fruit is medium sized, obovate, and has a pebbled, dark orange rind. The flesh is orange-colored, tender, and exceptionally juicy. When cross-pollinated, the fruit is seedy. The flavor is rich and sweet. Ortanique fruit mature in the spring in California, and the fruit holds very well on the tree.

Page (VI 58), a cross between Minneola tangelo and Clementine mandarin, was developed by William Gardner, Philip Reece, Frank Gardner, and Jack Hearn of the USDA starting in 1942. The fruit are medium sized and round with a slightly pebbled orange-red rind. The flavor is rich and sweet. Page blossoms are self-incompatible and must be cross-pollinated by a suitable pollinator to ensure good fruit set. The season of maturity is November through January.

Sidi Aissa (VI 508) is a Clementine selection of unknown origin introduced from Morocco in 1987. The fruit are reported to have good coloration at the beginning of the season in Morocco and larger fruit with a thicker rind than Cadoux. Sidi Aissa was selected for inclusion in a new multilocation replicated trial of several mandarins to evaluate yield and other fruit qualities due to good-quality fruit characteristics in California.

W. Murcott (W. Murcott Afourer) (VI 462) (fig. 4.12) is a tangor of unknown parentage that was imported to California from Morocco in 1985. The tree is moderate in size and vigor. Fruit are usually flattened, with a thin, smooth, orange rind that is easy to peel. The fruit is low seeded in the absence of cross-pollination but seedy when cross-pollinated. The flesh is orange and juicy, with a rich, sweet flavor. As with most mandarins, W. Murcott is susceptible to alternate bearing. The fruit matures in February and holds on the tree very well.

Mandarins that produce seedy fruit

These varieties produce seedy fruit even when grown in single-variety blocks or under isolated conditions.

Daisy (VI 382), named after Daisy Young of Young's Nursery in Thermal, California, is a hybrid between Fortune and Freemont that hybridized in 1963 and was developed and named by John Carpenter of USDA in Indio. In 1980 it entered the Citrus Clonal Protection Program. Daisy is grown in the desert in California but is not extensively planted, despite its excellent flavor. The fruit mature in late December to January.

Dancy (VI 386) mandarin traces its origin to a seedling tree growing in the orchard of Colonel G. L. Dancy in the 1860s. Its parent was a mandarin known as the Moragne tangerine, reported to have been growing in the orchard of N. H. Moragne as early as 1843, that is believed to have been introduced from Tangiers, Morocco, by Major Atway, the previous owner of the Moragne property. Regardless of this variety's tangled history, its reputed origin in Tangiers gave rise to the term "tangerine," which today may refer to any mandarin. The Dancy tree is a large, vigorous, densely foliated tree with a tendency to alternate bearing. The fruit is usually medium sized and oblate to obovoid. The thin, smooth rind is reddish orange at maturity and easily peeled. The flesh is a deep orange, with a rich flavor. The fruit usually contain a moderate number of seed. Dancy fruit mature in midseason and do not hold well on the tree, although they do store quite well after harvest. Dancy has traditionally been an important mandarin variety in Florida. In California, fruit are generally small and overly acidic, except when produced in the desert valleys region.

Ellendale (VI 464) tangor originated in Queensland, Australia, where it was found growing on the Ellendale property of E. A. Burridge as early as 1878. Today, it is an important variety in parts of Australia and other citrus-growing regions of the world. The tree is medium to large, spreading, and has a tendency to limb breakage as a result of weak crotches. Ellendale fruit are large and oblate, with a thin, smooth, orange-red rind that is easily peeled. The flesh has an attractive dark orange color. The seed are monoembryonic, and fruit grown in solid-block plantings are low seeded, whereas those grown in mixed-block plantings are very seedy. Ellendale fruit are especially juicy, and the flavor is sweet and rich. They mature in the late midseason, but the rind is sometimes slow to develop full color.

Fairchild (VI 246) mandarin is a hybrid of Clementine mandarin and Orlando tangelo made by J. R. Furr at the U.S. Date and Citrus Station at Indio, California, and released in 1964. It has proven to be particularly well suited to the California and Arizona deserts, where it provides an early-season fruit for the market. Fairchild trees are vigorous, with a wide-spreading habit. For best fruit production, a pollinator is necessary. The fruit is medium sized, oblate, and has a thin, slightly pebbled, dark orange rind. Fairchild is not especially easy to peel, and the fruit typically contain many seed; but the flesh is quite juicy, and the flavor is rich and sweet.

Honey (VI 133) is often referred to as California Honey to distinguish it from the variety Murcott that is marketed by Florida growers as Honey. California Honey, a hybrid of King and Willowleaf mandarins, is one of the early mandarin varieties bred by H. B. Frost at the UC Citrus Experiment Station. The fruit are small and oblate, with lightly pebbled rind. The flesh is light yellowish orange, with a rich flavor and early maturity. The trees have vigorous growth and there is a tendency to alternate bearing.

Kinnow (VI 1), a hybrid between King and Willowleaf mandarins, was developed at the UC Citrus Experiment Station by H. B. Frost in 1915 and released in 1935. Kinnow is the most widely planted mandarin in Pakistan. The tree grows vigorously and has an upright form, with a strong tendency to alternate bearing. The fruit is oblate, with a smooth orange rind that does not peel especially well for a mandarin. The flesh is orange, seedy, and has a rich, distinctive flavor. Kinnow matures in midseason and holds well on the tree.

Figure 4.12 W. Murcott (W. Murcott Afourer) (VI 462) tangor. *Photo:* D. Karp.

Murcott tangor (VI 147) is the same variety marketed by Florida growers as the variety Honey. It is believed to have originated in the USDA citrus breeding program in Florida in the early 1900s. Murcott trees are moderate in size and vigor, with a somewhat upright growth habit and willowy branches. The fruit tends to be borne near the outside of the tree. Alternate bearing often occurs in this variety, and the fruit burden is so excessive in the on-year that the tree may die due to carbohydrate depletion, commonly referred to as Murcott collapse. The fruit is medium sized when the tree is carrying a moderate fruit load. The orange rind is thin, smooth, and peels moderately well. Murcott differs from the variety W. Murcott in a number of fruit characteristics such as fruit shape, size, flavor, and especially seediness, since Murcott fruit will be seedy even when grown in single-location blocks or isolated from cross-pollination.

Nova (VI 282), a sibling of the Lee, Osceola, and Robinson mandarins, is a hybrid of Clementine mandarin and Orlando tangelo. The tree grows vigorously and can be thorny. The medium-sized fruit is usually subglobose, with a slightly pebbled, thin, reddish orange rind that is easily peeled. The flesh is dark orange, fine textured, and juicy with a sweet, rich flavor. In mixed plantings, Nova fruit are moderately seedy, but in isolation they are seedless but less plentiful. Nova matures in early winter and holds well on the tree.

Ponkan (VI 311) is probably the most widely grown mandarin in the world, especially in China, India, and Brazil, but it is not commonly grown in California. Ponkan trees are vigorous, upright, and medium sized. In California, Ponkan fruit are usually oblate, with a loosely adherent, thick, orange-yellow rind, and they ripen in midseason.

Sunburst (VI 377) is a hybrid Clementine and Orlando tangelo developed by P. C. Reece in Orlando, Florida, in 1961. Grown extensively in Florida, the fruit of Sunburst has a deep reddish rind that is very thin and fine textured. The fruit are low seeded if grown in isolation but seedy if cross-pollination occurs. The harvest season begins in early November in Florida.

Temple (VI 330), which is also called Temple Orange and Royal mandarin, is a tangor with a convoluted history and origin that are somewhat unclear, dating to before 1896 in Jamaica. The tree is spreading and bushy. Trees are more cold sensitive than other mandarins and sweet oranges and have a high heat requirement, which causes this variety to have a limited range of adaptability. The fruit are medium large and very broadly obovate to slightly subglobose, sometimes with a neck. The rind is deep reddish orange, medium thick, and somewhat pebbled in texture but still peelable. The fruit flesh is orange and moderately juicy with a rich flavor. When grown in Florida, Temple has excellent fruit quality but produces satisfactory quality only in the hottest interior region of California. Temple is medium to late in maturity.

USDA 88-3 (VI 488) is a recently released mandarin hybrid of Robinson and Lee developed at the USDA station in Orlando, Florida. The trees are moderately large at maturity. The fruit are low seeded to seedless in isolation but can be quite seedy, with approximately 15 seed per fruit, when cross-pollination occurs. The fruit is medium sized, and the rind is medium orange and slightly pebbled. The harvest season starts in November, and the fruit hold well on the tree.

Willowleaf mandarin (VI 118) is known under many other regional names, the most common ones being Mediterranean and Avana mandarin. This cultivar is known to have been growing in the Mediterranean since the very early 1800s. It is distinctive in both tree and fruit characteristics. The tree is of medium size, broadly spreading and pendant. The leaves are narrow and thin, suggestive of willow leaves, and fragrant. The fruit is medium sized, oblate, with a smooth, loose, orange rind that contains distinctively fragrant oil. The flesh is light orange, tender, juicy, and sweet. Willowleaf matures in winter, but the acidity drops, the rind puffs, and fruit quality decreases if the fruit is held on the tree past maturity.

Grapefruit and Hybrids

Grapefruit are thought to have originated as a natural hybrid between a pummelo and sweet orange in the Caribbean region during the early eighteenth century (Gmitter 1995). The grapefruit was unknown outside the Caribbean until the early nineteenth century, when it was introduced into Florida. Since its introduction into Florida, a number of new cultivars have been produced by natural and induced mutations. These new cultivars differ from the original type, which was white fleshed and seedy, in that they are seedless and have white or pink flesh. The pink-fleshed cultivars, often referred to as red grapefruit, contain the pigment lycopene. Grapefruit trees have the tendency to set fruit in clusters, unlike pummelos, which set fruit singly. They are known to be vigorous and large and are also highly resistant to heat and have similar medium-high cold tolerance to sweet oranges. Grapefruit have a high heat requirement to obtain fruit maturity and high quality, which explains why most grapefruit production in California is in the southern desert valleys.

Chironja (VI 436 and VI 559) grapefruit hybrid was a seedling tree discovered in 1956 in a mountainous section of Utuado, Puerto Rico. Although the parentage is unknown, the fruit resembles both a sweet orange and a grapefruit and is thought to be a hybrid of the two. The name reflects the local term for sweet orange *(china)* and the Spanish word for grapefruit *(toronja)*. The trees are vigorous, large, and spreading. The leaves have broad wings, are cupped, and have margins that irregularly undulate. The fruit are grapefruit sized and broadly obovoid to pyriform, with a bright yellow rind at maturity. Fruit are borne singly rather than in clusters as is typical of grapefruit. The flesh is yellowish orange and very juicy, with a mild flavor that lacks the typical bitterness of grapefruit, and mature from January to May.

Duncan (VI 660) grapefruit is believed to be the oldest grapefruit variety grown in the United States. Although not named until 1892, it was growing as early as 1830 near Safety Harbor, Florida. The tree is large, spreading, and grows vigorously. The fruit is almost round and has a smooth, light yellow rind. The flesh is very pale yellow, juicy, and seedy. The flavor is considered by many to be superior to that of similar seedless grapefruit varieties. Duncan is considered to mature somewhat early, between January and May. The fruit holds well on the tree.

Flame (VI 470) grapefruit is reported to originate from seed of a sport of Ruby Red grapefruit in the Houston, Texas, orchard of C. Henderson. C. J. Hearn made the final selection, and Flame was released as a new variety in 1987. Flame trees grow vigorously to a large size and are reported to be more cold tolerant than Star Ruby. The fruit has a smooth, yellow rind and usually has a pink blush. The flesh is tender and juicy and has an internal color almost as dark as Star Ruby. Flame is a mid- to late-season variety, maturing from February to June, yet the fruit holds well on the tree, with some fading of the internal color when held past maturity.

Marsh (VI 135 and VI 142) grapefruit originated as a chance seedling around 1860 in Lakeland, Florida. Because it was the first seedless variety to be promoted, it soon became the most widely planted grapefruit variety. The tree grows vigorously to a large size. The fruit is almost round, with a smooth light yellow rind. The flesh is pale yellow, juicy, and tender, with a good flavor. The fruit matures late and holds well on the tree. Several selections of Marsh have been released, including Frost Marsh (VI 29), Brown Marsh (VI 135), Reed Marsh (VI 142), and Miami Marsh (VI 279). Only Brown Marsh and Reed Marsh are registered bud source trees at the CCPP.

Melogold (VI 323) is a hybrid of Siamese Sweet pummelo, an acidless pummelo, and a white tetraploid grapefruit. It was developed at UCR by J. Cameron and R. Soost. A sibling of Oroblanco, it was released in 1986. Melogold trees grow vigorously to a large size with a somewhat spreading form. Melogold fruit are oblate, with a slightly flattened base. The rind is smooth and medium to dark yellow at maturity and thinner than Oroblanco. The large fruit typically average 1 pound. The flesh is pale yellow, seedless, tender, and juicy. The flavor is mild, sweet, and reminiscent more of pummelo than grapefruit. Melogold matures early and holds well on the tree. Due to its low-acid pummelo parent, Melogold can reach maturity in climatic regions that are marginal for the production of other grapefruit.

Oroblanco (VI 309) (fig. 4.13) is a hybrid of Siamese Sweet pummelo and a white tetraploid grapefruit developed at UCR by J. Cameron and R. Soost. The cross was made in 1958, and the Oroblanco selection was released in 1980. Oroblanco trees grow vigorously to a large size with a somewhat spreading form. The fruit has a smooth, greenish yellow rind at maturity. Oroblanco fruit are oblate, with a flattened base that has a pronounced indentation. The rind is thicker than the typical grapefruit, and both the rind and membranes exhibit the bitterness characteristic of Oroblanco's pummelo parentage. The flesh is very pale yellow and seedless. The flavor is mild and sweet. Oroblanco matures from early December to March in Riverside and from late November through February in the San Joaquin Valley; it holds very well on the tree. Due to

Figure 4.13 Oroblanco (VI 309) grapefruit. *Photo:* D. Karp and T. Siebert.

its low-acid pummelo parent, Oroblanco can reach maturity in climatic regions that are marginal for the production of other grapefruit.

Rio Red (VI 440) grapefruit originated in the breeding program of R. A. Hensz at Texas A&M University in Weslaco, Texas. It originated as a bud sport of a tree produced from irradiated budwood of a Ruby grapefruit seedling. The tree grows vigorously to a large size. Plantings of Rio Red grapefruit have performed quite well in the Coachella Valley desert region of California. The fruit is large, with a slightly pebbled surface. The rind frequently has a pink blush, especially where two fruit grow against each other. The flesh is juicy and well pigmented, with the darkest color next to the segment membranes. Rio Red matures from midseason to late season.

Ruby and Redblush (VI 31) grapefruit are red-fleshed sports of the Thompson grapefruit, which is a pink-fleshed sport of the white-fleshed Marsh. Ruby was discovered in the Lower Rio Grande Valley of Texas around 1929. It was patented in 1934, the first citrus to receive a plant patent. Redblush was discovered in Donna, Texas, in 1931 and introduced in 1934. The two are indistinguishable and are usually considered identical. Because of the attractive flesh color and pink rind blush, they were planted extensively, first in Texas, then in other U.S. citrus-growing areas. The trees are large, spreading, and vigorous. The fruit is almost round, medium sized, and has a smooth, light yellow rind that is blushed with pink. The flesh is dark pink, tender, and juicy. The fruit matures from midseason to late season and holds well on the tree, although the flesh color fades somewhat when held past maturity.

Figure 4.14 Star Ruby (VI 355) grapefruit. *Photo:* D. Karp and T. Siebert.

Star Ruby (VI 355) grapefruit (fig. 4.14) is the result of irradiated Hudson grapefruit seed from the breeding program of R. A. Hensz of Texas A&M University. Star Ruby is more difficult to grow than other grapefruit varieties. It has exhibited greater susceptibility to Phytophthora diseases and other pest problems, nutrient deficiencies, and cold temperatures. It does not appear to grow as vigorously as other grapefruit varieties, and the fruit are often smaller. Nevertheless, when grown well, Star Ruby produces attractive red-blushed fruit with a smooth, yellow rind. The flesh is very darkly pigmented, juicy, and low seeded or seedless. It matures in mid- to late season and holds well on the tree, with some loss of flesh color as the season progresses.

Pummelos and Hybrids

Pummelos, or shaddocks, are considered one of the three primordial citrus species. They originated in more tropical areas than most other citrus types. Pummelos have the largest leaves, flowers, and fruit of all citrus species. The trees of different cultivars vary in size from being fairly small to some that are the largest of all citrus trees. Most pummelo trees are drooping in habit. Pummelo fruit range in shape from subglobose to subpyriform. In addition, the fruit have very thick rinds, with numerous carpels (sections) that have tough carpel membranes. The juice vesicles that make up the flesh are larger than any other citrus species; in most cultivars, the juice vesicles do not adhere to each other, which produces a grainy texture to the flesh. The flesh can range from white to pale green to pink due to the pigment lycopene in the pink-fleshed cultivars. The flesh of pummelos can also vary in the level of acid. Acidless, or sweet, types such as the cultivar Siamese Sweet have a sweet but insipid flavor. All pummelos appear to be self-incompatible, which means that if they are self-pollinated they do not produce seed. However, if they are surrounded by other types that can cross-pollinate them, they produce seedy fruit, since they have numerous viable ovules. Pummelos also generally have plentiful viable pollen that can cross-pollinate other citrus types.

Chandler (VI 11) pummelo (fig. 4.15) is a hybrid of Siamese Pink pummelo and Siamese Sweet pummelo, released in 1961 by J. Cameron and R. Soost at UCR. The tree grows vigorously to a large size with pendulous branches. The fruit are large to very large and almost round, with a thick, yellow rind that is occasionally blushed with pink. The flesh varies from light pink to very dark pink, depending on where it is grown. The flesh texture is somewhat grainy but still juicy for a pummelo. The fruit mature early, and the flavor is pleasant and sweet. Chandler fruit are seedy when cross-pollinated by citrus fruit with viable pollen and seedless when grown in a solid block or in proximity to citrus fruit that are pollen sterile, such as navel oranges and Satsuma mandarins.

Cocktail (Mandalo) (VI 127) is not a grapefruit but rather a hybrid of Siamese Sweet pummelo and Frua mandarin (Frua mandarin = King × Dancy). The cross was made at Riverside in the 1920s and released in the 1950s. Cocktail grapefruit trees are large and vigorous. The fruit can vary from the size of an orange to the size of a grapefruit and has a thin, smooth, yellow rind. The flesh is very seedy, yellow-orange, and exceptionally juicy. The flavor is pleasantly subacid. The fruit mature in early winter and hold well on the tree, puffing when they become very old but not desiccating. Due to its seediness and juiciness, Cocktail grapefruit is best used as a juicing fruit for the home.

Pomelit (VI 566) pummelo was selected from the Djeroek Deleema Kopjor of Indonesia. The tree grows vigorously but is reported to be susceptible to branch breakage under heavy fruit loads. The fruit is round, with a slightly flattened base. The rind is thin for a pummelo, greenish yellow to yellow, and smooth. The pink flesh is tender and juicy, with a finer texture than is typical of a pummelo. The fruit are seedy but of good flavor. Pomelit matures early and holds well on the tree.

Reinking (VI 274) pummelo was initially reported to be a hybrid of Kao Phuang pummelo and Shamouti sweet orange made in 1948 by J. R. Furr at the U.S. Date and Citrus Station at Indio, California. However, it is similar in appearance to pummelo, with little or no suggestion of sweet orange in its background, and subsequent examination of the Reinking variety by USDA personnel and more recent DNA-level research have cast doubt upon its parentage; it is now believed to be of pure pummelo parentage. The tree grows well to a large size with big leaves and drooping branches. The fruit is large and pear shaped, with a slightly flattened bottom and a slightly pebbled, thick, yellow rind. The flesh is light yellow, grainy, but juicy. The flavor is good, but the fruit are seedy. Reinking matures early and holds fairly well on the tree.

Sarawak (VI 414) pummelo is sometimes referred to as Tahitian pummelo. The tree is large and vigorous. The fruit is round, with a flattened bottom, and has a greenish yellow rind that is thinner than the typical pummelo. The flesh is greenish, juicy, and sweet, with a hint of melon or lime. The fruit matures from early season to midseason and holds well on the tree. Thong **Dee** (VI 569) pummelo is a Siamese variety that was imported from Florida in the early 1990s. VI 569 is actually a selection of Thong Dee made in Florida for breeding purposes, with superior combining ability, a specific type of cold hardiness, and excellent fruit quality. Thong Dee has medium large fruit, with light yellow rind at maturity. Under favorable conditions both the albedo and the flesh of the fruit are pink tinged, with the flesh coloration occurring in streaks. The carpel membranes are thin and tough and readily separate from the pulp, which has large vesicles. The fruit is juicy and a pleasant flavor. The fruit matures in midseason.

Figure 4.15 Chandler (VI 11) pummelo. *Photo:* T. Siebert.

Lemons

Current research indicates that lemons appear to be a hybrid between citron and sour orange. Since sour orange is considered to be a hybrid of pummelo and mandarin, lemons are considered a complex natural hybrid that arose in the remote past. Lemon trees are generally vigorous, upright, spreading, and open. The fruit are oval, with an apical mammilla, or nipple, at the stylar end of the fruit, and the rind is yellow when mature, tightly adherent, and fragrant. Lemon fruit are more resistant to cold and heat than limes and citrons, but they are much less resistant than most other commercial types of citrus. Most commercial lemons have high levels of acid, but some lemon cultivars have low levels of acid and are commonly called sweet lemons. Limettas are similar to lemons and have both acid and low-acid forms. The low-acid forms are more common and are often called sweet lemons. Lemon hybrids such as Meyer lemon are also commonly considered to be lemons.

Allen Eureka (VI 227 and VI 368) lemon consists of two selections. VI 227, the Allen Young clonal line, was discovered at the Willits and Newcomb Ranch in Thermal, California, in 1963. VI 368 is a nucellar seedling from Utt Ranch in Oxnard, California, that was brought to the UC Citrus Experiment Station in 1949 and to the CCPP in 1977. The Allen Young clonal line is more commonly grown in California than VI 368.

Dr. Strong Lisbon (VI 345 and VI 371) has two selections: the nucellar seedling selection VI 345 was received as budwood by Domingo Hardison, La Campana Ranch, Fillmore, California, in 1966. This Lisbon selection tended to have strong juvenile characteristics, so when the shoot-tip grafting technique became available 6 years later, VI 371 was selected through this process. Although the CCPP has not seen significant differences between the two budlines in fruit characteristics, they are keeping both available, since some nurseries have a definite preference of one over another.

Eureka lemon is one of the most commonly grown lemons in California and a major lemon type throughout the citrus-growing world. It originated in a group of seedling trees grown in Los Angeles in the 1850s. The seed that produced these trees were of Italian origin. As is typical of lemon trees, Eureka trees grow vigorously with a spreading, open form. Eureka lemon trees are slightly less cold hardy and are less thorny than Lisbon lemon trees. The new growth and flowers are tinged with purple. Eureka lemon trees bear multiple crops per year, but the main season is late winter, spring, and early summer. Eureka lemons are not compatible with Trifoliate Orange *(Poncirus trifoliata)* or Trifoliate Orange hybrids (i.e., Carrizo and Troyer Citrange) rootstocks. Rootstocks such as Volkamer Lemon *(C. volckameriana)* and Alemow *(C. macrophylla)* are recommended for Eureka trees. The fruit are borne on the outside of the canopy and are frequently held in clusters. Eureka lemons are medium-small, elliptical, and sometimes longitudinally ridged. The fruit usually have a slight neck and a short nipple. The rind is yellow at maturity, has sunken oil glands, and is slightly textured. The flesh is pale greenish yellow, low seeded, and very acidic. There are many named selections of Eureka lemon with characteristics that distinguish them from the original cultivar. Three of the better known selections of Eureka lemon are Allen Eureka, Frost Eureka, and Variegated Pink-Fleshed Eureka lemon.

Femminello (VI 692) lemons are the most important lemon group in Italy. They grow vigorously and are of moderate size at maturity. The trees tend to have few thorns and flower and set fruit throughout the year. In Italy, Femminello lemon trees are culturally managed by a process called *forzutura* to produce four crops per year. The autumn crop is called *primofiore,* the winter to spring crop is called *limoni,* the spring crop is called *bianchetti,* and the summer crop is called *verdelli*. There are a number of named selections within the Femminello group, and the primary harvest seasons and fruit characteristics vary with the selection. The medium-sized fruit are elliptical to oblong, with a neck that varies from short to long and a nipple that varies from short and blunt to long and tapered, depending upon the selection. The rind is medium thick, finely pitted with sunken oil glands, and yellow at full maturity. The flesh is pale greenish yellow, low seeded to seedless, juicy, and very acidic.

Frost Eureka (VI 21) lemon is a nucellar seedling selection of Eureka lemon selected by Howard Frost, the first citrus breeder at the Citrus Research Center, around 1915. Frost Eureka was the standard Eureka selection grown for many years, but Allen Eureka is now more common.

Limoneira 8A Lisbon (VI 380) lemon (fig. 4.16), a vigorous Lisbon selection, originated from the Limoneira Ranch, Ventura County. This selection produces more blossoms than other Limoneiras, which results in a longer harvesting season. The fruit are similar in size and seediness to Eureka lemon. This selection is the most popular Lisbon lemon planted in California and Arizona today.

Limonero Fino (Limonero Fino 49), the principal winter lemon in Spain, was one of promising selections originating from dooryard trees in Murcia, Spain. This selection is reported to be vigorous, thorny, and highly productive. In Arizona, this selection is reported to be an early producer with uniform yield. Fruit are spherical to oval, with a smooth rind and a relatively short nipple. The fruit have relatively high acid and about 5 seed per fruit. Based on a trial of a number of Eureka and Lisbon selections in Arizona, Dr. G. Wright (2010) found that Limonero Fino 49 lemon had statistically similar yields to Limoneira 8A, but Limonero Fino 49 fruit were generally larger.

Lisbon lemon is one of the most widely grown lemons in California and is planted extensively throughout the citrus-growing regions of the world. It is believed to be a Gallego seedling selection of Portuguese origin. Lisbon lemon trees grow vigorously into large, thorny, upright, spreading trees. Due to their vigorous growth habit, the fruit are usually found within the canopy of the trees. The new leaves and flowers are tinged with purple. Lisbon trees produce several crops per year, but the main crop is in winter and early spring. The fruit are medium sized and oblong, with a prominent nipple. The rind is slightly textured and yellow at full maturity. The flesh is pale greenish yellow, low seeded, and very acidic. There are many named selections of Lisbon lemon; two are well-known commercial Lisbon selections (Dr. Strong Lisbon and Limoneira 8A Lisbon), and the third is an introduction that is demonstrating promise. Additional information and fruit quality data about other commercial selections are available at the CCPP (2013) and CVC (2013) websites.

Meyer (VI 319) lemon (fig. 4.17) is believed to be a hybrid of lemon and orange parentage. Frank Meyer, a plant explorer of the USDA, brought the tree to the United States from Beijing, China, in 1908. The tree is moderately vigorous and cold hardy, shrubby, and relatively small at maturity. It is an attractive garden tree. Meyer lemon flowers intermittently throughout the year, but the main season is in the spring. The fruit is medium sized and short elliptical, with a smooth, thin, yellow-orange rind. The flesh is light orange-yellow, moderately seedy, juicy, and acidic. The aroma and flavor of Meyer lemon are distinctive, and many people find it especially desirable. Meyer lemon is sometimes also listed as Improved Meyer lemon in California, but the "improved" refers to the fact that the original introduction of Meyer lemon was a symptomless carrier of citrus tristeza virus (CTV) and citrus tatterleaf virus. Today, all Meyer lemon trees are tested to be free of these viruses.

Millsweet (VI 591) lemon is a limetta of very low acidity. Although strongly resembling lemons, limettas have distinctive characteristics that set them apart. The tree grows as vigorously as the common lemon and has a similar form, but the leaves are more oval, round pointed, and cupped. The new-growth leaves and blossoms are faintly purple. The fruit is medium sized and round, with a distinctively flattened blossom end. The nipple is prominent and surrounded by a deep areolar furrow. The rind is slightly bumpy, pitted with sunken oil glands, and is yellowish orange at maturity. The flesh is pale yellow, low seeded, and low in acidity, giving the juice a sweet taste. Millsweet flowers and sets fruit throughout the year, but the main flowering season is in the spring. The fruit hold on the tree well. A similar limetta is Limonette de Marrakech (VI 573), which has highly acidic fruit. Externally, the fruit are indistinguishable, and the trees are also similar, except that the new growth and flowers of the Millsweet limetta are not tinged with purple.

Figure 4.16 Limoneira 8A Lisbon (VI 380) lemon. *Photo:* D. Karp and T. Siebert.

Figure 4.17 Meyer (VI 319) lemon. *Photo:* D. Karp and T. Siebert.

Seedless Lemon (VI 492) was imported to California sometime before 1985 from Lassocock's Nursery in South Australia under the name Seedless Lisbon, but DNA fingerprint data indicated that this lemon is not a Lisbon. The fruit average fewer than 1 seed per fruit and are very close to seedless when grown in Riverside.

Variegated Pink-Fleshed Eureka (VI 486) lemon, sometimes sold under the name Pink Lemonade, is a sport of the conventional Eureka lemon. It originated in a home garden in Burbank in 1931. The leaves are variegated green and white, making the tree quite ornamental, and the tree grows a little less vigorously than a typical Eureka. The rind is striped green and cream and is sometimes rougher than conventional Eureka. When fully ripe, the stripes fade, and the rind turns yellow with distinct pink oil glands. The flesh is light pink at full maturity, low seeded, and very acidic. The juice is the color of standard lemon juice, not pink.

Limes

Recent studies indicate that limes appear to be a hybrid of citrons and *Citrus micrantha* or a similar type in the *Papeda* subgenus of *Citrus*. Lime cultivars vary greatly and are generally divided into two groups, the acid (sour) limes and the acidless (sweet) limes. The acid limes are further divided into the small-fruited cultivars, which include Mexican, Key, and West Indian limes, and the large-fruited cultivars, which include Persian (also called Bearss, or Tahiti) limes. Although there are similarities between the small- and large-fruited acid limes, the differences are more pronounced. West Indian and Mexican lime trees are less vigorous and less robust and are more cold sensitive than the Persian limes. In addition, Mexican lime trees have finer stems, many more thorns, and smaller, paler leaves than Persian limes. The small-fruited West Indian and Mexican cultivars are seedy, and the flavor has a strong pungent aroma. The large-fruited Persian limes are seedless triploids, with a much milder aroma and flavor than the small-fruited limes. Sweet lime trees have larger leaves than the sour limes, and the fruit has very smooth rind and insipid flavor due to the low acid content.

Figure 4.18 Mexican (VI 419) lime. *Photo:* D. Karp and T. Siebert.

Mary Ellen (VI 625) sweet lime *(C. limettioides)* is a recently introduced lime originating in Mexico, where similar fruit are referred to as *lima dulce*. The mature tree is medium large, with a spreading habit. The flowers are pure white, and the new leaves are green, without the purple shading associated with acid limes. The fruit are small, round to slightly oval, and have a smooth pale green to greenish yellow rind at maturity. The flesh is pale yellow, tender, and juicy, with few seed. The flavor is insipid due to the lack of acidity in the fruit, but it is appealing to some.

Mexican (VI 419) lime *(C. aurantifolia)* (fig. 4.18) is also known by many other names such as Key lime, Bartender's lime, and West Indian lime. The trees are moderately sized and bushy, almost shrublike, and the leaves are distinctively aromatic when crushed. Some selections are quite thorny, while other selections are thornless (see the description of Thornless Mexican lime, below). Mexican lime trees are sensitive to cold. The fruit are small, approximately 1.5 inches in diameter and almost round, with a thin, smooth, greenish yellow rind at maturity that is especially fragrant. The flesh is greenish yellow, seedy, and highly acidic, with a fine texture. Once Mexican limes reach full maturity, usually in autumn to early winter, they drop from the tree. Various selections of Mexican limes have been imported into California but are not currently registered as bud sources by CCPP, including Castelo (VI 637), Chulo (VI 659), and Miranda (VI 670).

Palestine (VI 81) sweet lime *(C. limettioides)* is also known as Indian sweet lime. The tree is medium large, with an irregular, spreading form. The flowers are pure white, and the new growth is green; neither flowers nor new leaves are purple tinged, as acidic limes are. The fruit are small and round to slightly oblong, and they have a thin, smooth rind and prominent oil glands. At maturity, the rind is pale green to orange-yellow. The flesh is pale yellow, tender, and juicy, with some seed. The flavor is insipid due to the lack of acidity in the fruit, but it is appealing to some. Palestine sweet lime is also used as a citrus rootstock.

Persian (VI 358) lime *(C. latifolia)* (fig. 4.19) is also known by many other names, such as Tahiti lime and Bearss lime. The nearly thornless trees grow vigorously to a medium large size, with a spreading form. Persian lime trees are more cold hardy than Mexican lime trees and should grow well in areas where lemons grow well. The fruit are larger than Mexican limes, approximately 2 to 2.5 inches in diameter, with a thin, smooth, light yellow rind at full maturity. The seedless flesh is pale greenish yellow, acidic, juicy, and fine textured. Once Persian limes reach full maturity, usually late autumn to early winter, they drop from the tree. Most Persian lime selections, including VI 358, have the genetic disorder wood pocket, which causes tree decline and death in 7 to 15 years, depending upon climate. A purportedly wood pocket–free selection from Florida, Persian lime SPB-7 (VI 708), was recently introduced from Florida and should be planted instead of VI 358. At this time, the actual performance of VI 708 regarding wood pocket is under evaluation in California.

Thornless Mexican (VI 502) lime, which is probably a bud sport selection of Mexican lime, was introduced into the UCR Citrus Variety Collection in 1937 as budwood from the Marcy Ranch in Tustin, California. The original Marcy Ranch trees were purchased from Frank May, a nurseryman who reportedly imported the budwood from Mexico. The fruit characteristics are virtually indistinguishable from Mexican lime, yet the trees are almost completely thornless and more upright in their growth habit than the standard Mexican lime.

Figure 4.19 Persian (VI 358) lime. *Photo:* D. Karp and T. Siebert.

Citrons

Citrons are considered one of the three primordial citrus species and were the first to be brought under cultivation and to reach the Mediterranean. Citron trees are shrubs or small trees with straggly growth, soft wood, and short, stout thorns. The trees are highly sensitive to frost. The fruit are generally very large, often weighing 2 to 3 pounds, and variable in shape but usually oblong. A noted exception is the cultivar Buddha's Hand, or Fingered citron, which resembles a hand. The smooth or bumpy rind of the fruit is very thick, tightly adherent, and highly aromatic and sometimes sweet; it is the part of the fruit that is commonly used commercially for its scent or candied for cakes and candies. The flesh is not very juicy, can be high acid or low acid (commonly called sweet), and has an insipid flavor. Some citrons have small amounts or an absence of flesh.

Buddha's Hand (Fingered) (VI 369) citron is a unique citrus grown mainly as a curiosity, but it has been used in floral decorations, and the fruit have a strong, pleasing scent for use in interior design. The tree is small and shrubby with an open habit. It is very frost sensitive. Citron leaves are distinctive in form, being oblong and somewhat rumpled, with serrate margins. The flowers and new-growth leaves are heavily tinged with purple. The very small immature fruit may also be flushed with purple. The 6- to 12-inch fruit are apically split into a varying number of segments that resemble a human hand with many fingers. The rind is yellow and highly fragrant at maturity. The interior of the fruit is solid rind, with no flesh or seed. Fingered citron fruit usually mature in late fall to early winter and hold moderately well on the tree, but not as well as other citron varieties.

Etrog (Ethrog) (VI 426) citron, although often treated as a single cultivar, may refer to a dozen or more distinct cultivars or genotypes. Etrog is commonly used in the Jewish Sukkot harvest festival. Etrog fruit used in this ceremony must be free of blemishes, symmetrical, have a persistent style (portion of flower), and must come from trees that have not been grafted. Etrog trees are small and shrubby with an open growth habit. The new growth and flowers are flushed with purple, and the trees are sensitive to frost. The leaves are oblong but slightly pointed and somewhat rumpled, with serrate margins. The aromatic fruit is shaped somewhat like an oblong lemon but is considerably larger. The yellow rind is glossy, thick, and bumpy. The flesh is pale yellow and acidic, but not very juicy. Citron rind is traditionally candied for use in holiday fruitcake. The fruit hold well on the tree.

Kumquats and Hybrids

Kumquats are acidic fruit that are popular in Asian countries, especially China. They are commonly considered to be citrus fruit, but they are actually classified in the genus *Fortunella*, which is closely related to the genus *Citrus*. Although they are markedly similar in appearance to citrus fruit and can hybridize with them, kumquats were placed in a separate genus because of differences in fruit and flower structure. Kumquats are distinctive due to their small fruit with tart flesh and a sweet, edible peel. The trees are relatively small, symmetrical, and can become 25 to 30 feet tall. The attractive foliage is dark green on the upper surface and light green on the under surface. One other notable characteristic of kumquats is their outstanding cold hardiness. They are more cold hardy than all *Citrus* types including Satsuma, but less hardy than cultivars in the genus *Poncirus,* which are commonly called trifoliate oranges. Some kumquats and hybrids do not perform well or do not produced long-lived trees when propagated onto Trifoliate Orange or Trifoliate Orange hybrid rootstocks. As an alternative the use of Calamondin rootstock is recommended. Kumquat trees are especially susceptible to zinc deficiency, which can cause small leaves and reduced internode distance.

Calamondin (VI 408) is an acidic fruit that is most commonly grown in the Philippine Islands, although it is probably of Chinese origin. It is believed to be a natural hybrid of kumquat and probably an acid mandarin. In the Philippines, it is sometimes called *calamonding* or *calamansi*. In the United States, it is primarily planted as an ornamental, growing either in the ground or in a container. Calamondin is cold resistant. The tree is upright and columnar, and the leaves are small and dense, giving the tree a fine-textured appearance. The fruit are very small, round, and orange at full maturity. The orange flesh is acidic, juicy, and contains a few seed. Calamondin trees flower and set fruit intermittently throughout the year, adding to their decorative appeal. A variegated form (VI 475), with marbled leaves and faintly striped fruit, is sometimes marketed under the name Peters.

Centennial (VI 594) kumquat hybrid was released by the USDA Agricultural Research Service in 1993. It originated as a bud sport on a seedling tree resulting from an open pollination of Nagami kumquat. Centennial is a small, attractive, shrubby, thornless tree with an upright, fine-textured growth habit that appears to grow well on Trifoliate Orange rootstock. The leaves are variegated pale yellow and cream, and the immature fruit are striped light green and light yellow. At maturity, the fruit acquire an orange color, and the striping becomes less pronounced. The fruit are oval and necked, up to 2.5 inches long, with a smooth rind. The flesh is acidic, juicy, light orange, and low seeded. The fruit matures in winter and holds well on the tree.

Changshou kumquat *(Fortunella obovata)* is also called the Fukushu kumquat in Japan. The naturally small tree is very ornamental, has a spreading form, and is thornless. The leaves are larger and broader than other kumquat species. The fruit are also larger and slightly different in shape, about 1.5 inches long and oval, with a depressed apex. The rind is orange and thinner than Nagami or Meiwa. The flesh is orange, acidic, and contains a few seed.

Meiwa (VI 277 and VI 306) kumquat *(Fortunella crassifolia)* is a lesser-known species of kumquat. The two selections VI 277 and VI 306 are very similar. The tree is similar to Nagami in appearance, but it cannot be budded onto all the same rootstocks as Nagami. Trifoliate seems to be the best rootstock choice for Meiwa. As with Nagami, Meiwa trees are semidormant in winter, allowing them to withstand temperatures below freezing. The flowering season is in summer, and the fruit mature in late winter. The almost-round fruit are orange at maturity and up to 1 to 1.5 inches in diameter. The sweet rind is thicker than the rind of Nagami, making it seem sweeter than Nagami. The flesh is light orange, contains a few seed, and is acidic.

Nagami (VI 276) kumquat *(Fortunella margarita)* is the most commonly grown type of kumquat. The tree is small to medium sized with a dense and somewhat fine texture, and it is remarkably cold hardy due to its tendency to become semidormant from late fall to early spring. The typical Nagami fruit is oval, 1.25 inches long and 0.75 inch wide. The entire fruit can be eaten; the orange rind is sweet, and the light orange flesh is acidic. Each fruit contains about 5 or 6 seed. The flowering season for kumquats is summer, and the fruit mature in late winter, holding well on the tree.

Nordmann (VI 592) is a newly released entirely seedless selection of Nagami kumquat discovered by George Otto Nordmann on a Nagami seedling in 1965 in DeLand, Florida. The tree is similar to the standard Nagami kumquat, small to medium sized with a fine, dense texture. The tree becomes semidormant in winter, making it relatively cold hardy. The fruit are orange at maturity, up to 2 inches long and 0.75 inch wide, teardrop shaped, and frequently hanging in ornamental clusters. The fruit are seedless, and the rind is sweet while the light orange flesh is acidic. Preliminary data indicate that Nordmann appears to be incompatible with Carrizo rootstock but compatible with Rough Lemon, Yuma Ponderosa Lemon, and Alemow *(C. macrophylla)*. The flowering season is in summer, and the fruit matures in winter, holding well on the tree.

Papedas and Hybrids

Swingle classified papedas in the separate subgenus *Papeda* rather than as *Citrus* due to a number of distinctive flower, fruit, seed, and leaf characteristics. Papeda fruit have high concentrations of droplets of acrid oil in the pulp vesicles that cause them to taste bitter and unpleasant. Although the fruit are seldom eaten fresh, they have many other uses in Southeast Asia. Papeda leaves have distinctive large, winged petioles. The leaves of one species, *Citrus hystrix,* which is commonly called Kaffir lime or Kuffre lime, is used as a condiment in Southeast Asian foods.

Citrus hystrix (VI 463) is sometimes referred to as Kaffir lime, or more correctly, Kuffre lime, but it is not a lime. The tree is small and shrubby, with distinctive leaves that have a petiole almost as large and wide as the leaf blade. It is the pungent leaves, and not the fruit, of this species that are commonly used in Thai and Indonesian cooking. The fruit is small, oval to short-pyriform, and has an irregularly bumpy rind. When the fruit reach full maturity in late winter to early spring, the rind turns yellow and the fruit fall from the tree. In some places, the fruit is used to make a shampoo that repels insects.

Sudachi (VI 693) is believed possibly to be a hybrid of *C. ichangensis* × *C. reticulata*. It is a small acidic citrus fruit that is similar in use to Yuzu. Sudachi trees are medium sized and spreading. The fruit have a pungent aroma that is quite distinctive. They are used in Japanese cuisine, both in cooked dishes and as a food preservative, and as a garnish. In Tokushima, Japan, where Sudachi is grown and especially prized, the fruit are available primarily in summer and are still dark green at harvest.

Yuzu (VI 619) *(C. junos* or *C. ichangensis* × *C. reticulata* var. *austera)* is a small acidic citrus fruit that originated in China but is now more widely used in Japan. The tree is medium sized and spreading and is quite cold hardy. The fruit are small and round to slightly flattened, with a rind that is thick, rough, and bumpy. Yuzu fruit are normally harvested in autumn while still green, although the fruit eventually turn a dark yellow color. The flesh is yellow, acidic, bitter, and seedy. The most notable characteristic of Yuzu is its pungent aroma. Both the rind and the juice of Yuzu find use in Japanese cuisine, and dried Yuzu is used as a spice. The fruit are also used in baths at Lunar New Year, which is believed to ensure good health.

Commercial Variety in a Related Citrus Genus

Australian Finger Lime (VI 697) *(Microcitrus australasica)* (fig. 4.20) was released for propagation in 2008. The California selection of Australian Finger Lime was received as budwood from J. Furr of the U.S. Date and Citrus Station in Indio in 1968; the original source of this selection was seed collected by W. V. Mangomery, Maryborough, New South Wales, Australia. The fruit of this citrus relative are elongate, like a small chili with dark purplish rind. The juice vesicles that make up the flesh are not fused to each other and are round with a stalk, so they resemble light pink to clear fish eggs. Although not a lime, the flavor is acidic, like a lime. The California selection has much lighter-colored juice vesicles than some of the deep-colored selections available in Australia. The trees are very small, slow growing, and have many small thorns and small leaves. Trees flower from February to April; although fruit are available year round, the peak production period is November through December.

Acknowledgments

This chapter would not have been possible without advice and suggestions from a number of individuals with expertise on citrus diversity. We would like to thank John Bash, Norm Ellstrand, Ben Faber, Louise Ferguson, David Karp, Robert Krueger, Richard Lee, Mikeal Roose, Toni Siebert, and Glenn Wright for their support.

Figure 4.20 Australian Finger Lime (VI 697). *Photo:* D. Karp.

References

Barrett, H. C., and A. M. Rhodes. 1976. A numerical taxonomic study of affinity relationships in cultivated *Citrus* and its close relatives. Systematic Botany 1:105–136.

CCPP (Citrus Clonal Protection Program). 2013. University of California, Riverside, Department of Plant Pathology CCPP website, http://www.ccpp.ucr.edu.

Chao, C. T. 2005. Super early season Satsuma mandarins for San Joaquin Valley and California. Topics in Subtropics Newsletter 3(2): 3–5.

Chao, C. T., J. J. Fang, and P. S. Devanand. 2005. Long-distance pollen flow in mandarin orchards determined by AFLP markers: Implication for seedless mandarin production. Journal of the American Society for Horticultural Science 130(3): 374–380.

CVC (Citrus Variety Collection). 2013. University of California, Riverside, Department of Botany and Plant Sciences, View Citrus Varieties website, http://www.citrusvariety.ucr.edu.

Gmitter, F. G., Jr. 1995. Origin, evolution and breeding of the grapefruit. In J. Janick, ed., Plant breeding reviews. Vol. 13. Hoboken, NJ: Wiley. 345–363.

Hodgson, R. 1967. Horticultural varieties of citrus. In W. Reuther et al., eds., The citrus industry. Vol. 1. Rev. ed. Berkeley: University of California Press. 431–591.

Kahn, T. L., O. J. Bier, and R. J. Beaver. 2007. New late-season navel orange varieties evaluated for quality characteristics. California Agriculture 61(7): 138–143.

Kahn, T. L., T. Siebert, Z. Zheng, and K. H. Xu. 2013. Fruit quality evaluations of introduced Satsuma selections for California. Citrograph 4(2): 34–51.

Marais, L. J., and N. V. O'Connell. 2002. Fukumotos and some rootstocks don't mix. Citrograph 87(2): 5, 11–13.

NASS (USDA National Agricultural Statistical Service). 2010. California citrus acreage report. NASS website, http://www.nass.usda.gov/Statistics_by_State/California/Publications/.

Nauer, E. M., W. P. Bitters, D. Cole, D. R. Atkin, and T. L. Carson. 1985. Lindcove Navel Strain Trials after nine years. Citrograph 71:31–37.

Parsons, C. Market prices separating mandarins, July 2, 2010. Western Farm Press website, http://www.westernfarmpress.com/citrus/market-prices-seperating-mandrins-0705/index.html.

Platt, R. G. 1973. Planning and planting the orchard. In W. Reuther et al., eds., The citrus industry. Vol. 3. Rev. ed. Berkeley: University of California Press. 8–81.

Roistacher, C. N. 2007. History of the parent Washington navel orange tree. Ecoport website, http://ecoport.org/ep?SearchType=slideshowViewSlide&slideshowId=79.

———. 2009. The parent Washington navel orange tree. Part 1: Its first years. Topics in Subtropics Newsletter 7(2): 3–5.

———. 2010. The parent Washington navel orange tree. Part 2: Its decline and recovery. Topics in Subtropics Newsletter 8(1): 6–14.

Roose, M. L., R. K. Soost, and J. W. Cameron. 1995. Citrus. In J. Smartt and N. W. Simmonds, eds., Evolution of crop plants. 2nd ed. Essex, UK: Longman Scientific and Technical. 443–448.

Roose, M. L., T. E. Williams, J. W. Cameron, and R. K. Soost. 2000. 'Gold Nugget' mandarin, a seedless, late-maturing hybrid. HortScience 35(6): 1176–1178.

Saunt, J. 2000. Citrus varieties of the world. 2d ed. Norwich, UK: Sinclair International.

Scora, R. W. 1988. Biochemistry, taxonomy and evolution of modern cultivated Citrus. Proceedings of the 6th Congress of the International Society of Citriculture 277–289.

Soost, R. K., and M. L. Roose. 1996. Citrus. In J. Janick and J. N. Moore, eds., Fruit breeding. Vol. 1. Hoboken, NJ: Wiley. 257–323.

Swingle, W. T., and P. C. Reece. 1967. The botany of *Citrus* and its wild relatives. In W. Reuther et al., eds., The citrus industry. Vol. 1. Rev. ed. Berkeley: University of California Press. 190–430.

USDA Agricultural Marketing Service. Fruit and Vegetable Market News. USDA AMS website, http://www.marketnews.usda.gov/portal/fv.

USDA Agricultural Research Service. National Plant Germplasm System website, http://www.ars-grin.gov/npgs/acc/acc_queries.html.

Wright, G. 2010. Results of new cultivar selection trials for lemon in Arizona, 2008-09 and 2009-10. Arizona Department of Agriculture website, http://www.azda.gov/CDP/NewCBC/Final_08-03_and_08-05c.pdf.

5 Rootstocks

Mikeal L. Roose

Choosing a Rootstock

Choosing an appropriate rootstock for a new planting is among the most important decisions a grower makes because it lasts for the life of the planting. Growers should systematically consider the factors that affect rootstock performance on the specific site before selecting a rootstock. This chapter reviews the main factors that influence rootstock performance and summarizes the characteristics of the commonly used California rootstocks.

Once the decision is made on which scion variety to use, incompatible rootstocks can be eliminated. The fruit quality of the scion can also influence rootstock selection. Rootstocks that reduce solids and acid content (e.g., the Rough Lemon group) should not be used with a scion that also has low solids and acids, but they may be acceptable for a higher-quality scion. Similarly, with most scions, rootstocks that increase fruit size are desirable, but late navel oranges tend to have such large fruit sizes that a further increase may decrease crop value.

Rootstocks can strongly influence a tree's response to a variety of stress factors. Important soil factors include salinity, poor drainage, and calcareous soils. Climatic stresses such as the probability of freeze should be considered. In California, the major diseases and pests that influence rootstock choice are Phytophthora gummosis and root rot, citrus nematode, and citrus tristeza virus (CTV). Some other diseases, including dry root rot and Diaprepes weevil, are either rare or affect all rootstocks similarly. Holding fruit for late harvest can also be considered a stress on the tree and fruit. Rough Lemon rootstocks cause high levels of granulation in late-harvested navel oranges, including Lane Late. Choosing one of these rootstocks limits the harvest window available to the grower.

After eliminating unacceptable choices based on these considerations, the grower can choose among the remaining rootstocks. Using a standard rootstock with a long history in California is less risky than using a newer rootstock. However, newer rootstocks may turn out to have higher yield or longevity. Risk can be reduced by using knowledge about the performance of various rootstocks on the planting site. Local experience is often the best guide, so consider planting a small area with some newer rootstocks to evaluate their performance at your site. The experience of others on nearby sites with similar soils and scions can also be very helpful. For a summary of rootstock characteristics, see table 5.1.

General Notes on Rootstock Descriptions

Origin: Information on the origin of the rootstock.

Nursery behavior: Seediness of the rootstock fruit, degree of uniformity of seedlings, and ease of pushing buds of common scions under greenhouse conditions.

Tree size:. The effect of the rootstock on tree size in locations where it is well adapted, not locations where it is stressed by soil, disease, or crowding. In general, comments apply to trees of any age. Exceptions, such as trees that typically grow rapidly when young but more slowly later, are noted. For all scions except Eureka lemon, trees on Troyer or Carrizo are considered standard size. Trees more than 20% larger than the standard (in canopy volume) are considered to be large, while those more than 20% smaller are semidwarf ("intermediate" in table 5.1). Most data on tree size are for oranges; the limited data on lemon

Table 5.1. Effects of citrus rootstocks on tree performance, disease resistance, and fruit quality

Rootstock	Yield/tree	Tree size	Tristeza declines	Phytophthora	Citrus nematode	Calcareous soils
Carrizo/Troyer	I-H	I-Lg	T	I	S/T	P-I
C35	I-H	I	T	T	T	P
Swingle	I	I	T	T	T	P
Trifoliate	I-H	I-Lg	T	T	S/T	P
Cleopatra	L-I	Lg	T	S	S	I
Sour Orange	I-H	I-Lg	S	T	S	G
Rough Lemon	I-H	I-Lg	T	S	S	G
Volkameriana	H	Lg	T	S	S	G
Macrophylla	H	I	S/T	T	S	I

Rootstock	Salinity	Poorly drained soils	Freeze tolerance	Incompatible scions*	Fruit quality
Carrizo/Troyer	P	I	G	EL, FN?	G
C35	P	I	I	EL?, FN?	G
Swingle	P	P	G	EL, O?	G
Trifoliate	P	G	G	EL, FNN	G
Cleopatra	G	P	I		I
Sour Orange	G	G	G		G
Rough Lemon	I	I	P		P
Volkameriana	I	I-G	P		P
Macrophylla	I	?	P	EL,LL†	P

Key: G = good, H = high, I = intermediate, L = low, Lg = large, P = poor, S = susceptible, T = tolerant.

Notes:

*EL = Eureka lemon, FN = Fukumoto navel, FNN = Frost nucellar navel, LL = Lisbon lemon, O = oranges (all).

†Eureka lemon often declines by about 10 years, Lisbon lemon by 15 to 20 years.

and grapefruit trees indicates generally similar behavior. Mandarins are less well characterized.

Yield: Actual yield (pounds of fruit per tree) of trees planted at standard density (about 100 trees per acre for oranges) in an area where the rootstock is well adapted. Since yield changes over time, the evaluation given is approximately the cumulative yield of trees during the first 10 to 20 years. In many orchards, trees will have reached full size by this time. In any case, little data on performance of trees older than 20 years is available.

Yield efficiency: Yield per unit of canopy volume, a ratio used to predict the productivity of a planting in which tree density is optimized for tree size. Rootstocks producing smaller trees should be planted at a higher density to achieve this. Tree cost is not considered, so this is not an economic efficiency measure. A vigorous rootstock planted at the same density will require extensive pruning or tree removal earlier than a less vigorous rootstock. The grower must determine the optimal planting density for a particular scion-rootstock combination. A rough guideline for tree spacing is that spacing for a rootstock classed as intermediate should be about 75% of that for a rootstock classed as large. For example, on sites where trees on Carrizo produce a large tree and a standard spacing is 20 feet, trees on C35 should be planted at a spacing of 15 feet.

Internal fruit quality: Solids, acid, juice content, and rag are considered. High quality for oranges would combine the following: a solids to acid ratio of at least 10:1 or a California standard value of 99, high levels of solids and acids (richer flavor), at least moderately high juice content, and low rag.

External fruit quality: High external quality would be a smooth, firm rind that is thin or of moderate thickness, with intense rind color. If rootstocks are known to affect color break, this is noted here.

Fruit size: The average size of fruit in locations where the rootstock is well adapted. This is primarily based on packline sizing of navel and Valencia oranges from trials at Lindcove Research and Extension Center. When these results do not agree with others reported in the literature, a comment is made.

Tristeza tolerance: A summary of the performance of CTV-infected orange varieties on the rootstock. Only tolerance to common California CTV isolates is considered. Trees with good tolerance do not show quick decline, are no more than slightly stunted, and have fruit of normal size.

Phytophthora tolerance: Tolerance to both gummosis and root rot are covered, and both *Phytopthora parasitica* and *P. citrophthora* are included. Gummosis is almost never observed on a rootstock rated as resistant. For root rot, the main character is tolerance—that is, the ability of the tree to grow normally in locations where root rot is common for susceptible rootstocks. The standard of susceptibility is sweet orange.

Citrus nematode resistance: Level of resistance to citrus nematode races found in California. There are at least four distinct races of citrus nematode in California. All can infect Sweet Orange, three can attack Troyer and Carrizo, and one can attack some Trifoliate Orange selections. Responses to each race are discussed separately when available. The distribution of these races in California has not recently been characterized. The rating given is based primarily on seedling tests in a greenhouse, but available evidence indicates that it correlates well with performance of trees in the field. Rootstocks with excellent resistance should be little damaged by citrus nematodes in the field.

Soil adaptation: For many rootstocks, their adaptation to various soils is rather anecdotal in that few trials have compared rootstocks on different soil types. Salinity tolerance is fairly well known for many rootstocks, as is the extent of their iron chlorosis on calcareous soils. It is difficult to translate a soil analysis into guidelines that accurately predict performance of all rootstocks. Major soil problems in California are poor drainage (poor aeration), salinity (generally chlorides), and calcareous soils that lead to iron deficiency, typically called iron chlorosis. The best guide to rootstock responses to these stresses is performance of a standard rootstock (Carrizo or Troyer) on this soil. If these rootstocks perform well, most rootstocks will be suitable. If there are problems with Carrizo, more careful investigation will be needed.

Freeze tolerance: Rootstocks affect freeze tolerance of both trees and fruit. For trees, tolerance is primarily due to the degree of dormancy induced by cool temperatures before the freeze occurs. This is different than the common situation in Florida and Texas, where fall and winter temperatures are warmer than in California and trees do not become as dormant. Therefore, freeze tolerance ratings from Florida and Texas do not always apply to trees grown in California. The ratings given are primarily based on observation of trees in rootstock trials that were subject to natural freezes. Rootstocks can also confer some degree of freeze tolerance to fruit, probably through their effects on soluble solids content. Fruit with high soluble solids are somewhat more freeze tolerant than those with low soluble solids content.

Incompatibility: Compatibility with various scion cultivars, summarized from all available evidence; ratings assume that no pathogens such as exocortis viroid are present. The major practical concern is delayed incompatibility, which becomes evident when trees decline after 10 to 15 years. This is usually characterized by a bud union with a severe shoulder in which the scion is much smaller than the rootstock (see fig. 5.1). Ratings are based on available evidence from California and elsewhere, but many rootstock-scion combinations may not have been tested. In general, varieties in the major fruit types (oranges, lemons, grapefruit, and mandarins) behave similarly, but exceptions to this may occur: for example, the differences in compatibility of Eureka and Lisbon lemons with Trifoliate Orange hybrid rootstocks.

Overall: A judgment on the overall value of the rootstock and specific environments in which it performs well or should not be grown.

Carrizo and Troyer Citranges

Origin: Troyer and Carrizo rootstocks originated from a single hybrid seedling of Washington navel orange × Trifoliate Orange (parental cultivar unknown) made in 1909; the hybrid was named Troyer in 1934. The identity of seed planted at Carrizo Springs, Texas, was apparently lost, and this source was named Carrizo in 1938. While there are occasional reports of differences in performance between Troyer and Carrizo, there is no statistically convincing evidence that either is superior.

Figure 5.1 Bud unions of 25-year-old Washington navel orange on selected rootstocks at Lindcove Research and Extension Center. *Photos:* C. Federici.

Nursery behavior: Rootstock fruit are very seedy, and the seedlings are genetically very uniform (at least 98% nucellar). Buds push fairly well and trees grow well in the nursery.

Tree size: Tree size is standard with nearly all scions. Trees continue to grow for many years and may eventually become quite large if not controlled by pruning or crowding.

Yield: Trees on Troyer or Carrizo typically have good to excellent yields.

Yield efficiency: Moderate.

Internal fruit quality: Good to excellent with all common scions. Overall, fruit quality is good.

External fruit quality: Peel quality is generally fair to good, but creasing can be a problem with oranges. Peel thickness is medium, and texture is intermediate.

Fruit size: Fruit size is generally average in locations where trees are well adapted.

Tristeza tolerance: Good tolerance to typical California isolates.

Phytophthora tolerance: Excellent resistance to gummosis. Moderate tolerance to root rot.

Citrus nematode resistance: Resistant to some citrus nematode races, susceptible to others. The frequency and distribution of resistance-breaking races in California is not clear.

Soil adaptation: Well adapted to loam, sandy loam, and sandy soil. May perform poorly on very heavy soils with poor drainage. Poor tolerance to salinity (chlorides) and to highly calcareous soils, but among the best Trifoliate hybrids. Usually acceptable on slightly calcareous soils (e.g., pH 7.8).

Freeze tolerance: In California, trees on Carrizo or Troyer generally have relatively good freeze tolerance, because our climate induces dormancy. However, they are typically somewhat more susceptible than those on Trifoliate. Fruit also have relatively good freeze tolerance. In climates with warmer winter temperatures, such as Florida and Texas, trees often do not become dormant and are not particularly freeze tolerant.

Incompatibility: Good compatibility with all oranges, grapefruit, and Lisbon lemons. Considered incompatible with Eureka lemon, although the age at which incompatibility becomes evident varies considerably over locations and perhaps among Eureka selections. Compatibility with mandarins is complex and not well understood. Nearly all mandarins perform well for at least 10 to 15 years, but many eventually develop bud union crease and decline.

Overall: Excellent rootstocks that have been standards in California for many years. Good balance of traits and few major defects. Except for sites with calcareous soils, a very low risk choice. In specific locations, other rootstocks may perform better, but these are less broadly adapted than Carrizo or Troyer and therefore are more likely to have problems in a new location.

C35 Citrange

Origin: A hybrid seedling of Ruby orange × Webber-Fawcett trifoliate made in 1951 and released by UC Riverside in 1986.

Nursery behavior: Seed source trees generally produce fair to good crops, fruit have 8 to 12 seed per fruit, and the seedlings are moderately uniform (about 85% nucellar). Buds push fairly well and trees grow well in the nursery.

Tree size: Somewhat reduced with nearly all scions. Typically, young trees grow well, so that size differences do not become apparent until trees have borne several crops. Relative to Carrizo, canopy volume has varied from 54% in a replant trial with Valencia to 127% with Lane Late in a replant trial of 8-year-old trees. Over 11 trials, tree size averaged about 83% as large as the trees on Carrizo. Tree size seems to depend more on the site than on the scion.

Yield: Trees on C35 typically have good to excellent yields for their size, but absolute yields are sometimes lower because of reduced tree size. Over 11 trials, the yield per tree ranged from 75% of that of Carrizo in a trial with Lane Late scion to 159% in a trial with Washington navel. The average yield over all trials was 106% of that of trees on Carrizo. Trials in which trees on C35 had higher yields than those on Carrizo or Troyer were replant trials in which the yield of Troyer or Carrizo was low or moderate.

Yield efficiency: High, mainly because trees on C35 are usually as productive as those on Carrizo or Troyer but are smaller in size.

Internal fruit quality: Good to excellent with all common scions, generally similar to the fruit quality of trees on Carrizo. Overall, fruit quality is good.

External fruit quality: Peel quality is generally fair to good. Peel thickness is medium, and texture is also intermediate. With some scions, color break may be delayed by a few days relative to Carrizo, but this is not always observed.

Fruit size: Generally average in locations where trees are well adapted.

Tristeza tolerance: Good tolerance to typical California isolates.

Phytophthora tolerance: Excellent resistance to gummosis and very good tolerance to root rot caused by both *P. parasitica* and *P. citrophthora*.

Citrus nematode resistance: Resistant to all citrus nematode races found in California.

Soil adaptation: Well adapted to loam, sandy loam, and sandy soil. May perform poorly on very heavy soils with poor drainage, but not much data on this is available. Poor tolerance to salinity (chlorides) and to calcareous soils, where it is somewhat worse than Carrizo.

Freeze tolerance: In California, trees on C35 generally have relatively good freeze tolerance, because our climate induces dormancy. However, they are typically somewhat more susceptible than those on Trifoliate. Fruit also have relatively good freeze tolerance. In climates with warmer winter temperatures, such as Florida and Texas, trees may be much less freeze tolerant.

Incompatibility: Good compatibility with all oranges, grapefruit, and Lisbon lemons. Fukumoto navel on C35 is subject to a decline of unknown cause that may be a type of incompatibility. Appears to be strongly incompatible with Eureka lemon, but data is limited. Compatibility with mandarins is complex and not well understood. Nearly all mandarins perform well for at least 10 to 15 years, but many eventually develop bud union crease and decline.

Unknowns: While trees on C35 have performed very well in most trials, performance of Valencia trees became very poor after about 10 years in one trial in San Diego County. The cause of this poor performance is not yet known, but the use of any relatively new rootstock inevitably involves some risks.

Overall: An excellent rootstock where it is well adapted. To take advantage of its high yield potential, trees should be planted at about 30% higher density on average than those on Carrizo. Good balance of traits and few major defects. A slightly greater risk of poor tree performance than with Carrizo, but also a good chance of higher yield in a higher-density planting that needs less tree size management than an equivalent planting on Carrizo.

Swingle Citrumelo (4475)

Origin: A hybrid seedling of Duncan grapefruit × Trifoliate (parental cultivar unknown) made in 1907. It was named Swingle and released in 1974. Before release, it was originally sent to California as seed, identified by its temporary number CPB 4475, and some nurseries may still use this designation.

Nursery behavior: Rootstock fruit are seedy, and the seedlings are moderately uniform (about 90% nucellar). Buds push fairly well under optimal conditions, but these conditions are more restrictive than for Carrizo or Troyer. Trees grow well in the nursery.

Tree size: Tree size is generally similar to that of trees on Carrizo but can vary considerably with location, ranging from 65 to 127% over 11 trials. Over all 11 trials, the average tree size was 102% of that of trees on Carrizo.

Yield: Trees on Swingle typically have good to excellent yields, sometimes superior to those of trees on Carrizo. Over 11 trials, the average yield was 106% of that of trees on Carrizo. Trees on Swingle had higher yields than those on Carrizo in 8 of the 11 trials. Swingle had lower yields in two trials with Lane Late and one with Rio Red and Redblush grapefruit.

Yield efficiency: Moderate.

Internal fruit quality: Good to excellent with all common scions. Overall, fruit quality is good.

External fruit quality: Peel quality is generally fair to good. Peel thickness is medium, and texture is also intermediate.

Fruit size: Generally average in locations where trees are well adapted.

Tristeza tolerance: Good tolerance to typical California isolates.

Phytophthora tolerance: Excellent resistance to gummosis and root rot.

Citrus nematode resistance: Resistant to some citrus nematode races, susceptible to others. The frequency and distribution of resistance-breaking races in California is not clear.

Soil adaptation: Well adapted to loam, sandy loam, and sandy soil. Performs poorly on heavy soils. Poor tolerance to salinity (chlorides) and to calcareous soils.

Freeze tolerance: In California, trees on Swingle generally have good freeze tolerance, but they are typically somewhat more susceptible than those on Trifoliate. Fruit also have relatively good freeze tolerance. Swingle is also considered freeze tolerant in Florida.

Incompatibility: Bud union typically shows more extreme shoulder development than do trees on Trifoliate, and eventual compression girdling seems likely. Declines apparently due to this incompatibility have been observed in Washington navel orange trees at Lindcove (Tulare County) that were 22 to 25 years old. Considered incompatible with Eureka lemon. Compatibility with mandarins has not been studied. Most mandarins will probably perform well for at least 10 to 15 years but will eventually develop bud union crease and decline.

Overall: Excellent rootstock where it is well adapted. Disease resistance makes it a good choice for replant situations. Somewhat narrower soil adaptation than Carrizo. Incompatibility with Washington navel at 20 to 25 years predicts short tree life, but it is not yet known whether these declines occur with all scions and locations.

Rough Lemon

Origin: Apparently a hybrid that occurred naturally in India, where there are several very distinct forms. Several selections have been made, including the Florida selections Estes and Milam, with burrowing nematode resistance, and a California selection, Schaub, with better Phytophthora root rot tolerance. Several other selections are available, but no remarkable differences in tree performance on these have been documented. Rough Lemon is similar to but distinct from Volkameriana (see below), which is also from India.

Nursery behavior: Rootstock fruit are seedy, and the seedlings are genetically very uniform (about 98% nucellar). Buds push well, and trees grow well in the nursery.

Tree size: Generally similar to that of trees on Carrizo but can vary considerably with location, ranging from 54 to 129% over 11 trials. Over all 11 trials, the average tree size was 96% of that of trees on Carrizo. Overall, trees are more vigorous than those on Carrizo, but tree size is more frequently reduced by Phytophthora root rot, citrus nematode, and freeze damage.

Yield: Trees on Rough Lemon typically have good to excellent yields, sometimes superior to those of trees on Carrizo. Over 11 trials, the average yield was 96% of that of trees on Carrizo. Trees on Rough Lemon had higher yields than those on Carrizo in 5 of the 11 trials. Rough Lemon had much lower yields than Carrizo in all three trials with Lane Late navel scion.

Yield efficiency: Moderate.

Internal fruit quality: Poor to fair with all common scions. In comparison with fruit from trees on Carrizo, fruit from trees on Rough Lemon have 1 to 2% lower soluble solids, lower acid, and a lower juice percentage. The fruit are more susceptible to freeze damage, and late-harvested oranges granulate earlier. Lemon quality is good.

External fruit quality: Peel quality is generally fair to good. The peel is often thicker than that of fruit from trees on Carrizo, and the texture is sometimes rougher.

Fruit size: Generally average in locations where trees are well adapted.

Tristeza tolerance: Good tolerance to typical California isolates.

Phytophthora tolerance: Poor tolerance to gummosis and root rot. Schaub is slightly better than other Rough Lemon selections in most greenhouse tests, but field performance has not been much different.

Citrus nematode resistance: Susceptible to all citrus nematode races.

Soil adaptation: Well adapted to sandy and sandy loam soils. Often performs poorly on heavy soils. Fair tolerance to salinity (chlorides) and good tolerance to calcareous soils.

Freeze tolerance: In California, trees on Rough Lemon generally have poor freeze tolerance. Fruit also have poor freeze tolerance.

Incompatibility: No common incompatibilities but often produces many suckers. Removal of these creates wounds that can lead to gummosis infection.

Overall: Fair rootstock where it is well adapted, particularly on sandy soils. Should not be used for late-harvested oranges or in cold locations. Poor choice for scions with naturally low solids and acid content.

Volkameriana

Origin: Apparently a hybrid that occurs naturally in India. Also known as Volk and Volkamer Lemon, it is similar to Rough Lemon in many characteristics.

Nursery behavior: Rootstock fruit are seedy, and the seedlings are fairly uniform (about 80% nucellar). Buds push very well, and trees grow well in the nursery. Testing in California began only in about 1989, so less data is available than for some other rootstocks.

Tree size: Young trees are usually very vigorous, and trees can be large. In a lemon rootstock trial near Santa Paula, Volk with Lisbon scions produced the largest trees in the trial, but Eureka trees on Volk were only moderate in size (similar to Eureka lemon on Macrophylla rootstock). Yield and yield relative to tree size were moderate. With Lane Late navel scion, 11-year-old trees at two locations were much smaller than those on Carrizo. In a Valencia trial in Pauma Valley, trees on Volk were slightly smaller than those on Troyer.

Yield: Trees on Volk typically have average yield for their size.

Yield efficiency: Moderate.

Internal fruit quality: Fair with all common scions. In comparison with fruit from trees on Carrizo, fruit from trees on Volk have 1 to 2% lower soluble solids, lower acid, and a lower juice percentage. The fruit are more susceptible to freeze damage, and late-harvested oranges granulate earlier. Lemon quality is good.

External fruit quality: Peel quality is generally fair to good. The peel is often thicker than that of fruit from trees on Carrizo, and the texture is sometimes rougher.

Fruit size: Generally average in locations where trees are well adapted.

Tristeza tolerance: Good tolerance to typical California isolates.

Phytophthora tolerance: Moderate tolerance to gummosis and root rot, somewhat better than that of Rough Lemon, but worse than Carrizo. Susceptible to *P. citrophthora*.

Citrus nematode resistance: Susceptible to all citrus nematode races.

Soil adaptation: Well adapted to sandy and sandy loam soils. Often performs poorly on heavy soils. Fair tolerance to salinity (chlorides) and good tolerance to calcareous soils.

Freeze tolerance: In California, trees on Volk generally have poor freeze tolerance. Fruit also have poor freeze tolerance.

Incompatibility: No common incompatibilities.

Overall: A fair rootstock where it is well adapted, particularly on sandy soils. Should not be used for late-harvested oranges or in cold locations. A poor choice for scions with naturally low solids and acid content. Overall, similar to Rough Lemon, but with slightly better Phytophthora tolerance.

Trifoliate Orange

Origin: A citrus relative that grows in China. Many different selections have been made. Those commonly used as rootstocks in California fall into three groups: large-flowered types (Pomeroy), small-flowered types (Rubidoux and Rich 16-6), and Flying Dragon (a distinctive small-flowered type).

Nursery behavior: Rootstock fruit are seedy, and the seedlings are moderately uniform (about 80 to 90% nucellar). Buds are somewhat difficult to push, and trees grow more slowly in the nursery, sometimes taking a year longer than Carrizo. Flying Dragon is particularly difficult in the nursery, with a higher percentage of zygotic seedlings and slower growth than other Trifoliate types.

Tree size: With Rubidoux or Rich 16-6, trees are generally similar in size to trees on Carrizo but can vary considerably with location, ranging from 53 to 151% over eight trials. Over all eight trials, the average tree size was 91% of that of trees on Carrizo. Trees on Pomeroy are typically somewhat larger than those on Rubidoux or Rich 16-6, and those on Flying Dragon are strongly dwarfed, orange trees not reaching more than about 7 feet tall after 20 years.

Yield: Trees on Trifoliate typically have good to excellent yields, sometimes superior to those of trees on Carrizo. Over eight trials, the average yield was 91% of that of trees on Carrizo. Trees on Trifoliate had higher yields than those on Carrizo in three of the eight trials. Yields of trees on Pomeroy are typically somewhat greater than those of trees on Rubidoux or Rich 16-6 but are proportional to tree size.

Yield efficiency: Moderate.

Internal fruit quality: Good to excellent with all common scions. Fruit typically have high solids, acids, and juice content.

External fruit quality: Peel quality is generally good. Peel thickness is medium and texture is smooth.

Fruit size: For Rubidoux and Rich 16-6, fruit size is generally average in locations where trees are well adapted. Trees on Pomeroy often produce larger fruit than those on Carrizo.

Tristeza tolerance: Good tolerance to typical California isolates.

Phytophthora tolerance: Excellent resistance to gummosis and root rot.

Citrus nematode resistance: Resistant to some citrus nematode races, susceptible to others. The frequency and distribution of resistance-breaking races in California is not clear.

Soil adaptation: Well adapted to loam, sandy loam, and clay soils. Can also perform well on sandy soils, but only if irrigation is managed very carefully because roots are shallow and therefore trees on Trifoliate are quite susceptible to drought. Poor tolerance to salinity (chlorides) and to calcareous soils.

Freeze tolerance: In California, trees on Trifoliate generally have the best freeze tolerance of any rootstock. Fruit also have relatively good freeze tolerance.

Incompatibility: Bud union typically shows considerable shoulder development, which sometimes leads to compression girdling. Rich 16-6 was selected to be more compatible with Frost nucellar navel orange and generally has somewhat less shoulder development than other selections. Trifoliate is incompatible with Eureka lemon. Most mandarins perform well for at least 10 to 15 years but eventually develop bud union crease and decline.

Overall: An excellent rootstock where it is well adapted. Disease resistance makes it a good choice for replant situations. Narrower soil adaptation than Carrizo.

Cleopatra Mandarin

Origin: A high-acid mandarin type that apparently originated in India.

Nursery behavior: Rootstock fruit are seedy, and the seedlings are uniform (about 98% nucellar). Seedlings grow slowly at cool temperatures. Buds are sometimes difficult to force.

Tree size: Generally similar to that of trees on Carrizo. For trials of 22 to 25 years with navel, Valencia, and Minneola scions, the average tree size was 101% of that of trees on Carrizo.

Yield: Trees on Cleopatra typically have relatively poor yields when young but gradually improve to levels similar to that of trees on Carrizo by age 15 to 20. Over three trials, the average yield was 96% of that of trees on Carrizo.

Yield efficiency: Moderate.

Internal fruit quality: Good, and similar to that of trees on Carrizo with common scions.

External fruit quality: Peel quality is generally good. The peel thickness is medium, and the texture is intermediate.

Fruit size: Generally small to average.

Tristeza tolerance: Good tolerance to typical California isolates.

Phytophthora tolerance: Susceptible to gummosis and root rot.

Citrus nematode resistance: Susceptible to citrus nematodes.

Soil adaptation: Well adapted to all except calcareous soils, where it is moderately tolerant. Good tolerance to salinity.

Freeze tolerance: In California, trees on Cleopatra generally have good freeze tolerance, but they are typically somewhat more susceptible than those on Trifoliate. Fruit also have relatively good freeze tolerance.

Incompatibility: Considered incompatible with Eureka and Lisbon lemons.

Overall: Slow growth of young trees and delay in bearing generally mean that there are better choices, but good where salinity problems are expected.

Macrophylla

Origin: Apparently a hybrid that occurs naturally on Cebu Island, Philippines. Also known as Alemow, it seems closely related to limes.

Nursery behavior: Rootstock fruit are only moderately seedy, but the seedlings are genetically very uniform (about 98% nucellar). Buds push well, and trees grow well in the nursery.

Tree size: With lemon, the only scion for which it is widely used in California, trees are generally 10 to 20% smaller than trees on Carrizo.

Yield: Lemons on Macrophylla typically have good to excellent yields during the first 10 to 15 years, often superior to those of trees on other rootstocks. However, as trees age, they typically develop health problems and yields decrease.

Yield efficiency: Good.

Internal fruit quality: Not used with oranges because internal quality is usually poor, with lower soluble solids, lower acid, and a lower juice percentage. These are not important issues for lemon.

External fruit quality: Peel quality is generally fair to good. The peel is often thicker than that of fruit from trees on Carrizo.

Fruit size: Generally average in locations where trees are well adapted. Little data is available on this characteristic.

Tristeza tolerance: Susceptible to typical California isolates, and lemons budded on it will be stunted if the rootstock becomes infected. The reaction is due to stem pitting of the rootstock (not quick decline). Isolates that do not cause stem pitting of most cultivars will do so with Macrophylla. Therefore, rootstock seedlings are generally grown in CTV-free conditions. Budded trees typically perform well in areas with CTV, perhaps because lemon is a relatively poor host.

Phytophthora tolerance: Good tolerance to gummosis and root rot.

Citrus nematode resistance: Susceptible to all citrus nematode races.

Soil adaptation: Well adapted to sandy and sandy loam soils. Often performs poorly on heavy soils. Fair tolerance to salinity (chlorides) and to calcareous soils.

Freeze tolerance: Trees on Macrophylla have poor freeze tolerance, much worse than that of trees on any other commonly used rootstock.

Incompatibility: Trees on Macrophylla, particularly lemons, are generally short-lived (< 20 years) due to rootstock necrosis. This may not be a true incompatibility, but its effect is similar. Little is known about compatibility with other scions.

Overall: The standard rootstock for lemons in many areas due to high yields. Should not be used in cold locations. There has been some recent use of Macrophylla with mandarins in Spain and elsewhere.

Sour Orange

Origin: A natural hybrid, probably of pummelo and mandarin parentage.

Nursery behavior: Rootstock fruit are seedy, and the seedlings are fairly uniform (about 85 to 95% nucellar). Buds push well under cooler conditions, and trees grow well in the nursery.

Tree size: Tree size is generally large, but in five trials it ranged from 64 to 155% of that of trees on Carrizo. The mean tree size was 105% of that of Carrizo trees. Variation seems more attributable to location than scion, since results with similar-age Lane Late navel trees ranged from 65 to 111% of that of Carrizo.

Yield: Yields in five trials ranged from 70 to 118% of those of trees on Carrizo, with an average of 88%. Superior to Carrizo in locations where Carrizo performs poorly because soils are calcareous or poorly drained.

Yield efficiency: Generally very similar to that of trees on Carrizo.

Internal fruit quality: Sour orange produces fruit of high internal quality.

External fruit quality: Peel quality is generally good.

Fruit size: Fruit size is generally average, usually slightly smaller than fruit from trees on Carrizo.

Tristeza tolerance: Susceptible to many isolates of CTV, with orange, grapefruit, and mandarin trees showing quick decline and death within a few years after infection. Sour orange can be used with lemons in the presence of CTV.

Phytophthora tolerance: Excellent tolerance to gummosis and root rot.

Citrus nematode resistance: Susceptible to all citrus nematode races.

Soil adaptation: Well adapted to a broad range of soil types, including sand, sandy loam, loam, and clay soils. Good tolerance to salinity and to calcareous soils.

Freeze tolerance: In California, trees on Sour Orange generally have good freeze tolerance. Fruit have intermediate freeze tolerance, showing more damage than those from trees on Trifoliate or Carrizo, but less than fruit from trees on Rough Lemon.

Incompatibility: No common incompatibilities, except with kumquat and perhaps Eureka lemon, where trees decline at 12 to 15 years.

Overall: An excellent rootstock with broad soil adaptation, Phytophthora tolerance, and positive fruit quality effects. As tristeza becomes more widespread in California, all trees on Sour Orange (except lemons) are likely to be killed.

Other (Minor) Rootstocks

African Shaddock × Rubidoux Trifoliate (ASRT): A hybrid from the USDA (Indio) breeding program. It has excellent tolerance or resistance to Phytophthora and citrus nematode and typically produces large trees with excellent yields. However, recent studies indicate that trees are stunted by some mild Florida isolates of CTV and some common California isolates. Its susceptibility to CTV makes use of ASRT quite risky.

Bitters Trifoliate hybrid: A hybrid of Sunki mandarin and Swingle trifoliate from the USDA Indio breeding program released by UC Riverside in 2009. It was tested under the code C22 and may be referred to by this code by some nurseries. It is tolerant to CTV, moderately tolerant to Phytophthora, but fairly susceptible to citrus nematode. Bitters produces a semidwarf tree that often has high production for its size. It is very tolerant to calcareous soils. Fruit quality appears to be similar to that of trees on Carrizo.

C32 Citrange: A hybrid of Ruby orange and Webber-Fawcett trifoliate (a sister of C35 Citrange) from the UC Riverside citrus breeding program. It has good disease resistance and produces a large, vigorous tree with high yields in most locations. However, the rootstock is difficult to propagate because rootstock trees produce few fruit and the fruit have few seed. The seedlings are also somewhat variable.

Gou Tou: A Chinese variety that has been used some as a rootstock in Florida. It is tolerant to CTV, but susceptible to Phytophthora and citrus nematode. Gou Tou is somewhat tolerant to calcareous and wet soils. Fruit quality is relatively poor. Several different types have been imported into California under this name, but none have yet been tested as rootstocks.

Rangpur (Rangpur Lime): The standard rootstock in Brazil because it is tolerant to CTV and performs well without irrigation. It is very susceptible to Phytophthora and has performed poorly in several trials in California. Not recommended.

Sun Chu Sha (Sun Chu Sha Kat): A mandarin type that has been considered promising as a rootstock in Florida. Limited testing (three trials with lemons and Lane Late navel) in California has been disappointing. Yields have been lower than those of trees on Carrizo, but tree size is similar or larger. In Florida it is susceptible to Phytophthora and citrus nematode, tolerant to CTV, and fairly tolerant to calcareous soils. There is probably little reason to use this rootstock except on calcareous soils. Testing to evaluate performance on calcareous soils in California was initiated in 2001; it does not appear to be very tolerant in these tests.

Sweet Orange: Can be used as a rootstock for most varieties, but its susceptibility to Phytophthora and citrus nematode often limit the productivity and lifespan of trees. In nearly all recent trials, yield, tree size, and tree health ratings have been considerable lower than those of trees on Carrizo. These trials have all been on ground previously planted with citrus where diseases are more likely to limit productivity. Trees on Sweet Orange have poor drought tolerance and are very susceptible to Phytophthora, so irrigation must be carefully managed.

Taiwanica: Previously, Taiwanica was considered similar to Sour Orange, but with CTV tolerance. Several trials during the last 20 years indicate that it usually performs poorly as a rootstock. Internal fruit quality is typically similar to that of Rough Lemon, and yields have been low. There is little reason to consider this rootstock.

Trifeola. A Minneola × Trifoliate hybrid from South Africa. There is currently little information on its performance in California, but it is included in several trials planted in 1997 and 2001.

X639. A Cleopatra × Trifoliate hybrid from South Africa. There is currently little information on its performance in California, but it is included in several trials.

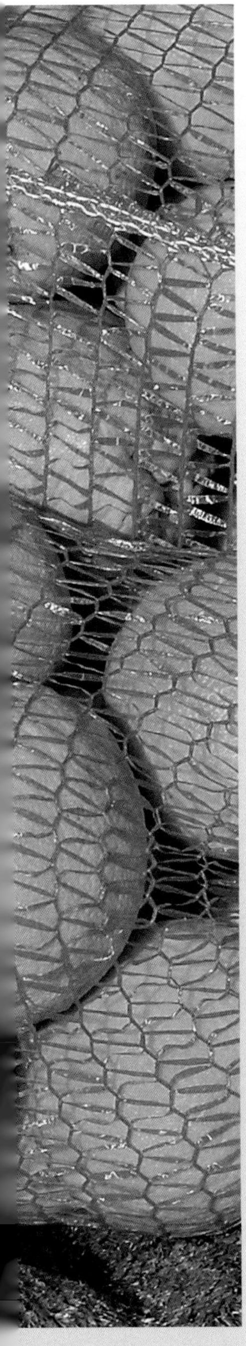

6 Commercial Production of Container-Grown Nursery Trees

Timothy M. Spann and Louise Ferguson

Virtually all the trees in California's citrus orchards are produced on seedling rootstocks selected for their desired characteristics. Citrus trees are not germinated from the seed of the desired scion cultivars (cultivated varieties), as most citrus cultivars with good fruit characteristics do not produce roots resistant to common soilborne diseases or do not have other desired rootstock traits. Therefore, citrus trees for both commercial and landscape use consist of a scion selection propagated onto a rootstock seedling.

The seed used for most citrus rootstocks have one interesting characteristic: they are polyembryonic; that is, they have multiple embryos. In addition to the normal (zygotic) embryo produced by the sexual process of pollination and fertilization of the citrus flower, they produce additional embryos asexually from the cells of the nucellus, the embryo lining. These additional nucellar embryos, unlike the zygotic embryos, are genetically identical to the mother tree. Most of the popular citrus rootstocks have a high degree of nucellar embryony and produce a high percentage of seed genetically identical to their mother tree. Thus, when germinated, citrus seed often produce more than one seedling. Zygotic seedlings are easily identified by their different morphology and can be rogued, or removed, from the germination bed early in rootstock production. Among the advantages of using a rootstock are, in order of importance

- climatic adaption
- disease and nematode resistance
- adaption to soil type
- enhancement of the horticultural characteristics of yield and fruit size and quality

These factors, particularly soilborne disease resistance, have directed rootstock selection and improvement programs for the past century.

The seedling rootstock is budded in the nursery with buds of the desired citrus cultivar. A selected bud grows into the top, or aboveground, portion of the tree bearing the fruit and is called a scion. The scion, because it is vegetatively propagated, will consistently produce the desired fruit.

The origin and source of California citrus scion budwood is highly regulated to prevent the spread of disease and ensure trueness to type. Over 30 citrus pathogens can be transmitted by infected scion budwood via budding, mechanical transmission (pruning tools), and insect vectors. To prevent the spread of bud-borne pathogens, the California citrus industry and the University of California jointly formed and administer the California Citrus Clonal Protection Program (CCPP). A full description is available at their website: http://www.ccpp.ucr.edu/budwood/budwood.php. For more information, see chapter 4, "Scion Cultivars," and chapter 7, "The California Citrus Clonal Protection Program."

Citrus trees can be produced in several ways. An earlier method was to transplant rootstock seedlings to the field, bud them there, and later dig out the tree, wrap the roots in burlap, and transport them to the orchard. This was known as the balled and burlapped method. In container-grown production, the common commercial practice today, the seed are germinated in the same fashion but transplanted to pots, budded, and raised in a greenhouse. When sold, the trees are transferred to an enclosed truck that remains closed until delivery to the grower.

In 2008, the Asian citrus psyllid was found in California. Asian citrus psyllid is the insect that vectors the devastating bacterial disease huanglongbing (HLB). To prevent the spread of this disease via new orchard establishment, all new citrus tree nursery production must be done inside insect-excluding greenhouses. Previously, few citrus growers attempted to produce their own trees for new orchards, and virtually none will do so now.

Greenhouse Requirements

Currently there are no mandatory requirements for where and how commercial California citrus nurseries must be located and constructed. However, legislation is now pending for all California citrus nursery production to be done within protective, insect-proof structures to exclude insect vectors. The protective structures must have double entryways with positive air pressure to the outside and connecting hallways for transporting nursery stock, equipment, and supplies. Insect exclusion screening must be semipermeable with no gaps. Air holes in the screens must be small enough to exclude the tiny cotton aphid. A regulatory official must approve new structures prior to planting or moving stock. Hopefully, existing structures meeting these performance standards can be grandfathered into the new mandatory regulations. These structures will probably have semiannual inspections and periodically inspected insect traps.

Germinating Rootstock Seedlings

Most nurseries purchase seed from nurseries that maintain rootstock seed trees. This ensures viable, true-to-type seed with a high germination percentage. Establishment of seed source trees and seed extraction are specialized procedures designed to increase germination percentage through proper harvest times, decrease variability by seed grading, eliminate fungal problems using heat and fungicide treatment, and decrease seed deterioration with appropriate cold storage. A few larger nurseries maintain blocks for producing their own rootstock seed. The process of producing large volumes of rootstock seed includes harvesting the fruit at normal maturity, crushing the fruit in water, stirring the resulting mixture with pectinase enzymes at specified temperatures to separate the seed from the pulp, and washing out the seed. The washed seed is surface-sterilized with 125°F water for 10 minutes, dipped in 1% 8-hydroxyquinoline sulfate, air-dried on screens in the shade, and packaged in plastic bags. When storing, it is important to leave these storage bags open initially until the seed equilibrate with the refrigeration temperature. If the bags are closed before refrigeration, the undissipated heat and humidity may generate condensation inside the bags, which promotes fungal pathogens. Fresh seed has the highest germination percentage; however, seed can be stored at 40°F for as long as 6 months with little loss of viability. Seed stored at room temperature looses viability quickly.

Generally, citrus rootstock seed is purchased by weight. There are from 2,000 to 4,000 seed per pound. Once seed is purchased it can be sealed in plastic bags that permit gas exchange and stored at 40° to 45°F. Prior to germination, nursery staff visually cull the seed for off-types, which includes misshapen, off-color, or dissociated seed or seed that appears desiccated or is much lighter or darker than other seed. Generally, the number of seed sown should be double the number of rootstocks required.

Germination

Germination is the most carefully monitored phase of rootstock production. Rootstock seed can be germinated anytime during the year in a greenhouse. Some nurseries prefer to peel seed (remove the seed coat) before sowing to reduce time to germination and increase germination percentage and uniformity. This is particularly helpful with older seed and certain varieties that may have poor germination. Some nurseries also find that soaking seed in clean water for 24 hours prior to sowing improves germination. Seed are sown in a variety of trays, flats, tubes, or fused plastic cell trays. The larger rectangular trays hold multiple seed. Seed are planted singly in trays with individual cells or in plastic tubular cells that taper to pointed ends with air holes for root pruning. All contain a sterilized soilless medium that may contain peat moss or pine bark, perlite, vermiculite, Styrofoam, and other inert ingredients, with dolomitic limestone to maintain a pH of 6 to 7 and controlled-release fertilizers and micronutrients. Seed should be planted at 0.25 to 0.50 inch deep. The trays or containers of seed should be placed on raised benches at least 3.5 feet above the ground to encourage root pruning and avoid contamination from splashing. Polystyrene trays should be treated with sealer to avoid root penetration of the polystyrene, as this can be a source of contamination when the trays are reused. If roots have penetrated the trays, the trays should be sterilized with heat or fungicide before reuse. Humidity should be maintained between 80 and 90%, temperatures should be above 85°F, and light should shine for 16 hours per day, though some nurseries maintain that light is not important to the quality of germination. Germination is further enhanced if the seed trays can be heated from the bottom to maintain the 85°F temperature in the media. Irrigation is generally by overhead sprinklers as needed to maintain a moist medium. Germination beds should be visually inspected daily.

Transplanting

Citrus seed germinate rapidly but unevenly within a seed lot, under controlled conditions. Seedlings planted in trays can be 2 inches tall with three to five leaves within 10 to 14 days after planting, though it can take as long as 40 days for all the seed to germinate. They may be transplanted when the first full leaf is fully expanded. However, seedlings may be held longer if sown in individual cells of sufficient size. At transplanting, seedlings should be individually examined and graded. Nonuniform seedlings should be discarded at this point. Seedlings with curved or crooked lower stems at or below the soil line should be discarded; they tend to outgrow the deformity slowly. Stunted and albino seedlings should also be discarded. Any seedling with irregular stem length or girth or an off-color leaf, stem, or root should be discarded. The remaining seedlings should be classified into no more than two size groups and transplanted together by size to enhance uniform growth within a bench. Once planted, the seedlings should be allowed to grow unstaked until they have at least a caliper of 0.15 inch at 6 inches from the soil line.

The final containers into which citrus rootstock seedlings are transplanted range from 0.75 to 1.5 gallons, 4 to 5 inches wide, and up to 14 inches deep. They generally fall into two types: plastic sleeves and rigid containers. The former are cheaper, lighter, more easily removed at planting, and less of a disposal problem for the grower. The rigid containers are more expensive, sturdier, and could be reused but seldom are as they must be collected, transported, and sterilized. In some of today's more advanced nurseries, rigid pots are necessary with automated equipment (e.g., pot fillers). Finally, the rigid pots can have vertical ridges or slots that help discourage circling root growth, which can result in slow root spread after planting.

These containers should be housed in a bench frame that supports them but allows free drainage. The density of containers on a greenhouse bench and the width of the aisles between benches is dictated by how the benches are constructed. The most modern citrus nurseries use rolling bench technology adopted from the ornamental industry. Once a bench is loaded with transplanted container trees, the trees remain on the same bench, and the bench can be transported about the greenhouse via a system of rails until loaded on the delivery truck. This system eliminates the need for aisles between each bench; only one aisle, wide enough to accommodate one worker with tools, is needed. When access is needed, the benches can be moved, opening an aisle where it is needed. In this system, the large, bicycle-wheeled carts needed to transport trees are unnecessary, and the entire bench can be tagged instead of individual trees. Less-mechanized greenhouses with stationary benches require wider aisles to accommodate carts for container transport, and individual rootstock seedlings must be coded to ensure identification.

As with the germination trays, the containers on both rolling racks and stationary benches should be elevated at least 18 inches above the ground to encourage root pruning and avoid contamination from splashing. A raised bench also facilitates budding, pest scouting, weeding, and spraying. The arrangement of trees on a bench varies from nursery to nursery. For practical reasons, the number of rows of trees on a bench should not be more than that which allows for the middle tree to be reached from either side of the bench. Even following this rule of thumb, the inner rows of trees will often be thinner and taller than the trees in the outer rows. This is partly due to airflow along the outside edges of the bench, which causes the pots along the edge to dry out more quickly, as well as shading toward the middle of the bench as trees grow. To counteract this, some nurseries initially place the thicker trees on the inside two rows; however, some grading and moving of trees is virtually unavoidable as they grow. Rows should run north-south to maximize light use, and attention should be paid to the orientation of the greenhouse structure itself to minimize shading due to support beams or other structural components.

A major objective of nursery production is optimal utilization of space; this is particularly important now that new pest exclusion regulations are significantly increasing the cost per square foot of construction and operation. A nursery producer's main objective should be to produce the highest-quality tree in the least amount of space in the least time possible.

Budding Rootstock Seedlings

Budwood Source

Budwood production in California is carefully

controlled and monitored to prevent the spread of pathogens that once devastated the California citrus industry and to ensure that budwood is true to type. This program monitors the budwood from the original parent source tree through a series of tests (indexing) and cleanup procedures to produce a California Department of Agriculture (CDFA) registered foundation block maintained in a pest exclusion screenhouse at the UC Lindcove Research and Extension Center. From these foundation trees, nurseries obtain buds to develop CDFA-registered budwood source trees and nursery increase blocks for budding rootstock seedlings. A complete discussion of the California Citrus Clonal Protection Program, including how to obtain budwood, including early release of new cultivars, is detailed at their website: http://www.ccpp.ucr.edu/budwood/budwood.php. For more information, see chapter 7, "The California Citrus Clonal Protection Program."

Budding Time

Budding can be done throughout the year as long as bark is slipping. The rootstock seedlings should have a caliper the size of a pencil, 0.15 to 0.25 inch at a height of 6 inches. Anything smaller will flush slowly and have stunted growth; seedlings budded lower will have greater potential for Phytophthora infection. Nurseries producing trees in greenhouses generally bud from early spring through summer.

Budwood Collection and Storage

All implements used in budwood collection and budding should be sterilized with a 10% sodium hypochlorite (household bleach) solution to eradicate any mechanically transmissible viroids, bacteria, or fungal pathogens.

Budwood is generally collected from the shoot behind the current growth flush or from the current growth flush after it has begun to harden or mature. The wood should be straight, round, and as near in diameter to the rootstock seedling as possible. It should be fully developed with mature leaves. The bark should be tender, green, and not woody. A few longitudinal gray lines on the green bark are a good indicator of proper maturity. The buds in the leaf axils should be well formed. The sticks should be trimmed immediately after harvest to 8 to 10 inches long, and the leaves should be removed, leaving about 0.8 inch of petiole to protect the bud. The two blind, insufficiently developed buds at the base should be removed. Young, recently hardened, rounded buds are the first to grow out and generate the best take. The budwood should be identified with cultivar, date, and source.

The budwood should be washed with running water, dipped for 60 seconds in 10%, 3.5 to 5% a.i. bleach solution, and shaken to remove the excess. Some nurseries treat budwood with fungicides at this point. Air-dry the budwood on a rack for no more than 15 minutes at 64°F. Precool the budwood at 40°F, seal it in a plastic bag, and store it at the same temperature. Inspect the bags for condensation within 2 days; if there is more than a fine mist, dry the inside of the bag with paper towels. Desiccation is the major problem in stored budwood. After this initial inspection, inspect weekly for abscised petioles and decayed wood, removing both. If storage is longer, repeat the sanitation and surface-drying procedure detailed above at 2 months. However, most nurseries prefer to use budwood cut the same day.

Budding

In California and the other citrus-producing states in the United States, citrus is most commonly propagated by T-budding, or more precisely inverted T-budding. The technique of budding is relatively simple and requires only a few tools, although the art of budding takes some time to master. The most important tool for budding is a quality, sharp budding knife. Budding knives are designed with thin blades that are flat on one side and ground to the cutting edge on the other edge. By grinding the edge only on one side, the knives more easily glide just beneath the bud when cutting it from the bud stick. The knives are also available in right- and left-hand versions. Another feature of budding knives is a bump on the back edge of the blade. This bump is designed for opening the flaps of the T to allow the bud to be inserted more easily. The other materials needed for budding are budding tape, a suitable rootstock seedling, and budwood.

Most citrus trees are budded about 6 inches above the soil. Seedling rootstocks to be budded should have a relatively straight section at least 1 inch long where the bud is to be inserted. The leaves and thorns should be clipped off a couple of inches above and below the insertion point for easy budding. It is important to clip the leaves and thorns rather than strip them in order to prevent bark tears and other jagged injuries that will be susceptible to disease. Make a shallow vertical cut, only as deep as the green bark, by inserting the point of the budding knife at the top of the section where the bud is to be placed and drawing it down for about 0.75 inch. Next, make a horizontal or cross cut about 0.2 to 0.25 inch long at the base of the vertical cut, thus forming an inverted T (fig. 6.1). The knife should be held at an upward angle when making this cut, and the blade should be rolled around the stem, cutting just through the bark. Keeping your budding knife razor sharp allows these cuts to be made very easily and creates clean edges that heal quickly. After making the cross cut, open the two flaps either by using the bump on the back edge of the blade or, as most experienced budders do, simply rock the blade after making the cut, making the cut and opening the flaps in one motion.

The next step in budding, cutting the bud (fig. 6.2), is the most difficult and can be the difference between success and failure. Using a prepared bud stick as described above, place the flat edge of the knife against the bud stick about 0.4 inch above the bud to be cut. Holding the blade at about a 45° angle with the blade facing toward you, draw the knife under the bud, exiting about 0.4 inch below the bud. The knife must be razor sharp so that this cut can be made in one clean stroke with no rocking or sawing motion. The flat edge of the blade must be kept against the stem to prevent the knife from penetrating too deeply. The finished bud should be about 0.75 inch long to match the slit that was cut previously, and the bud should appear as a very thin slice of green bark with just a little white wood.

Insert the bud into the prepared rootstock by placing the upper end of the bud under the bark flaps at the bottom of the inverted T (fig. 6.3). Using your thumb, gently push the bud up under the bark, being sure that both edges of the bud remain under the bark flaps. Insert the bud until its full length is beneath the bark flaps. The bud should be centered in the slip between the flaps. If the bark of the rootstock is slipping properly this step will be easy.

Next, wrap the bud to ensure good contact between the cambial layers of the bud and the rootstock and to prevent the bud from drying out until it heals and can begin receiving water and nutrients from the rootstock (figs. 6.4 and 6.5). Begin wrapping from below the horizontal cut of the T and move upward, overlapping each turn slightly over the previous one. Continue wrapping to the top of the vertical cut. Finish the wrap by tying the tape by pulling the loose end under the final turn. As you wrap, maintain some tension on the budding tape by stretching it slightly. This will ensure that adequate pressure is applied to the bud to achieve good contact for rapid healing. Each tree should be clearly labeled at this time, indicating the rootstock and scion varieties and any other information that may be helpful or required by law, such as the nursery registration number or the location of the tree in the nursery (see fig. 6.6).

Figure 6.1 The vertical cut for a T-bud (shown) is followed by a horizontal cut at the top of the vertical cut. The bump on the top of the knife is used to pry the vertical cut open enough to insert the bud. *Photo:* G. S. Brar.

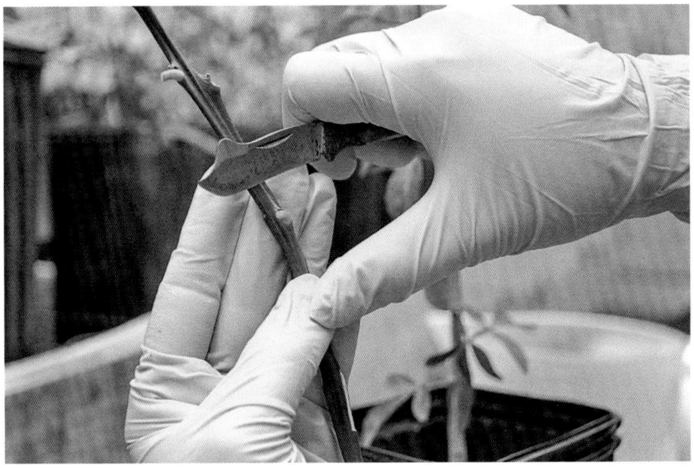

Figure 6.2 A vegetative scion bud is cut immediately prior to insertion into the T-bud of the rootstock seedling. *Photo:* G. S. Brar.

Figure 6.3 Inserting the vegetative scion bud into the seedling rootstock. Holding the bud by the subtending trimmed leaf stem ensures the correct upward bud orientation. *Photo:* G. S. Brar.

Figure 6.4 Wrapping the bud. *Photo:* G. S. Brar.

Figure 6.5 Fully wrapped bud with thorn growth. *Photo:* G. S. Brar.

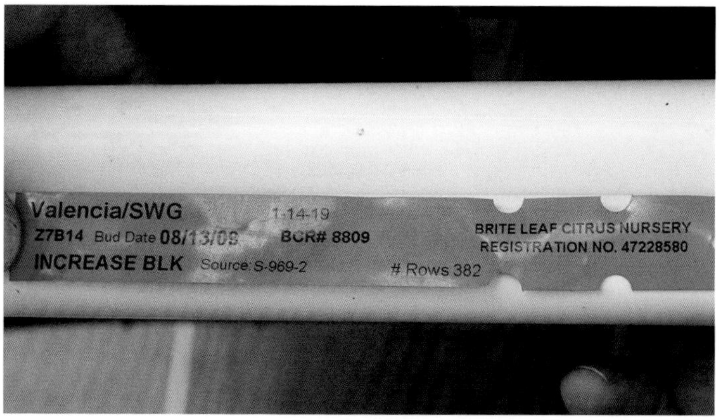

Figure 6.6 A tag designating rootstock variety and scion cultivar is put on each tree after budding. *Photo:* L. Ferguson.

The inserted bud should heal in 10 to 14 days, depending on environmental conditions and the growth status of the rootstock. To unwrap the buds, simply make a vertical cut through the wrapping on the side of the rootstock opposite the bud (don't cut through the bark). If the bud has taken, it will be bright green and will show the development of callus tissue along the edges of the flaps. If the bud is brown or black, it has failed. The rootstock can be rebudded in another location on the stem. With practice, an experienced budder can easily achieve 98% success.

Now that the bud is healed, it must be forced to grow. This is done by breaking the apical dominance of the rootstock seedling. Common methods of breaking this dominance are bending, lopping, and topping (cutting off). In bending, the rootstock stem above the bud is bent over in a loop and tied or otherwise secured to the base of the rootstock stem. A modification of bending, sometimes referred to as the Brazilian twist, is to twist the rootstock stem twice while bending it. This twist prevents the stem from standing back up and eliminates the need for tying. In lopping, a cut is made partway through the rootstock stem above the bud union, leaving it only partially attached, and the stem is bent over. In topping, the entire rootstock stem above the bud union is removed. Bending has the advantage that if a bud fails to grow or grows poorly, the rootstock can be rebudded. In general, bending and lopping result in better bud growth than topping, likely due to the continued supply of photosynthates from the upper part of the rootstock stem. Any competing sprouts must be removed from the rootstock as the young bud grows.

Once the first flush of growth from the new bud (fig. 6.7) has stopped, the rootstock stem above the bud can be completely removed. This should be done using a clean (sanitized in 10% bleach solution), sharp clipper about 0.5 inch above the bud. This is also a convenient time to stake and tie each tree. Stakes can be wood, metal, plastic, or bamboo; the growing scion should be secured to it about every 6 inches to ensure a straight stem. The trees should be ready for sale within 12 months from budding. A good container-grown nursery tree should have well-developed lateral branches and a well-developed root system that fills the container (figs. 6.8 and 6.9). When the trees are removed from their containers for transplanting, no more than 10% of the growing media should fall away. Trees should be transported from the nursery to the field in an enclosed truck or trailer, which should be loaded through an opening that ensures pest exclusion.

Nursery Cultural Practices

Irrigation Methods

Irrigation is the most critical, expensive, and labor-intensive practice in nursery production. Four major types of irrigation are used in container-grown citrus nurseries: hand-held hoses, micro and macro overhead sprinklers, drippers, and microsprinklers. Each has advantages and disadvantages.

Hand-held hoses are initially inexpensive and suitable for fertigation. However, the labor requirement is high, and the uniformity of water and fertilizer application is low. Leaching is rarely effective, particularly if the water-holding capacity of the medium is high or the containers are overfilled. Splashing and the muddied hoses can spread pathogens. The sprinkler head can also spread pathogens if it contacts the soil.

Overhead sprinklers are cheaper to install than drip or microsprinklers, but the wasted water soon negates this advantage. Further, fertigation is impractical and may lead to leaf toxicity damage. Finally, as the trees develop canopies the irrigation becomes nonuniform, leading to under- and overirrigation and possibly Phytophthora damage. Greenhouse structural members or other physical obstructions that deflect or block the water from reaching its intended target may further reduce the uniformity of overhead irrigation.

Drip irrigation is considered to be the best method for container-grown plants. The major disadvantages are the high initial cost and the maintenance cost to prevent the emitters from becoming clogged. Drip irrigation requires a good filtration system to decrease clogging, and adding phosphoric acid to fertigation water helps deter clogging. When selecting a drip system, consider delivery uniformity, ease of unclogging, and cost, in that order. The most precise (and therefore the most expensive) drippers may not be necessary because overirrigation is not common using drippers and occurs only when irrigation is too frequent.

Microsprinkler irrigation has high installation and maintenance costs. Its major advantages are water-use efficiency and the ability to observe operation, although determining uniformity requires testing. To be effective, microsprinklers should be situated beneath the seedling canopy and above the container lip. They should be placed around the outer edge of the benches, preferably alternating from one side of the bench to the other, to ensure uniform coverage. Specific placement and frequency is a function of wetting pattern. As with drip systems, microsprinklers require filtration systems, and

Figure 6.7 The vegetative scion shoot beginning to flush. *Photo:* G. S. Brar.

Figure 6.8 A good container-grown nursery tree should have well-developed lateral branches. *Photo:* L. Ferguson.

Figure 6.9 Good nursery trees should have a well-developed root system that fills the container. When the trees are removed from their containers for transplanting, no more than 10% of the growing media should fall away. *Photo:* L. Ferguson.

irrigation should begin and end with clean water to avoid plant damage and clogging. Of the two systems, drippers use water more efficiently. Because the growing media used in container nurseries is well drained, it is impractical to monitor irrigation with indicators such as tensionometers, gypsum blocks, or TDR probes. The frequency and volume of irrigation are best determined by observing the trees and knowing the water-holding capacity of the growing media and how quickly it is depleted.

Fertilization

Both insufficient and excess fertilization produce poor growth; the latter causes chlorosis and dieback through increased salinity. The method of fertilization depends on the growing media, irrigation method, and water quality. Even if a premixed growing medium is used, it is wise to test the pH of the soil-water extract initially at planting and periodically thereafter, as both nutrient uptake and seedling growth depend on pH. Preplant fertilization generally includes nitrogen, phosphorus, potassium, and trace elements in slow-release fertilizer.

Fertilizing with various irrigation systems

With in-line fertigation through a drip or low-rise microsprinkler system in individual containers, the only form of preplant fertilization needed may be pH adjustment for the water, because all the macro- and micronutrients can be supplied with every irrigation. One disadvantage of applying all the nutrients through the drip system is that all the elements cannot be combined in concentrated form without precipitation. This means two nutrient stock solutions must be stored separately and injected into the irrigation system simultaneously at different points by injection pumps. Alternatively, large-capacity storage is required to make a more diluted complete solution for injection. The disadvantage of the latter is its tendency to generate algae, which clogs drippers, but phosphoric acid can eliminate the algae growth. Fertigating also demands that a leaching fraction be calculated and included in every irrigation to prevent the buildup of salinity. This increases the cost of fertilizer and the pollution of the water source in nurseries that have closed water systems. An alternative method would be to fertilize the growing medium with phosphorus and slow-release trace elements before planting and use only nitrogen and potassium in the nutrient solution. If irrigation is by overhead sprinklers or hand-held hose, the preplant fertilizer should be a slow-release N-P-K, which is easier than individual application of each element. For more information, see chapter 12, "Fertigation."

Monitoring nutrient levels and leaching

Leaf analysis has not been demonstrated to be a reliable method of monitoring seedling nutrient status, and no nutrient standards have been established for citrus nursery production. Leaf nutrient status as indicated by visible symptoms is the most reliable and most often used method, but it appears too late to be an indicator of developing problems. Regularly monitoring the soil-water extract for cation exchange capacity (EC) and pH and keeping these two values within the ranges where nutrients are readily available is a better way to prevent deficiencies or excesses from developing.

Growing media should be tested every 6 months, particularly if the media is well drained, to check for decreasing phosphate levels. Single superphosphate has a rapid initial release rate followed by a drop to levels inadequate for growth within a year. Phosphorus should be applied every 6 to 8 months or added at a rate of 20 mg/l with every fertigation. Some growers collect weekly soil cores, squeeze out the leachate, and sample it.

Salinity in the media is a result of too much fertilization, fertilizing at the wrong time (e.g., when uptake is slow due to cold temperatures), or inadequate leaching. Salinity produces visible leaf toxicities and reduces growth. It also produces saturated conditions in the containers and increases the seedling susceptibility to Phytophthora. Therefore, even if leaching fractions are being calculated and applied, the EC of media should be regularly monitored along with the pH. There is no connection between the two parameters, so one cannot be used as an indicator of the other. Salts are easily leached from sand, but organic material in a medium retards salt leaching. Therefore, the EC of the medium should be measured monthly, but the EC of the leachate should be monitored weekly. For more information, see chapter 10, "Nutrient Deficiency and Correction."

Monitoring Salinity

Two methods are used to control salinity in fertigation. The first is to only replace the water lost since the previous irrigation through several consecutive irrigations; heavy leaching occurs when the EC_e reaches critical levels. The other, and less dangerous, method is to calculate a leaching fraction and leach it with every irrigation. This has the advantage of creating uniform nutrient availability in the media, but it wastes nutrients. Beginning and ending each fertigation with clean water decreases nutrient losses. Monitoring should continue even with regular leaching.

Pest Control

General Aspects of Pest Control

Any damage to the apical bud and leaves of a newly budded and growing citrus tree will result in decreased growth. Therefore, it is essential to control all pests. Among the pests common in California citrus nurseries are aphids, mealybugs, mites (including flat, red, bud, and red spider), orange dog and other caterpillars, red scale, soft brown scale, snails and slugs, thrips, leaf miners, and leafrollers. Pesticides, rates, and application regulations change frequently; consult current registrations, label rates, and labels before attempting to control any pest or disease. For more information, see the UC IPM Citrus Pest Management Guidelines, http://www.ipm.ucdavis.edu/PMG, and part 5 of this manual, particularly chapter 17, "Citrus IPM."

Integrated Pest Management for California Container-Grown Citrus Nurseries

Prophylactic organophosphate sprays at 7- to 14-day intervals were once a common practice in California citrus nurseries. In the early 1980s, thrips developed resistance to the most effective pesticides due to repeated applications on nonbearing nursery trees. Now, integrated pest management (IPM) developed for orchards is increasingly implemented in nurseries. The emphasis is to monitor and spray only when necessary and only for what is present. Nurseries must develop regular pest monitoring programs to determine when to spray. For more information, see chapter 17, "Citrus IPM."

New Invasive Pests and Diseases

Citrus nurseries play a pivotal role in disease and pest control. Because budwood and budwood sources are introduced into nurseries, and because budded trees are distributed from nurseries, citrus nurseries are in a unique position to vector diseases. The discovery that budwood was the vector for the viral and bacterial diseases that decimated the California citrus industry in the 1950s led to the formation of the California Citrus Clonal Protection Program (CCPP) and the California Citrus Tristeza Eradication Agency (CCTEA). The CCPP oversees a series of monitoring, testing, and control measures that certify that the budwood used by a nursery is true to type and disease-free. The CCTEA monitors and controls for the citrus tristeza virus (CTV) in producing orchards with the objective of preventing its spread to other producing trees and potential budwood sources. The CCPP will most likely play a role in determining the regulations that will be formulated to meet the threat posed by huanglongbing disease (for more information, see chapter 7, "The California Citrus Clonal Protection Program").

Pathogen Control

The primary method of pathogen control in nurseries should be prevention. The nursery should be fenced or enclosed, with copper footbaths placed at all entrances. All tools and equipment should be treated with a 5-minute dip in a 10% chlorine solution. The major pathogens in citrus nurseries are *Phytophthora* species. The *Phytophthora* status of water, roots, and growing media should be continually monitored to avoid any buildup. Of particular importance is the drainage of the media in pots. Anything that impedes drainage from a pot, whether the quality of the media, the drainage hole placement, or benches that impede water flow from the bottom of the pot, can produce conditions that exacerbate *Phytophthora* buildup. The time between irrigations should be as long as possible and the duration of irrigation as short as possible. Generally, *Phytophthora* develops only under suitable conditions. Therefore, even if *Phytophthora* is present, it will not develop into root rot unless the conditions are amenable. If *Phytophthora* is detected in a nursery, the source must be isolated and eradicated. This is far preferable to fungicide treatment, which may mask the root rot symptoms until the tree reaches the orchard. As with pesticides, repeated applications of systemic fungicides eventually cause resistance. Generally, it is better to protect a nursery from pathogen entry than to suppress it once detected.

Nursery Management and Administration

A nursery is a business, and all businesses require record keeping. Records should include sufficient detail to determine the origin of all seed and budwood, all treatment of the seed and subsequent rootstock and budded seedlings, and the final destination of the sold tree. Regular records of facilities maintenance and improvements should be kept, as well as records of treatment of media, water, grounds, and equipment. Financial records should be maintained in the same fashion as for any farming operation, with three levels of financial management and analysis: production costs, cash overhead costs, and noncash overhead costs. These financial records must be analyzed to determine whether the operation is truly profitable. For more information, see chapter 2, "Costs of Establishment and Production."

References

Baker, K. R., and C. N. Roistacher. 1957. Treatment of nursery containers. In K. F. Baker, ed., The U.C. system for producing healthy container-grown plants. Berkeley: University of California Division of Agricultural Sciences, Agricultural Experiment Station; repr. 1985, Chipping Norton, New South Wales, Australia: Surrey Beatty and Sons. 210–216.

CCPP (Citrus Clonal Protection Program). 2012. University of California, Riverside, Department of Plant Pathology CCPP website, http://www.ccpp.ucr.edu.

Handreck, K. A., and N. D. Black. 1984. Growing media for ornamental plants and turf. Kensington, New South Wales, Australia: New South Wales University Press.

Lee, A. T. C., and K. Roxburgh. 1993. Guidelines for the production of container-grown citrus nursery trees in southern Africa. London, UK: Outspan Publishing.

Matkin, O. A., P. A. Chandler, and K. F. Baker. 1985. Components and development of mixes. In K. F. Baker, ed., The U.C. system for producing healthy container-grown plants. Berkeley: University of California Division of Agricultural Sciences, Agricultural Experiment Station; repr. 1985, Chipping Norton, New South Wales, Australia: Surrey Beatty and Sons. 86–107.

Sauls, J. W. 2008. Citrus nursery production. Texas citriculture and subtropical fruit website, http://aggie-horticulture.tamu.edu/citrus/nursery/L2305.htm.

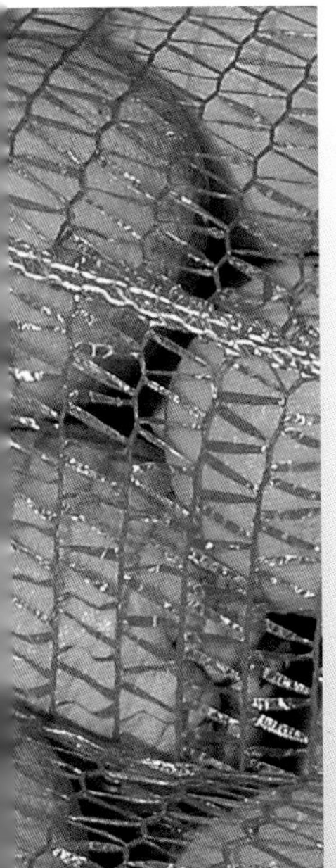

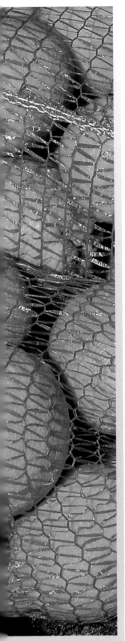

7 The California Citrus Clonal Protection Program

Georgios Vidalakis, David J. Gumpf, MaryLou Polek, and John A. Bash

The world's leading citrus-producing countries established their industries by importing, maintaining, and improving varieties obtained from countries where citrus was native, principally in Australasia and later Asia and Africa (Wallace and Drake 1959; Bayer et al. 2009; Pfeil and Crisp 2008). As citrus was moved from these areas to the new citrus-growing areas, so were pathogens and insect pests that in many cases were also vectors of pathogens. This is especially true of viral pathogens, which were unrecognized during these early years of citrus movement. Most of the older citrus varieties that were moved around the world were infected with one or more pathogens, including those that cause psorosis, concave gum, exocortis, tristeza, cachexia, and vein enation. As a result, these diseases have become widely distributed and are present in virtually every citrus-growing area of the world (Wallace 1978; Wallace and Drake 1959; Childs 1978). Today, a number of serious prokaryotic diseases, including stubborn, huanglongbing (HLB, or citrus greening), and variegated chlorosis, as well as diseases of unknown etiology such as impietratura and cristacortis, are common in certain geographic areas of the world but absent in others. Other problems and diseases related to bud union disorders, (e.g., tatter leaf/citrange stunt) and various tree declines (e.g., citrus sudden death) are present in only one or two citrus-growing areas. Many of these diseases were spread around the citrus-growing world when movement of plant materials was virtually free (Timmer et al. 2000; Wallace 1978; Roistacher 1991).

About 1930, most citrus-growing countries realized the need to initiate quarantine regulations to halt unrestricted importation of plants and plant-propagating materials from other citrus-growing areas. These restrictions were initially designed to prevent the introduction of obvious diseases (e.g., fungal or bacterial) and insect pests. However, as information and understanding of the diseases and their causal agents increased, quarantine regulations were modified and expanded to cover the entire spectrum of pests, including viruses and bacteria, that were not visible or easily detected. California adopted such regulations when it was discovered that certain viral diseases could be transmitted through budwood. From the 1930s until the 1960s, the importation of new varieties into California (as well as a high percentage of the existing propagating materials for oranges and lemons) was restricted to seed (Childs 1978; Wallace and Drake 1959; Reuther 1959; Hiltabrand 1959; Calavan et al. 1978; Krueger 1997). The seed method of introduction was considered safe because studies had shown that even if a mother tree were infected with a viral disease, seedling progeny were healthy (Weathers and Calavan 1959, Cameron et al. 1959). Introducing new varieties by seed, however, has a number of disadvantages, even though most citrus varieties produce nucellar seedlings that are genetically similar to the mother plant: nucellar seedlings are often very thorny, and fruit production may be delayed for years, so that determining trueness to type is difficult. Also, desirable cultivars may not produce nucellar seedlings or may be seedless. These problems are not encountered if the variety is introduced as budwood (Wallace and Drake 1959; Cameron et al. 1959).

The long-term health, survival, and sustainability of a country's commercial citrus industry require frequent budwood importations of new scion and rootstock varieties that provide disease resistance or improved fruit quality and production. It is also very desirable for a citrus-growing country to have a germplasm collection available to researchers developing new varieties through classical breeding or by genetic engineering. Introduction of varieties as budwood requires that importing countries be able to test the newly imported materials for the presence of graft- or bud-transmissible diseases and clean up the imported varieties so they do not present a threat to the existing citrus industry (see Frison and Taber 1991; Wallace and Drake 1959; Reuther 1959; Calavan et al. 1978; Roistacher 1991; Lee 2003; Lee et al. 1999; T. Kahn et al. 2001; Vidalakis et al. 2010; Gumpf 1999).

The California Situation

The citrus industry in California, as in many other citricultural areas, has long imported budwood of citrus varieties from around the world. In the past, problems such as tristeza, rootstock-scion abnormalities, citrus nematodes, and replant problems were just a few of the conditions that compelled an ongoing search for and importation of new citrus materials for study and evaluation. More recently, California growers have demanded the importation of new and different citrus varieties to satisfy the changing tastes of consumers and to fill specific marketing niches. Varieties such as very-early-maturing and very-late-maturing navel oranges and a vast array of easy-peeling mandarin oranges that mature at different times have been in great demand for importation. Fortunately, this need to expand the diversity of citrus varieties in California was recognized not only by growers and citrus researchers at the University of California, Riverside (UCR), but also by the United States Department of Agriculture (USDA) and the California Department of Food and Agriculture (CDFA). These groups took action when they recognized the potential problems that could result from the importation of new pathogens and pests into the United States and California along with the citrus budwood. In 1954, a national cooperative program was initiated for the importation of citrus scion and rootstock selections from other citrus-growing regions of the world. It was fortunate that this program had built-in safeguards (such as fumigation) against insect pests and propagation and observation for a long period (a minimum of 2 years and in many cases 5 or more) of the imported material under quarantine at the USDA Plant Introduction Station in Glenn Dale, Maryland (Roistacher and Nauer 1968; Reuther et al. 1972; R. Kahn et al. 1967; Wallace and Drake 1959; Gumpf 1999).

About this same time, growers and researchers in California also recognized the importance of pathogen-tested budwood in the establishment and maintenance of a viable and productive citrus industry (Reuther 1959; Hiltabrand 1959). Viral, viroid, and bacteria-like pathogens in propagation budwood can reduce tree survival and fruit production and quality. Since the principal means of spread of these diseases in California occurred through propagation from infected stock, a comprehensive program was initiated to produce, maintain, and distribute budwood that has tested negative for all known bud-transmitted diseases. This program was originally called the Citrus Variety Improvement Program (CVIP). In 1978, the name was changed to the Citrus Clonal Protection Program (CCPP) to more accurately reflect a major change from therapy and maintenance of existing California varieties to include the direct importation and quarantine of new varieties from all over the world (Calavan et al. 1978; Nauer et al. 1967; Reuther 1981; T. Kahn et al. 2001). This change was aided when the CCPP director was issued a permit from the USDA Animal and Plant Health Inspection Service (APHIS) to import foreign citrus varieties and hold them in quarantine while performing the testing and cleanup necessary for release into the field.

The CCPP as it exists today is a cooperative venture involving UCR, USDA, and CDFA, along with the citrus industry of California as represented by the Citrus Research Board (CRB), a box-research marketing order, and the California Citrus Nursery Board (CCNB), a handler marketing order. UCR, CRB, and CCNB provide the financial support of the CCPP and advise the program through a 12-member committee made up of growers and nursery professionals. Currently, UCR is developing a CCPP advisory committee; members under consideration include Agricultural Experiment Station (AES) scientists and representatives of the California citrus industry, state and federal regulatory agencies, as well as members of the UCR Chancellor's Agricultural Advisory Council.

In 2009, the CCPP was recognized as a regional center of excellence by the National Clean Plant Network (NCPN).The NCPN partners with plant centers for disease diagnosis and pathogen elimination in the production and maintenance of clean propagative material for U.S. specialty crops such as grapes, peaches, berries, citrus, and other fruit. The CCPP plays a leading role nationally in the introduction of the highest-quality, true-to-type propagative citrus materials that are free of targeted citrus pathogens and pests (Vidalakis et al. 2010a, 2010b).

The California Citrus Clonal Protection Program

The CCPP is the basic element of a long-term multilevel program whose objectives include avoiding or restricting the spread of bud- or graft-transmitted pathogens of citrus in support of a profitable, competitive, and sustainable citrus industry. Such programs should contain the following components:

- a quarantine to prevent the unrestricted movement of citrus propagative material such as budwood
- the elimination of pathogens from infected budlines
- the registration of sources of propagative material
- a variety improvement program that includes the introduction and development of superior budlines and the constant evaluation of their disease status and horticultural characteristics
- the eradication of all infected trees and their replacement with certified ones when relevant and worthwhile

In California, the CCPP is pivotal in providing all of these components, as is reflected in its stated objectives:

- To provide a safe mechanism for the introduction into California of citrus varieties from any citrus-growing area of the world for purposes of research, variety improvement, or use by the commercial industry of California.
- To maintain primary sources of pathogen-tested and true-to-type citrus propagative material of all important fruit and rootstock varieties in protected foundation blocks for budwood distribution to the California citrus industry and citrus researchers.

The Functional Pillars of the Citrus Clonal Protection Program

The steps and procedures in the following discussion are employed by the CCPP in its commitment to safeguarding the citrus industry of California (for more information, see Gumpf 1999; T. Kahn et al. 2001; Gumpf et al. 1996; Bash 1997; Vidalakis et al. 2010a, 2010b).

Introduction of Citrus Varieties under Quarantine

The citrus quarantine in California is a cooperative venture involving federal, state, and county departments of agriculture, along with the University of California. The federal government, represented by the USDA Animal and Plant Health Inspection Service Plant Protection and Quarantine (USDA-APHIS-PPQ) is concerned with preventing citrus pests from foreign sources from entering the United States. Federal quarantines also exist between states. The director of the CCPP has a permit issued by USDA-APHIS-PPQ to import citrus budwood from foreign countries (one of few such permits issued for citrus introduction into the United States). The permit spells out specific stipulations regarding handling and treatment of materials in quarantine that must be followed when citrus material enters the CCPP Rubidoux Quarantine Facility at Riverside. These stipulations are also required by the state of California and are enforced by CDFA. Under California law, all citrus budwood is prohibited from entry into California unless authorized under a CDFA or USDA permit such as the one issued to the CCPP director.

All citrus germplasm entering the United States first goes through the federal quarantine facility in Beltsville, Maryland. Upon arrival, the budwood is visually inspected and treated for any pests or any other potentially biohazardous materials (e.g., soil). It is then sent to the CCPP Rubidoux Quarantine Facility. The Rubidoux facility consists of an insect-resistant greenhouse, with the temperature and lighting controls that are required for biological indexing; a screenhouse; and a modular office and laboratory area. The facility is located within the city of Riverside adjacent to the original location of the Citrus Experiment Station (1907) and is isolated from the nearest commercial citrus and UCR experimental orchards by about 3 miles. The CCPP quarantine and campus laboratory facilities are part of the space allocated by the UCR Department of Plant Pathology and Microbiology of the College of Natural and Agricultural Sciences.

Typically, when a new import is received by CCPP at the Rubidoux facility, propagations are made onto Rough Lemon rootstocks to preserve the budline and to produce budwood for future indexing or therapy. These propagations are placed in a warm greenhouse for maximum rapid growth. In addition, some of the imported budwood is cultured in vitro, and the meristems of the young flushes produced are used for in vitro micropropagations using shoot-tip micrografting. The remaining portion of the original imported budwood is used to graft-inoculate indicators and for laboratory testing for the pre-index screening. The pre-index will indicate whether the imported budline contains regulated pathogens such as tristeza and psorosis viruses, citrus viroids, or the hunglongbing associated bacteria. (fig. 7.1). Imported budwood is frequently infected with one or more of these bud-transmissible pathogens, especially if collected from a field tree rather than a disease-tested citrus collection.

Disease Diagnosis and Pathogen Detection

The CCPP program of importation, production, and distribution of pathogen-tested propagative materials is based on a comprehensive indexing (testing) program to detect graft-transmissible diseases and pathogens that may arrive in imported budlines. Graft-transmissible diseases may be caused by viruses, viroids, or other pathogens (e.g., bacteria, phytoplasmas) and are vegetatively transmitted with an infected budline. Graft-transmissible diseases can seriously affect fruit quality, production, and tree

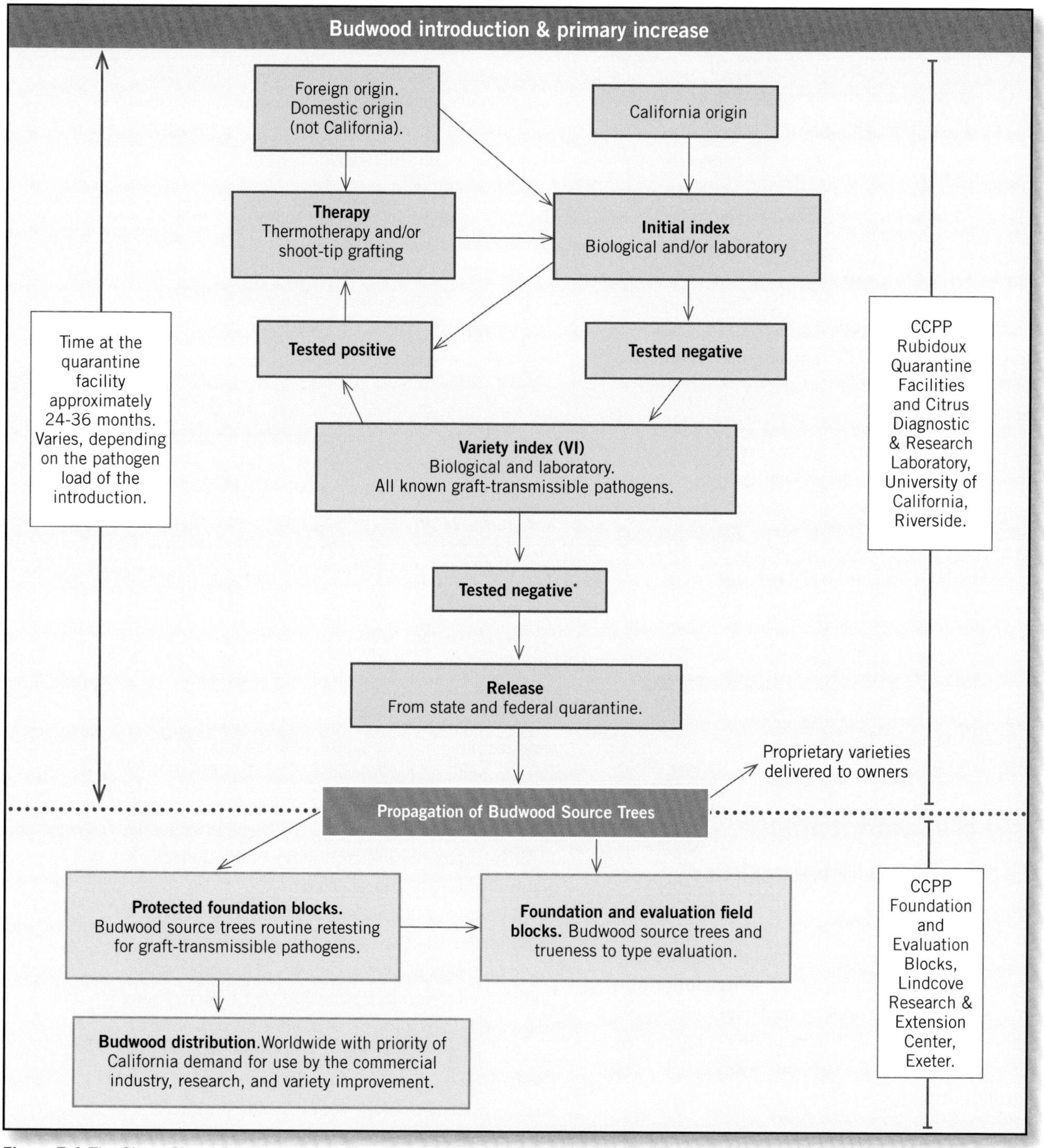

Figure 7.1 The Citrus Clonal Protection Program variety introduction and distribution scheme. *Note: In order to fulfill movement requirements between quarantine zones, all materials must undergo therapy regardless of the pre-index results.

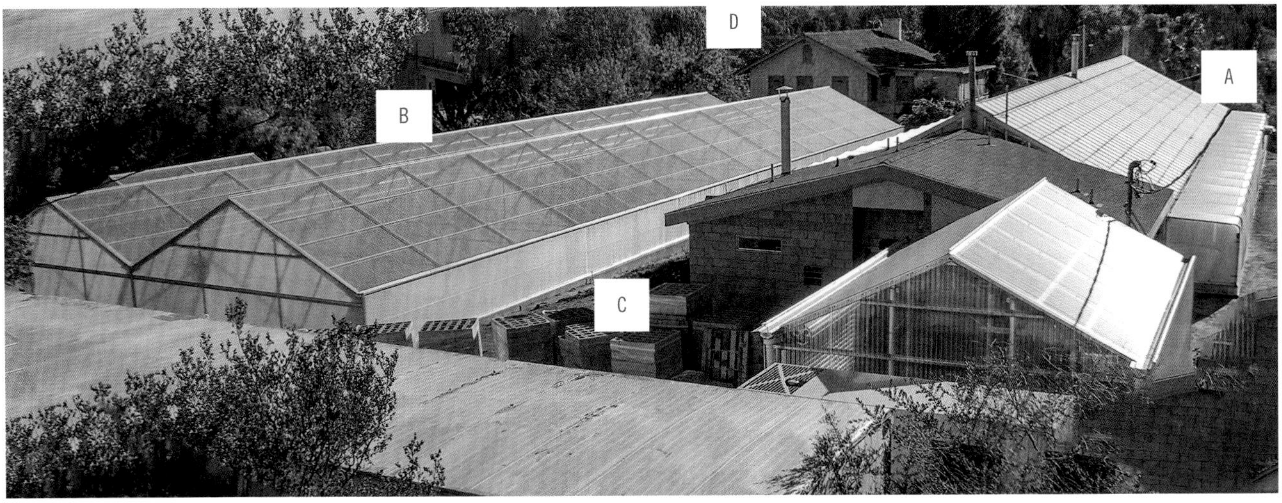

Figure 7.2. Panoramic view of the CCPP Rubidoux Quarantine Facilities at Riverside, California. Quarantine glasshouse (5,000 sq ft) (A) and screenhouse (9,000 sq ft) (B) for biological indexing and maintenance of citrus germplasm, respectively. The original structures were built from 1959 to 1962 and were fully renovated from 2010 to 2012. The Delfino Family Plant Laboratory (750 sq ft) (C) for citrus therapy and diagnostics was constructed from 2010 to 2011. The original Citrus Experiment Station (1907), the precursor of UC Riverside, has housed the CCPP offices (D) since October 2013. *Photo:* G. Vidalakis.

health and longevity. Additionally, diseases from infected field propagations may be spread to neighboring orchards by insects or farming equipment (Timmer et al. 2000; Wallace 1978).

Disease diagnosis and pathogen detection take place at the Rubidoux quarantine facility in the insect-resistant greenhouse and the Delfino Family Plant Laboratory (fig. 7.2) as well as in the UCR Citrus Diagnostic Laboratory. The detection of graft-transmissible citrus diseases is based primarily on biological indexing, which is accomplished by grafting tissue of the imported budline onto citrus indicators (fig. 7.3). Specific citrus indicators are used to detect specific diseases. Indicator varieties have been selected for sensitivity to diseases and ability to express symptoms. In each index, adequate positive controls (disease-infected indicators along with healthy controls of each indicator) are held under the same environmental conditions as test indicators. Controls are used as a comparison with the test source and also as confirmation that environmental conditions in the greenhouse are optimal for plant growth and symptom expression (Roistacher 1991; Childs 1978; Vidalakis et al. 2004).

Laboratory tests are also part of pathogen detection. These include tests such as enzyme-linked immunosorbent assay (ELISA), sequential polyacrylamide gel electrophoresis (sPAGE), reverse transcription (RT) followed by conventional or quantitative polymerase chain reaction (PCR), hybridization, and culture in growth media (see, e.g., Roistacher 1991; Palacio-Bielsa et al. 1999; Garnsey et al. 2002; Sieburth et al. 2002).

Indexing occurs at various times. A small-scale pre-index takes place at the time of introduction. The results of the pre-index indicate whether and which therapy may be needed; another pre-index indicates whether the therapy was successful (figs. 7.4 and 7.5). If indexing indicates that therapy was successful, the budline enters a full-scale VI index. The VI index lasts 6 to 8 months and consists of bioindexing onto a broad range of indicators, including a range of negative and positive disease controls (mild, moderate, and severe isolates), along with the most current laboratory tests.

Pathogen Elimination Therapy

If the pre-index indicates the presence of a disease or pathogen, the budline must be subjected to therapy that can eliminate the disease agent. The CCPP employs two methods of therapy, thermaltherapy (or thermotherapy) and shoot-tip micrografting. UCR and CCPP scientists have been instrumental in the development, validation, and employment of both techniques since the early 1970s (Calavan et al. 1972; Roistacher and Calavan 1972; Murashige et al. 1972; Navarro et al. 1975; Roistacher et al. 1976; Roistacher and Kitto 1977).

Thermaltherapy is performed by taking buds from the infected budline and grafting them onto Citrange (Sweet × Trifoliate Orange) seedlings. The infected bud is tightly wrapped with budding tape so that the bud will not flush during thermaltherapy. The Citrange seedlings with an infected bud grafted on them are placed into a hot greenhouse for 30 days, with temperatures maintained at 82° to 104°F during the day and approximately 77°F at night for preconditioning. The plants are then placed into a controlled-temperature growth chamber for a period of 3 months that is set for 16-hour days at 104°F and 8-hour nights at 86°F (fig. 7.5). Upon removal of the plants from the growth chamber, the buds are unwrapped, the rootstock seedling is bent over, and the top of the seedling is pushed into the potting soil so the grafted bud will become the terminal bud. The plants are then placed in the greenhouse until sufficient budwood is produced from the grafted bud for further indexing. This method does not effectively eliminate viroids, and its ability to eliminate certain viruses (e.g., citrus tatter leaf) has been reported to be variable. All imports received by CCPP originating outside the United States are routinely subjected to thermaltherapy as a precaution (Calavan et al. 1972; Roistacher 1977; Roistacher and Calavan 1972; Roistacher and Calavan 1974; Zhang 1996; Koizumi 1984).

Shoot-tip micrografting (STG) is the other form of disease therapy employed by the CCPP. Some pathogens, particularly citrus viroids (e.g., causal agents of exocortis and cachexia diseases), are difficult or impossible to eliminate by thermaltherapy and are much more readily eliminated by

Figure 7.3 Biological indexing of graft-transmissible diseases of citrus. Branch of healthy citrus (A), dead branch/shock reaction of virus-infected citrus indicator (B), wood of healthy citrus (C), gumming and wood discoloration of viroid-infected citrus (D), leaf of healthy citrus (E), leaf deformation of virus-infected citrus (F). *Photos:* G. Vidalakis (A, B, E, F); J. Semancik (C, D).

STG. In STG, several new growth tips slightly less than 1 cm long are taken from one of the original infected imported propagations. Under a dissecting microscope, an apical meristem of about 0.15 mm is removed from the infected growth tip and grafted onto a rootstock seedling grown in vitro (see fig. 7.4). If small enough when removed from the growth tip, the apical meristem of the infected imported variety is not yet developed enough (i.e., has not developed phloem tissues) to contain pathogens, and the micrografted propagation will be free of pathogens and will not express any disease. The propagations are then returned to glass tubes and placed under light in a culture chamber. When the scion of the micrografted propagation reaches about 2 cm long, it is regrafted onto a clean Rough Lemon seedling and moved to the greenhouse. STG is effective against all graft-transmissible agents, including viroids (De Lange 1978; Murashige et al. 1972; Nacer et al. 1983; Navarro 1981; Navarro et al. 1975; Navarro and Juarez 1977; Roistacher and Kitto 1977; Roistacher et al. 1976).

After any therapy procedure, all propagations produced during therapy undergo indexing in order to determine their disease status. If this indexing indicates that a disease or pathogen is still present, the plant material is subjected to therapy again. This cycle of therapy and testing continues until all tests are negative. If propagation of a budline tests negative in the indexing after therapy, it may then enter the full-scale VI index (see fig. 7.1).

Figure 7.4 In vitro growth of shoot-tip-grafted plants in artificial media. *Photo:* G. Vidalakis.

Figure 7.5 Thermaltherapy chamber. *Photo:* G. Vidalakis.

Quarantine Release

If an introduced budline is shown to be free of known diseases in the VI index, it is then considered ready for release from quarantine. The CCPP must first obtain a release from CDFA by outlining the testing procedures and test results. Once released by CDFA, an application for federal quarantine release is sent to USDA-APHIS containing the testing information and a copy of the letter of approval by CDFA for release from quarantine. The distribution of citrus material that has been released from quarantine is also a highly regulated and carefully executed procedure that involves close interaction among CDFA, CCPP, and citrus nurseries and growers.

The above-described testing, therapy, and release from quarantine is available to private entities that wish to import patented or other proprietary varieties into California. The propagator or owner must sign an agreement with UC Riverside for the recovery of the cost of the testing and therapy procedures. When such varieties are released from quarantine, they are delivered to the owner and are no longer maintained in the CCPP foundation blocks.

Maintenance

The Lindcove foundation and evaluation block

Newly introduced varieties that have passed testing and have been released from state and federal quarantine are propagated at the Rubidoux facility on suitable rootstocks for field planting in the CCPP foundation and evaluation blocks. These blocks are located at the UC Lindcove Research and Extension Center (REC) near Exeter in the San Joaquin Valley (figs. 7.6 and 7.9).

The foundation and evaluation block is planted on fumigated soil and has a wide planting distance between rows and trees to allow for better visual evaluation of each tree. Each tree of the block is examined several times per year by CCPP and interested university and industry personnel for horticultural trueness to type, fruit quality, freedom from budsports and chimeras, spontaneous genetic disorders, and symptoms of disease. Until 2007, each tree was biologically reindexed annually for tristeza and was tested up to five more times annually by ELISA during budwood distribution. Any tree that showed abnormal growth characteristics or tested positive for a disease was immediately removed from the block, and the soil was fumigated before planting another tree.

In 2006 and 2007, a significant number of trees at the Lindcove REC tested positive for tristeza, and some of these were located in the foundation and evaluation block. Therefore, after almost 50 years of budwood distribution from this outdoor field block, the CCPP began budwood distribution exclusively from the indoor protected foundation block. The outdoor block is now used only as an evaluation block for trees that have been propagated from the indoor block and have been evaluated for trueness to type (see fig. 7.6).

The Lindcove protected foundation block

Because of the increasing threat of natural spread of tristeza around the Lindcove REC in 1998, the CRB sponsored the construction of the protected foundation block, a protective screenhouse that was completed in two phases from 1998 to 1999 (see fig. 7.9). In 2007, after positive tristeza findings in the foundation and evaluation block, the protected foundation block became the primary source of budwood. In 2008, the CRB committed the necessary funds for the expansion of the protected block. Construction was completed in 2010, and tree plantings began in the spring of 2010. In 2011, the CRB funded the construction of a positive-pressure greenhouse for the support and expansion of the CCPP foundation operations. Inside these structures, the CCPP maintains both potted and in-ground trees for the varieties whose budwood has traditionally been in high demand. Every tree is tested regularly for regulated diseases such as tristeza, psorosis, and exocortis. The testing requirements of trees registered for budwood distribution changed in May of 2010, when diseases such as huanglongbing (HLB) were added in the list and the frequency of the various tests was revised.

Registration of trees and budwood availability

Before any budwood is distributed from foundation block trees, the trees must be registered as budwood source trees with CDFA.

Prior to May 2010, registration by CDFA required mandatory testing for tristeza and voluntary testing for citrus viroids and psorosis. If these tests were negative, the tree was assigned a CDFA registration number, which accompanied any

Figure 7.6 The CCPP field evaluation block located at the UC Lindcove Research and Extension Center in Tulare County. *Photo:* D. Gumpf.

budwood distributed from any foundation tree (fig. 7.7). In order to remain registered budwood sources, CCPP retested these foundation block trees annually for tristeza, every 3 years for citrus viroids, and every 5 years for psorosis.

In May 2010, CDFA filed regulations for a mandatory Citrus Nursery Stock Pest Cleanliness Program as an emergency action. Under this new program, registered budwood source trees must be tested annually for tristeza and HLB, every 3 years for citrus viroids, and every 6 years for psorosis and psorosis-like diseases. In addition, since January 1, 2012, all registered scion mother trees must be maintained in insect-resistant structures (for details and updates, see the CDFA website, http://cdfa.ca.gov, and search for "Section 3701").

When registration of source trees is completed, budwood from the Lindcove foundation block is distributed to California nurseries and growers. Limited quantities of budwood are available, and recipients normally use foundation block budwood to produce their own nursery- or grower-owned registered trees or nursery increase blocks, which are also regulated by CDFA. Limited amounts of early-release budwood from container-grown trees of selected newer varieties may also be distributed. After release from quarantine, these early-release trees are maintained in a protected screenhouse at the Lindcove REC. Early-release trees have not fruited but have otherwise undergone all the required indexing and are registered with CDFA. Recipients of small lots of early-release budwood understand that the fruiting characteristics of that budline have not yet been evaluated by CCPP. A waiver of liability for budwood that may not be horticulturally true to type or that may contain budsports must be signed prior to receiving early-release budwood. For an overview of budwood movement before and after quarantine release, see figure 7.8.

Budwood from the foundation block typically is cut in January, June, and September of each year. In June 2013, the CCPP initiated a pilot program for monthly budwood distribution, which proved to be successful and continues to this day. The CCPP primarily serves California citrus nurseries and growers, but clients outside the state can also request budwood. Budwood cut dates, order forms, prices, and other information is available at the CCPP website, http://ccpp.ucr.edu.

Outreach

The CCPP makes all information related to budwood distribution, variety evaluation, citrus disease, and management available to the public via its website and by publication in agricultural magazines, peer-reviewed journals, and scientific conferences. Yearly foundation evaluation block inspections, or walk-throughs, are scheduled by the CCPP for the benefit of citrus nursery personnel and interested growers. The CCPP also is present at many local and international agricultural shows, with booths displaying the citrus varieties of its collection and distributing information about the program and its objectives.

CCPP Sanitation Procedures

To prevent the introduction and spread of pests into and within its facilities, a number of safeguards and precautions have been instituted as part of the normal CCPP operating protocols:

- External entrances to the quarantine facilities and the protected foundation block have a double-door system to protect against the entrance of insect pests (see fig. 7.9B).
- Other entrance or exit ports, such as ventilation ducts and fans, are protected with filters, screens, or self-closing louvers (see fig. 7.2A).
- Potting soil, pots, plant stakes, and potting tools are steam-sterilized before entering the facilities.
- All other tools used for plant maintenance, budwood collection, and grafting, such as clippers and knives, are sterilized by dipping or spraying with 1% sodium hypochlorite before each use and between each new task.

Figure 7.7 A CDFA tree registration tag (A) and a budwood label, which accompany every budwood shipment from the CCPP (B). *Photo:* G. Vidalakis.

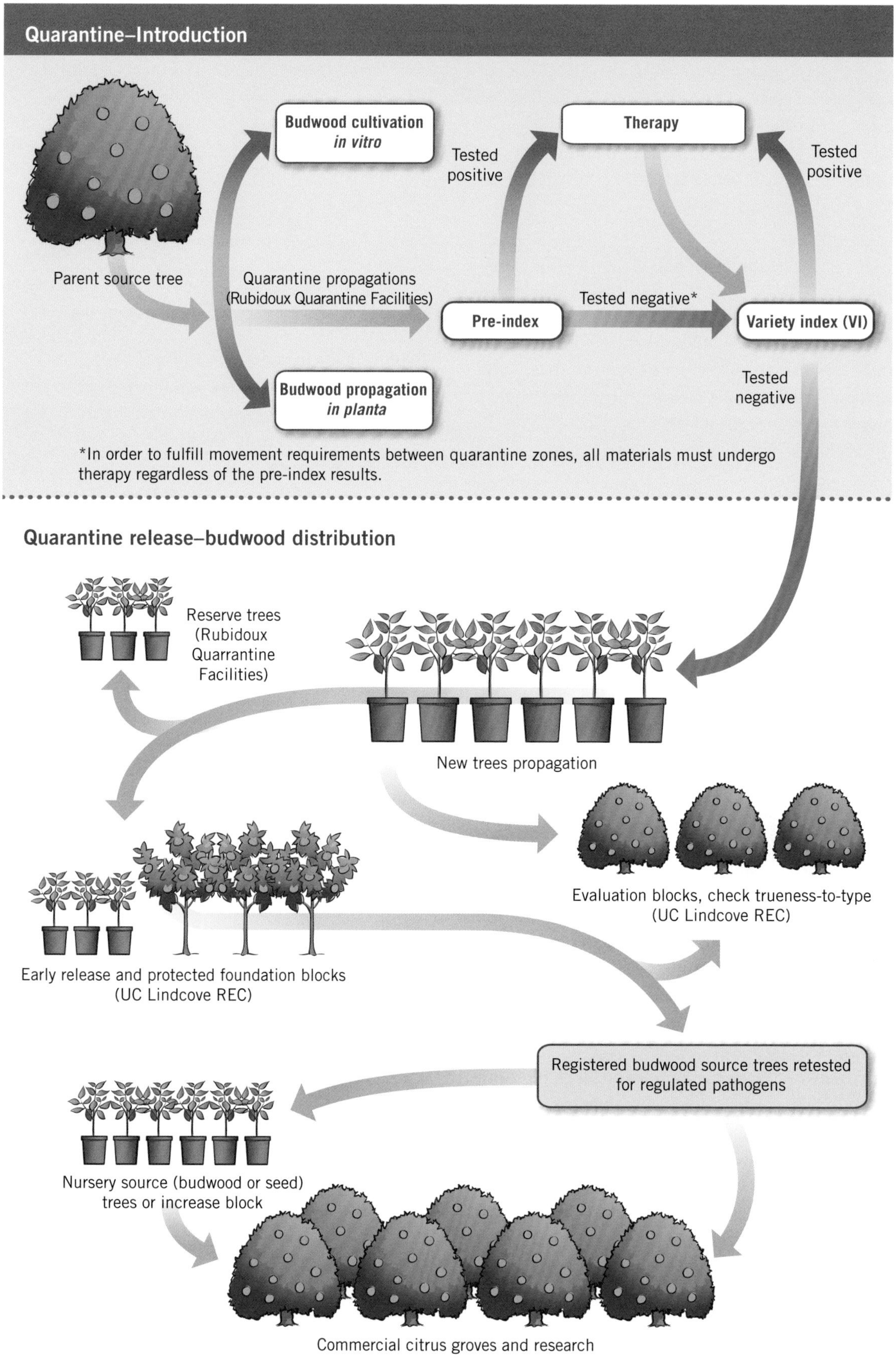

Figure 7.8 Schematic representation of budwood pathways at the CCPP.

- Under no circumstances are bare-rooted or potted plants allowed to enter the quarantine facilities.
- In the quarantine facility, all plants are kept in containers on benches or raised runners that are routinely disinfected with copper naphthenate.
- To prevent the spread of soilborne diseases, the ends of irrigating hoses are not allowed to touch the ground, personnel are not allowed to stand or rest their feet on the benches, and potted plants that have come in contact with the ground are not placed on benches.
- Regular inspections and treatments for insects and mites are performed.
- Budwood entering the CCPP program should be collected from disease-tested sources when possible, or at the least from healthy-appearing trees that are free of pests and without a history of any symptoms on the fruit, foliage, limbs, or trunk.

CCPP Facilities and Equipment

Rubidoux Quarantine Facility, Riverside

The Rubidoux Quarantine Facility (see fig. 7.2) is equipped with all equipment necessary for the molecular, serological, and biological detection of citrus pathogens and the execution of citrus therapeutic protocols, including

- approximately 5,000 square feet of insect-resistant greenhouse, with temperature and light controls required for biological indexing
- approximately 9,000 square feet of screenhouse for tree source maintenance of positive disease controls, original introductions, and foundation budwood sources
- an office and a laboratory area of approximately 750 square feet each

Figure 7.9 Panoramic view of the Citrus Clonal Protection Program foundation block operations in the Lindcove Research and Extension Center in Tulare County. The first screenhouse (40,000 sq ft) was constructed from 1998 to 1999 (A); the second screenhouse (30,000 sq ft) was completed in 2010 (B); and the positive-pressure greenhouse (5,700 sq ft) was completed in 2011 (C). The interior of the screenhouses can accomodate container (D) and in-ground (E) planted budwood source trees. *Photos:* (clockwise from top) E. Grafton-Cardwell; T. Kapaun; D. Gumpf.

University of California, Riverside

The UCR Citrus Diagnostic and Research Laboratory consists of different laboratory spaces of several thousand square feet that allow sensitive diagnostic and research activities such as working with quarantined plant materials, nucleic acid extraction, setting up diagnostic reactions, performing final tests, and storing plant tissue and nucleic acid samples. The laboratory is equipped with all equipment necessary for the molecular and serological detection and characterization of graft-transmissible pathogens of citrus and is approved for working with recombinant organisms according to the National Institute of Health and UC guidelines. In addition, the laboratory has access to UCR core facilities for DNA sequencing and other genomics as well as protein and cellular analysis (UCR Institute for Integrative Genome Biology), the Statistical Consulting Collaboratory, and the greenhouse and experimental field plot infrastructure of the UCR Department of Plant Pathology and Microbiology, the Citrus Research Center (CRC), and the AES.

Lindcove Research and Extension Center, Exeter, CA

The CCPP Foundation Block Operations at the Lindcove REC (fig. 7.9) consist of

- an evaluation block of approximately 25 acres that currently contains over 1,300 trees representing over 300 different scion and rootstock varieties (see fig. 7.6)
- protected foundation blocks with approximately 76,000 square feet of protective structures (screen- and greenhouses) currently containing 1,000 trees that represent over 350 scion and rootstock varieties
- a citrus fruit evaluation laboratory with all equipment necessary for data collection, analysis, and online publication

Acknowledgments

We wish to thank the UCR, CRB, CCNB, CDFA, USDA, and NCPN as well as all the past and present staff and researchers of the CCPP and Lindcove REC for their invaluable and continuing support to the CCPP.

References

Bash, J. A. 1997. The California Citrus Clonal Protection Program. Proceedings of the 5th World Congress of the International Society for Citrus Nurserymen 323–327.

Bayer, R. J., D. J. Mabberley, C. Morton, C. H. Miller, I. K. Sharma, B. E. Pfeil, S. Rich, R. Hitchcock, and S. Sykes 2009. A molecular phylogeny of the orange subfamily (Rutaceae: Aurantioideae) using nine cpDNA sequences. American Journal of Botany 96(3): 668–685.

Bernad, L., and N. Duran-Vila. 2006. A novel RT-PCR approach for detection and characterization of citrus viroids. Molecular and Cell Probes 2:105–113.

Calavan, C. E., S. M. Mather, and E. H. McEachern. 1978. Registration, certification, and indexing of citrus trees. In W. Reuther, C. E. Calavan, and G. E. Carman, eds., The citrus industry. Vol. 4. Oakland: University of California Division of Agricultural Sciences. 185–222.

Calavan, E. C., C. N. Roistacher, and E. M. Nauer. 1972. Thermotherapy of citrus for inactivation of certain viruses. Plant Disease Reports 56:976–980.

Cameron, J. W., R. K. Soost, and H. B. Frost. 1959. The horticultural significance of nucellar embryony in citrus. In J. M. Wallace, ed., Citrus virus diseases. Berkeley: University of California Division of Agrcultural Sciences. 191–196.

Childs, J. F. L., ed. 1978. Indexing procedures for 15 virus diseases of citrus trees. Washington, DC: USDA Agricultural Research Service Agriculture Handbook. 333.

De Lange, J. H. 1978. Shoot-tip grafting: A modified procedure. The Citrus and Subtropical Fruit Journal 539:13–15.

Frison, E. A., and M. M. Taber, eds. 1991. FAO/IBPR Technical guidelines for the safe movement of citrus germplasm. Rome: FAO.

Garnsey, S. M., D. L. Zies, M. Irey, P. J. Sieburth, J. S. Semancik, L. Levy, and M. E. Hilf. 2002. Practical field detection of citrus viroids in Florida by RT-PCR. Proceedings of the 15th Congress of the International Organization of Citrus Virologists 219–229.

Gumpf, D. J. 1999. Citrus quarantine, California. In R. P. Kahn and S. B. Mathur, eds., Containment facilities and safeguards for exotic plant pathogens and pests. St. Paul, MN: APS Press. 151–156.

Gumpf D. J., J. A. Bash, G. Greer, J. Diaz, R. Serna, J. S. Semancik, and R. Gonzales. 1996. The California Citrus Clonal Protection Program. Proceedings of the 8th Congress of the International Society of Citriculture 1:445–447.

Hiltabrand, W. F. 1959. Certification program for maintenance of virus-free propagation sources of citrus in California. In J. M. Wallace, ed., Citrus virus diseases. Berkeley: University of California Division of Agrcultural Sciences. 229–231.

Kahn, R. P., et al. 1967. Incidence of virus detection in vegetatively propagated plant introductions under quarantine in the United States, 1957–1967. Plant Disease Reporter 51:715–719.

Kahn, T. L., R. R. Krueger, D. J. Gumpf, M. L. Roose, M. L. Arpaia, et al. 2001. Citrus genetic resources in California: Analysis and recommendations for long-term conservation. Report of the Citrus Genetic Resources Assessment Task Force. Oakland: University of California Division of Agriculture and Natural Resources Genetic Resources Conservation Program Report 22. GRCP website, http://grcp.ucdavis.edu/publications/doc22/contents.pdf.

Koizumi, M. 1984. Elimination of tatter leaf-citrange stunt virus from Satsuma mandarin by shoot-tip grafting following pre-heat-treatment. Proceedings of the 9th Conference of the International Organization of Citrus Virologists. 229–233.

Krueger, R. R. 1997. The California citrus certification program. Proceedings of the 5th ISCN International Congress. 303–305.

Lee, R. F. 2003. Certification programs for citrus. In S. A. M. Naqvi, ed., Management of fruit and vegetable diseases: Diagnosis and management. Vol. 1. Dordrecht, Netherlands: Kluwer Academic Publishers. 291–305.

Lee, R. F., P. Lehmann, and L. Navarro. 1999. Nursery practices, budwood and rootstock certification programs. In L. W. Timmer and L. Duncan, eds., Citrus health guide. St. Paul, MN: APS Press. 35–46.

Murashige T., W. P. Bitters, T. S. Rangan, E. M. Nauer, C. N. Roistacher, and P. B. Holliday. 1972. A technique of shoot apex grafting and its utilization towards recovering virus-free Citrus clones. HortSci 7:118–119.

Nacer, E. M., C. N. Roistacher, T. L. Carson, and T. Murashige. 1983. In vitro shoot-tip grafting to eliminate citrus viruses and virus-like pathogens produces uniform budlines. HortSci 18:308–309.

Nauer, E. M., E. C. Calavan, C. N. Roistacher, R. L. Blue, and J. H. Goodale. 1967. The Citrus Variety Improvement Program in California. California Citrograph 52:133, 142, 144, 146, 148, 151, 152.

Navarro, L. 1981. Citrus shoot-tip grafting in vitro (STG) and its applications: A review. Proceedings of the 4th Congress of the International Society of Citriculture 452–456.

Navarro, L., and J. Juarez. 1977. Elimination of citrus pathogens in propagative budwood. II. In vitro propagation. Proceedings of the 2nd Congress of the International Society of Citriculture 3:973–987.

Navarro L., C. N. Roistacher, and T. Murashige. 1975. Improvement of shoot-tip grafting in vitro for virus-free citrus. Journal of the American Society of Horticultural Science 100:471–479.

Palacio-Bielsa, A., X. Foissac, and N. Duran-Vila. 1999. Indexing of citrus viroids by imprint hybridization. European Journal of Plant Pathology 105:897–903.

Pfeil, B. E., and M. D. Crisp. 2008. The age and biogeography of *Citrus* and the orange subfamily (Rutaceae: Aurantioideae) in Australasia and New Caledonia. American Journal of Botany 95(12): 1621–1631.

Reuther, W. 1959. A program for establishing and maintaining virus-free citrus stock. In J. M. Wallace, ed., Citrus virus diseases. Berkeley: University of California Division of Agricultural Sciences. 215–217.

———. 1981. The Citrus Clonal Protection Program. California Agriculture 35:30–32.

Reuther W., E. C. Calavan, E. M. Nauer, and C. N. Roistacher. 1972. The California citrus variety improvement program after twelve years. Proceedings of the 5th Conference of the International Organization of Citrus Virologists 271–278.

Roistacher, C. N. 1977. Elimination of citrus pathogens in propagative budwood. I. Budwood selection, indexing and thermotherapy. Proceedings of the 2nd Congress of the International Society for Citriculture 3:965–972.

———. 1991. Graft-transmissible diseases of citrus: Handbook for detection and diagnosis. Rome: FAO.

Roistacher, C. N., and E. C. Calavan. 1972. Heat tolerance of preconditioned citrus budwood for virus inactivation. Proceedings of the 5th Conference of the International Organization of Citrus Virologists 256–261.

———. 1974. Inactivation of five citrus viruses in plants held at warm glasshouse temperatures. Plant Disease Reporter 58.

Roistacher, C. N., and S. L. Kitto. 1977. Elimination of additional citrus viruses by shoot-tip grafting in vitro. Plant Disease Reporter 61:594–596.

Roistacher, C. N., and E. M. Nauer. 1968. Frequency of virus infection in citrus budwood introduced into the United States. Proceedings of the 4th Conference of the International Organization of Citrus Virologists 386–391.

Roistacher, C. N., L. Navarro, and T. Murashige. 1976. Recovery of citrus selections free of several viruses, exocortis viroid, and *Spiroplasma citri* by shoot-tip grafting in vitro. Proceedings of the 7th Conference of the International Organization of Citrus Virologists 186–193.

Sieburth P. J., M. Irey, S. M. Garnsey, and R. A. Owens. 2002. The use of RT-PCR in the Florida citrus viroid indexing program. Proceedings of the 15th Conference of the International Organization of Citrus Virologists 230–239.

Timmer, L. W., S. M. Garnsey, and J. H. Graham, eds. 2000. Compendium of citrus diseases. St Paul, MN: APS Press. 51–69.

Vidalakis, G., J. V. da Graca, W. N. Dixon, D. Ferrin, M. Kesinger, R. R. Krueger, et al. 2010a. Citrus quarantine, sanitary and certification programs in the USA: Prevention of introduction and distribution of citrus pests. Part 1: Quarantine and introduction programs. Citrograph 1(3): 26–35. ———. 2010b. Citrus quarantine, sanitary and certification programs in the USA: Prevention of introduction and distribution of citrus pests. Part 2: Certification schemes and national programs; highlights of new California citrus nursery regulations. Citrograph 1(4): 27–39.

Vidalakis, G., S. M. Garnsey, J. A. Bash, G. D. Greer, and D. J. Gumpf. 2004. Efficacy of bioindexing for graft-transmissible citrus pathogens in mixed infections. Plant disease 88, 1328–1334.

Wallace, J. M. 1978. Virus and viruslike diseases. In W. Reuther, C. E. Calavan, and G. E. Carman, eds., The citrus industry. Vol. 4. Berkeley: University of California Division of Agricultural Sciences. 67–184.

Wallace, J. M., and R. J. Drake. 1959. An indexing program to avoid viruses in citrus introduced into the United States. In J. M. Wallace, ed., Citrus virus diseases. Berkeley: University of California Division of Agricultural Sciences. 209–214.

Weathers, G., and C. E. Calavan. 1959. Nucellar embryony: A means of freeing citrus clones of viruses. In J. M. Wallace, ed., Citrus virus diseases. Berkeley: University of California Division of Agricultural Sciences. 197–202.

Zhang T. 1996. Effective methods for the elimination of citrus tatter leaf virus by thermotherapy and shoot-tip grafting. Proceedings of the 13th Conference of the International Organization of Citrus Virologists 310–312.

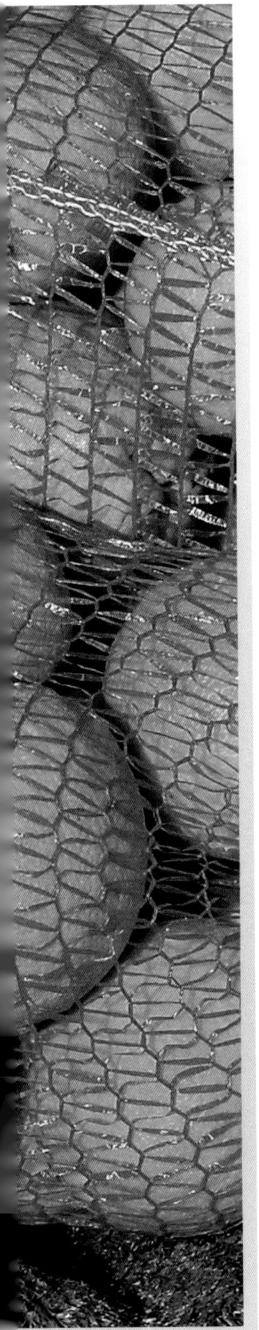

8 Soil and Water Analysis and Amendment

Blake Sanden, Allan Fulton, and Louise Ferguson

Three important factors that can seriously decrease citrus growth and yield can be evaluated only through a soil and water sampling program:

- potentially excessive levels of salinity
- specific ion toxicity, usually sodium, chloride, or boron, even though general salinity may be acceptable
- poor infiltration, leaching, or drainage

This chapter describes why these factors are important, how to conduct an effective sampling program, how to diagnose the results, and amendment strategies to correct potential problems.

Soil and Water Analysis

Sampling Philosophy

When sampling soils and water to diagnose and manage salinity problems, sampling can best be conducted prior to planning an orchard and continued annually or biannually for routine monitoring, or it can be conducted after the orchard has been established and only when necessary to troubleshoot problems. Sampling prior to planting an orchard is like doing a title search before you buy a house. It should always be done to avoid unwanted complications down the road.

Regardless of which approach you choose, soil and water sampling must be representative for the results to be of value. Although obtaining representative composite samples involves some effort and expense, it should not require more than about 6 hours of labor and $1,000 (about $12 per ac) for an 80-acre orchard.

Soil Sampling Procedure

Soil types, nutrients, and salts can vary considerably throughout an orchard. Sampling in 1-foot increments—that is, taking separate samples at depths of 0 to 1 foot, 1 to 2 feet, 2 to 3 feet, and so on—at the same time each year in the area of the orchard floor wetted by the irrigation system helps achieve informative, representative sampling. The most important factor when evaluating soil quality and managing salinity is consistency of sampling. You should take at least one composite soil sample for several depths in each area in the orchard having a similar soil type. USDA soil surveys are good starting points for targeting sampling areas. A minimum of one composite sample for each 40 acres should be taken. Figure 8.1 illustrates the possible soil variability for a 160-acre field in southern Tulare County. Excavating a backhoe

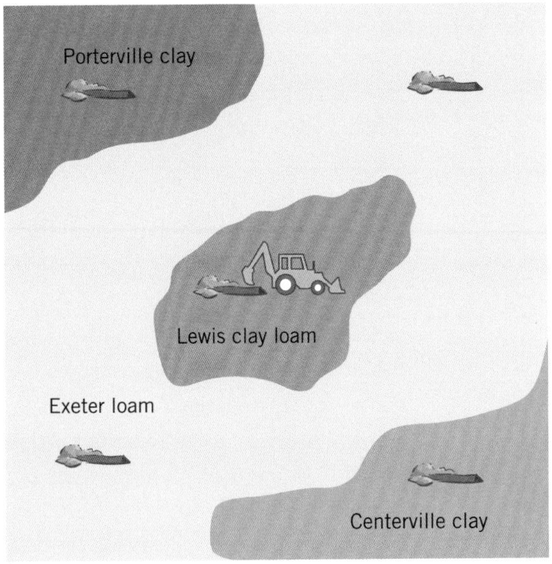

Figure 8.1 Possible soil variability and required sampling areas for developing a 160-acre citrus orchard.

pit to 6 or 7 feet deep allows a visual examination of the soil profile in each area to assess potential drainage problems and the depth of ripping required. Soil samples for each 1-foot increment can be taken from these pits but should be composited with other subsamples for a given soil type as described below.

Begin by sampling the soil in 1-foot increments to at least a depth of 4 feet in 6 to 12 locations (each location 20 to 60 ft apart) over similar soil types. A 2- to 3-inch auger works best. Use four buckets, one bucket for each depth. Mix all samples from the same depth in the bucket to form one sample for that depth. The result will be composite samples for that soil type. Take about 1.5 pounds of a moist sample to the laboratory the same day. If you have to wait a few days to submit your samples, air-dry the sample by rolling back the bag opening and setting it outside in a dry area exposed to the wind. Compositing minimizes the number of soil samples requiring analysis while achieving the most representative sample. If the soil is fairly uniform, then sampling 20-inch instead of 12-inch intervals is sufficient and reduces the total number of samples sent to the lab.

Repeat the soil sampling in existing orchards at the same time each season every couple of years to evaluate salinity and amendment management strategies with respect to rainfall patterns, irrigation scheduling, and water distribution around the tree. This ensures that salinity distribution and accumulation reflect the long-term trends occurring in the orchard and not just different levels of applied water and crop evapotranspiration (ET, or water use) for that year. Sampling after harvest provides a salinity assessment in an orchard when root zone salinity is usually highest. Irrigation is commonly delayed or deficit during harvest, allowing salts to accumulate in the root zone. Also, fall sampling gives the most advance notice that additional irrigation water may be needed for salinity control during the winter season, when the ground is cold and extra irrigation for leaching is less likely to cause problems with Phytophthora root rot.

The method of irrigation and the ability to apply water uniformly must be considered when collecting representative soil samples. In flood-irrigated citrus, sampling 5 to 10 feet to the side of a tree row is the best location. In sprinkler-irrigated orchards, sample across the sprinkler pattern to ensure that two-thirds of the composite locations are soils from the center of the wetting pattern, which receive the most applied water, and about one-third from the edges of the wetting pattern, where salts tend to accumulate. The same strategy applies to microsprinklers and drip irrigation.

Variable soil textures contribute to nonuniform salinity levels in an orchard. Sandy loam tends to have lower salinity than silt loam and clay loam because infiltration rates are often higher and leaching is greater in sandy loam. Sampling from similar soil types reduces variability in salinity levels. Uniform growth is a good indicator of uniform soil type and water availability.

It is important to take samples from several depths, because each depth reveals different information. If infiltration is a problem, a sample from 0 to 3 inches may be needed to diagnose crusting problems. Surface soils containing salinity and sodium levels far in excess of levels in the irrigation water strongly indicate that soil crusting may restrict water infiltration. Thus, the combination of irrigation water quality and soil type may present a problem. Sampling subsoils in increments of 12 to 20 inches to 4 or 5 feet usually reveals whether water penetration is a limiting factor by indicating a zone of much greater salt accumulation. If no such zone is found, the salinity over all depths can be averaged to determine the overall root zone salinity. This will provide an indication of the impact of salinity on ET and possible specific ion toxicities.

Water Sampling Procedure

Sampling irrigation water to assess salinity is much simpler than sampling soils. First, rinse a plastic container in the water that is to be sampled. Collect a small sample (4 to 8 oz). Completely fill the container with water; this eliminates air, which would otherwise promote calcium carbonate precipitation.

Before taking a sample from a well, let the pump run for at least 30 minutes. This should be sufficient time to flush the well of static water and establish the pumping water elevation that represents the primary water-bearing strata. If the orchard is surrounded by many other deep wells, sample in August, when ground water pumping levels tend to be deepest. Wells located in stable aquifers may not require annual sampling, but wells in declining aquifers require frequent sampling. If ground water depths are declining, collect a representative water sample for laboratory analysis to establish a baseline. Invest in an inexpensive portable electrical conductivity (EC) meter (about $70) and monitor the total salinity of the well water. Submit a new sample for analysis when the total salinity (EC_w) increases about 20%. Otherwise, analysis once every 3 years is sufficient.

To establish a baseline for surface water, take samples from water flowing in canals or ditches. Submit the sample for analysis and use this as a reference point. If possible, submit a water sample for analysis on the same day that it is collected. If the sample must be stored, refrigerate it to minimize changes in salinity. Storage at room temperature allows calcium (Ca) and bicarbonate (HCO_3) to precipitate and lower the total salinity level of the water. Usually, irrigation districts test for water quality and can supply you with the lab results. As described for ground water monitoring, use a portable electrical conductivity meter to determine how often the surface water should be analyzed.

Laboratory Analytical Reports

An analysis of an irrigation water sample or a saturated soil extract (the most common method of analyzing soil salinity in the western United States) measures dissolved salts, not exchangeable ions that are attached to clay particles in the soil. Exchangeable ions mostly affect soil structure. Laboratories can do a separate analysis for exchangeable ions, but most often they simply calculate the exchangeable sodium percentage (ESP) from the soluble analysis; the ESP is used in calculating the gypsum requirement. Table 8.1 gives the determinations usually provided in an analytical report relating to irrigation water or a soil extract. The following sections define terms that may appear in analytical results.

Saturation percentage (SP)

After the composite soil sample is dried, distilled water is added to make a saturated paste. The saturation percentage is the weight of water required to saturate the pore space divided by the weight of the dry soil. Saturation percentage is useful for characterizing soil texture. Very sandy soils have SP values of less than 20%; sandy loam to loam soils have SP values between 20 and 35%; and silt loam, clay loam, and clay soils have SP values from 35 to over 50%. Also, salinity measured in a saturated soil can be correlated to soil salinity at different soil-water contents measured in the field. As a rule, the SP soil-water content is about two times higher than the soil-water content at field capacity. Therefore, the soil salinity in a saturation extract (the water extracted from the sample and used for all the lab analyses described in table 8.1) is about half of the actual concentration in the same soil at field capacity.

pH

The pH of a soil or water measures hydrogen ion concentration (activity). Although pH is closely related to bicarbonate concentration and the availability of some macro- and micronutrients, it does not correlate with total salinity. It is, however, important when selecting the most appropriate soil amendments.

Table 8.1. Laboratory tests and units for evaluating irrigation water quality and soil salinity

Test	Symbol	Units	Water	Soil
saturation percentage	SP	%	no	yes
acidity/alkalinity	pH	none	yes	yes
electrical conductivity	Ec_e, Ec_w	dS/m	yes	yes
total dissolved solids	TDS	mg/l	yes	no
calcium	Ca^{2+}	meq/l	yes	yes
magnesium	Mg^{2+}	meq/l	yes	yes
sodium	Na^+	meq/l	yes	yes
bicarbonate	HCO_3^-	meq/l	yes	yes
carbonate	CO_3^{2-}	meq/l	yes	yes
chloride	Cl^-	meq/l	yes	yes
sulfate	SO_4^{2-}	meq/l	yes	yes
boron	B	mg/l (ppm)	yes	yes
nitrate-nitrogen	NO_3-N	mg/l (ppm)	yes	yes
sodium adsorption ratio	SAR	none	yes	yes
adjusted sodium adsorption ratio	SAR_{adj}	none	yes	no
exchangeable sodium percentage	ESP	%	no	yes
gypsum requirement	GR	ton/ac-6 in	no	yes
lime requirement	LR	ton/ac	no	yes
lime percentage	$CaCO_3$	%	no	yes

Electrical conductivity (EC)

The electrical conductivity (EC_e for extracts from soil, EC_w for irrigation water) is a measure of total salinity based on how easily an electric current passes through the extract, but it does not give any indication of the salt composition. However, EC is one of the most important numbers on an analysis because nearly all crop salt tolerance levels are based on EC. The internationally accepted reporting unit for EC is deciseimens per meter (dS/m). This unit is equal to millimhos per centimeter (mmhos/cm), which is still used by some labs. Many labs that do environmental and agricultural testing often report EC_w in micromhos per centimeter (μmhos/cm); divide these values by 1,000 to obtain dS/m or mmhos/cm.

Total dissolved solids (TDS)

The total dissolved solids (TDS) is the weight of all soluble salts in milligrams per liter of water (mg/l; about 640 mg/l = 1 dS/m). TDS is not useful in evaluating salinity, because crop tolerance thresholds are correlated with EC_e and EC_w rather than with TDS.

Salts

Salts such as sodium chloride (NaCl) and calcium sulfate ($CaSO_4$) consist of positively charged cations and negatively charged anions bonded by opposing charges. In irrigation water or soil-water, many of the bonds are broken, and the water salinity consists of individual cations and anions. To understand the impact of salinity on soil structure and crop tolerance, soil-water and irrigation water samples must be analyzed for these soluble cations and anions. Calcium (Ca^{2+}), magnesium (Mg^{2+}), and sodium (Na^+) are the major cations in soil extracts and irrigation water. Although soluble potassium (K^+) is important as a nutrient, it is usually a very minor component of salinity. Bicarbonate (HCO_3^-), sulfate (SO_4^{2-}), and chloride (Cl^-) are the main anions in most soil extracts and irrigation water. Along with these anions, boron (B) and nitrate-nitrogen (NO_3-N) are commonly reported in lab results. Boron does not contribute significantly to total salinity and the osmotic effects on soil-water availability, but it is important in diagnosing specific ion toxicity. Knowing the nitrate-nitrogen content of the soil and irrigation water is valuable in making fertilizer decisions, but nitrate does not usually contribute significantly to the total salinity.

Salinity (meq/l)

The preferred unit for reporting individual cations and anions is milliequivalents per liter (meq/l). This unit is used specifically in salinity evaluation and reporting. Many agriculturists who work with pesticides, fertilizers, and tissue analyses are familiar with parts per million (ppm) and milligrams per liter (mg/l), which are equal but may be less familiar with meq/l. Reporting all ions in meq/l provides the best comparison of the reactive strength of the different cations and anions. Clay particles in soil are negatively charged and adsorb the positively charged cations. It is the concentration of the charges, not the weight of the ions, that affects soil structure and eventually the decisions you will make on soil amendments. Reporting cation and anions in meq/l is one of the hallmarks of a quality lab. These numbers are easily converted back to the weight of the various salts, which is needed when calculating the tonnage of required amendments (table 8.2).

Table 8.2. Conversion factors from meq/l to mg/l or ppm

Cation/anion	Symbol	Conversion factor
bicarbonate	HCO_3^-	61
calcium	Ca^{2+}	20
carbonate	CO_3^{2-}	30
chloride	Cl^-	35
magnesium	Mg^{2+}	12
sodium	Na^+	23
sulfate	SO_4^{2-}	48

Note: mg/l = meq/l × conversion factor.

SAR

The unadjusted sodium adsorption ratio (SAR), adjusted sodium adsorption ratio (SAR_{adj}), and exchangeable sodium percentage (ESP) are calculated from the individual cation and anion determinations. These indices must be used along with the EC values to evaluate salinity and sodicity accurately.

The unadjusted SAR indicates the ratio of sodium to calcium and magnesium in a soil-water extract or irrigation water sample. An increasing SAR value indicates higher levels of sodium in comparison with calcium and magnesium. Rising levels of sodium reduce soil stability, decrease water infiltration, and increase the chance that sodium will accumulate to toxic levels in leaf tissue. Use the SAR, rather than sodium cation levels, to evaluate sodicity problems. More sodium can be tolerated in a soil extract or water sample when calcium increases proportionally with sodium.

The SAR_{adj} is calculated and reported only for water samples. This index predicts the reaction of bicarbonate with calcium when water is applied to the soil. Irrigation water with low levels of bicarbonate or carbonate anions usually has a SAR_{adj} that is very similar to its unadjusted SAR. This water will very slowly dissolve lime from the soil and contribute calcium to offset sodium in the soil-water. Irrigation water high in bicarbonate or carbonate usually has a SAR_{adj} higher than its unadjusted SAR. Such water precipitates calcium with bicarbonate and forms lime, which reduces free calcium levels in the soil-water and increases the proportion of sodium. Prior to 1988, SAR_{adj} was calculated according to an empirical equation using pH constants (pH_c). This has since been proven to overestimate SAR_{adj}. The new procedure is based on the proportion of calcium and bicarbonate in a water sample. Debate still exists over the real value of using SAR_{adj} and many labs now use only SAR.

ESP is closely related to SAR. These ratios differ in that SAR is an index of the water-soluble sodium compared with water-soluble calcium and magnesium, but the ESP is an indicator of exchangeable sodium bound to the clays in the soil. Today, most laboratories do not measure ESP directly because this would require the extra expense to measure the cation exchange capacity and the exchangeable sodium content. Instead, most laboratories report an estimated ESP based on a correlation between SAR and ESP. This correlation is fairly accurate for most soils, but it can substantially underestimate ESP for very sandy, low-salinity soils.

Gypsum requirement (GR)

Most laboratory reports of a soil analysis provide either a gypsum requirement (GR) or a lime requirement (LR). A gypsum requirement is usually provided for alkaline soils with pH above 7 and SAR above 10 to 15. The most common method of determining the gypsum requirement is the Schoonover test. It measures how much soluble calcium must be added in the form of gypsum to replace nearly all of the sodium on the soil exchange sites. A lime requirement is made only for acid soils with a pH less than 7.

Verification of Laboratory Results

Check the quality of an analytical report before making management decisions based on the results. A good report will give the cation and anion concentrations in meq/l, which is a measure of the reaction potential and not the mass of the salt. The accuracy of the analysis can be evaluated by checking that the total meq/l of cations roughly equals the total anions and that the total cations or anions should be about 10 times the EC. Table 8.3 is a sample analytical report. The next two paragraphs refer to table 8.3 to show how to check the quality of an analysis.

Table 8.3. Sample irrigation water quality analysis

Analysis	Example
pH	8.4
EC_w	1.0 dS/m
Ca	0.5 meq/l
Mg	0.1 meq/l
Na	9.6 meq/l
HCO_3	4.2 meq/l
CO_3	1.0 meq/l
Cl	4.6 meq/l
SO_4	0.1 meq/l
B	0.7 mg/l
NO_3	5.2 mg/l
SAR	17.5
SAR_{adj}	16.6

Check 1: Cation-anion balance method

Salts such as NaCl, $NaHCO_3$, and $CaSO_4$ consist of cations and anions bonded together by electrical charges. For each cation there is an equivalent charge (meq) of an anion bonded to form the salt. This is referred to as the cation-anion balance. Using table 8.3,

$$Na + Ca + Mg \approx HCO_3 + CO_3 + SO_4 + Cl$$

$$10.2 \text{ meq/l} \approx 9.9 \text{ meq/l}.$$

When dissolved in a water sample or soil extract, the bonds are broken and the salts exist as individual cations, anions, or neutral ion pairs. The individual cations and ions must be reported in meq/l to perform this check. Omit boron and nitrate when conducting this check because they are reported in mg/l and are usually an insignificant amount of the total salinity.

Check 2: Comparing total salinity with the sum of cations or anions

In a valid analysis, the salinity level (EC_w) multiplied by a factor of 10 about equals the sum of the cations or anions. Using table 8.3,

$$EC_w \times 10 \approx (HCO_3 + CO_3 + SO_4 + Cl)$$

$$1.0 \times 10 \approx (4.2 + 1.0 + 0.1 + 4.6)$$

$$10 \approx 9.9.$$

Beware of a report in which the EC multiplied by 10 exactly equals the sum of either the cations or anions, or one in which the cations exactly equal the anions. Such a result may indicate that one of the individual cations or anions was estimated by subtraction rather than determined by direct measurement. Sulfate and sodium are the most likely components to be estimated, because measuring them requires additional analytical steps and expense.

Diagnosing Salinity Problems

As mentioned at the start of this chapter, salinity analyses are used to diagnose three types of conditions in the field: excess root zone salinity, specific ion toxicity, and poor water infiltration.

Citrus Salt Tolerance and Potential Yield Loss

A high EC_e or EC_w value indicates high salinity. Excessive salts reduce the amount of water that plant roots can absorb from the soil and thus reduce growth. Trees grown in saline soil may show symptoms of water stress even though the soil may appear or feel as though it contains sufficient water. Crop evapotranspiration (ET) is directly related to vegetative growth. As water vapor leaves the open stomata in the leaf surface, carbon dioxide (CO_2) enters for the production of carbohydrates. When levels of salts in the root zone are too high, the trees may display inadequate shoot growth, reduced fruit size, excessive leaf drop, and an unthrifty appearance. Brown to black necrotic or dead tissue along leaf tips and margins usually indicates excess salt accumulation.

Citrus is in the sensitive category for salt tolerance. The most commonly quoted threshold soil EC_e where yield decline begins is 1.7 dS/m (1.1 dS/m for EC_w), with a 16% relative yield loss for each additional 1 dS/m above that level (Maas and Hoffman 1977). Shalhevet and Levy (1990) compiled salt tolerance data from nine studies from California, Israel, and Australia and got a very similar result. Figure 8.2 shows that both results indicate a 50% decline in yield at a soil EC of around 5 dS/m. To estimate the relative yield (RY) where soil EC_e equals 2.5 dS/m,

$$\begin{aligned} RY\ (\%) &= 100 - \text{decline}\ \% \times (EC_e - \text{threshold}) \\ &= 100 - 16 \times (2.5 - 1.7) \\ &= 87.2\%. \end{aligned}$$

These limits assume that irrigation is managed to provide an annual leaching fraction to prevent increasing salt accumulation (for more information on the leaching fraction, see "Maintaining Acceptable Root Zone Salinity," below).

Toxic Accumulation of Specific Ions

The accumulation of an ion to the point of toxicity can take several years. In this regard, sodium, chloride, and boron are the primary ions of concern. Trees grown on soils with an excess of these elements can accumulate ions in the woody tissue and eventually in the leaves. These symptoms and critical levels for citrus have been well documented for decades. Burn on leaf margins often means excess chloride or sodium in leaf tissue. The margins of foliage containing excess boron may develop leaf burn that expands into interveinal necrosis, with twisting and curling and often a black tip at the end of the leaf. Gummosis, the oozing of sap in excessive amounts from the trunk of the tree, can also occur. Accumulation of sodium, chloride, or boron ions is likely to reduce production of necessary plant hormones and contribute to nutritional disorders.

In general, citrus is in the sensitive category for specific ion toxicity. However, citrus rootstocks vary in their ability to exclude sodium, chloride, and boron, and some rootstocks are more tolerant than others. The most tolerant rootstocks may withstand up to 20 meq/l chloride or sodium in the soil saturation extract, while the most sensitive might show symptoms at 8 to 10 meq/l. Citrus is extremely sensitive to boron, and tolerance in rootstocks varies from about 0.5 to 1.0 ppm boron in soil concentrations (sprinkling citrus foliage with irrigation water at these concentrations would cause significant burn). Table 8.4 shows the tolerance to these ions for some common citrus rootstocks. Keep in mind that tolerance to alkaline pH levels and the ability of roots to extract micronutrients from the soil is different from chloride, sodium, and boron tolerance. For example, Trifoliate Orange does not perform well on soils with pH greater than 7.5 even if salinity is not a problem.

It is important to diagnose ion accumulation before levels become elevated in the woody and leaf tissues. Once ions accumulate there, the trees have no mechanism to rapidly expel them. Correcting the toxicity in the root zone may require several seasons of proper irrigation management. Analysis of soil, irrigation water, and leaf tissue can diagnose conditions in which toxic ion effects may become a problem.

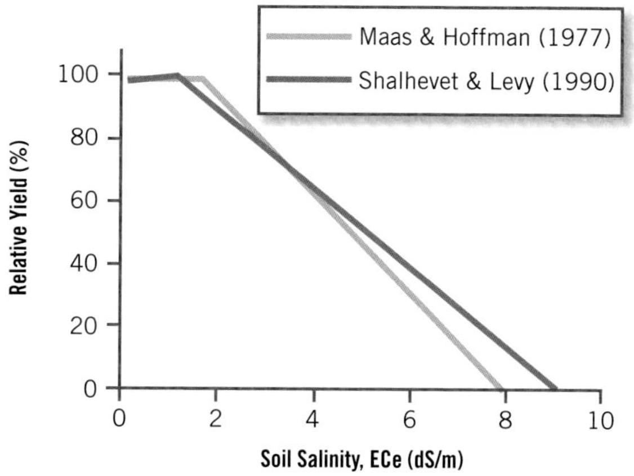

Figure 8.2 Relative yield (RY) of selected citrus varieties and rootstocks as a function of soil EC_e.

Table 8.5 gives guidelines for ranges of soil and water salinity and sodium, chloride, and boron. If soil or water analysis indicates levels in the "Increasing" range, carefully monitor leaf tissues. It may be only a matter of time before one or more ions accumulate to toxic levels in the trees. Soil and water analyses over time should show declining or steady levels of these ions. If they increase in concentration, rethink your amendment strategy and irrigation scheduling. Table 8.6 gives critical levels for sodium, chloride, and boron in leaf tissue.

Impacts on Crop Evapotranspiration

Water can enter plant roots only through the process of osmosis, where the concentration of solutes in the root sap is higher than the salt concentration of the soil-water. As the salt concentration in the root zone increases, so does osmotic stress (a moist saline soil can produce the same level of plant stress as a drier nonsaline soil). Consequently, crop water use, or evapotranspiration (ET), can be substantially reduced when irrigating with marginally saline water. This reduction may be 5 to 40%, depending on the soil, type of salt, and irrigation system. Fine-tuned irrigation scheduling is even more important when using saline irrigation water (for more information, see chapter 11, "Irrigation").

Nitrate-Nitrogen Toxicity

Nitrate (NO_3^-) toxicity becomes a concern when too much nitrogen fertilizer is applied. If overapplication is severe, the first effect may be defoliation, but the trees will most likely regrow with tremendous vigor. Foliage may grow so large, in fact, that the leaves curl. Use soil and water analyses to avoid overuse of fertilizer and to achieve efficient nitrogen management practices.

Most labs report the amount of nitrogen in the form of nitrate, hence the term is written NO_3-N. This makes it easy to convert the number on the report into the pounds of nitrogen potentially available for crop nutrition. For irrigation water, a nitrate-nitrogen level from 0 to 3 mg/l is considered low; from 3 to 10 mg/l is moderately low to moderately high; and above 10 mg/l is high. For soil samples taken from 0 to 1 foot, a nitrate-nitrogen level from 0 to 10 mg/l is considered low; 10 to 20 mg/l is moderately low to moderately high; and a level exceeding 20 mg/l is high.

To convert the level of nitrate-nitrogen in a water analysis to pounds of nitrogen per acre-foot of water, multiply the concentration reported in mg/l by 2.7 (2.7 million lb of water in 1 ac-ft.) For example, if the analysis reports that a sample of water contains

Table 8.4. Citrus rootstocks ranked in order of decreasing tolerance to chloride, sodium, and boron accumulation in scion

Tolerance	Chloride	Sodium	Boron
	Cleopatra	Cleopatra	Sweet Orange
higher	Rough Lemon	Rough Lemon	Troyer Citrange
	Citrumelo 4475	Sweet Orange	Cleopatra
	Trifoliate Orange	Citrumelo 4475	Trifoliate Orange
	Sweet Orange	Troyer Citrange	Citrumelo 4475
lower	Troyer Citrange		Rough Lemon
	Carrizo Citrange		Sweet Orange

Sources: Adapted from Maas 1992 and Ayers and Westcott 1985.

Table 8.5. Guidelines for soil and water salinity and specific ions for mature citrus trees; values assume foliage is not sprinkled with irrigation water

Criterion	Unit	Degree of restriction for oranges		
		None	Increasing	Severe
root zone EC_e	dS/m	< 1.7	2–5	> 5
irrigation water EC	dS/m	< 0.8	0.8–2.0	> 2
irrigation water Na	meq/l	< 5	5–25	> 25
irrigation water Cl	meq/l	< 5	5–25	> 25
irrigation water B	meq/l	< 0.5	0.5–1.5	> 1.5

Source: Adapted from Ayers and Westcott 1985.

Table 8.6. Critical levels of specific ions in citrus leaf tissue

Specific ion	Degree of possible toxicity		
	None	Increasing	Excess
sodium (%)	< 0.16	0.17–0.24	> 0.25
chloride (%)	< 0.3	0.4–0.6	> 0.7
boron (mg/l)	< 100	101–260	> 260

Source: Reisenauer 1978.

2.3 mg/l of nitrate-nitrogen, the sample contains 6.2 pounds of nitrogen per acre-foot of water.

Similarly, you can convert the level of nitrate-nitrogen reported in a soil analysis to pounds of nitrogen per acre-foot of soil. In the lab report, find the level of nitrate-nitrogen (ppm) in a composite sample that represents all samples taken to a depth of 1 foot. Multiply the number by 4 (about 4 million lb of soil in 1 ac-ft). For example, if the analysis reports that the sample contains 9.7 mg/l of nitrate-nitrogen, the sample contains about 39 pounds of nitrogen per acre-foot. Make sure that the lab is reporting nitrate-nitrogen as ppm in dry soil and not the nitrate-nitrogen concentration of the saturation extract. The latter requires a different calculation to estimate the pounds of nitrogen per acre-foot of soil.

Reclaiming Saline Soils and Managing Salinity

The first part of this discussion deals with reclamation, correcting an existing salinity problem, assuming that an orchard or potential orchard site has excess salinity, toxic levels of specific ions, poor infiltration rates, or some combination of these problems. The second part of this discussion focuses on maintaining acceptable levels of soil salinity, assuming that the orchard has been established on a soil already suitable for citrus production. The goal is to manage the soil and irrigation water to avoid salinity buildup and the accumulation of toxic ions and maintain acceptable infiltration rates. Whether the objective is reclamation or maintenance, proper management of irrigation water is the key. In some situations, application of soil and water amendments will also be needed. The following general concepts are critical to effective reclamation or salinity maintenance:

- **Drainage.** At least 3 feet of aerated, low-salinity root zone is required for optimal citrus production. This is most easily maintained where there is no impermeable clay layer within 8 to 12 feet of the surface that might prevent drainage below the root zone. Hardpans at 3 to 4 feet are acceptable as long as they have some fracturing and allow for drainage. Salinity control is more difficult on poorly drained soils due to waterlogging and the much slower movement of salts below the root zone. Impermeable clay layers can create shallow, or perched, water tables that retain salt and usually rise and fall throughout the season. As a result, salinity that may be leached below the root zone in fall by fresh water irrigation during the growing season often "subs" back into the root zone in the spring when the water table rises. Citrus should not be grown on ground subject to perched water tables.
- **Recharging root zone to field capacity.** The soil-water content must exceed field capacity throughout the root zone before significant leaching can occur. Irrigation that simply replaces ET does not transport salts below the root zone.
- **Impact of irrigation frequency and method.** Small quantities of water (a depth of 1 to 2 in per application) applied frequently over several days are more effective for moving salinity below the root zone than an equal depth of water applied in one large application. Such one-time applications tend to percolate rapidly through the larger cracks and pores in the soil; much less water percolates through the small pores, and the result is a much higher salt concentration remaining in the small pores. Sprinkler, microirrigation, or rainfall allow more time for salts to diffuse from the small pores into the water moving through the big pores and ultimately to leach out of the root zone. Leaching by this method is most effective during the winter, when surface evaporation from the soil and crop transpiration are the lowest. Drip irrigation that wets a minimum of soil on the orchard floor can be effective for leaching during the summer, but it can build salt "shells" at the outer limits of the drip subbing pattern.

Reclamation: Reducing Soil Salinity

The depth of leaching required for reclamation is an estimate of the amount of water needed to reduce soil salinity to a level that will not cause yield loss. It is commonly expressed as inches of water required per foot depth of soil in the root zone. Reclamation is needed when excessive salts restrict production in existing citrus orchards or the salinity is higher than acceptable tolerance levels in a soil being considered for development of a new planting. Table 8.7 shows the depth of leaching required to reclaim various soils. The guidelines in table 8.7 assume that the moisture profile is full at the start of leaching and that the leaching requirement is applied in several small irrigations or rains separated by 2 or more days of drainage. Any water lost to evaporation must be estimated and added to the amounts shown in the table. The guidelines also apply to leaching chloride and sodium. Leaching boron is a special case, requiring as much as five times (depending on soil type) the water needed to leach other salts. If leaching is attempted in one irrigation with a large application of water, the leaching efficiency will decline, and it may require more than three times the quantity of water to achieve the same level of reclamation. If the field overlies a perched water table, care must be taken to avoid complete saturation of the root zone. When saturation occurs it is easier for salts from the lower depths to rise back up into the root zone through capillary action as evaporation pulls water and salts back toward the soil surface.

Example: Calculating water rates

Assume that a grower is considering a parcel of land for orchard establishment. Laboratory analyses of the irrigation water indicate excellent quality, with an EC of 0.4 dS/m. The average root zone salinity prior to any land preparation is 6 dS/m. The grower would like to have a soil salinity level of less than 1.0 dS/m to a depth of at least 5 feet. Table 8.7 shows that 6.0 inches of water per foot of root zone (in addition to the water needed to bring the soil to field capacity) is required to drop the salinity to 1 dS/m, or 30 inches total to reclaim all 5 feet of soil in the root zone. A root zone EC_e of 1.5 dS/m is probably acceptable. Reclamation to this level only requires 3.6 inches per foot, or 18 inches, for the 5-foot root zone. In practice, some of this leaching can be done after planting (assuming good drainage) in the form of supplying excess water with each irrigation. This can be especially effective with drip irrigation, where only a fraction of the whole root zone (usually 20 to 40%) is being reclaimed.

Maintaining Acceptable Root Zone Salinity

Root zone salinity increases when salts are transported into the orchard with irrigation water. The only way of decreasing or maintaining salinity levels is by moving salts out of the root zone with deep percolation. This important function of irrigation is referred to as leaching.

The leaching requirement for maintaining an acceptable salinity level is the fraction of infiltrating water that is not used to refill root zone water storage or for crop ET but instead percolates below the root zone. It is expressed as a percentage rather than as a specific quantity so discussion of the leaching fraction (LF) can be applied to orchards with various water requirements and water qualities. As the quantity of applied water increases, or as the concentration of the salts in the water increases, more salinity is transported into the orchard. Therefore, more leaching is required to push salts below the root zone. Variations in irrigation water quality and soil salinity create the need for different leaching fractions from one orchard to the next. Table 8.8 gives leaching fractions required for irrigation water qualities from 0.4 to 1.6 dS/m to adequately maintain two root zone salinity levels.

As an example of calculating the volume of irrigation required to supply a given leaching fraction, assume that mature citrus in the San Joaquin Valley on a nonsaline soil in a clean-cultivated orchard consumes about 34 to 36 inches of water annually. The irrigation water supply has an EC_w of 0.8 dS/m, and the goal is to maintain an average root zone EC_e of 1.5 dS/m. Table 8.8 shows that a leaching fraction of 15% is required.

Table 8.7. Depth of leaching water required per foot of root zone to be reclaimed given the initial average soil salinity and final desired salinity

Desired root zone salinity (dS/m)	Water required (in/ft of root zone) to leach initial salinity (dS/M)			
	2 dS/m	4 dS/m	6 dS/m	8 dS/m
1.0	1.2	3.6	6.0	8.4
1.5	0.4	2.0	3.6	5.2
2.0	0.0	1.2	2.4	3.6

Source: Adapted from Hoffman 1986.

Note: Applicable for all irrigation water less than 1.0 dS/m.

$$\text{Required seasonal irrigation} = ET \times (1 + LF)$$
$$= 35 \times (1 + 0.15)$$
$$= 40.2 \text{ in.}$$

If the EC_w of the irrigation water were 1.6 dS/m, the leaching fraction would be about 45%,

$$\text{Required seasonal irrigation} = 35 \times (1 + 0.45)$$
$$= 50.8 \text{ in.}$$

In practice, adding irrigation water to make up for the inefficiencies and nonuniformity of the irrigation system will produce a leaching fraction of 10 to 15%. If the trees have a full postharvest irrigation, then winter rainfall can supply another 5 to 10%.

Unless your trees begin to decline in the middle of the season and your soil and irrigation water salinity are high (> 1.5 dS/m), it is usually sufficient to apply water to meet normal crop water use (ET) plus some extra for irrigation system nonuniformity. For a new border system with tailwater return, this may be 15 to 25% more water than actual ET; for drip irrigation, 10 to 17%; and for microsprinkler, 6 to 15%. (For more information on distribution uniformity, see chapter 11, "Irrigation.") When using this approach, root zone salinity may increase some during the season, but it should return to acceptable levels with postharvest irrigation, winter rainfall, and a light spring irrigation. Experience has shown this to be the best program when combined with a continuing sampling program as outlined below.

The Need for Resampling

Every orchard has many unique factors that impact tree response to water and salinity management. These include nonuniformity of applied irrigation, imperfect scheduling, variable ET of the trees, extended harvest cutoff, and variable interaction of soils, irrigation water quality, and fertilizer application. Thus, the effectiveness of the above guidelines can be verified only by resampling the soils on an annual or biannual basis. Repeated sampling will confirm that your salinity management strategy is on track or indicate that adjustments must be made. Monitoring field moisture levels during the season should always be your first checkpoint.

Table 8.8. Leaching fraction required to maintain a specific level of root zone salinity with increasing salinity in the irrigation water

Irrigation water EC (dS/m)	Leaching fraction (%) required to maintain root zone EC_e of	
	1 dS/m	2 dS/m
0.4	7	2
0.8	23	7
1.2	44	14
1.6	70	23

Source: Adapted from Hoffman 1996.

Note: These leaching fractions are suited for soils that have already been reclaimed and have been under continuous cultivation for several years. The leaching requirement is then basically the same as the leaching fraction when crop salt tolerance is known.

Water and Soil Amendments

Soils with poor infiltration rates can usually be improved when treated with amendments. For these soils, the primary purpose of soil amendments is to improve soil structure for better percolation of water, leaching of salts and toxic ions such as sodium and boron, and improving aeration. Adjusting soil pH is also an important consideration for many applications. Although organic matter and crop rotation play a significant role in improving soil structure and fertility, the following discussion of soil amendment focuses on inorganic amendments such as gypsum that make calcium available to improve soil structure. Water amendments are discussed later in this section.

Improving Water Penetration

Amendments can be applied to the soil to increase the concentration of calcium ions attached to exchange sites in the soil in order to displace sodium and, in some instances, magnesium and sodium. Sodium causes swelling and dispersion of soil aggregates during irrigation. Exchangeable calcium stabilizes and improves aggregation, increasing soil porosity and improving infiltration. Soils that seal up are usually dominated by sodium. Inorganic soil or water amendments such as gypsum contain free calcium (and usually sulfate), which can displace excess sodium when dissolved by irrigation water. Acids and sulfur provide free calcium by dissolving existing lime in the soil.

The balance between EC and SAR determines at what level sodium becomes a problem. As EC increases, higher SAR values (more sodium) can be tolerated before causing excessive soil dispersion and sealing (fig. 8.3). SAR values of 15 to 25 are often found on saline Westside cotton soils irrigated with brackish water; infiltration is not reduced on these soils. Citrus, however, requires much lower levels of salinity and also a lower SAR.

Irrigation water that is too low in salts (EC < 0.2 dS/m) can often cause problems with dispersion and soil sealing (fig. 8.4). With water above this level, a good rule of thumb is to keep the SAR less than five times the EC; if this condition is met but the root zone salinity is still excessive, leaching may be the most appropriate first step toward correcting the salinity problem before adding amendments to improve the balance of salts and increase infiltration rates. If improving infiltration is your only concern, additional leaching is unnecessary, but if the soil EC_e is too high, leaching must follow the amendment application.

Water Amendments

The two general types of amendments are calcium salts and acid-forming amendments. Calcium salts are direct suppliers of calcium, and acid-forming amendments are indirect calcium suppliers through the breakdown of native soil lime that usually exists in high-pH soils. Common calcium salts include gypsum ($CaSO_4 \cdot 2H_2O$), lime (also called calcite, $CaCO_3$), dolomite ($CaMg(CO_3)_2$), calcium chloride ($CaCl_2$), and calcium nitrate ($Ca(NO_3)$). Each salt has a specific solubility in water. Calcium nitrate and calcium chloride are highly soluble; gypsum is moderately soluble; and dolomite and lime are only slightly soluble when pH is greater than 7.2. Calcium chloride can be used on sandy soils where chloride is less than 4 meq/l, but other calcium sources are usually cheaper.

Applying the highly soluble salts through the irrigation water is convenient but typically more expensive. High-grade product and special injection equipment are usually needed for this method (figs. 8.5 and 8.6). Gypsum is reasonably simple to add to irrigation water and is less expensive than calcium chloride and calcium nitrate per unit of calcium. Lime and dolomite are unsuitable as a water-run amendment, as they are nearly insoluble. Gypsum, calcium chloride, or calcium nitrate have negligible effects on soil pH. Lime or dolomite is used to increase the pH of acid soils.

Sulfur (S), sulfuric acid (H_2SO_4), urea sulfuric acid (N-pHuric, $H_2NCONH_2 \cdot H_2SO_4$), Nitro-Sul ($(NH_4)_2S_2$), and lime sulfur ($CaS_x + CaSO_x \cdot XH_2O$) are common acid-forming amendments used in salinity management. Since all contain sulfur or sulfuric acid but no calcium, they supply exchangeable calcium indirectly by dissolving lime that is native to the soil. The sulfur compounds undergo microbiological reactions that oxidize it to sulfuric acid. The acid dissolves soil lime to form a calcium salt (gypsum), which then dissolves in the irrigation water to provide exchangeable calcium. The acid materials do not have to undergo the biological reactions but react immediately with soil lime on application. Acid-forming amendments can also increase the

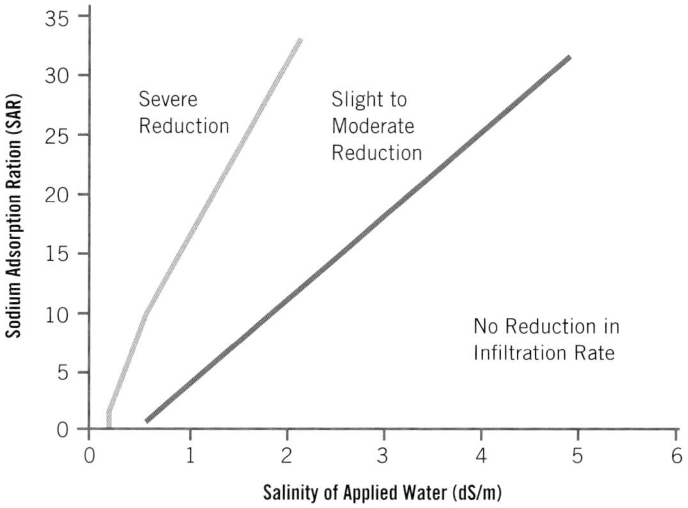

Figure 8.3 Effect of total salinity as EC and the SAR of applied water on infiltration. Figure based on Ayers and Westcott 1985.

Figure 8.4 Surface sealing of Wasco sandy loam due to very low salinity in soil and irrigation water. *Photo:* B. Sanden.

Figure 8.5 Gypsum injection machine attached to large silo for automatically adding bulk solution-grade gypsum. *Photo:* B. Sanden.

Figure 8.6 This batch mixer with vertical mixing bar is Suitable for dissolving some fertilizers for injection as well as solution-grade gypsum. *Photo:* B. Sanden.

availability of calcium in irrigation water by neutralizing bicarbonate and CO_3 carbonate that otherwise tie up some of the calcium to form lime precipitates. Since these amendments form an acid in the soil reaction, they can reduce soil pH if applied in sufficient quantity.

Water Amendment Selection

Selecting a soil amendment to reclaim a new orchard or maintain acceptable infiltration in an existing orchard depends largely on the level of lime in the soil and the relative cost of the materials. As long as lime is abundant in the soil (particularly the surface soil), consider a calcium salt or an acid-forming amendment. The choice of amendment depends on the following parameters:

- cost
- how quickly a response is desired
- the depth to which reclamation is desired
- the degree of soil pH correction desired

Another factor influencing the choice of an amendment is the anion that will be added to the root zone. Amendments can add sulfate ($SO^{2-}{}_4$), chloride (Cl^-), or nitrate (NO_3^-); the nitrate and chloride content of these materials limits the amount that can be added to a soil. The amount of nitrogen added should not exceed annual crop needs. Adding amendments is not likely to increase concentrations of chloride and sulfate to toxic levels in field soil, but it may in nursery plantings.

Do not use acid-forming materials if the soil zone lacks significant amounts of lime. Such a soil is neutral or acidic, so only calcium salts should be used. As soil acidity increases, adding lime or dolomite is preferable to adding gypsum and calcium chloride, especially where the pH is less than 6.8. Acid-forming amendments reduce soil pH when applied in sufficient quantity. The amount of amendment needed to do this will depend largely on the percentage of lime in the soil. Micronutrient nutrition can be compromised when the soil pH exceeds 8. Acid-forming amendments may be more effective on alkaline soils (those with a pH above 8) than calcium salts.

Table 8.9 lists the amounts of selected amendments needed to supply equal amounts of sodium to soil and calcium to irrigation water.

Water Amendment Rates

Amendments are most often added to water to improve water infiltration into the surface soil. Amendment rates from 1.0 to 3.0 meq/l calcium are considered low to moderate; rates from 3.0 to 6.0 meq/l calcium are considered moderate to high. For example, table 8.9 indicates that an application rate of 234 pounds of pure gypsum, or 133 pounds of pure sulfuric acid, per acre-foot of water supplies the equivalent of 1.0 meq/l calcium, assuming that enough lime is in the surface soil to react with the sulfuric acid.

Example: Calculating gypsum rates

To demonstrate how to calculate the rate of gypsum to apply to supply the equivalent of 1 meq/l calcium, we will use the water analyzed in table 8.3 (see also fig. 8.7). This is not the best water for irrigating citrus, but the EC, boron, and chloride levels are manageable with a sufficient leaching fraction, as discussed in the previous section. To achieve good infiltration, some of the sodium must be offset with calcium. The water should be treated by injecting gypsum using the following four steps:

1. Determine the purity of the gypsum and the actual pounds per acre-foot needed to supply 1 meq/l calcium. Assume that the solution gypsum purity is 92%; table 8.9 indicates that 234 pounds of gypsum will be required to supply 1 meq/l:

$$234 \text{ lb/ac-ft} \div 0.92 = 254 \text{ lb/ac-ft per 1 meq/l calcium}$$

2. Use the desired application rate to calculate the EC of the additional calcium and new water:

$$500 \text{ lb/ac-ft} \div 254 \approx 2 \text{ meq/l}$$
$$\text{New EC} = 1.0 + 0.2 = 1.2 \text{ dS/m.}$$

3. Calculate the new SAR:

$$\text{SAR} = \text{Na} \div [(\text{Ca} + \text{Mg}) \div 2]^{0.5}$$
$$= 9.6 \div [(2.5 + 0.1) \div 2]^{0.5}$$
$$= 8.4.$$

Table 8.9. Amount of amendments required for calcareous soils to replace 1 meq/l of exchangeable sodium in the soil or to increase the calcium content in the irrigation water by 1 meq/l

Chemical name	Trade name, composition	Weight of amendment required to replace 1 meq* exchangeable sodium (lb/ac-6 in. in the soil)†	Weight of amendment to obtain 1 meq/l free calcium (lb/ac-ft)†
sulfur	100% S	321	43.6
gypsum	$CaSO_4{\cdot}2H_2O$, 100%	1,720	234
calcium polysulfide	lime sulfur, 6% Ca, 23% S	1,151	156
calcium chloride	Hi-Cal, 13% Ca	3,076	418
potassium thiosulfate	KTS, 25% K_2O, 26% S	1,890‡ 3,770	256‡ 513
ammonium thiosulfate	Thio-Sul, 12% N, 26% S	807§ 2,470**	110§ 336**
ammonium polysulfide	Nitro-Sul, 20% N, 40% S	510§ 1,000**	69§ 136**
monocarbamide dihydrogen sulfate/sulfuric acid	N-pHuric 10/55, 10% N, 18% S	1,090§ 1,780**	148§ 242**
sulfuric acid	100% H_2SO_4	981	133

Source: Adapted from Oster et al. 1992, p. 113.

*The meq of exchangeable sodium to replace = initial ESP – desired ESP × cation exchange capacity (meq/100 grams soil).

†Salts bound to the soil are replaced on an equal ionic charge basis and not equal weight basis. Depending on initial and final desired ESP, laboratory tests show an extra 14 to 31% of the amendment above the amount shown is needed to completely displace the Na.

‡Assumes 1 meq K beneficially replaces 1 meq Na in addition to the acid generated by the S.

§Combined acidification potential from S and oxidation of N source to NO_3 to release free Ca from soil lime. Requires moist, biologically active soil.

**Acidification potential from oxidation of N source to NO_3 only.

4. Locate the intersection of the new SAR and EC on the infiltration chart (fig. 8.8).

You can see that increasing the application to 750 pounds per acre-foot (a 50% increase) gives a very small additional infiltration benefit and is not cost effective.

Acidifying Water Amendments

The water used for the above example would be a good candidate for acidifying amendments. Starting in the late 1950s, the Harmon SO_2 Generator (manufactured in Bakersfield) was the first production agriculture "sulfur burner" made available to farmers. These machines feed ground elemental sulfur into a small furnace. The burning produces sulfur dioxide, which combines with water trickled through the machine to make sulfurous acid; the acid is then injected into the irrigation water. In recent years, better pumps and safeguards have been developed to inject concentrated sulfuric acid directly.

Returning to our water in the example above, the pH is quite high (8.4) and the bicarbonate level is 4.2 meq/l. Adding gypsum to this water and running it through a drip system will significantly increase the chances of plugging the system with lime precipitate. The soil to be irrigated with this water is likely also to be alkaline. If the soil pH is greater than 7.7, acidification of this water and/or the soil may be beneficial to crop growth. Neutralizing the bicarbonate will definitely increase free calcium in the soil-water and improve infiltration. Table 8.9 shows that it takes 133 pounds per acre-foot of 100% pure sulfuric acid to release 1 meq/l calcium (assuming that the acid contacts lime in the soil neutralizing the carbonate molecule and releasing the calcium). This is the same amount of acid required to neutralize 1 meq/l of bicarbonate in the water; for the example water, the amount would be 532 pounds per acre-foot of 100% sulfuric acid. Because the additional acid will rapidly drop the pH, request a lab to run a "pH breakthrough" curve to show how much acid can safely be added to the water. Brass valves, transite pipe, and some membranes and plastics in older irrigation systems (pre-1992) are sensitive to pH less than or equal to 4.5. Newer all-plastic

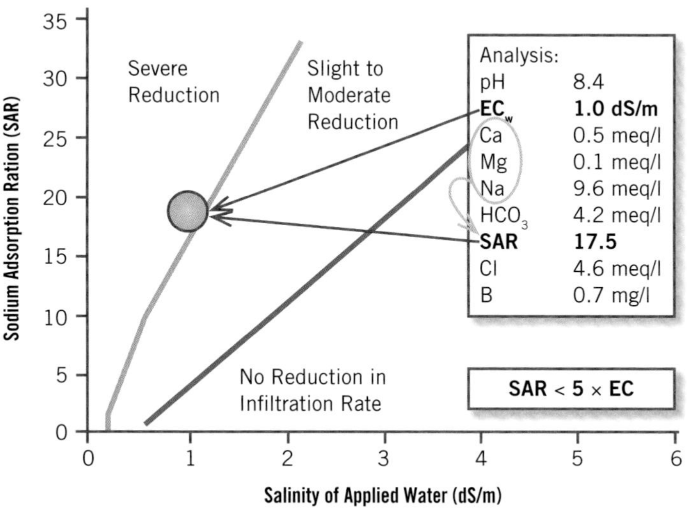

Figure 8.7 Estimating potential infiltration problems and determining amendment options from an irrigation water analysis.

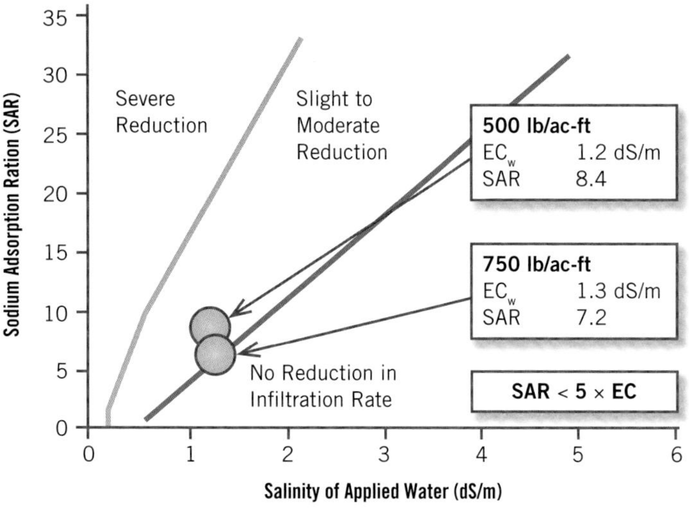

Figure 8.8 Revised infiltration potential after injecting gypsum into irrigation water at rates of 500 and 750 lb/ac-ft.

systems with fiberglass or epoxy-lined filters are generally safe to a pH of 2.5.

Although a plastic venturi injector can be used to inject acid, noncorroding solid displacement pumps are preferred for accuracy and safety. As a general rule of thumb, however, do not allow the water pH to drop below 5.5. In the final analysis, a cheaper and less-involved alternative is to band ground sulfur on drip lines or under microsprinklers.

Soil Amendment Rates

Amendment rates for soil are considerably higher than those for water. The purpose of applying a soil amendment is to reduce the exchangeable sodium throughout the root zone, not just at the soil surface. For soil that may be used for citrus production, the 100% gypsum requirement (GR) may range from 1 to 8 tons per acre (0.25 to 2 ton/ac-ft of soil over a 4-ft root zone). Depending on the choice of amendment, table 8.10 shows the cost can range from $27 to $766 per acre. If higher amendment rates are needed, the soils may be too costly to reclaim and will not be suitable for citrus.

Amendment Application Methods

Water Application of Amendments

Adding amendments directly to the water is ideal for managing soils with infiltration problems caused by surface crusting. On many soils, research has shown that as little as 0.5 inch of irrigation or rainfall can cause a crust that restricts infiltration. Such a crust is only at the soil surface and is often as thin as 0.25 to 0.5 inch. Improving the quality of irrigation water with water-run amendments puts the amendments at precisely the point where they are most needed: the soil surface. Soils with such problems need relatively small amounts of amendment applied frequently. Water application makes this convenient and accurate. Before applying an acid-forming amendment via the water, be certain that lime is present in the surface soil or, as in the previous example, that the water contains high levels of calcium and bicarbonate. Amendments in a liquid formulation are the easiest to apply in this manner and are usually most effective in drip or microsprinkler irrigation systems. Solution-grade gypsum, injected as a slurry, is probably the most common amendment applied by this technique (fig. 8.9).

Broadcasting Then Irrigating (No Till)

Broadcasting amendments such as gypsum onto the soil surface and irrigating the amendment into the soil is an alternative to water-run applications. The primary advantage to broadcasting is that the gypsum used is less expensive. However, for surface applications to be as effective, the application must be properly timed.

If infiltration is a problem in the summer, the amendment should be applied at the onset of summer and not during the preceding fall or winter. Applying the amendment too early (winter or spring) often makes it no longer effective when it is most needed (July and August). Surface applications are most effective when applied monthly during June, July, and August at rates equivalent to 250 to

Table 8.10. Approximate bulk purity and moisture content, field tons required, and applied cost/acre for selected calcium-supplying amendments to provide a 100% pure gypsum requirement of 1 to 4 T/ac

Amendment	Purity (%)	Moisture (%)	Approx. cost/T ($)	Field tons and total cost/ac to meet the below 100% gypsum requirements*					
				1 T/ac		2 T/ac		4 T/ac	
				T	$	T	$	T	$
Westside pit gypsum	50	8	22	2.17	95	4.35	183	8.70	365
Lima gypsum (Ventacopa)	75	4	32	1.39	80	2.78	146	5.56	289
bulk solution gypsum† (delivered)	92	3	130	1.12	146	2.24	291	4.48	583
ground wallboard	92	5	27	1.14	63	2.29	111	4.58	215
Hi-Cal (delivered)†	Ca-13	6	220	1.79	394	3.58	788	7.16	1,575
soil sulfur (granular)‡	99	2	190	0.19	54	0.38	94	0.77	172
sulfuric acid (applied)‡	98	NA	220	0.58	152	1.16	288	2.33	562
Thio-Sul†‡ (delivered)	N-12, S-26	NA	280	0.47	131	0.94	263	1.88	526
lime sulfur‡§	Ca-6, S-23	NA	240	0.67	186	1.34	356	2.68	697
N-pHuric 10/55†‡ (delivered)	N-10, S-18	NA	380	0.63	241	1.27	482	2.54	964
beet lime**	60	10	25	1.08	58	2.15	101	4.31	194

Notes:

Costs determined for Kern County, May 2008. Soils with extremely high silt content (> 50%), mica, or zeolite clays often do not respond predictably when applying amendments. Consult local experts before investing in amending these soils.

*Price assumes freight at $15/T and spreading (where applicable) at $15/ac to spread 3 T/ac.

†Chemigation, no application cost.

‡Free lime must be present in soil.

§Some free calcium, but soil lime is needed for complete reaction.

**Acid soil only.

1,000 pounds of 100% gypsum per acre; for more consistent results, broadcast finer, evenly graded gypsum. If only one application is to be made in the dormant season, apply coarse material, as the large chunks will require several irrigations to dissolve. Waste wallboard gypsum is ideal for this application. For orchards using alfalfa-type valves, it is possible to place coarse gypsum in a 200- to 300-pound pile next to the valve and let the water carry it down the field. Growers often find midseason broadcasting of gypsum to be a nuisance (or a cause of crop damage) and prefer to add amendments to the water.

Broadcasting Followed by Incorporation

Land application of amendments (i.e., the equivalent of 3 to 6 tons pure gypsum per acre) is most appropriate when the objective is to reclaim a saline or sodic root zone and not just the surface of the soil. It is usually most cost effective to broadcast coarser, less-refined amendments during orchard site preparation. Incorporating the amendment by plowing, shanking, or slip-plowing speeds up reclamation by quickly getting the amendment to the deeper soil so the exchange reaction can occur. Note that organic matter is also often spread and incorporated during site preparation. The benefit of organic amendments cannot be calculated in the same manner as for inorganic amendments, but the combination of gypsum and manure or compost always provides a much greater benefit then either amendment by itself. Orchard site preparation also often uses the banding technique described below to incorporate acid or sulfur down the tree row prior to planting.

Figure 8.9 Mobile gypsum solutionizer machine with horizontal mixing bar for injecting gypsum slurry into drip system. *Photo:* B. Sanden.

Banded Surface Application

Banding acid-forming amendments is most often done to correct a micronutrient deficiency in alkaline soils by lowering the soil pH to increase nutrient availability. The major concern is crop fertility, but the practice usually benefits soil tilth and infiltration as well due to the release of free calcium from the soil lime. When preparing ground prior to planting, effective rates of acid applied in the treated band range from 1 to 4 tons of acid (500 to 3,000 lb of soil sulfur) per applied acre, depending on the lime content of the soil. Banding is most often followed by a slip plow. For deeper incorporation of soil sulfur, the sulfur can be banded over the slip plow trench and followed by a 3-foot chisel shank or 2-foot furrowing shovel with the wings bent back. In established orchards, however, applications should not exceed 1,500 pounds per applied acre (a 3- to 4-ft spray band), or crop damage may result. Even smaller amounts of acid, fine sulfur, or gypsum shanked into the soil or banded along drip hoses can be effective.

Apply these more expensive amendments into areas of active rooting; it is not essential to modify the entire orchard floor. It takes 0.98 pound of pure sulfuric acid to neutralize 1 pound of lime. Thus it would take 20 tons of pure acid to neutralize a 1% lime content in 1 acre of soil 1 foot deep! However, it is necessary only to dissolve all the lime in a small area of rooting to increase the availability of micronutrients. When coupled with microirrigation and water-run amendments, this technique can be more economical and highly effective.

References

Ayers, R. S. 1977. Quality of water for irrigation. Journal of Irrigation and Drainage Engineering 103:135–154.

Ayers, R. S., and D. W. Westcott. 1985. Water quality for agriculture. Rome: United Nations FAO Irrigation and Drainage Paper No. 29, Rev. 1.

Hanson, B., S. R. Grattan, and A. Fulton. 2006. Agricultural salinity and drainage. Oakland: University of California Agriculture and Natural Resources Publication 3375.

Hoffman, G. J. 1986. Guidelines for reclamation of salt-affected soils. Applied Agricultural Research 1(2): 65–72.

———. 1996. Leaching fraction and root zone salinity control. In K. K. Tanji, ed., Agricultural salinity assessment and management. ASCE Manuals and Reports on Engineering Practice 71. New York: Macmillan. 237–247.

Maas, E. V. 1992. Salinity and citriculture. Proceedings of the 7th Congress of the International Society of Citriculture 3:1290–1301.

Maas, E. V., and J. G. Hoffman. 1977. Crop salt tolerance: Current assessment. Journal of Irrigation and Drainage Engineering 103(IR2): 115–134.

Oster, J. D., M. J. Singer, A. Fulton, W. Richardson, and T. Prichard. 1992. Water penetration problems in California soils. University of California Kearney Foundation of Soil Science.

Reisenauer, H. M., ed. 1978. Soil and plant-tissue testing in California. Berkeley: University of California Division of Agricultural Sciences Bulletin 1879.

Shalhevet, J., and Y. Levy. 1990. Citrus trees. In B. A. Stewart and D. R. Nielsen, eds., Irrigation of agricultural crops. Agronomy Monograph 30. Madison, WI: American Society of Agronomy. 951–986.

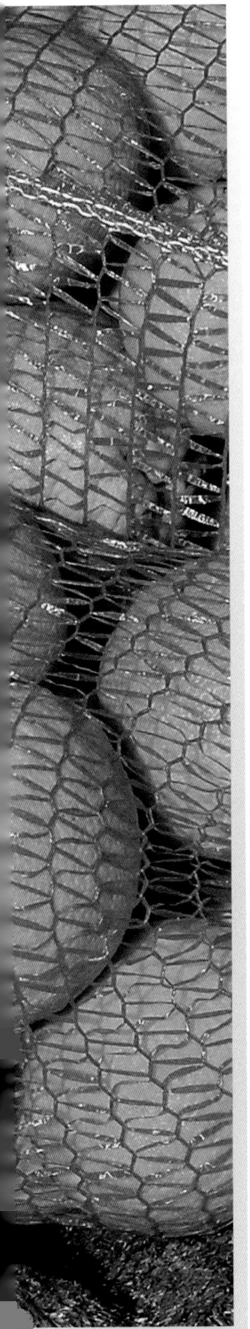

9 Establishing the Citrus Orchard

Gary S. Bender

The success of an orchard will depend largely on how well the site is selected. Trees on the correct rootstock can be planted properly, the roads can be engineered correctly, and the irrigation system installed by a qualified irrigation engineer, but if the site is located in a chronically frosty or windy area, the orchard will likely be a failure. In this chapter, several important elements to consider are discussed, including site selection, orchard preparation, and orchard design, spacing, and planting. For information on irrigation system design and installation, see chapter 11, "Irrigation"; for more complete information on sizing of pipe, friction losses, effect of gravity on water pressure, and so on, consult a qualified irrigation engineer.

Site Selection

Site Characteristics

Selecting the proper site for the orchard is critical for success. The ideal site would have the following characteristics.

Good soil drainage

Soils that do not drain well due to soil compaction or clay layers may cause roots to die of asphyxiation. Root rot caused by the fungus *Phytophthora* is much worse in poorly drained soils. As a rule of thumb, if an 18-inch-deep hole is filled with water and water is still in the hole the following day, the site is not suitable for planting. Sandy loam soils that are reasonably deep are usually considered best for citrus, but only if the subsoil has good drainage. Citrus can be grown on soils that have a high clay content, but good water management will be critical. Root and tree growth can be restricted if the clay content of the soil is over 50%. When replanting an orchard that has had root rot or poor drainage, deep ripping is usually recommended to break through hardpans and clay layers.

Suitable climate

Perhaps the greatest danger in citrus production is the potential for frosts or freezes. Most citrus production areas in California are subject to some degree of frost and moderate freezes, but if the location is chosen wisely, severe freeze damage will be avoided. Because most types of citrus fruit show internal freeze damage when temperatures remain below 26° to 29°F for a few hours, the site must be located in a warm zone with good air drainage at night. Most frost damage is caused by temperature inversions, in which warm air rises to a layer just above the trees during a cold night and cold air settles toward the ground. In this situation, wind machines are useful to stir the air and push the warm air back down. Cold air flows downhill, collecting in low areas. If the potential orchard is to be planted in a low-lying area, the temperature at the site should be recorded for at least 3 years prior to the planting, and neighbors and farm advisors should be consulted to determine whether the location is too risky.

Evaluate the slope and elevation. Cold air flows if the land is sloped, and on the hillsides cold air is usually not standing long enough to create a cold atmosphere, unless the elevation is high enough that snow is common and would remain for a period of time—long enough to lower the temperature well below freezing. Most successful orchards in California are located at elevations between 230 feet below sea level to approximately 2,000 feet above sea level. On flatter ground, small increases in elevation compared with surrounding areas can make quite a difference; it was reported in Florida that an elevation of just 4 to 5 feet can cause an increase of from 2° to 5°F on calm, clear, cold nights.

Is there a basin? In evaluating the frost hazard of an area, determine whether the area forms a basin. This is the case with most valley bottoms. It could be a thousand-acre area, yet if all the slopes come down continuously around the area and form a basin with no open sides, cold air will fill the basin and create a frost hazard for the trees planted there. If there is an opening in the basin, the cold air may be able to flow out. Dense stands of trees (as can be found along creeks and rivers), buildings, elevated freeways, and small hills can also form basins that can cause cold air to accumulate. Spots where water does not flow out can be spots that are too cold for citrus.

Temperature records

Given the sizable investment required for development of a citrus orchard, it is wise to consult specific temperature data, especially if citrus is not grown in the surrounding area. Digital recording thermometers that can be downloaded into a computer are better than minimum-recording thermometers because they give the duration of a frost. This is important because it is generally believed that if a frost below 29°F lasts for over 4 hours, severe damage to lemon fruit will occur, and if a frost below 26°F lasts for 4 hours, damage to oranges will occur. Minimum-recording thermometers do not give the duration of low temperatures, but they do give a good indication of frost potential. A problem with weather data is that it can be misleading if only 1 year of data has been collected. It is for this reason that 3 years of data is needed to say with some confidence that an area may be safe for citrus.

Not too hot

Citrus growth may be slowed and major fruit drop will occur if the summer temperatures have prolonged periods over 122°F, and young succulent foliage may be damaged if the temperatures are over 104°F, especially if the trees are water stressed. Damage to Valencia orange fruit (in the form of sunburned peel, dried flesh, reduced fruit size, and increased granulation) in the Coachella Valley was noted when average temperatures were 112°F with relative humidity at 20%, but no damage occurred at 105°F. Temperatures over 104°F at a sensitive time immediately after fruit set have been known to exacerbate the typical June drop. A rapid increase in temperature accompanied by low humidity (less than 20%) and desiccating winds has been known to drop almost an entire crop. Sensitivity to heat varies among the types of citrus; Valencia oranges, grapefruit, lemons, and most mandarins produce well in areas of high temperatures, but navel oranges and Satsuma mandarins are poor producers when high temperatures occur during bloom and fruit set.

Match the variety to the climate

Lemons can be grown in all the California citrus production zones from the coast to the desert, but the time of harvest will be different. Lemons in the desert are harvested as one crop in the fall and early winter, whereas lemons along the coast are almost everbearing. Navel oranges need a relatively high mean temperature for best quality and are usually grown in the warm interior valleys. Yield and quality of navels in the desert and on the coast are poor, yet yield and quality of Valencia oranges in the desert and along the coast are satisfactory to excellent. Grapefruit quality is best under high mean temperatures, and desert conditions produce excellent-quality winter fruit.

Soil pH

Citrus trees grow best at soil pH 5.5 to 7.0 due to optimal availability of nutrients in the soil. Citrus will grow at soil pH 7.5 to 8.5 if the proper rootstock is chosen. At the higher soil pH, iron is often limiting, and trees show chronic iron deficiency symptoms unless the soil pH is lowered by applying soil sulfur, or a micronutrient spray is applied to the foliage every year (see chapter 8, “Soil and Water Analysis and Amendment”). If the soil pH is below 5.5, aluminum toxicity (causing a tip burn in the leaves) may be a problem and liming the soil may be necessary (see chapter 10, “Nutrient Deficiency and Correction”).

Slope

Citrus grows well on slopes, but steeper slopes make pest control, harvest, and fruit removal more difficult. The ground should be flat enough for pest control equipment and pruning equipment to move efficiently through the orchard. A westward-facing slope usually has more heat units, and fruit from these orchards often mature earlier. On the other hand, an eastward-facing slope is beneficial for growers trying to hold fruit on the trees.

Wind

Care should be taken to locate the orchard in an area that is not facing constant wind. Prevailing winds can cause young trees to grow lopsided and can also cause peel damage to fruit from mature trees. Chilling winds from the ocean reduce, and sometimes stop, tree growth. These winds can also reduce temperature during bloom such that pollination is negatively affected. High winds also cause fruit scarring, leaf drop, fruit drop, and a dramatic increase in water usage. In southern California, some east-facing slopes will be subject to easterly Santa Ana winds, which usually develop in the fall but can appear in the spring or winter as well. These winds have been known to cause almost complete fruit drop. Santa Ana winds are characterized by low humidity and high temperature, with average speeds of 35 miles per hour and gusts up to 50 miles per hour. Occasionally winds at much higher speeds will occur; the author measured a Santa Ana wind at 70 to 100 mph in the Pauma Valley region of San Diego County in early 2003 that stripped all the leaves and most of the fruit off trees at the edges of Valencia citrus orchards. Fruit harvested from those orchards in the summer of 2003 had extensive peel scarring, regreening, and small fruit size. Windbreaks have proven to be effective in protecting citrus trees from milder chilling winds from the ocean, as well as prevailing winds. Windbreaks can be constructed with rows of eucalyptus (the 60 to 100 ft blue gum, *Eucalyptus globulus,* is most commonly used), Lombardy poplar (*Populus nigra* L. var. *italica* Du Roi), or Australian pine (*Casuarina* spp.). It should be remembered that these trees must be watered to become established, and as they mature they will steal water and fertilizer from the first few rows of citrus in the orchard. Windbreaks have become less popular in recent years due to the belief that the extra space taken by the trees, and the consumption of water and fertilizer, do not make up for the benefit of wind protection. Careful analysis of the effect of wind at a site should be made before deciding whether to plant windbreaks.

Water supply

Most of the rainfall that occurs in California comes during the winter, when water consumption by citrus is low. Therefore, citrus must be irrigated during the dry months. The lack of rainfall in the summer months is actually an advantage for California because the fruit peels are not susceptible to fungal diseases common in areas with higher humidity and summer rainfall. District water (water from the Colorado River, the Delta region in northern California, and perhaps local water sources) is piped to most orchards in southern California, and most orchards in the Central Valley have access either to ground water or to canal water provided by the local water district. Growers in remote areas may not have access to district water and must use only well water. Due to the high cost of district water (especially for orchards in southern California), it may be preferable to use well water for all or part of the irrigation. The economics of the citrus industry dictate that irrigation should come from ground water if at all possible; most orchards in the Central Valley get their water from wells, which allows them to stay more economically viable. When choosing a site for an orchard and sizing the irrigation pipes, calculate the quantity of water required for the mature orchard so that production from a well or other sources matches the irrigation requirement. Assuming that production of water from the well remains constant, this calculation will indicate how many acres can be planted. However, due to drought cycles in California, be aware that water production from wells rarely remains constant.

Water quantity and quality

Most of the smaller orchards (especially those in southern California) were planted with the notion that the water would be supplied from the local water district. However, with the cost of water increasing and the supply susceptible to cutbacks, many landowners are interested in drilling for water to supply part or all of the water needs of the orchard. Careful planning should be done before the orchard is planted to determine the water-use demands of a mature orchard in the summer, whether all or part of the water can be supplied from a well, and whether a reservoir is required for storage. The water requirements for citrus vary considerably, depending on the location of the orchard in California. For more information on irrigation quantities, see chapter 11, "Irrigation."

Citrus is sensitive to salinity in the irrigation water; water that is too saline will hurt growth and production. Knowing the quality of the water supply is crucial, and water samples from wells should be analyzed by a lab before the trees are planted. Important mistakes have been made when growers calculated that the water output from a well was sufficient to meet the water requirement of the orchard, but the water quality was too saline for citrus production. In these cases, the growers failed to add an extra quantity of water for a leaching fraction. For more information on salinity management, see chapter 8, "Soil and Water Analysis and Amendment."

Access

In some cases, orchards are established in areas where the only access is through private property. Make sure there is an understanding (hopefully a legal agreement) that will provide access for all farming operations for the life of the orchard, including pickers and transportation of the fruit on an orchard road.

Location of infrastructure

It is less expensive to operate a citrus orchard when the orchard is located reasonably close to the packinghouse and other important components of the operation, including labor supply, farm supply stores, orchard managers, and trucking. During the harvest season, most large orchards have a truck coming from the packinghouse every day to pick up fruit; this trucking cost can become expensive for orchards in remote locations. Also, if you are relying on a professional orchard manager but you want to plant your orchard an hour away from the nearest orchard manager, you must be prepared to manage the orchard yourself.

Size of the orchard

Orchards should be large enough for labor contractors to move their picking crews around efficiently. The larger packinghouses will usually not send a crew out to pick the same variety of citrus fruit if the orchard is less than 5 acres.

County permits

Last on this list, but quickly becoming one of the most important of the criteria for choosing a site is the necessity for obtaining a grading permit if working in an area that hasn't been farmed in the last 5 years. In addition, approval by the county is required if you want to clear and plant in a zone that might be covered by a multi-species habitat plan. Most counties are developing plans that will make it very difficult if not impossible to plant orchards in certain areas of the county due to the necessity to preserve habitat for endangered species. Some counties require permits to drill wells, and in most counties it is now illegal to cut down an oak tree. Before purchasing a site for orchard development, it is very important that you check with the county department of planning and land use for any restrictions that might come with the property.

Orchard Preparation

Variety and Rootstock Selection

One of the first steps is to determine which variety and rootstock to plant. Trees are not always immediately available at the nursery; major plantings are often ordered from the nursery sometimes 2 years in advance of the actual planting. Not only must the grower consider the type of citrus that will grow successfully in the area, but there must be a consideration for what will be desirable at the market for many years in the future. If a minor type of citrus is to be grown, such as blood oranges, the grower must expect to play a major role in market development—the job doesn't end after the fruit is harvested.

Market trends, consumer consumption, foreign competition, and local competition also enter into the decision as to which variety to plant. In the early 1990s, growers were impressed by high market prices for blood oranges and kumquats. Unfortunately, too many people planted these, and there was little effort to increase consumer demand. Prices went from $45 per carton for blood oranges in 1990 to the point that much of the fruit could not be sold in 2000. Major varieties such as grapefruit have seen falling consumer demand; Valencia oranges have faced a declining market due to many factors, including competition from summer navels imported from Australia.

In order to choose a variety, the grower should check with the local packinghouses to see what can be packed and sold from that house and what are the trends for the future. Farm advisors, produce buyers for chain stores, and bankers should also be consulted. For more information, see chapter 4, "Scion Cultivars," and chapter 5, "Rootstocks."

Permits

The local county department of planning and land use must be consulted before any clearing or grading is done on the land. Many counties are developing plans to conserve endangered species of animals and plants, and if the proper permit is not obtained before clearing, the grower may be subject to fines and replanting the land with native plant species. Do not drill a well or cut down an oak tree without checking with the county.

Land Clearing

When given permission by the county to clear the land, most growers brush off the land before building the roads because in thick brush it may be difficult for the tractor operator to see the terrain. Several options are available for brushing the land. On flat ground and gentler slopes, the land can be cleared by a dozer; the brush can be cut with a heavy-duty brush mower pulled by a dozer or tractor; or the brush can be crushed by a "sheeps-foot" (without the water) pulled by a dozer. On the steeper slopes, the brush must be cleared by hand with a crew using machetes, chain saws, and brush hoes. Sometimes a chain dragged behind two tractors, with the chain between the tractors, can clear brush from gullies. Doing the work with machinery is the most economical method.

Brush can either be chopped into small pieces and left on the ground (to reduce soil erosion) or stacked into piles and burned on burn days. (A burn permit as well as permission to burn on a burn day are required.) Fire safety rules must be closely adhered to because if the fire gets away, the grower may be liable for damages and all costs in putting out the fire.

Deep ripping

Examine the soil in various places around the potential orchard for hardpan layers. For the examination, it is easiest to use a backhoe or a power auger to dig holes about 5 feet deep. If a hardpan is located anywhere from the 1-foot to 4-foot depth, the area should be deep ripped to break through the hard layers before planting. The hardpan will not reform, and the improved drainage could make the difference between success and failure of the orchard. Sometimes a plowpan caused by years of soil compaction from wheeled traffic or disks is found at 6 to 20 inches deep. Chiseling 20 inches deep with shanks 18 to 24 inches apart can break up plowpans. On slopes, the hard layer below the surface may be granite rock; this cannot be ripped, but water can usually drain through cracks and fissures in the rock.

Leveling

With the advent of drip and microsprinkler irrigation systems, flood- or furrow-irrigated orchards have become much less common in California. If an orchard is to be developed with flood or furrow, the land should be leveled with a laser leveler to provide an irrigation gradient. It is not necessary to level orchards irrigated by drip or microsprinklers, but it is a good idea to pull out as many rocks and stumps as possible before the trees are planted to make it easier for equipment to move through the orchard.

Eradicating weeds

Certain weeds such as johnsongrass and bermudagrass compete strongly with new citrus trees and should be eradicated before tree planting. The weeds can be disked repeatedly during the summer, then irrigated and the new growth sprayed with a systemic herbicide. If the orchard is to be nonorganic, a preemergent herbicide can be worked into the soil in the late fall. For more information, see chapter 16, "Integrated Weed Management."

Soil fumigation

Soil fumigation is usually not necessary when planting virgin land. Most land planted to citrus, however, has had a previous crop of citrus, and high populations of nematodes and soil fungi (such as *Phytophthora* and *Fusarium*) can severely damage the young trees. Methyl bromide has been commonly used in the past to control pathogens and weed seed, but this material is now under restriction in the United States and is in the process of being phased out. Preplant nematicides are available but these, like methyl bromide, are under restriction. Consult the local county agricultural commissioner before using these materials. Soil solarization, a method of heating the soil with clear plastic tarps on the ground during the summer, is effective in reducing nematode and fungal populations.

Road Systems

The main purpose of roads in the orchard is to facilitate efficient removal of bins of fruit. Traditionally, most citrus orchards were located on fairly flat ground, and road construction was easy. In many citrus-producing counties of California, however, most available flat ground has been taken for use by housing and other development, leaving the steeper slopes for orchards.

On flat ground, a dozer is used to build a road completely around the future square or rectangular block of trees, with drive roads through the orchard about every 20 trees. On hillsides, two types of road networks are used. If there is a hill in the middle of the orchard, a continuous primary road is built upward around the hill with a loop at the top of the hill. Secondary roads are built from the primary road; these roads run on the contour of the hill, usually looping completely around the hill and rejoining the primary road. If the orchard is located on a hillside, a primary road is built up the hill with switchbacks at the end of each grade. Secondary roads are built from the primary road, usually on the contour of the hill; these usually have dead ends, with a wide spot at the end for turning around.

Topographical maps are essential when planning roads with 10% grades. The maps do not show outcroppings of rocks and other obstacles, therefore walking the orchard and marking the location of roads with flags is necessary to get the final adjustments for the exact road locations. Contour roads should be spaced with six rows of trees between them on the gentler slopes and four rows between them on the steeper slopes. In the case of hillside orchards, fruit is carried down the hill to be placed in bins on the orchard roads. For orchards on flat ground, bins are usually driven into the orchard by forklifts and often placed next to or near the tree being harvested.

Erosion control on access roads

Grading roads for any development can be the largest source of water runoff and erosion. The key to proper road design and drainage is to frequently divert runoff water from the road before it causes erosion on the road itself or on the slope below. Roads that are properly designed, constructed, and maintained will avoid long-term costs of erosion and grading. Initially, the landowner must have the necessary permits from the regulatory agencies before commencing work. This will require the submission of an engineering plan for the roads along with specifications and an environmental assessment.

Road design and construction

When determining the design of an access road system, several guidelines should be followed:

- Main roads should be a minimum of 14 feet wide for two-way traffic. Haul or pick roads can be 10 feet wide for single-lane traffic. Avoid slopes in excess of 66% (1.5:1, horizontal: vertical), which is the natural angle of repose of soil or a pile of sand. Steeper slopes require excessive cut and fill, and this consumes usable land.
- A large dozer is used initially to build the roads. Slope boards are used to create the inner slope. A smaller (less expensive) dozer is then used to clean the dirt berm at the base of the slope. The inner edge of the road needs to be clean or the dirt will flow into the culvert and clog it during the first rain.
- For roads with less than 8% gradient, the road should be sloped slightly away (or out-sloped) at 2% from the cutbank and toward the outside edge of the road. Do not leave a berm along the outside of the road because the flow off the road is designed to sheet evenly onto a well-vegetated fill bank, reducing the erosive action of runoff. The optimal gradient for haul roads is 0 to 2%; main access roads should be 2 to 10%. If gradients exceed this, surfacing with rock, asphalt, or concrete should be considered.
- Roads with greater than 8% gradient (usually the main access road) should be sloped slightly into the cutbank (in-sloped) at 2%. The collected water can then be conveyed safely under the road with culverts into stable material or existing drainage ways. Provide rock energy dissipaters as needed so that additional gullies are not created.
- Final road grading should be done after mid-April to reduce the chances of exposing loose soil to rainfall.
- Both cut and fill banks should be at a slope of 2:1 if practicable. On steeper slopes, this is not possible, and they must be graded according to existing terrain.
- In addition to installing culverts and drainpipes under roads (see below), it may be necessary to build small "Arizona drains" to move small amounts of water across roads where it is difficult to build culverts. These drains are open drains built with rock and concrete across dips in the road, draining water into gullies.
- In most large orchards, the main entrance to the primary road is paved with asphalt to reduce the dust in the most heavily traveled areas. Paving also makes it easier for the boom trucks from the packinghouse to gain entrance on rainy days.

Waterbreaks

Waterbreaks, also called water bars, effectively divert accumulated water from the road surface onto the vegetated fill banks or toward the cutbank on gradients over 8%. The following points must be considered:

- Install waterbreaks before the middle of November.
- Space waterbreaks as indicated in table 9.1.
- The optimal size of an earthen waterbreak is 6 inches above the road surface and 6 inches below the road surface. If it is smaller, it may be less effective in diverting water, will require increased maintenance, and will probably break down faster.
- Do not allow diverted water to flow directly into unstable areas, septic fields, or natural watercourses.
- Avoid placing waterbreaks in swales, gullies, or low areas where water has no escape. Place waterbreaks above these areas whenever possible.
- Place waterbreaks above changes in grade to minimize water flowing down steeper portions of the road.
- Place waterbreaks above road intersections and curves in the road.
- Runoff water from waterbreaks should not be directed onto fill material unless a downspout or other conveyance measure (e.g., rocks, pipe, chute) is provided and the water is drained away from the fill. Compacted berms should be placed along the outside bank to divert the concentrated water to the outlet.
- Each waterbreak should have a continuous, firm berm of soil at least 6 inches high and 6 feet wide that is parallel to the waterbreak cut.
- All waterbreaks must be open at the lower end so the water can run off easily.

Table 9.1. Recommended spacing of waterbreaks for orchard roads

Road gradient (%)	Interval between breaks (ft)
5	125
10	78
15	58
20	47

- Avoid allowing water to pond behind a waterbreak. The ponded area may fill with sediment, causing the waterbreak to fail as runoff flows over it into the road.
- Waterbreaks should be installed at an angle of between 30° and 45° downhill from the perpendicular to the road; this will catch and direct runoff water to the outlet.
- Keep the waterbreak and outlet clear of debris and sediment so that water drains freely.

Culverts

Culverts will be needed at all natural drainages. Install them at right angles to the road (fig. 9.1). Determine the size of the culverts based on the amount of water produced from the watershed. Reduce plugging by installing catch basins or debris racks several feet upstream of the culvert entrance. Drop inlets to culverts create proper hydraulics and also reduce siltation.

Roadway vegetation

When building roads on slopes, all exposed soils, including the roadbed, should be seeded with annual ryegrass at 24 pounds per acre (0.56 lb per 1,000 sq ft) and mulched with straw (not hay) at 2 tons per acre (1.5 bales per 1,000 sq ft) tacked or tucked into the soil. Fertilizer is also needed, such as ammonium phosphate (16-20-0) at 500 pounds per acre (11.5 lb per 1,000 sq ft). Vegetation should be established by mid-October on exposed areas, including the surface of roads (the main access road may be excluded). The procedure is to seed first, apply straw, and fertilize as the seedlings emerge through the straw.

Road maintenance

Maintenance is very important. Several actions can be incorporated into most road maintenance programs that reduce the long-term cost and need for annual road maintenance:

- Restrict access during wet weather. Driving on the roads aggravates erosion and drainage problems. Roads usually dry out within a day or two in coarse-textured soils.
- Make adequate drainage more important than vehicle speed.
- Inspect and clean out culverts the first time after the rain creates surface flows.
- Avoid excessive road maintenance. Whenever possible, retain stabilizing vegetation in road ditches by removing small blockages by hand. Avoid undercutting cut banks and retain bank sloughage when adequate road width remains.
- Regrade only to remove deep ruts or areas damaged during severe storms. Regrading should be done in the spring when the soil moisture allows for good compaction of graded material.

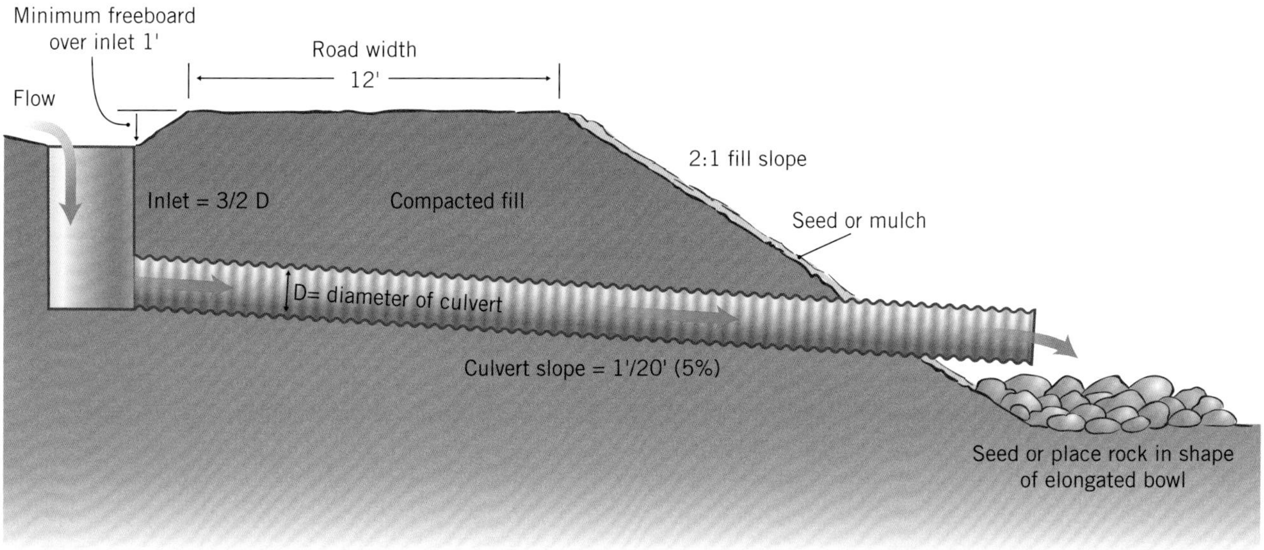

Figure 9.1 Profile of culvert installation.

- Inspect the road system at regular intervals and immediately apply soil erosion and water control measures necessary to prevent damage.
- For larger operations, have a night watchman during a major rainfall to ensure that all drainage systems are functioning properly.
- Abandon or relocate road sections that have repeated high maintenance.
- Consider mowing roads or "driving out" vegetation across the full width of the roads if the primary purpose for conducting road maintenance is to reduce fire hazard and minimize soil disturbance.

Erosion Control

Erosion control practices are a necessary part of citrus orchard development. Citrus orchards have traditionally been developed on flat ground, but as land has become more valuable for houses, citrus orchards are being developed more on steeper ground. This ground is initially cheaper to purchase, but more expensive to develop. Steep ground is more prone to erosion problems. The following erosion control measures should be taken in addition to controlling erosion on access roads, as discussed above.

Vegetated filter strips

In addition to controlling erosion on slopes below the road, it is important to prevent eroded sediment in runoff from entering natural watercourses. A vegetated filter strip is an area of planted or resident vegetation below a road that can catch sediment before it reaches a stream. Design a drainage system so that each water bar or culvert releases water onto an energy dissipater, and each energy dissipater releases water onto a filter strip that is adequately vegetated to trap sediment. If necessary, seed the filter strip area with suitable grasses to improve its stability and sediment-trapping capability.

Figure 9.2 Cover crops planted between young trees on a gentle slope. *Photo*: G. Bender.

Critical area planting

Use plants to control erosion in critical areas of the orchard such as slopes and road banks. Native annuals are best, but they must be seeded and mulched if the seed base is buried when land is cleared with equipment (a good operator can clear only the brush and leave the topsoil and seed base). If the seed base contains annual broadleaf weeds instead of grasses, it may require additional herbicide control. Even if native or resident vegetation is used, plant a seed mix for the first few years as a vegetative cover until the tree canopy shades the ground. Avoid planting in the sprinkler pattern of the tree. The use of erosion control nettings over a straw mulch is an effective alternative but may be more costly. Nettings are tucked in or pinned and rolled down the slopes.

Cover crops

Cover crops can replace critical area plantings as erosion control, and they also can improve traction, soil structure, soil fertility, and water infiltration; they can, however, increase water use, frost hazard, pests, and overall costs (for more information on cover cropping, see Ingels et al. 1998). Establish planted annual grasses or natural vegetation on the contour between tree rows to control sheet erosion (fig. 9.2). The types of equipment used and the cultural operations performed should maintain a population of planted or desirable resident nonweedy species that provides at least 60% ground cover during the erosive period. Control the plant height by mowing. As the trees grow, shade may change the makeup of the cover crop or eliminate it entirely. Do not allow the cover crop to grow around sprinklers, as this will disrupt the spray pattern and result in poor water distribution to the root zone.

Mulch

Clean organic material such as straw can be applied around trees for erosion control in the area of the sprinkler pattern, since this area will not be planted to a cover crop. Mulching 2 to 4 inches deep reduces weed growth, conserves moisture, and improves soil tilth. Do not incorporate the mulch into the soil. Mulch should be kept about 8 inches away from the trunk to avoid wetting the lower trunk for long periods of time. Mulch laid against the trunk has been known to increase the infection rate of Phytophthora gummosis and other trunk diseases.

Underground drainpipes

A properly sized underground pipe can carry surface runoff down steep terrain to a controlled outlet or natural drainage way to control gully erosion. The amount of water produced in the watershed and its drainage patterns must be evaluated to size the pipe and design its installation.

Irrigation Pipelines

The final step in orchard development is installation of the main irrigation lines. The irrigation system should be designed by a qualified irrigation engineer. Many factors are considered in the design, including pressure and flow rates from the wells or the district connection, slope, number of trees in a block, emitter discharge rates, and so on. Careful consideration as to tree spacing is made at this point (see below). If larger submain pipes and other main pipelines are backfilled during installation, it is usually necessary to control erosion on the exposed soils with critical area planting. Seeding, mulching, and fertilizing are done to the same specifications as for controlling sheet erosion.

Orchard Design, Spacing, and Planting

Orchards should be designed to maximize light interception by the tree canopy, minimize light striking the ground between trees, and provide enough space between the rows for orchard sprayers, pruning equipment, and bin forklifts. One of the difficulties in an orchard design is spacing the trees so as to give maximum yield with good fruit quality and at the same time be economically efficient for the life of the planting, not just in the early years.

In California, as in Florida, tree spacing has become closer for a number of reasons. Before 1940, most citrus orchards were spaced at 22 by 22 feet, with some Valencia orchards spaced at 24 by 24 and some lemon orchards in the desert spaced at 30 by 30. In the 1960s, tree spacing for new plantings was usually from 20 by 20 to 25 by 25. With the advent of plastic irrigation pipe for minisprinkler irrigation systems, which came in 20-foot lengths, most growers settled on a spacing of 20 by 20, but orchards were also planted at 15 by 20, with 15 feet between the trees in the row and 20 feet between the rows. Also in the 1960s, growers experimented with closer spacings: higher-density orchards were planted at 10 by 20, 9 by 22, and 10 by 18. These were grown as hedgerows, or every other tree was removed at the point of crowding to give a spacing of 20 by 20, 18 by 22, or 20 by 18.

Closer spacing down the row is possible if the citrus variety is grafted on the dwarfing Flying Dragon rootstock (a Trifoliate selection). With this rootstock, a tree spacing of 6 to 8 feet down the row might be possible, but apparently it is not economically viable since the practice has not caught on with growers. The standard Rubidoux Trifoliate rootstock (considered a semidwarf rootstock) is often used in plantings that are 15 feet between trees in the row. Neither of these rootstocks is used with lemons. The following must be taken into consideration when considering planting on close spacing:

- Higher tree densities improve early yield per acre because light strikes more leaf area in each acre of the orchard. This might be an advantage if an early cash flow is important.
- Higher tree densities cost more to develop—not just in tree cost but also in planting, irrigation installation, and tree care.
- Although high-density plantings provide an earlier yield, at maturity the planting bears about the same volume of fruit as normal-density planting because the bearing canopy surface (leaf surface per acre) is about equal. Shading may reduce yield in a high-density planting if the orchard is not pruned. Shading may also delay color development in fruit and contribute to smaller fruit size.
- In high-density plantings, fruit within 7 to 8 feet of the ground can be harvested without a ladder. This fruit can be harvested faster and easier than fruit higher in the tree. In the future, direct labor costs and other labor issues (including costs for workers' compensation insurance) may necessitate that the trees be pruned at a lower height. Trees that are kept at a lower height (less than 10 ft) can be kept on close spacing within the row, but there must be at least 18 feet between the rows for equipment to move through the orchard. This will require annual pruning and the extra expenses associated with it.
- Lower-vigor scions and rootstocks should be used in high-density plantings.

Tree Density and Height

With a conventional spacing, the trees should be no higher than twice the distance of the clear row middle. Usually a row middle 8 feet wide is the minimum necessary for equipment to pass between the trees; thus, the trees should be no taller than 16 feet. This allows adequate light to reach the lower canopy of neighboring trees. Side pruning and topping will be necessary every 3 to 4 years in oranges and grapefruit, and usually every 1 to 2 years in lemons and limes.

Fruit location in trees on closer spacing can be different from fruit location in trees on a wider spacing. In Florida, researchers determined that 71% of the fruit was located from 8 to 16 feet high in trees on a spacing of 10 by 15 feet, but only 47% of the fruit was located in this same range in trees on a spacing of 20 by 25 feet (Whitney and Wheaton 1984). Also, trees on the closer spacing essentially had no inside fruit (fruit located more than 2 feet from the outside canopy surface), whereas trees on the wider spacing averaged 14% inside fruit over a 5-year period.

These same researchers in Florida reported that after topping and pruning began in the 8th to 10th year, the yield per acre in some orchards declined, sometimes to half of the ideal yield. Their explanation is that excessive vigor in the rootstock, and perhaps too much irrigation and fertilization, created an imbalance in the natural growth of a tree constrained by the space allocated to it. Too much wood (including fruiting wood) is pruned off. The excessive vigor apparently increases yield in the early years but decreases it in later years. Therefore, if too much wood is continually being pruned, a reduction in the cultural inputs may be necessary.

Spacing Research in California

Several spacing trials were begun in the 1960s under the direction of University of California, Riverside, specialist S. B. Boswell. A Washington navel trial on Troyer Citrange rootstock was planted in 1961 in Kern County at 11 different densities, ranging from 9 by 11 to 22 by 22 (Boswell et al. 1970). In the 1964–65 harvest (the first harvest), trees spaced at 9 by 11 (440 trees per ac) bore 431 boxes per acre (50 lb boxes). Yield was less in the other spacings (in relation to the number of trees per acre), with the least yield at 90 boxes per acre for the trees spaced at 22 by 22. However, the yield per acre on the closest spacing declined until the trees were thinned in 1967, after which yields increased. The best yield per acre for the first 5 years of harvest was from the plots at 11 by 11 (thinned to 11 by 22 in 1967): the 5-year average was 422 boxes per acre. The plots at 22 by 22 averaged the lowest yield per acre in the first five harvests, bearing an average of 265 boxes per acre. The highest-producing spacings in the 5th year of harvest were the 11 by 22 (677 boxes per ac), 11 by 11 (thinned to 11 by 22 the previous year and yielding 652 boxes per ac), and 11 by 18 (651 boxes per ac).

Spacing also had an effect on fruit quality. The deepest orange color in fruit peels was from trees spaced at 22 by 22. Full color development of fruit in the high-density plantings was delayed as much as 45 days after the fruit reached legal maturity, a problem if the grower is trying to harvest to an early market. Fruit size was also affected: fruit from the 22 by 22 spacing averaged one size larger than fruit from trees at the higher densities.

After the first 5 years of harvest, the best net income was from the trees spaced at 11 by 22 (+ $515 per ac); the 9 by 11 and 11 by 11 spacings lost money due to extra costs for tree planting and removal.

In the second 5 years of the experiment (1970–74), yield on the trees at wider spacings increased as trees continued to fill into their spaces, and during this time, production per acre was highest at a spacing of 22 by 22, and least on trees that began at a spacing of 9 feet in the rows (Boswell et al. 1975). These results were opposite those obtained in the first 5 years of the trial. During the second 5-year period, trees at 22 by 22 produced an average of 13,805 pounds per acre per year, and trees at 11 by 22 produced the second-best yield, averaging 12,648 pounds per acre per year.

For the entire 10-year trial, there were no significant differences in average yield per year from the 11 spacings used in the trial. However, trees spaced at 11 by 22 had the highest overall average yield (16,389 lb per ac per year). At the conclusion of the trial, a cost study revealed that the highest net income came from trees spaced at 11 by 22 (+ $7,335), and the least net income came from trees originally planted at 9 by 11 (+ $2,704). The researchers concluded that, especially in close spacings, pruning of vigorous trees to keep them within an allotted space is apparently detrimental to fruit production (Boswell et al. 1970).

Conclusions

It is important to note that the trees grown in the trial described above were removed by the grower after 10 years of production. In general, there is a lack of long-term information on various spacings of citrus in California. More studies are needed, especially on the interactions of various types of citrus, cultivars, rootstocks, soils, and climate. It would appear from the research by Boswell and colleagues that under the conditions of their experiments, moderate spacings provide the best balance between early yields, early returns, and long-term sustainability of economically successful fruit production. Spacing suggestions made in 1966 by University of California Cooperative Extension Farm Advisors R. M. Burns and B. W. Lee for orchards planted in Ventura County are still applicable today (see table 9.2; for the number of trees per acre in selected spacings, see table 9.3).

Table 9.2. Planting distance (ft) for citrus trees in California

Citrus type	In the row	Between the rows
	Lemons*	
Eureka	16–18	20–22
Lisbon	18–20	22–24
	Oranges†	
Valencia	9–18	20–22
Navel	9–18	20–22

Source: Burns and Lee 1966.

Notes:

*These spacings are designed to account for individuality in trees: Lisbon trees are often more vigorous, larger trees and should be planted farther apart than Eureka.

†With the closer spacings (9–11 ft), every other tree should be removed when crowding occurs. If a grower is not concerned with greater early production, a spacing of 18 by 20 or 22 feet would be suggested.

Table 9.3. Number of trees per acre at selected spacings

Distance between trees in the row (ft)	Distance between the rows (ft)		
	20	22	24
9	242	214	201
10	217	198	181
11	198	180	165
12	181	165	151
14	156	141	130
16	136	124	113
18	121	110	100
20	109	99	91
22	99	90	83
24	91	82	76

Note: Number of trees per acre = 43,500 square feet per acre ÷ (distance between rows × distance between trees).

Row Orientation

In California there is apparently no published data on north-south versus east-west row orientation in citrus. On slopes, the trees should be oriented along the contour of the slopes to stabilize irrigation pressure in the lines. However, on flatter ground, the trees should be oriented in the north-south direction, especially if the trees are to be kept in a hedgerow (Wheaton et al. 1978). An advantage to north-south orientation is that light is distributed more evenly on both sides of the trees during the day. Row orientation should be determined at each particular site, in relation to slope, erosion control, air drainage, and convenience for equipment.

Orchard Layout

Once the soil has been prepared and the decision has been made on planting distances and orientation, the orchard spacing is marked out. Generally, a planting wire or string marked with the correct spacing is stretched across the field, starting at a baseline (usually a road). A flag, stake, or a handful of gypsum is placed at the site of each tree. The irrigation system should be completed before the trees are planted. For more information, see chapter 11, "Irrigation."

Planting the Trees

The best time to plant citrus trees in California is from mid-February to early May, after the danger of frost has passed and before the onset of high summer temperatures. Generally, nursery trees that are dug and balled earlier in the year are semidormant and are less subject to shock. As the season progresses and trees become active, digging the tree results in the loss of a lot of active roots, and the trees may sit for some time without growing; if the temperature suddenly becomes hot, many young trees can be lost. Trees that are grown in plastic pots tend to have less shock and loss, but if the trees are planted properly both types of trees usually have equal growth rates at the end of the year.

Trees come from the nursery balled (root ball and soil wrapped in burlap) or in plastic or tar paper pots. It is critical that balled trees and trees in pots be kept moist while waiting for planting. It is likely that, in the near future, all nursery trees will be grown in greenhouses and screenhouses due to the potential threat from the Asian citrus psyllid, and balled trees or trees in plastic pots grown outdoors will no longer be sold.

It is best to moisten the soil before digging the holes. Holes are usually dug with an auger mounted on a tractor, but they can leave a slick side to the hole, making it difficult for roots to penetrate, especially if the ground was not prepared well by ripping. Many growers still use shovels to dig the holes with the belief that roots make better contact with the soil.

Dig the hole slightly larger than the diameter of the root ball and deep enough so that the tree, when placed in the hole, is at the same height as it was in the nursery. This means that the top of the tree ball will be just even with the surrounding ground. It is very important that the tree does not settle so that the bud union is near or below the soil surface. If the scion above the bud union comes in contact with soil, it will likely be susceptible to infection by *Phytophthora* fungi. Watch the planting crew to verify that trees are not planted too deep, or too shallow, or pushed into the hole by foot, fracturing the root ball.

For balled trees, backfill the hole about three-quarters full, tamp the soil well to eliminate air pockets, and cut the twine around the trunk. Fold the burlap into the hole, then fill and tamp the soil. Water the soil immediately; the best method is to form a basin and pour 5 gallons of water into the hole to provide enough moisture for the trees until the planting is completed and the irrigation system can be turned on.

For trees grown in open-bottomed containers, lower the container into the hole while supporting the bottom of the root ball by hand. Slit the container vertically with a knife while in the hole. Pack 6 inches around the base of the container and lift the container out. This process reduces fracturing of the lower ball and roots. For trees in closed-bottom containers or pots, roll the container on the ground to loosen roots from the side of the container. Then gently slide the root ball of the container onto a piece of cardboard. Using the cardboard to support the root ball, gently place the tree into the hole and backfill with tamping, as described above.

References

Ayers, R. S. 1977. Quality of water for irrigation. Journal of the Irrigation and Drain Division 103:135–154.

Bender, G. S., V. W. Smothers, and L. Francis. 2003. Site selection, road engineering and erosion control. In G. S. Bender, ed., Avocado production in California. Book one: A cultural handbook for growers. University of California Cooperative Extension, San Diego County. 57–68.

Boswell, S. B., L. N. Lewis, C. D. McCarty, and K. W. Hench. 1970. Tree spacing of 'Washington' navel orange. Journal of the American Society for Horticultural Science 95(5): 523–528.

Boswell, S. B., C. D. McCarty, K. W. Hench, and L. N. Lewis. 1975. Effect of tree density on the first ten years of growth and production of 'Washington' navel orange trees. Journal of the American Society for Horticultural Science 100(4): 370–373.

Burns, R. M., and B. W. Lee. 1966. Planting distances for citrus: Citrimation for Ventura County. University of California Cooperative Extension Ventura County.

Ingels, C. A., R. L. Bugg, G. T. McGourty, and L. P. Christensen. 1998. Cover cropping in vineyards: A grower's handbook. Oakland: University of California Agriculture and Natural Resources Publication 3338.

Wheaton, T. A., W. S. Castle, D. P. H. Tucker, and J. P. Whitney. 1978. Higher density plantings for Florida citrus: Concepts. Proceedings of the Florida State Horticultural Society 91:27–33.

Whitney, J. D., and T. A. Wheaton. 1984. Tree spacing affects citrus fruit distribution and yield. Proceedings of the Florida State Horticultural Society 97:44–47.

Part 4 Citrus Orchard Management

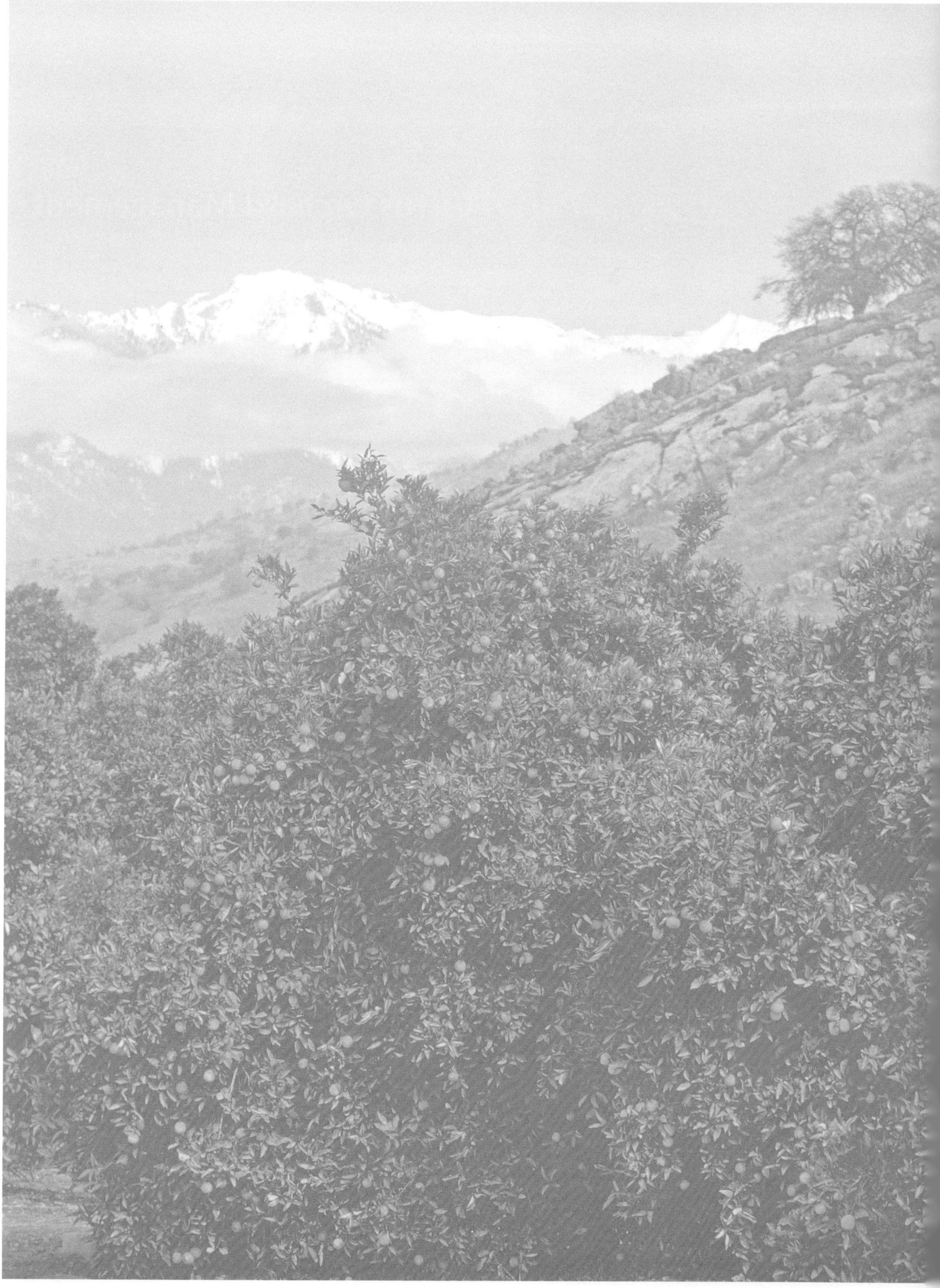

10 Nutrient Deficiency and Correction

Carol J. Lovatt

Essential Elements

Plants are unique among most other organisms because they are autotrophs, organisms that can produce their own food. In the process of photosynthesis, plants use simple inorganic molecules of carbon dioxide (CO_2) and water (H_2O) and light energy captured by chlorophyll to synthesize energy-rich organic food molecules. Plants must break down the energy-rich organic food molecules by the process of cellular respiration to obtain the energy to do cellular work (metabolism) and to obtain carbon, hydrogen, and oxygen to use as building materials for new and different molecules.

In addition to organic molecules, plants require mineral nutrient elements that are essential for their existence, health, and well-being. Plants and animals have a common requirement for many nutrients, making plants a rich source of these nutrients for animals. An element is considered an essential nutrient if the organism cannot complete its life cycle if this element is lacking, the element cannot be replaced by another element, and the element has a distinct function (Arnon and Stout 1939). For example, zinc is an essential cofactor for the enzyme alcohol dehydrogenase in both plants and humans, and magnesium is essential to plants as the central atom of the chlorophyll molecule.

Plants require 17 essential nutrient elements: carbon (C), oxygen (O), hydrogen (H), nitrogen (N), phosphorus (P), potassium (K), calcium (Ca), magnesium (Mg), sulfur (S), iron (Fe), chlorine (Cl), manganese (Mn), boron (B), zinc (Zn), copper (Cu), molybdenum (Mo), and nickel (Ni). Nutrient elements required by plants in large quantities (C, O, H, N, P, K, Ca, Mg, and S) are referred to as macronutrients; those nutrients required in smaller quantities (Fe, Cl, Mn, B, Zn, Cu, Mo, and Ni) are referred to as micronutrients. Table 10.1 illustrates the average relative concentrations of the essential nutrient elements in dry plant material at levels considered adequate. Note that the macronutrient required at the lowest concentration, sulfur, is still required at a concentration 10-fold greater than the concentration of the micronutrient required in the greatest amount, chlorine. The optimal proportion of one nutrient to another nutrient (nutrient balance) varies among plants, including citrus. It is important to maintain nutrient balance to optimize yield parameters of economic value (see the section "Leaf Analysis").

Table 10.1. Relative concentrations of nutrient elements in plant material at levels considered adequate

Element	Chemical symbol	Concentration in dry matter (μmole/g)
hydrogen	H	60,000
carbon	C	40,000
oxygen	O	30,000
nitrogen	N	1,000
potassium	K	250
calcium	Ca	125
magnesium	Mg	80
phosphorus	P	60
sulfur	S	30
chlorine	Cl	3
iron	Fe	2
boron	B	2
manganese	Mn	1
zinc	Zn	0.3
copper	Cu	0.1
molybdenum	Mo	0.001
nickel	Ni	0.0002

Source: Epstein 1972.

Information regarding the role of each essential nutrient in plants in general as well as in citrus physiology is increasing. Research results on the typical mode of uptake and the efficacy of foliar uptake are also available. All nutrients must be in solution to be taken up by the cells of any plant. Ion channels in plant cell membranes facilitate the uptake of some essential elements. Plants contain highly selective channels for the cations H^+, K^+, and Ca^{+2}. When the soil nitrate concentration is high, nitrate is taken up by a low-affinity transport system, which may be a carrier protein or anion channel, through which chloride is also taken up. When soil nitrate availability is low, a high-affinity nitrate-proton cotransport protein works with an ATPase proton pump, which catabolizes ATP to obtain the energy required for the uptake of nitrate and other anions, including chloride. Since nitrate and chloride are taken up by the same carrier systems, increasing the concentration of one can reduce the uptake of the other. In contrast, ammonium is taken up by high- and low-affinity ammonium-specific carrier systems. Nutrients taken up by the roots are moved in solution to other parts of the plant via the transpiration stream that flows through the conducting cells of the xylem transport tissue. Once in the plant, nutrients can also move from one part of the plant to another by translocation in the conducting cells of the phloem transport tissue, but the rate of this transport depends on the nutrient. The relative mobility of each of the essential nutrient elements is also presented below, along with the consequences of nutrient mobility versus immobility on the development of symptoms of deficiency or toxicity. The relative mobility of each nutrient and relative rates of foliar uptake for each nutrient are summarized in table 10.2.

Macronutrients

Carbon

Carbon is taken up as carbon dioxide (CO_2) by leaves and as bicarbonate (HCO_3^-) by roots. Carbon is essential to all major classes of molecules found in living cells: sugars, amino acids, fatty acids, nucleotides, and hormones.

Oxygen

Oxygen is absorbed as O_2 by leaves and roots and is also a component of all major classes of molecules found in living cells: sugars, amino acids, fatty acids, nucleotides, and hormones.

Hydrogen

Hydrogen is absorbed as part of the water molecule (H_2O) and is also a constituent of the major molecules of life: sugars, amino acids, fatty acids, nucleotides, and hormones.

Table 10.2. Relative absorption by leaves and mobility in the phloem

Nutrient	Absorption	Mobility	Leaves showing first symptoms of deficiency
boron	moderate	partial	young
calcium	moderate	immobile	young
chlorine	rapid	mobile	mature
copper	slow	partial	young
iron	slow	partial	young
magnesium	rapid	mobile	mature
manganese	moderate	partial	young
molybdenum	slow	partial	young
phosphorus	slow	mobile	mature
potassium	rapid	mobile	mature
sulfur	moderate	mobile	young
urea nitrogen	rapid	mobile	young
zinc	moderate	partial	young

Nitrogen

Roots absorb nitrogen as nitrate (NO_3^-) or ammonium (NH_4^+). For citrus leaves, urea-nitrogen uptake is greater than ammonium or nitrate uptake. Even mature, hardened citrus leaves can take up urea-nitrogen. Once in the plant, nitrogen is very mobile (see table 10.2). Nitrate must be reduced to ammonium (NH_4^+) in order to be used by plant cells. The reduction of nitrate to ammonium requires energy and thus is more costly to the plant than the metabolism of urea or ammonium nitrogen fertilizers. Ammonium is incorporated into amino acids, which are the basic units of proteins, and also into nucleotides, which are the building blocks of DNA and RNA. In chlorophyll, the green pigment of the plant that is essential for photosynthesis, four nitrogen atoms hold the central magnesium atom in place, and 75% of the nitrogen in a leaf is contained in ribulose-1,5-bisphosphate carboxylase-oxygenase, the first enzyme in photosynthesis. When nitrogen is limiting, this enzyme and chlorophyll are degraded and photosynthesis is reduced. Several plant hormones contain nitrogen. Nutrients that are highly mobile in the phloem, like nitrogen, typically exhibit deficiency symptoms in older leaves before new leaves, but because nitrogen is used in the synthesis of complex molecules essential for growth, new leaves develop symptoms of nitrogen deficiency first (see table 10.2) (figs. 10.1, 10.2, and 10.3). Nitrogen deficiency of new citrus leaves is frequently observed in early spring when soils are too cold or wet to adequately supply nitrogen to the new foliage. Foliar urea-nitrogen applications are an excellent means to meet the tree's demand for nitrogen at this time.

Phosphorus

Roots absorb phosphorus mostly as a dihydrogenphosphate ion ($H_2PO_4^-$) because of its greater availability in the soil solution, but roots also take up monohydrogenphosphate (HPO_4^{-2}). Foliar uptake of dihydrogenphosphate is limited for most plant species, especially citrus, but it is taken up in its less-oxidized state as dihydrogenphosphonate (phosphite) ($H_2PO_3^-$). Once in plant tissues, phosphorus is very mobile in the phloem and is moved from old tissues to new (see table 10.2), so old tissues show deficiency first (figs. 10.4 and 10.5). Phosphorus is one of three quantitatively prominent elements to be absorbed as complex ions, the other two being nitrate and sulfate (SO_4^{-2}). However, phosphate is not reduced to a lower oxidation state in the cell, whereas nitrate is reduced to ammonium and sulfate to hydrogen sulfide (SH_2).

Phosphate plays a key role in energy metabolism. ATP is the universal "energy currency" for all living cells. ATP, ADP, AMP, sugar phosphates, and inorganic phosphate concentrations regulate enzymes and metabolic pathways, including photosynthesis. Phosphorus is found in phospholipids, including those that make up membranes, sugar phosphates, and nucleotides forming DNA and RNA. Phosphorus deficiency inhibits the export of sugar from the chloroplast, reduces the translocation of sucrose to other parts of the tree, and alters both carbohydrate and nitrogen metabolism. Due to the key role that phosphorus plays in regulating the cell's energy and biosynthetic reactions, it is important in protecting the plant from stress. Phosphorus deficiency is as disastrous as nitrogen deficiency. Flowers or fruit may abscise prematurely. For sweet orange cultivars, phosphorus deficiency results in rough peels (see fig. 10.5).

Phosphorus deficiency can occur when soil pH is low (< 4), when the phosphate molecule becomes chemically bonded to soil organic matter. Phosphorus must be applied close to the plant's roots in order for the plant to use it. Large applications of phosphorus without adequate levels of available zinc can cause zinc deficiency.

Potassium

Potassium is absorbed as K^+ both by roots and leaves. Its foliar uptake is considered rapid (see table 10.2). Once in the plant, potassium is very mobile in the phloem. Potassium is removed from older tissue; therefore older tissue becomes deficient first. Potassium is involved in enzyme activation and helps in

Figure 10.1 Grapefruit leaves with high (left), low (middle), and deficient (right) nitrogen concentrations. *Photo:* T. W. Embleton.

Figure 10.2 Nitrogen-deficient leaves from 25-year-old Marsh grapefruit in Corona, California. *Photo:* W. P. Bitters.

Figure 10.3 Nitrogen-sufficient (left) and -deficient (right) lemon leaves. *Photo:* T. W. Embleton.

Figure 10.4 Phosphorus-deficient lemon leaves. *Photo:* T. W. Embleton.

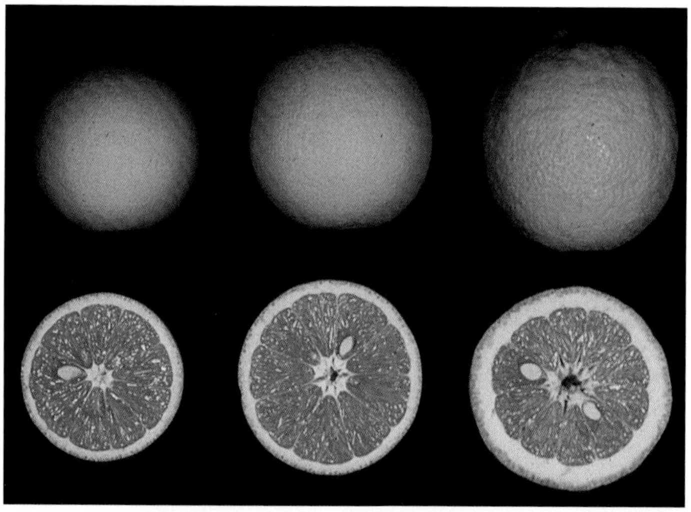

Figure 10.5 Valencia fruit from trees with leaf phosphorus concentrations that are, left to right, in excess (0.18% P), normal (0.13–0.14% P), and deficient (0.11–0.12% P, note rough peel). *Photo:* T. W. Embleton.

Figure 10.6 Potassium-deficient grapefruit leaves. *Photo:* T. W. Embleton.

the building of proteins, photosynthesis, and fruit quality by enhancing the flavor and color of fruit and vegetable crops, and it also helps protect plants from diseases by making them "tougher." Potassium also improves cold hardiness. It has a well-known role in regulating stomatal opening and closing by acting as an osmoticum in the guard cells, the two cells surrounding each stomate (opening in the leaf). Open stomates have guard cells with high concentrations of potassium per guard cell; closed stomates have low concentrations of potassium per guard cell. As potassium moves into the guard cells, it lowers the water potential inside the guard cells, and water flows into them. This creates hydrostatic (turgor) pressure in the guard cells, causing them to bend. The bending of the guard cells opens the stomate. As potassium leaves the guard cells, water also flows out of them, reducing the water pressure and leaving the guard cells flaccid and collapsed on each other, which closes the stomata (Hopkins and Hüner 2004). Potassium is also important to the water balance and ionic balance of plant cells. Potassium deficiency results in low yields, mottled or curled leaves, or yellow or necrotic edges of leaves (figs. 10.6, 10.7, and 10.8). In soil, potassium is highly mobile and is easily displaced from the soil cation exchange capacity. However, soils with greater cation exchange capacity (high organic matter) are less apt to become potassium deficient.

Calcium

Roots or leaves absorb calcium as Ca^{+2} when supplied as a salt. The anion in the salt will have a

Figure 10.7 Potassium-deficient lemon leaves. *Photo:* T. W. Embleton.

Figure 10.8 Potassium-deficient lemon leaves and fruit. *Photo:* T. W. Embleton.

significant effect on its solubility and thus its availability for uptake by roots or leaves. Foliar uptake of calcium is considered moderate, but within the plant calcium is immobile, that is, not translocated in the phloem. Calcium moves freely into the conducting cells of the xylem only between the root apex and zone of root elongation, where the endodermis and Casparian strip have not developed. Once in the xylem, calcium moves in the transpiration stream. Thus, calcium deficiency results in serious injury to shoot apical meristems because they have limited rates of transpiration.

Calcium is a structural component of cell walls and occurs in considerable quantities in the cell walls and middle lamella as calcium pectate, the glue that holds cells together, and it apparently influences cell wall elasticity and rigidity by forming ionic bridges between cell wall components. Indole-3-acetic acid (IAA) (at a low pH of 3.0) leads to less calcium bonding between carboxyl groups of cell wall components, making the wall less rigid and more elastic and resulting in growth by cell expansion. In addition to cell expansion, calcium is also necessary for cell division. Calcium is important in maintaining membrane integrity, plays a role in the response of roots to gravity, and is an important second messenger in signal transduction; the activity of many enzymes is calcium dependent.

Calcium accumulates as calcium oxalate crystals and calcium carbonate in leaves and woody tissues of trees. Calcium deficiency leads to stem collapse due to loss of cell wall rigidity. Thus, calcium also plays a role in protecting plants from invasion by disease pathogens. Calcium has been implicated in membrane function for years. It imparts selectivity to membranes; reduces sodium uptake; prevents cell leakage, which is important in regulating root exudation; and protects plants from high salt concentrations, excess boron, and other toxic ions. Since calcium is easily leached from the soil and is immobile once deposited in plant tissue, it must be constantly supplied during periods of high demand to maintain normal growth and fruit quality. Foliar uptake of calcium is moderate; foliar calcium fertilization is efficient for correcting calcium deficiency quickly, but due to the need to constantly supply calcium, application of small amounts to the soil through irrigation is optimal for preventing deficiency. Calcium deficiency causes stunting of new growth of stems, flowers, and roots. Symptoms range from distorted new growth to black spots on leaves and fruit. Yellow leaf margins may also appear. Calcium deficiency has been implicated in citrus rind disorders.

Magnesium

Magnesium is absorbed as Mg^{+2} by roots or leaves if provided as a salt. Its uptake through foliage and mobility within the plant is considered moderate. Magnesium serves as a structural component and is involved as an enzyme cofactor in many enzyme reactions. The role of magnesium as the central atom in the chlorophyll molecule is well known, but the magnesium content of chlorophyll is only 10% of the total magnesium in a leaf. Magnesium is also required for ribosome integrity and thus protein synthesis. Almost every phosphorylating enzyme in carbohydrate metabolism requires magnesium; ATP must be complexed with magnesium (or manganese). Pyruvate dehydrogenase kinase and phosphatase require magnesium to inactivate the pyruvate dehydrogenase complex by phosphorylation and to reactivate this enzyme by removal of the phosphate. The pyruvate dehydrogenase complex regulates the Kreb's tricarboxylic acid (TCA) cycle that controls citric acid metabolism and ATP synthesis by cellular respiration, which occurs in the mitochondria of plant cells.

A common phenomenon in fruit trees is antagonism between magnesium and calcium or magnesium and potassium, so that increased soil fertilization of one leads to diminished uptake and deficiency of the other. Unlike calcium, magnesium is readily translocated from mature older leaves to young, actively growing regions of the plant (see table 10.2). As a result, mature leaves show deficiency symptoms first. Deficient plants appear chlorotic (show yellowing between veins of older leaves), and leaves may drop (figs. 10.9, 10.10, and 10.11). Magnesium is leached by watering and must be resupplied. It can be applied as a foliar spray to correct deficiencies; its uptake by leaves is rapid (see table 10.2).

Figure 10.9 Magnesium-deficient navel orange leaves. *Photo:* T. W. Embleton.

Figure 10.10 Magnesium-deficient lemon leaves (0.03–0.15% Mg). *Photo:* T. W. Embleton.

Figure 10.11 Magnesium-deficient leaves from lemon on Macrophylla rootstock. *Photo:* W. P. Bitters.

Sulfur

Sulfur is taken up as sulfate (SO_4^{-2}) by plants and has to be activated through phosphorylation via ATP in order to be reduced to sulfite (SO_3^{-2}) and then sulfide (S^{-2}) to be incorporated into cysteine and methionine, amino acids essential for protein synthesis. Reduction of sulfate to sulfide is a many step, energy expensive process, requiring not only ATP but also reducing power produced by photosynthesis. Sulfur is found in vitamin cofactors that regulate enzymes, including those essential for chlorophyll synthesis. Sulfur is a key constituent of glutathione and ascorbate, important antioxidants that protect plants from reactive oxygen species formed during stress. Sulfur-containing compounds impart flavor to many vegetables, but not fruit. Because sulfur is essential to protein and chlorophyll synthesis, as is nitrogen, sulfur deficiency results in symptoms similar to nitrogen deficiency: light green young leaves. Sulfur is readily lost by leaching from soils and should be applied. Foliar uptake of sulfur is moderate, but once in the plant, it is highly mobile. The use of gypsum increases soil sulfur levels. Some water supplies may contain sulfur; it is also an air pollutant that can be washed from the air onto leaves and soil for uptake in significant quantities in some locations. The greater purity of fertilizers has reduced the amount of sulfur that was once supplied through standard fertilization with lower-grade fertilizers. Once nitrogen, phosphorus, potassium, and calcium fertilization are optimized for a crop, the next greatest gain in yield for many crops has resulted from optimizing sulfur.

Micronutrients

Iron

Iron is absorbed by roots and leaves as a cation, either Fe^{+2} or Fe^{+3}. Approximately 75% of the total iron content of the leaf is in the chloroplast, where iron is required for the synthesis of chlorophyll, plays a role in the synthesis of chloroplast proteins, and is essential for electron transport and energy production in photosynthesis. Iron is also in the mitochondrial cytochromes involved in plant and animal cellular respiration and energy production.

Iron uptake through the leaves is slow, and it is only partially mobile in the phloem (see table 10.2). Organic acid levels, especially citric acid, influence translocation of absorbed iron. A complex of the two acts like a chelated species of iron, making it more soluble and reducing interactions with anions. Iron is poorly translocated out of older tissues, so deficiency symptoms show up in new tissue first. Young leaves and fruit are pale due to the inability to synthesize chlorophyll. Subsequently, the leaves, including large veins, and fruit turn yellow (figs. 10.12–10.15). Iron deficiency is one of the most common and conspicuous micronutrient deficiencies in trees. It occurs chiefly on alkaline and calcareous soils, where the higher pH prevents iron absorption.

Chlorine

Chlorine is absorbed as chloride (Cl^-). Chloride uptake is rapid, and within the plant, chloride is mobile in the phloem (see table 10.2). It is essential for the water-splitting oxygen-producing step in photosynthesis and is co-transported with potassium into and out of the guard cells in stomatal opening and closing, respectively. It also functions as a cofactor to several enzymes and has a role with other anions in balancing the ratio of cations to anions in cellular compartments. Deficiency is rare. Symptoms include wilting, stubby roots, chlorosis, and bronzing. Chlorine was the second to last element to be demonstrated as essential to plants. Chlorine excess is a major problem in irrigated crop production, and chloride can harm citrus trees (see chapter 8, "Soil and Water Analysis and Amendment").

Manganese

Manganese is absorbed by roots and leaves as Mn^{+2} if provided as a salt. Foliar uptake is moderate, and manganese is partially mobile in the phloem (see table 10.2). Some rootstocks oxidize Mn^{+2} to manganese dioxide (MnO_2), which remains at the root surface. Manganese is required for the water-splitting oxygen-evolving process in photosynthesis. It also plays a prominent role in many reactions of the Kreb's TCA cycle, which is important to both energy and carbon metabolism. Manganese is involved in oxidation-reduction reactions, decarboxylation reactions, and hydrolysis reactions, and it can replace magnesium in phosphate group transfers from ATP. It influences the level of IAA in plants by regulating IAA degradation.

Manganese-deficient young leaves may show a network of green veins on a light green background, similar to iron deficiency. In the advanced stages, the light green parts become white, and leaves are shed. Brownish, black, or grayish spots may appear next to the veins. Manganese deficiency occurs most frequently on neutral or alkaline soils. In highly acidic soils, manganese may be available to the extent that it causes toxicity, especially in evergreen tree crops due to accumulation in leaves that persist for multiple years. Manganese is toxic at all but low concentrations and can accumulate in leaves of evergreen tree crops to toxic levels. Excess manganese reduces IAA concentrations by increasing degradation and compromises physiological processes regulated by this

Figure 10.12 Iron-deficient orange leaves. *Photo:* T. W. Embleton.

Figure 10.14 Iron-deficient lemon leaves. *Photo:* T. W. Embleton.

Figure 10.13 Iron-deficient grapefruit leaves. *Photo:* T. W. Embleton.

Figure 10.15 Comparison of iron-sufficient (left) and -deficient (right) lemon leaves and fruit. *Photo:* T. W. Embleton.

hormone. High concentrations of manganese in the soil can induce iron deficiency.

Boron

Boron is usually found in soils as boric acid (H_3BO_3), borate ($H_2BO_3^-$), or sodium metaborate ($NaB(OH)_4$) and in the plant as boric acid. Boron foliar uptake is moderate (see table 10.2). Older work suggested that boron was immobile in the phloem, but more recent work suggests that it is partially mobile and that boron mobility is species dependent (Brown and Hu 1996). Boron complexes with sugar alcohols and is translocated efficiently in tree crops that translocate carbon predominantly as sorbitol or other sugar alcohol, rather than sucrose. Citrus species translocate sucrose.

Boron has been demonstrated to be required for cell division and normal nucleic acid and carbohydrate metabolism in cell wall formation, and membrane integrity. Boron deficiency increases the activity of the shikimic acid pathway, resulting in increased synthesis of phenolics, superhigh levels of indole-3-acetic acid, and increased lignification of cell walls. Boron's involvement in these fundamental processes impacts at least 16 functions in plants, including flowering, pollen germination and pollen tube growth, fruit set and development, root and shoot growth, water relations, synthesis of propenoid secondary metabolites, and hormone balance. Boron deficiency symptoms in citrus include corking of the veins on the underside of the leaf, the appearance of pinholes in the leaf, shoot tip and twig dieback, and gum pockets in juice vesicles closest to the central core of the fruit. Boron-deficient leaves are thick, curled, and brittle (fig. 10.16), with corking of the underside of the veins (fig. 10.17). Fruit and roots are discolored, cracked, and flecked with brown spots.

Boron must be available throughout the life of the plant. Its translocation is limited, and it is easily leached from soils, so deficiency is common on soils low in boron. Within the plant, the range in concentration from deficient to toxic is very narrow. For tomato, 0.1 ppm is optimal, 0.01 ppm is deficient, and 1.0 ppm is toxic. Trees are not quite as sensitive as tomato. For citrus, less than 30 ppm is deficient, 30 to 100 is optimal, and greater than 100 ppm is considered too high. Boron toxicity is a problem in irrigated crop production. Boron moves in the transpiration stream through the xylem and accumulates in leaf tips at the end of the transpiration stream. Leaves can show boron toxicity, while shoot apical meristems, which have low rates of transpiration, show deficiency. Deficiency kills the apical meristem of roots, resulting in short roots with many lateral roots that develop all the way to the root apex, and the shoot apical meristem of terminal buds, resulting in rosette growth of the plant.

In irrigated agriculture, excess boron can be a problem. The symptoms of boron excess are similar to those of excess salt and biuret toxicity: tip burn and necrosis that expands downward along the margin of the leaf. The leaf blade appears watery or yellow in areas (figs. 10.18–10.23).

Zinc

Zinc is absorbed as Zn^{+2} by roots and leaves when supplied as a salt. Zinc can be taken up through the leaves and is considered partially mobile in the phloem (see table 10.2). Zinc is required as an enzyme cofactor for tryptophan synthase in the synthesis of IAA and for alcohol dehydrogenase in glucose metabolism during hypoxia of flooded or waterlogged soils.

Zinc-deficient plants have mottled leaves with irregular chlorotic areas (figs. 10.24 and 10.25). Zinc deficiency can cause puckering (cupping) of the leaf blade. Zinc deficiency in tree crops sometimes produces symptoms similar to virus or bacterial diseases. Little leaf is probably due to a lack of IAA; frenching looks like a lack of GA—that is, very short internodes. Citrus trees require from 25 to 100 ppm of zinc for optimal growth. Deficiencies can be quickly corrected by foliar application of zinc sulfate ($ZnSO_4$, 300 ppm), zinc oxide (ZnO_2), or zinc chelate. Zinc deficiency leads to iron deficiency, which causes similar symptoms. Deficiency occurs on eroded soils. The solubility of zinc is highly dependent upon soil pH. Zinc is least available when soils are in the pH range of 5.5 to 7.0. Alkaline calcareous soils have low levels of available zinc due to the formation of zinc carbonate; similarly, high bicarbonate and carbonate concentrations in irrigation water can reduce zinc availability. Lowering the pH can render zinc more available, possibly to the point of toxicity if the soil zinc concentration is high.

Copper

Copper is absorbed by roots and leaves as Cu^{+1} or Cu^{+2}. Copper foliar uptake is considered slow, and it is only partially mobile in the phloem (see table 10.2). Copper is part of the proteins that transfer electrons in photosynthesis and cellular respiration leading to ATP synthesis and reducing power essential for normal metabolism. Copper is required in enzyme reactions where molecular oxygen is used directly for oxidation of the substrate by ascorbic acid oxidase, tyrosinase, cytochrome oxidase, or plastocyanin. It is also required by the enzyme superoxide dismutase, which protects the plant's cells from reactive oxygen species formed during abiotic and biotic stress. Copper is also essential for the biosynthesis of lignin, the three-dimensional carbohydrate mesh that gives wood its strength.

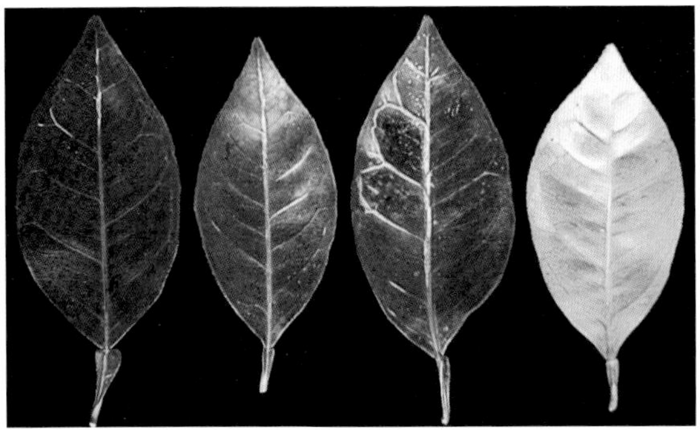

Figure 10.16 Boron-deficient leaves (5 ppm B) of navel orange on Trifoliate Orange rootstock. *Photo:* T. W. Embleton.

Figure 10.17 The corky vein symptom of boron-deficient navel orange leaves (5 ppm B). *Photo:* T. W. Embleton.

Figure 10.18 Boron-deficient leaves from 10-year-old lemon trees. *Photo:* C. J. Lovatt.

Figure 10.19 A boron-deficient lemon leaf with corking of the veins typical of boron deficiency. *Photo:* C. J. Lovatt.

Figure 10.20 Symptoms of boron excess in grapefruit leaves (760–1,150 ppm B). *Photo:* T. W. Embleton.

Figure 10.21 Symptoms of boron excess in leaves of navel orange. *Photo:* T. W. Embleton.

Figure 10.22 Symptoms of boron excess in lemon leaves. *Photo:* W. P. Bitters.

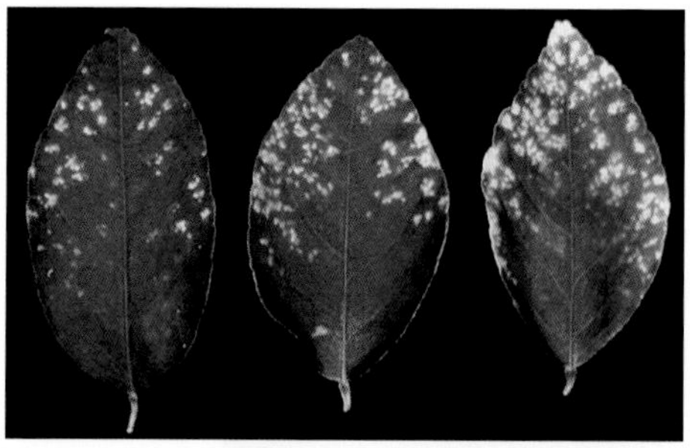

Figure 10.23 Symptoms of boron excess in leaves of Bearss lime. *Photo:* T. W. Embleton.

Figure 10.24 Zinc-deficient navel orange leaves. *Photo:* T. W. Embleton.

Figure 10.25 Zinc-deficient lemon foliage. *Photo:* T. W. Embleton.

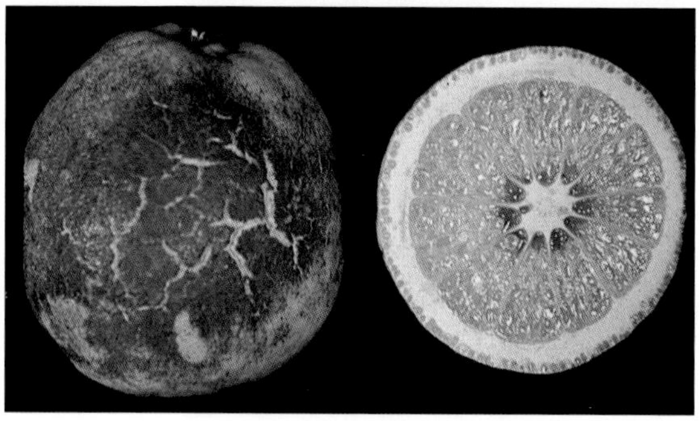

Figure 10.26 Copper-deficient fruit of navel orange on Troyer Citrange rootstock. *Photo:* T. W. Embleton.

Copper is required by citrus in low concentrations (5 to 16 ppm). It is bound tightly in organic matter and may be deficient in highly organic soils. It is not readily lost from soil, but it may be unavailable. Copper deficiency causes leaves at the terminus of the shoot to develop brown spots and abscise, and it causes shoot dieback in citrus (figs. 10.26–10.31). Twigs take on an S-shaped curve, and the tree eventually becomes twisted and malformed. Gum pockets form inside the shoots. Fruit from copper-deficient citrus trees are brown and cracked externally, with gum pockets in the juice vesicles near the central core. Copper-deficient trees have low levels of reducing sugars and high levels of organic acids.

Copper is toxic at high concentrations. Damage can result from foliar application of copper fungicide that contains too much dissolved or soluble copper. This damage appears as green to brown spots or patches on leaves or fruit (figs. 10.32, 10.33, and 10.34).

Molybdenum

Molybdenum is taken up as Mo^{+2} by roots and leaves. Molybdenum uptake through the foliage is slow, but it is partially mobile in the phloem once in the plant. Less than 1 ppm of molybdenum is sufficient for most plants (for citrus, 0.1 to 3.0 ppm is optimal). Molybdenum is part of the nitrate reductase complex that reduces nitrate to ammonium. Thus, molybdenum-deficient plants become nitrogen deficient if nitrate fertilizer is the only source of nitrogen. If urea- or ammonia-based fertilizers are the source of nitrogen, molybdenum deficiency does not affect nitrogen metabolism. Molybdenum is also a cofactor for several other enzymes, including those involved in purine metabolism, ABA biosynthesis, and sulfur metabolism. Molybdenum is necessary for pollen formation. Molybdenum deficiency is not common in tree crops, as soil typically provides sufficient molybdenum to meet the needs of the tree. Deficiency symptoms are pale green leaves with rolled or cupped margins. Molybdenum enhances iron uptake, and molybdenum-deficient plants can become iron deficient.

Nickel

Ni^{+2} is a cofactor for urease, the enzyme that catalyzes the breakdown of urea into ammonium and carbon dioxide, both of which are reused by the plant. Nickel was only recently added to the list of mineral nutrient elements essential to plants. It is required at even lower concentrations than molybdenum. Nickel deficiency is rare in tree crops.

Limiting Factors

Agronomic and horticultural crops can grow and produce fruit only to the level supported by the most limiting factor. This is known as Liebig's Law of the Minimum, after Justus von Liebig. In 1862, Liebig observed that the factor affecting growth the most tends to limit growth independent of the other factors. For example, if all nutrients are available to a tree at optimal levels except iron, which is available at half the optimal rate, and this lower rate of iron is known to reduce yield by 60%, the tree will produce only 40% of the yield produced by a tree with all nutrients available at optimal levels. In this scenario, iron is the limiting factor (fig. 10.35). Each of the 17 essential nutrients can be a limiting factor. In a situation where there are multiple limiting factors, the most limiting factor sets the upper limit for yield. When this factor is brought to an optimal level, yield increases to the degree permitted by the next limiting factor.

Different stages in the tree's phenology (e.g., time of flowering, fruit set, June drop, root growth) have greater or lesser demands for the essential nutrients. Nutrients must be available at each stage of the tree development at levels able to meet the needs of that stage. The seasonal cycle of flowering, fruit set, and fruit development for the Washington navel orange in California is depicted in figure 10.36. Fruit set (early fruit drop) is the most critical stage of fruit development from the grower's point of view. It is during this period that the greatest gains in fruit retention influencing final yield can be made. Events during this period also impact fruit size and quality (Lovatt 1999). In California, citrus flowering and fruit set, periods of high nutrient demand, occur when soil temperatures are low. Soil temperatures are generally lower than 55°F from January to April in the citrus-growing areas of California (Hamid et al. 1988). Low soil temperature reduces root metabolic activity, solubility of nutrients in the soil solution, and nutrient transport in the transpiration stream. Thus, the ability of trees to use nutrients applied to the soil depends on many factors unrelated to nutrient demand or the presence of adequate amounts of each nutrient in the soil. With increased use of sprinkler, drip, or microsprinkler irrigation, there is a growing trend to divide the annual amount of fertilizer to be applied into 6 to 12 small applications. While this strategy reduces the potential for ground water nitrate pollution, it ignores tree phenology and nutrient demand. Thus, it is clear that there are likely transient periods of nutrient deficiency during key stages in the tree's phenology that affect yield, fruit size, or fruit quality.

Figure 10.27 Copper-deficient Valencia orange fruit. *Photo:* W. P. Bitters.

Figure 10.28 Gum pockets in fruit caused by copper deficiency in Valencia orange. *Photo:* W. P. Bitters.

Figure 10.29 Copper deficiency in lemon. *Photo:* T. W. Embleton.

Figure 10.30 Copper deficiency in lemon fruit. *Photo:* T. W. Embleton.

Figure 10.32 Copper damage to Valencia fruit and leaves caused by a copper spray. *Photo:* T. W. Embleton.

Figure 10.33 Copper damage to a lemon leaf and fruit caused by a copper spray. *Photo:* T. W. Embleton.

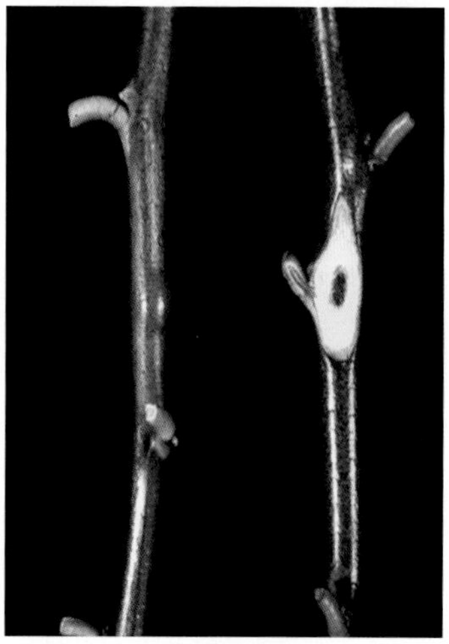

Figure 10.31 Gum pockets in shoots of copper-deficient lemon. *Photo:* T. W. Embleton.

Figure 10.34 Copper damage to a lemon fruit caused by a copper spray. *Photo:* T. W. Embleton.

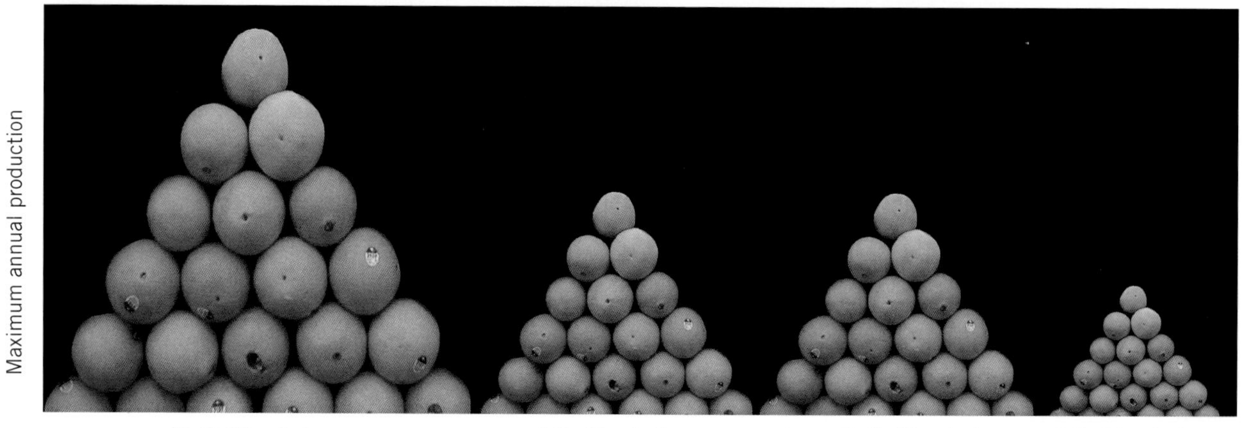

Figure 10.35 The concept of the limiting factor.

In addition to nutrient deficiencies, water, sunlight, temperature, disease, insect and nematode pests, salinity, nutrient toxicities, tree size and architecture, and pruning can become limiting factors. When yield is reduced by a limiting factor, all other factors supplied at the high rate required for maximum yield, fruit size, and quality, are, in part, wasted. In a well-managed orchard, these factors must be optimized, and eliminating temporary and early-stage nutrient deficiencies makes good economic sense.

Adequate Amounts of Essential Nutrients

Leaf Analysis

California citrus growers are fortunate that the optimal ranges for leaf concentrations of all essential nutrients supplied through the soil or applied as fertilizer are known, thanks to the considerable efforts of Homer D. Chapman, William W. Jones, Tom W. Embleton, Charles K. Labanauskas, and Robert G. Platt (Embleton et al. 1973, 1978). These ranges or threshold values are given in table 10.3. If leaf nutrient concentrations, as determined by laboratory analysis, are just below or above the optimal range, a deficiency or toxicity may be present or pending; similarly, leaf nutrient concentrations near the low or high end of the optimal range may indicate incipient deficiency or toxicity. Both low and high concentrations can have negative effects on yield, fruit size, or quality, without causing visible symptoms. Table 10.3 also provides threshold values that indicate sodium and chloride excess and toxicity (salinity).

Leaf nutrient values can also provide information about soil pH. High levels of iron, manganese, zinc, and aluminum, along with low levels of phosphorus, can indicate that the pH of the soil is dropping below 6.0 (fig. 10.37). Low levels of manganese, boron, copper, and zinc can indicate that the soil pH is approaching 4.5. Low levels of iron and manganese can also indicate that the soil pH is rising above 6.5; and low levels of boron, copper, and zinc can indicate that the pH is above 7.0. Test the soil pH to verify any of these conditions.

Leaf nutrient analysis does not reveal a transient nutrient deficiency. Leaves are analyzed at a specified time and do not reflect the availability of a given nutrient at other times in the tree's phenology. A good example of a transient nutrient deficiency in citrus is provided by the yellowing of leaves in the early spring when air temperatures are warm enough to promote shoot growth and flowering but soil temperatures are too low (or the soil is too wet) to support adequate nutrient uptake by roots to meet the demand of the shoots, inflorescences, and young fruit.

Frequency of leaf analysis

The longer or earlier in the season a tree's nutrient status remains at the low end or below the optimal range, the greater the negative effect on yield and the greater the response to fertilizer application. It is easier to correct an incipient deficiency than to overcome a severe deficiency. The citrus leaf analysis guidelines developed by Embleton et al. (1973, 1978) are best used for annual maintenance of tree nutrition (see table 10.3). Thus, leaf analysis should be done annually. Proper use of the leaf nutrient analysis guidelines will reduce the amount of fertilizer used, which saves the grower money and protects the environment.

Sampling for leaf analysis

The guidelines of Embleton et al. (1973, 1978, table 10.3) are applicable only to analysis of leaves sampled as follows. Use 5- to 7-month-old, spring flush, terminal healthy leaves from nonfruiting and nonflushing shoots for the sample. Leaves are typically sampled in September or October. The sample should consist of a minimum of 50 leaves collected at a height of 3 to 5 feet above the ground. The sample must adequately represent the block of trees for which the analysis is sought. The block should be visually evaluated for uniformity. Areas that differ in tree age, tree size, number of fruit set, foliage color or density, rootstock, or soil conditions should be sampled separately (as described below) and analyzed for the purpose of troubleshooting. Samples can be collected diagonally or in a U-shaped pattern across the block. Border trees should not be sampled. Collect two leaves from each selected tree from a quadrant in a pattern that advances by 90°. Thus, each tree quadrant will be equally represented in the sample. Five to 10 acres of trees that are uniform can be sampled as a single unit. If the results from 2 years of leaf sampling and analysis are similar for several uniformly fertilized units, the units can be combined for leaf sampling and analysis for the next 4 to 5 years and then sampled separately to determine whether any significant changes have occurred. Be sure to sample units individually after any significant change in fertilization or other cultural management practices. For troubleshooting, leaves from the trees in question should be collected separately using the sampling method provided above and analyzed for comparison with the uniform trees (Embleton et al. 1973, 1978).

Leaf samples should be promptly transported to the laboratory in clean plastic, paper, or porous bags in an ice chest. If ice is not available, they should be

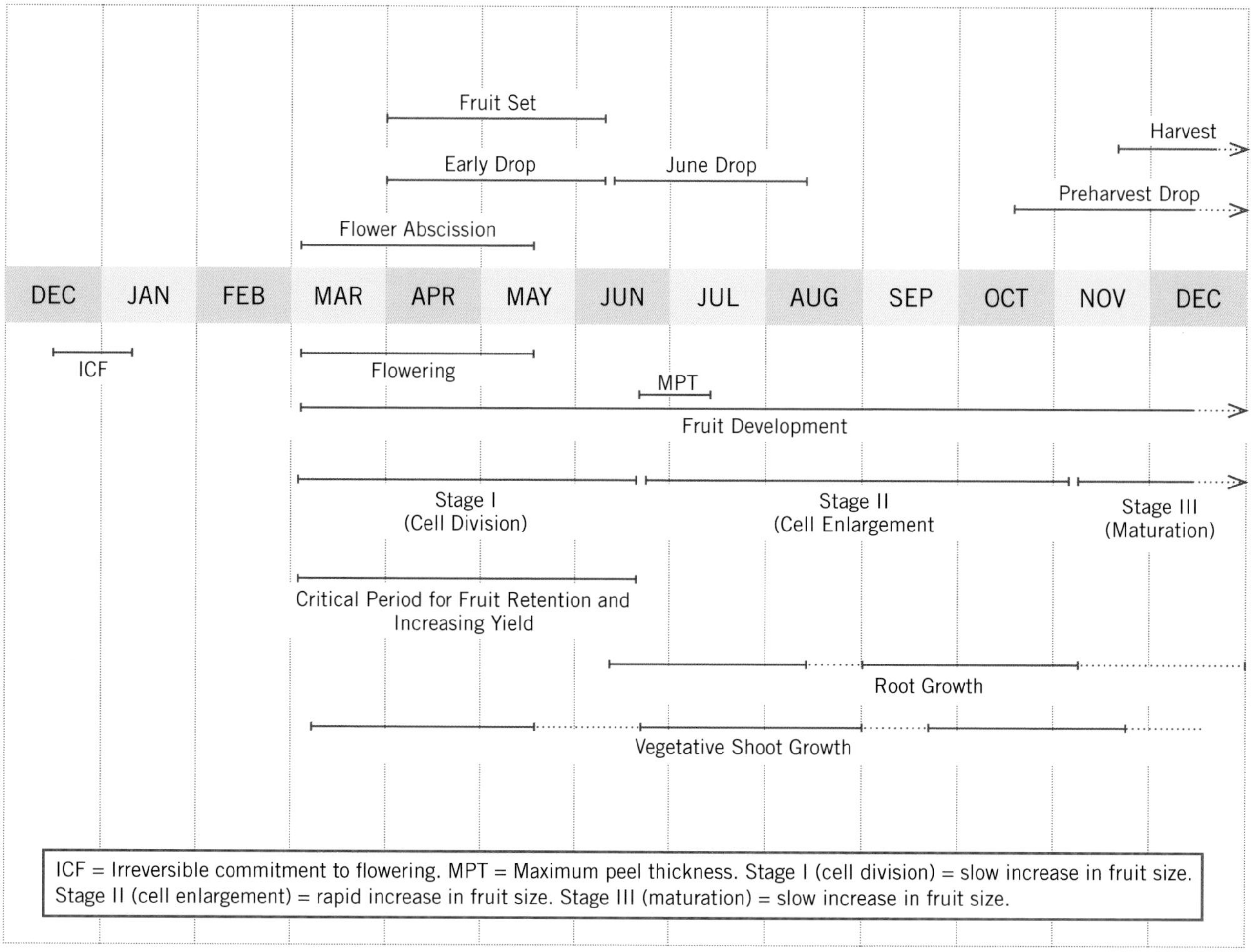

Figure 10.36 Phenology of the Washington navel orange in California. *Source:* Adapted from Lovatt 1999.

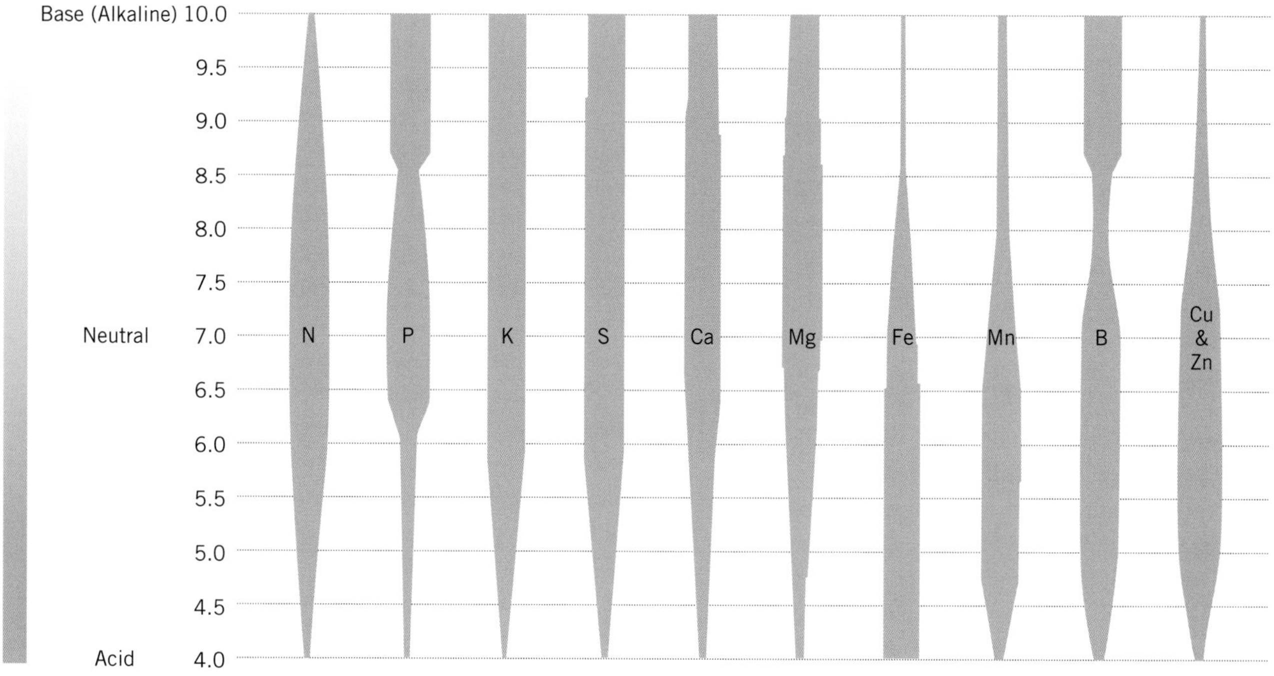

Figure 10.37 Effect of soil pH on nutrient availability (maximum availability is indicated by the widest part of the black bar).

transported in paper or porous bags and protected from the heat. Leaf samples can be kept overnight with refrigeration. At the lab, samples are washed in a detergent solution, rinsed with distilled or deionized water, dried at 140°F, and ground to pass through a 40-mesh screen. A high aluminum concentration in a leaf analysis without corroborating evidence of low soil pH, high aluminum content of the soil, or confirming analysis from resampling may be a good indication that the leaves were not washed thoroughly.

Soil and Water Analyses

Soil and water analyses are just as important as leaf analysis in producing optimal yields from a citrus orchard. For more information, see chapter 8, "Soil and Water Analysis and Amendment."

Correcting a Nutrition Problem

Quick Fix versus Long-Term Solution

A deficiency identified by visual symptoms or leaf analysis, even an incipient deficiency, should be corrected quickly. The longer a tree's nutrient status remains at the low end or below the optimal range, the greater the negative effects on yield and fruit size and quality. Foliar fertilization can successfully provide adequate amounts of certain essential nutrients such as iron much more rapidly than soil fertilization, and it is an excellent way to overcome these nutrient deficiencies quickly. For nitrogen, foliar-applied low-biuret urea can quickly correct nitrogen deficiency, but a minimum of three foliar

Table 10.3. Leaf analysis guide to nutrient status of mature Valencia and navel orange trees*

Element	Unit (dry matter basis)	Ranges†				
		Deficient	Low	Optimum	High	Excess
N	%	< 2.2	2.2 –2.3	2.4–2.6	2.7–2.8	> 2.8
P	%	< 0.09	0.09–0.11	0.12–0.16	0.17–0.29	> 0.30
K‡	%	< 0.40	0.40–0.69	0.70–1.09	1.10–2.00	> 2.30?
Ca	%	< 1.6?	1.6–2.9	3.0–5.5	5.6–6.9	> 7.0?
Mg	%	< 0.16	0.16–0.25	0.26–0.60	0.7–1.1	> 1.2?
S	%	< 0.14	0.14–0.19	0.2–0.3	0.4–0.5	> 0.6
B	ppm	< 21	21–30	31–100	101–260	> 260
Fe§	ppm	< 36	36–59	60–120	130–200?	> 250?
Mn§	ppm	< 16	16–24	25–200	300–500?	> 1,000?
Zn§	ppm	< 16	16–24	25–100	110–200	> 300
Cu§	ppm	< 3.6	3.6–4.9	5–16	17–22?	> 22?
Mo#	ppm	< 0.06	0.06–0.09	0.10–3.00	4–100	> 100?
Cl	%	unknown	unknown	< 0.3	0.4–0.6	> 0.7
Na	%	unknown	unknown	< 0.16	0.17–0.24	> 0.25

Source: Adapted from Embleton et al. 1978.

Notes:

*Modest adjustments to these levels were proposed by Embleton to increase the fruit size of the navel orange when size became a more dominant economic factor in marketing: he suggested that leaf nitrogen should be reduced to 2.4% and leaf K concentration should be increased to 1.2% (pers. comm.). Ranges are also applicable to mature Valencia orange trees and, with the exception of nitrogen values, applicable to grapefruit, lemon, and probably other commercial citrus cultivars. Question marks (?) indicate a value estimated but not proven.

†Leaves selected for analysis should be free of chlorosis, obvious tip burn, insect or disease injury, mechanical damage, and so on, and from trees not visibly affected by disease or other injury.

‡Potassium ranges are for effects on number of fruit per tree.

§These standards are not applicable for leaves sprayed or dusted with the particular element in question. Leaves sprayed or dusted with Fe, Mn, Zn, or Cu may analyze high or excessive in these elements, but the next growth cycle may have values in the deficient range.

#From fruiting shoots.

applications annually would be required to meet the total annual nitrogen needs of a citrus tree. Note that Embleton and Jones (1974) demonstrated that regardless of the fertilization method, maximum nutritionally attainable yields for sweet oranges required between 1 and 1.3 pounds of nitrogen per tree annually. Due to the potential for ammonia toxicity, a minimum of three applications must be made to supply the recommended rate of nitrogen. To provide nitrogen quickly, foliar-applied low-biuret urea should be applied; a long-term solution must include soil-applied nitrogen.

To develop a long-term solution to nutrition, soil pH, and salinity problems, soil fertilization or soil amendments must be used in addition to foliar fertilization. The great advantage of foliar fertilizers is that their application can be properly timed to key stages in the development of the tree when nutrient demand is high to stimulate a specific metabolic process to produce a positive effect on yield (Lovatt 1999, 2013). Applying small doses of fertilizer to the soil throughout the year can result in fertilizer being present when it is not needed, being leached below the root zone before it is used, or being immobilized. Dividing soil-applied fertilizer into many small doses might also result in too little fertilizer being applied when demand is high. Timing soil-applied fertilizers to stages in tree phenology that are characterized by high nutrient demand and rapid uptake and use is fundamental to fertilizer best management practices that strive to manage fertilizer applications to maximize crop yield (including fruit size and quality) and economic return while protecting water quality.

Foliar versus Soil-Applied Fertilizers

Foliar fertilization is an efficient way to supply nutrients to a crop when needed. In foliar fertilization, leaves of the crop or other target organs take up the nutrient. It is advantageous for a foliar-applied nutrient to be mobile in the phloem, but even immobile elements are beneficial to the metabolism of treated leaves or other organs. Relative rates of nutrient uptake and phloem mobility are given in table 10.2; actual rates of uptake measured for several nutrients are provided in table 10.4. In experiments comparing the efficiency of zinc, manganese, boron, and molybdenum applied to the foliage and to the soil, foliar fertilization was reported to be 4 to 30 times more efficient than soil fertilization, depending on the nutrient, crop, and soil (table 10.5) (PureGro, n.d.). Moreover, foliar fertilization can meet a tree's demand for a nutrient when soil conditions such as low temperature, low soil moisture, high temperature, pH, or salinity would render soil-applied fertilizers ineffective. Nutrients, especially phosphate, potassium, and trace elements dispersed in the soil, can become fixed and unavailable to plants through interactions with organic matter or other nutrients to form insoluble compounds. Applying essential nutrients directly to leaves, the major organ for photosynthesis, ensures that the plant's metabolic machinery is not compromised. It is important to note that foliar-applied mobile nutrients are translocated to all parts of the tree, including the smallest feeder roots. Foliar fertilizers reduce the potential for accumulation of nutrients in the soil, runoff water, surface water, and ground water, where they can contribute to salinity, eutrophication, and nitrate contamination, all of which have serious consequences on the environment and humans. Replacing soil-applied fertilizer with foliar-applied fertilizer when possible also contributes to fertilizer best management practices.

Efficiency of foliar fertilization

The best time to apply foliar fertilizers is early morning, when the plant's stomates are open. Temperature and relative humidity influence the uptake and resulting effect of foliar fertilizer applications. Temperatures should neither be too warm in summer nor too cold in winter. If temperatures are projected to reach approximately 85°F or higher during the day, applications should be made just after daybreak or after the temperature drops in the evening and the relative humidity increases. In addition, the increase in relative humidity in the field at night may induce renewed uptake of the nutrient from the dry spray residue and increase its effect. Winter applications should be made at midday, when temperatures are suitably warm. Foliar fertilizer treatments are more successful on healthy, well-watered trees.

The amount of water used is important for good results. Plant uptake is improved with dilution of the fertilizer with sufficient water to give full canopy coverage, and the material should be applied as fine droplets. Avoid spraying to runoff. Excessive liquid per tree causes pooling of the fertilizer at the tip of the leaf, which can result in tip burn; excessive misting will increase drift. For more information on spray application of fertilizers, see the UC IPM Citrus Pest Management Guidelines, http://www.ipm.ucdavis.edu/PMG/, and *The Safe and Effective Use of Pesticides,* 2nd ed. (O'Connor-Marer 2000).

Movement of nutrients through the leaf surface depends on the unique chemistry of water and the chemistry of the leaf surface. Water is bipolar, with areas of negatively and positively charged molecules. The several constituents making up the waxy surface of the leaf (cuticle) also make the leaf surface bipolar. Both the upper and lower surfaces of leaves contain transcuticular pores filled with water, which play a role in the movement of some dissolved nutrients from the leaf surface into the cells of the leaf. Additionally, leaves have lipohyllic transcuticular pores, which facilitate the uptake of lipid soluble (hydrophobic) compounds. Dissolved nutrients also move through open stomates on the underside of citrus leaves. In general, it is best to apply nutrients to the foliage in their most soluble form. Nutrient uptake is best when the application solution is at a pH from 5.5 to 6.5. For some nutrients, such as calcium, a chelated product is more soluble. For most cation nutrients (magnesium, potassium, zinc, manganese, and iron), application as a sulfate salt is equally effective as chelated products and less expensive. Sodium metaborate or boron hydroxide is commonly used to supply boron to the foliage. Timing of application is important. In general, application of foliar fertilizers just before flowering, or at 10% anthesis, prevents deficiencies that promote flower abortion and early fruit abscission and, thus, increase yield. Mature, hardened citrus leaves are able to take up many nutrients efficiently, including nitrogen as urea and magnesium, potassium, sulfur, zinc, calcium, manganese, and boron as soluble salts or chelates. Leaves are receptive to nutrients early in their growth cycle and at two-thirds full expansion. Developing leaves at one-half to two-thirds expansion provide sufficient surface area and a thin cuticle for efficient uptake and soon thereafter will serve as "exporters" of nutrients to the developing fruit. For optimal stages in citrus tree development when foliar-applied urea-nitrogen and phosphite phosphorus increased yield, fruit size, and fruit total soluble solids for Washington navel orange, see the sidebar. These results were shown to apply to mandarin cultivars when tested in California and Morocco (El-Otmani et al. 2003a, 2003b; Gonzalez et al. 2010; Zheng et al. 2013).

Foliar fertilizers have additional benefits. Some foliar fertilizers, such as sulfur, copper, iron, zinc, and manganese, are known to have fungicidal and bactericidal properties, whereas other foliar fertilizers are effective in stimulating the natural defense mechanisms of plants. Foliar sprays of zinc, copper, phosphite, or hydrated lime (calcium hydroxide) alone or combined with copper sulfate in Bordeaux mixture are effective treatments to reduce the symptoms of several diseases, including anthracnose, citrus blast, Botrytis diseases, brown rot, Phytophthora root rot and gummosis (see Lovatt and Mikkelson 2006 and the UC IPM Citrus Pest Management Guidelines, http://www.ipm.ucdavis.edu/PMG/), and most recently huanglongbing (Zhao et al. 2013). It has long been held that potassium and calcium increase plant toughness and resistance to the mechanical damage caused by insects.

Nutrients must be in solution before roots can take them up. The use of chelated forms or salts of calcium and magnesium or micronutrients (such as iron, manganese, or zinc) applied by drip irrigation to feeder roots increases the efficiency of soil application, since the nutrients are in solution and in close proximity to the feeder roots and not widely dispersed in the soil (for more information, see chapter 12, "Fertigation"). Chelation is especially effective in maintaining the solubility of the nutrient and preventing it from being fixed in the soil and becoming unavailable to the tree. Sodium metaborate or boron hydroxide is commonly used to supply boron to the soil.

Table 10.4. Relative absorption rates for foliar-applied nutrients

Nutrient	Time for 50% absorption
urea	½–2 hours
magnesium	2–5 hours
potassium	10–24 hours
calcium	1–2 days
manganese	1–2 days
zinc	1–2 days
phosphorus	5–10 days
iron	10–20 days
molybdenum	10–20 days

Source: PureGro, n.d.

Table 10.5. Efficiency of soil vs. foliar fertilization

Nutrient	Soil (kg/ha)	Foliar (g/ha)	Efficiency
zinc	10–12	500	> 20:1
manganese	25–30	750	> 30:1
boron	2–4	500	> 4:1
molybdenum	0.5	35	> 14:1

Source: PureGro, n.d.

Winter prebloom foliar applications of low-biuret urea (46-0-0; ≤ 0.25% biuret) or potassium phosphite (0-28-26) have been shown to increase yield (Ali and Lovatt 1992, 1994; Albrigo 1999). Lovatt (1999) demonstrated that summer applications of either fertilizer increased fruit size. Proper timing is important to achieve the desired outcome. The winter prebloom spray is designed to increase flower number and yield of commercially valuable-sized fruit (Lovatt et al. 1988; Ali and Lovatt 1994; Albrigo 1999) without reducing fruit size (Ali and Lovatt 1994). The most effective application time in California is from December 15 to February 15 in most years, but this window was extrapolated from research conducted in southern California. The timing of the winter prebloom fertilizer applications might be improved by determining experimentally when irreversible commitment to flowering occurs in orchards from the Arvin-Edison area to Madera. Later applications (March and April) are better than those that are too early (October and November). This is because once the opportunity to increase flower number has passed, late applications of either fertilizer will increase the retention of abscising reproductive organs. However, as time progresses from flowering through early fruit set to June drop, more flowers and fruit abscise. Hence, yield potential decreases and the contribution that either fertilizer can make to increasing yield diminishes. When used as a winter prebloom foliar spray, low-biuret urea is applied at the rate of 50 pounds in 200 gallons of water per acre. Lower volumes can be used as long as tree coverage is good, but high volumes, 500 to 700 gallons per acre, produce greater incidence of tip burn. The potassium phosphite formulation that has been used in all research trials reported in the literature thus far is Nutri-Phite (0-28-26). Nutri-Phite is applied for the winter prebloom spray at the rate of 2.6 quarts (0.64 gal) in 200 gallons of water per acre.

Summer applications of either low-biuret urea or potassium phosphite are designed to increase fruit size without increasing yield (Lovatt 1999). The time of application is important. The treatments are designed to extend the cell division stage of fruit development to achieve an increase in fruit size; the end of the cell division stage is characterized by maximum peel thickness. We experimentally determined that this period is between approximately June 11 and July 26 in California. Low-biuret urea is most effective when applied between July 1 and July 26. Applications of low-biuret urea that are too early (May and June) increase fruit retention and thus are less effective in increasing fruit size. Potassium phosphite is more effective in increasing fruit size when applied two times, May 15 ± 7 days and July 15 ± 7 days. It is not known whether low-biuret urea or potassium phosphite applications would be effective if applied later than the end of July. When applied in the summer at maximum peel thickness, low-biuret urea is applied as a single spray targeting July 15 ± 7 days at the rate of 50 pounds in 200 gallons of water per acre. To increase fruit size, potassium phosphite (Nutri-Phite, 0-28-26) is applied in two sprays at the rate of 2 quarts (0.49 gal) in 200 gallons of water per acre for each application. The first targets May 15 ± 7 days, and the second targets July 15 ± 7 days.

All applications should be made to give good canopy coverage, much like applying a pesticide or plant growth regulator. Solutions should be between pH 5.5 and 6.5. Applications of low-biuret urea or potassium phosphite should be made when the air temperature is below 80°F. Taking into consideration the potential maximum daytime temperature, applications should be made in early morning or late afternoon to early evening.

The information below is designed to help guide you in deciding whether to use foliar-applied low-biuret urea or potassium phosphite as part of your production management strategy. Keep in mind that this research is ongoing, and the information is the best interpretation of results to date.

Why should you use a winter prebloom foliar application of low-biuret urea or potassium phosphite?

» If your production goal for the year is to increase yield.

» If your current harvest is a heavy on-crop and you anticipate the next crop to be an off-crop.

» If you hang your fruit late and suspect a low return bloom.

» If you have a mild winter with insufficient chilling to induce a good bloom.

» If you have cold, wet soil, especially into February and March (late fertilizer applications are beneficial).

» If you wish to increase the total soluble solids in your fruit or total soluble solids per acre

Why should you use summer (maximum peel thickness) applications of low-biuret urea or potassium phosphite?

» If your production goal for the year is to increase fruit size.

» If you are carrying a heavy on-crop.

How do you decide whether to use low-biuret urea or potassium phosphite?

» If your production goal is to increase fruit size, do not use either fertilizer as a winter prebloom spray; the increased yield will make it more difficult to increase fruit size. Use low-biuret urea or potassium phosphite as summer applications only.

» If cost is your overriding consideration, low-biuret urea is the less expensive fertilizer, and only a single application is required to increase fruit size. However, potassium phosphite provides the secondary benefit of controlling *Phytophthora* and mitigating the effects of huanglongbing (Zhao et al. 2013).

» If you wish to harvest early, use potassium phosphite as a summer spray (May and July).

» If you wish to increase peel thickness, use potassium phosphite as a summer spray (May and July).

» For Valencia oranges, if the time of application is close to harvest, use potassium phosphite, NOT low-biuret urea, to prevent regreening.

» If crease is a problem, use potassium phosphite.

» If the fruit have rough, thick peels, use potassium phosphite instead of low-biuret urea as a winter prebloom spray while your current crop is still on the tree or as a summer maximum peel thickness spray.

» If soils are cold and wet in spring but the air temperature is periodically warm, use potassium phosphite to reduce the potential for crease.

» If soils are cold and wet and trees are yellow, use low-biuret urea after considering all the above.

» If trees have high leaf nitrogen concentrations (> 3.3% N), use potassium phosphite instead of urea.

Soil Amendments and Fertilizers

Leaf analysis is used in California citrus production to guide annual fertilization in a manner consistent with replacement fertilization: applying fertilizer to replace what the tree used to produce its current crop. This approach has two failings:

- It tends to ignore alternate bearing. Thus, by replacing what the mature on-year crop used when the tree is setting an off-crop, growers overfertilize the off-crop. Conversely, growers underfertilize the on-crop by replacing only what the off-crop used.
- It fails to recognize the times during tree development when nutrient demands are higher than at other times and therefore is not consistent with fertilizer best management practices.

Timing soil-applied fertilizer to key stages in tree phenology that are characterized by high demand and rapid uptake and use requires research to determine the seasonal patterns of nutrient uptake and how the on- and off-crop status of alternate-bearing cultivars affect nutrient uptake and partitioning in the tree. Thus, even for soil fertilization, time of application has an important effect on yield and fruit quality. This area of fertilization has been minimally investigated in citrus production. Soil-applied fertilizers, mulches, and other amendments can be used not only to supply essential mineral nutrient elements, but also to promote a pathogen-suppressive rhizosphere and to ameliorate soil problems such as salinity, poor soil structure, pH, and low water-holding capacity.

Few soil amendments provide as many benefits as lime (see Hardy et al. 2013). Lime not only raises soil pH, which improves the availability of calcium and magnesium and reduces the potential for aluminum and manganese toxicity in citrus, it also provides a more favorable environment for microorganisms. Soils with different textures and organic matter content require different amounts of lime to adjust the soil pH value. For more information, see chapter 8, "Soil and Water Analysis and Amendment."

Conclusions and Future Prospects

Optimizing soil and foliar fertilization is a cost-effective way to improve yield, increase fruit size, and enhance fruit quality to maximize the yield of commercially valuable fruit and increase economic return. Soil and foliar fertilizer applications must be properly timed to achieve this goal, and further research is under way on these practices. Preliminary results showed that foliar application of zinc sulfate at 10% anthesis significantly increased navel orange yield without reducing fruit size (Gonzalez and Lovatt 2010), and foliar applications of potassium and calcium in February reduced the incidence and severity of crease in navel orange cultivars. In Florida, fruit size of Sunburst tangerine *(C. reticulata* × *C. paradisi)* was increased with three foliar applications of potassium nitrate (KNO_3) made at dormancy (February), postbloom (~April), and during exponential fruit growth (July to August) (Boman 2002). This resulted in a significant increase in grower income because a greater number of fruit reached the minimum size and quality for early harvest and netted a higher market price than fruit on the untreated control trees. These results require verification in additional orchards in California. Application timing is also important for soil-applied fertilizers, and more research is needed on this practice as well.

Acknowledgments

The author wishes to acknowledge the critical importance of the fundamental research of Homer D. Chapman, William W. Jones, Tom W. Embleton, Charles K. Labanauskas, and Robert G. Platt to our understanding of citrus nutrition. The leaf analysis ranges for deficient, optimal, and excess concentrations of the essential nutrients and method of leaf sampling developed by these researchers became the standard for many citrus-producing countries. The author also wishes to note the pioneering efforts of William W. Jones and Tom W. Embleton in the field of foliar feeding. Through their efforts citrus was the model for foliar fertilization of other tree crops. The author also acknowledges Tom W. Embleton and Willard P. Bitters for the photographs of nutrient deficiencies and toxicities used in this chapter.

References

Albrigo, L. G. 1999. Effects of foliar applications of urea or Nutri-Phite on flowering and yields of Valencia orange trees. Proceedings of the Florida State Horticultural Society 112:1–4.

Ali, A. G., and C. J. Lovatt. 1992. Winter application of foliar urea. Citrograph 78:7–9.

———. 1994. Winter application of low-biuret urea to the foliage of 'Washington' navel orange increased yield. Journal of the American Society for Horticultural Science 119:1144–1150.

Arnon, D. I., and P. R. Stout. 1939. The essentiality of certain elements in minute quantity for plants with special reference to copper. Plant Physiology 14:371–375.

Boman, B. J. 2002. KNO_3 foliar application to 'Sunburst' tangerine. Proceedings of the Florida. State Horticultural Society 115:6-9.

Brown, P. H., and H. Hu. 1996. Phloem mobility of boron is species dependent: Evidence for phloem mobility in sorbitol-rich species. Annals of Botany 77:497–505.

El-Otmani, M., A. Ait-Oubahou, H. Gousrire, Y. Hamza, A. Mazih, and C. J. Lovatt. 2003a. Effect of potassium phosphite on flowering, yield, and tree health of 'Clementine' mandarin. Proceedings of the 9th Congress of the International Society of Citriculture 1:428–432.

El-Otmani, M., A. Ait-Oubahou, F.- Z. Taibi, B. Lmfoufid, M. El-Hila, and C. J. Lovatt. 2003b. Pre-bloom foliar urea application increases fruit set, size, and yield of Clementine mandarin. Proceedings of the 9th Congress of the International Society of Citriculture 1:559–562.

Embleton, T. W., and W.W. Jones. 1974. Foliar-applied nitrogen for citrus fertilization. Journal of Environmental Quality 3:338–392.

Embleton, T. W., W. W. Jones, C. K. Labanauskas, and W. J. Reuther. 1973. Leaf analysis is a diagnostic tool and guide to fertilization. In W. J. Reuther, ed., The citrus industry. Vol. 3. Berkeley: University of California Division of Agricultural Sciences Press. 183–211.

Embleton, T. W., W. W. Jones, and R. G. Platt. 1978. Leaf analysis as a guide to citrus fertilization. In H. M. Reisenauer, ed., Soil and plant tissue testing in California. Berkeley: University of California Division of Agricultural Sciences Bulletin 1879.

Epstein, E. 1972. The inorganic components of plants, In Mineral nutrition of plants: Principles and perspectives. New York: Wiley and Sons, Inc.. 63.

Gonzales, C., Y. Zheng, and C. J. Lovatt. 2010. Properly timed foliar fertilization can and should result in a yield benefit and net increase in grower income. Acta Horticulturae 868:273–286.

Hamid, G. A., S. D. Van Gundy, and C. J. Lovatt. 1988. Phenologies of the citrus nematode and citrus roots treated with oxamyl. Proceedings of the 6th Congress of the International Society of Citriculture 2:993–1004.

Hardy, D. H., M. R. Tucker, and C. E. Stokes. 2013. Crop fertilization based on North Carolina soil tests. Raleigh: North Carolina Department of Agriculture and Consumer Services, Agronomic Division. Circular No. 1. http://www.ncagr.gov/agronomi/obook.htm.

Hopkins, W. G., and N. P. A. Hüner. 2004. Introduction to plant physiology. Hoboken, NJ: Wiley & Sons, Inc.

Lovatt, C. J. 1999. Timing citrus and avocado foliar nutrient applications to increase fruit set and size. HortTechnology 9:607–612.

———. 2013. Properly timing foliar-applied fertilizers increases efficiency. HortTechnology 23:536–541.

Lovatt, C. J. and R. L. Mikkelsen. 2006. Phosphite fertilizers: What are they? Can you use them? What can they do? Better Crops 90:11-13.

Lovatt, C. J., Y. Zheng, and K. D. Hake. 1988. Demonstration of a change in nitrogen metabolism influencing flower initiation in citrus. Israel Journal of Botany 37:181–188.

O'Connor-Marer, P. 2000. The safe and effective use of pesticides. 2nd ed. Oakland: University of California Agriculture and Natural Resources Publication 3324.

PureGro Company. n.d. Soil vs. foliar. Sacramento: PureGro.

Zhao, H., R. Sun., U. Albrecht, C. Padmanabhan, A. Wang, M. D. Coffey, T. Girke, Z. Wang, T. J. Close, M. Roose, R. K. Yokomi, S. Folimonova, G. Vidalakis, R. Rouse, K. D. Bowman, and H. Jin. 2013. Small RNA profiling reveals phosphorus deficiency as a contributing factor in symptom expression for citrus huanglongbing disease. Molecular Plant 6(2):301–310.

Zheng, Y, T. Khuong, C. J. Lovatt and B. A. Faber. 2103. Comparison of different foliar-fertilization strategies on yield, fruit size and quality of 'Nules' Clementine mandarin. Acta Horticulturae 984:247–255.

11 Irrigation

Ben Faber and David A. Goldhamer

A large number of irrigation choices are available to citrus growers. In the past, the choices were limited to flood, furrow, and high-pressure sprinklers, but in the 1970s, drip irrigation became available, followed by a great number of other low-flow (low-volume) systems. Low-flow systems include drip, microsprinklers, fan sprays, and a host of new technologies such as vortex emitters and fan jets. Low-flow is distinctive in that emitter output is measured in gallons per hour (gph) rather than gallons per minute (gpm). Using low-flow emitters instead of high-flow impact sprinklers allows for a lower operating pressure, which permits the use of systems that are lighter, less expensive, more easily constructed, and more energy efficient.

Compared with high-flow full-pattern sprinklers or surface methods such as furrow irrigation, low-flow systems can improve the efficiency of irrigation and chemical application and can also reduce weed growth. These advantages, however, come at a cost. More maintenance is required on low-flow systems, and the root zone is restricted to the wetted pattern, reducing the soil volume available for water and nutrients.

Most citrus in California is grown on low-flow systems, so they are the focus of this chapter. However, the principles of water management are the same for all systems, so the ensuing discussion will be applicable to all systems. These principles strive to ensure a healthy, producing orchard.

Components of an Irrigation System

Before planting the citrus orchard, the irrigation system must be in place and ready for the trees. The various parts of the system can be assembled and installed by the grower (fig. 11.1), but it is usually best to have a qualified designer of low-flow irrigation systems create a master plan. The system should be designed to meet the water needs of the full-grown orchard during the peak irrigation periods. It should also be designed so that daily operation does not exceed 16 to 18 hours. Allowing some down time gives time for catch-up in case of a breakdown in the system, such as a pump that requires repair (see Schwankl et al. 1998).

Although many growers have water delivered by an irrigation district, in some instances it may be cheaper (or the only option) to drill a well. If a well is the source of water, select a pump and motor that will deliver the correct pressure and flow rate at the highest possible efficiency. The system designer determines the flow rate and pressure to be delivered by the pump, and the pump dealer matches the motor to the pump for the greatest efficiency.

A flowmeter and pressure gauges are critical parts of the system. The flowmeter indicates how much water is being applied, which is critical information for efficient irrigation and scheduling. For example, a decreasing flow rate measured at a given pressure might indicate clogging of the system; an increasing flow rate might suggest a leak in the system.

Valves help control the system. A main control valve is very important, particularly when using a well and pump, to prevent contamination of the water source or wellhead. A backflow prevention device should also be installed. Air or vacuum relief valves allow air to escape when the system is turned on and prevent air from entering when the system is shut down. Check valves prevent undesirable flow reversal in hilly terrain (see Brown 1972).

Filters should be used in low-flow systems because the emitter orifices have a great tendency to

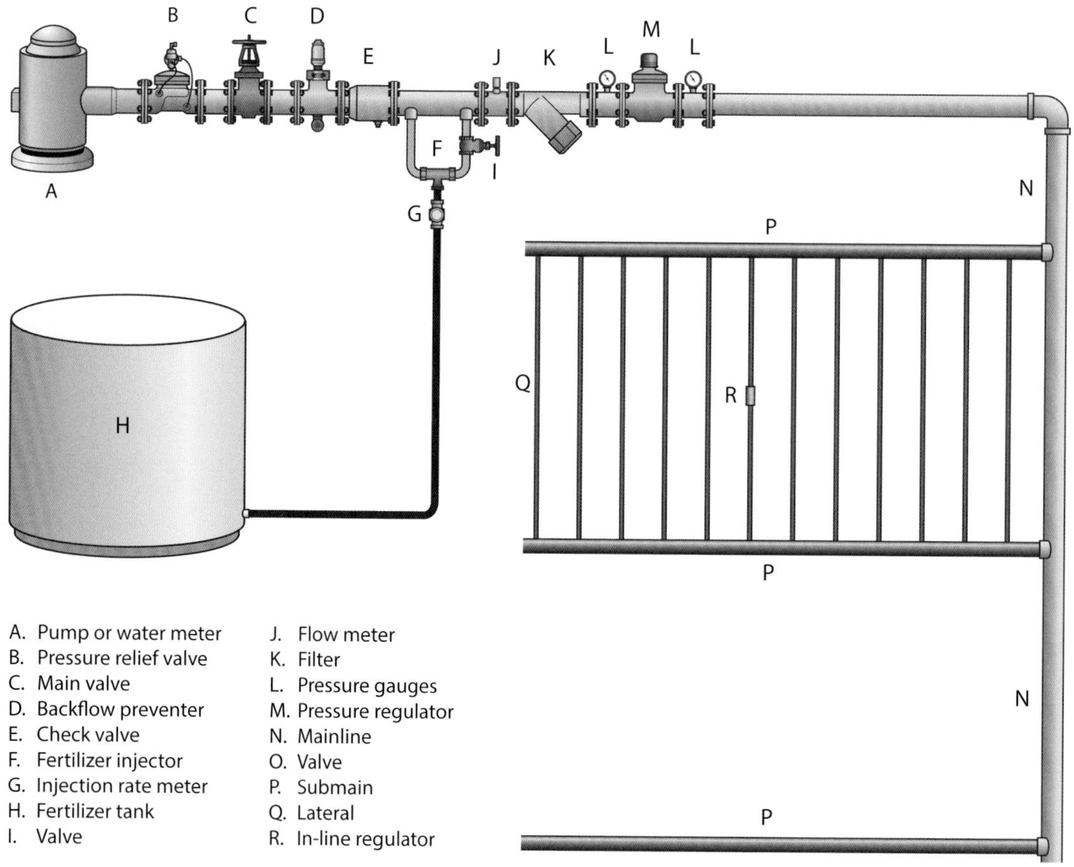

Figure 11.1 Parts of an irrigation system.

clog. Choices of filters include media (sand), disk, and screen (fig. 11.2). Typically, media filters are used with surface water that has a large load of organic sediment that would rapidly clog a screen filter. Screen filters are less expensive and should be used with well water that needs relatively less filtration. Disk filters with automatic backflushing are a happy medium between price and effectiveness for the removal of most sediments. If the water contains a high level of sand sediments, a sand or cyclone separator should be installed upstream from the filter.

Injection equipment is critical in preventing clogging of low-flow systems (fig. 11.3), but it is also a great convenience for fertilizer application (see Hanson et al. 2006). Differential pressure tanks, or batch tanks, are the simplest. Irrigation water flows passively in and out of the tank under line pressure. The major disadvantage of batch tanks is that as irrigation continues, the chemical concentration in the irrigation water decreases. If the chemical concentration must be kept constant, such as when injecting chlorine to prevent clogging, a batch tank should not be used; however, in most fertilizer applications, constant chemical concentration is not important. Venturi injectors, which are simple and inexpensive, rely on a pressure drop of 10 to 30% between the inlet and outlet of the injector. Venturis are better at maintaining a constant concentration of material than are batch tanks, but neither is as good as a positive displacement pump. These are powered by electricity, gasoline, or water and are the most complicated and expensive option. Most citrus growers find them to be unnecessary.

Mains and submains deliver water to the lateral lines and emitters. The size of the lines must be balanced between the cost of larger pipe versus the pressure loss when water moves through smaller lines. Lateral lines, usually polyethylene or PVC, deliver water to the emitters. The length and diameter of the laterals must not be too long or too small; if they are, the emitters may discharge water at different rates, resulting in nonuniform irrigation.

Pressure regulators can be installed as pressure-regulating valves after the filter at the head of the system, as preset regulators at the head of laterals, as in-line pressure regulators, or as a part of the

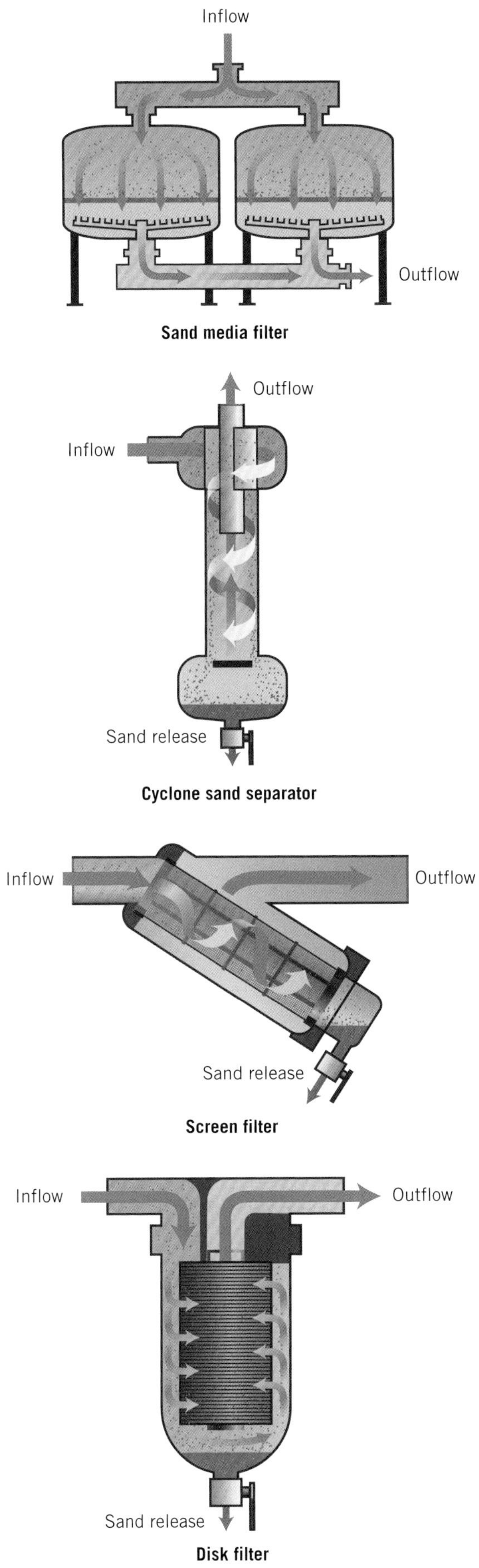

Figure 11.2 Various types of filters.

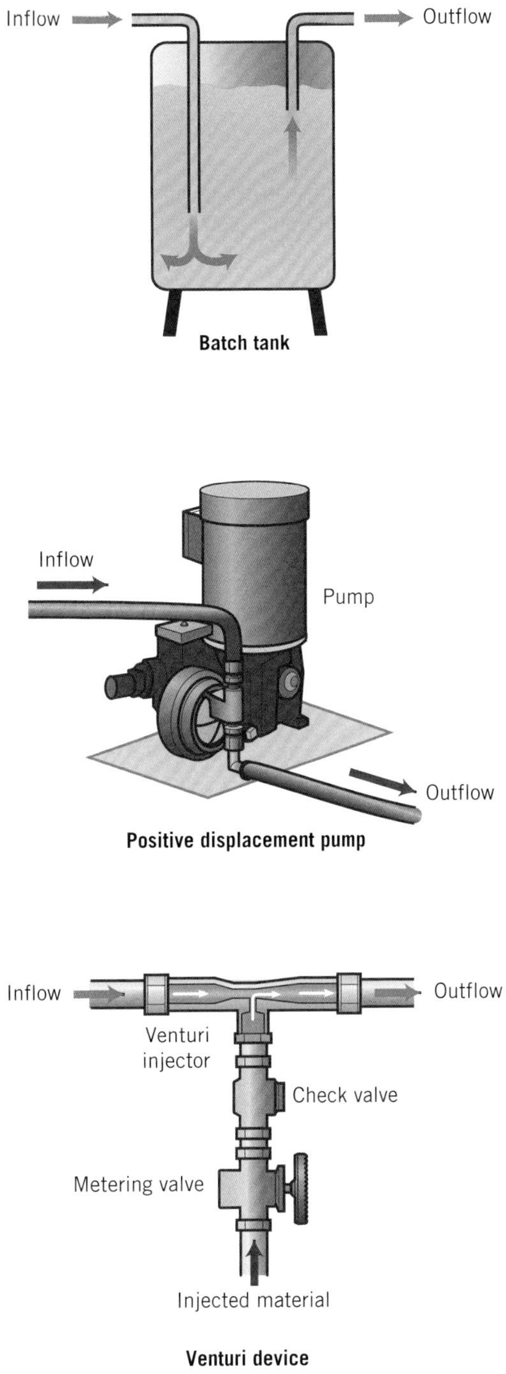

Figure 11.3 Various injectors.

emitter itself (pressure-compensating emitters). Pressure-compensating emitters are more expensive than standard emitters and may wear out sooner. However, pressure regulation is critical for uniform application of water, since the output of standard emitters varies with pressure.

Emitters come in many sizes and shapes, including drippers, microsprinklers, and fan sprays (fig. 11.4). They do not wet the entire orchard floor. Drip emitters with outputs of 0.5 to 4 gallons per hour wet a small spot at the surface, reducing weed growth and applying water only where it is needed. The wetted pattern enlarges below the soil surface; depending on the soil texture, this pattern can be a very bulbous onion shape (heavier soils such as clay) or more like a stove pipe (lighter, more sandy soils). Drippers are very good with young trees, since they wet only the area of the young roots. As the trees grow, more emitters and a second lateral should be added to accommodate the increasing water demand. Typically, six drip emitters should be able to meet the requirements of a mature citrus tree, depending on the soil type. However, drippers are notorious for their maintenance requirements. Tortuous path emitters have fewer problems, as they rely on a long, relatively large channel to reduce water flow, rather than just a small opening. Injecting chlorine or acid reduces clogging problems, but walking the lines must always be done.

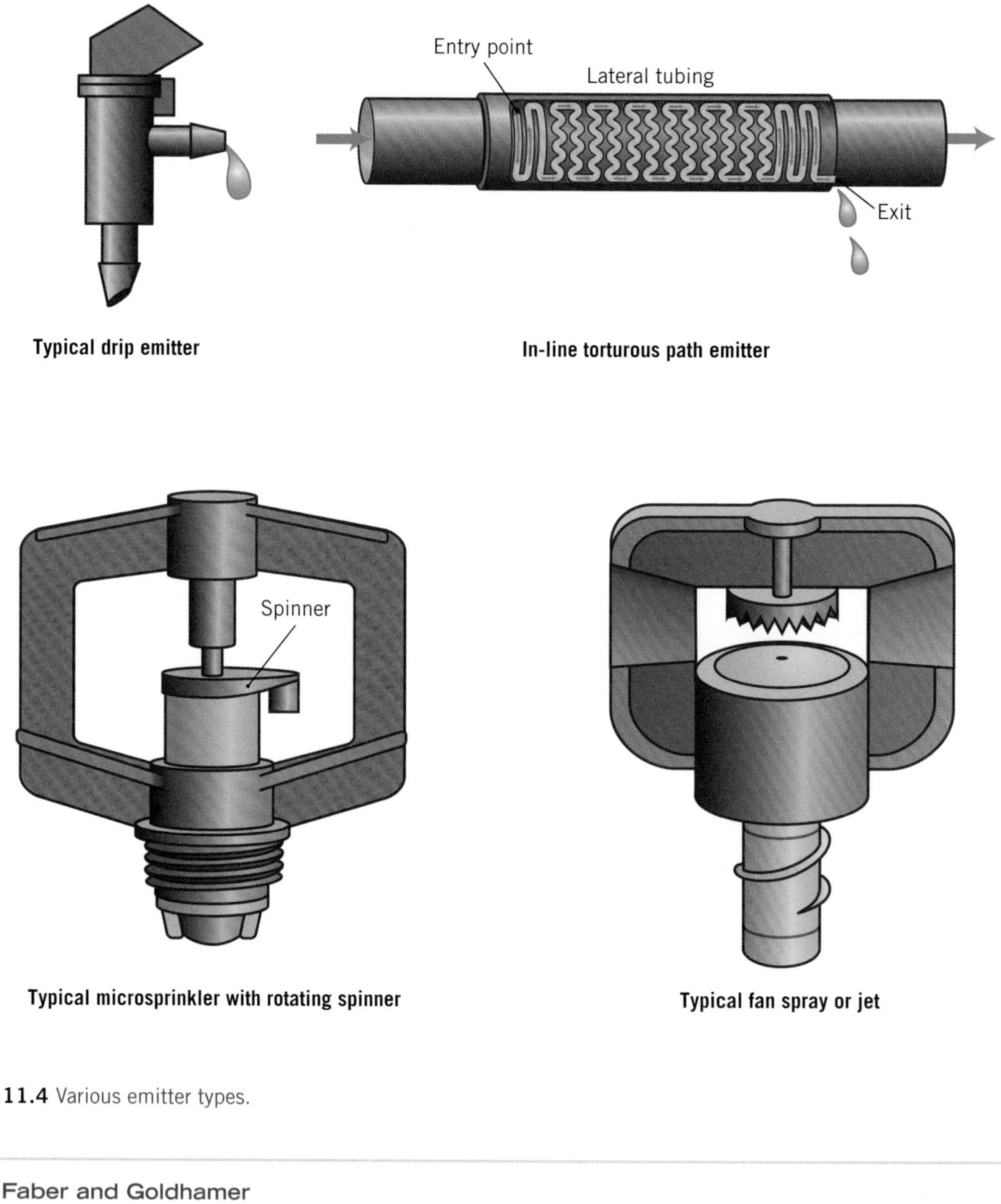

Figure 11.4 Various emitter types.

Microsprinklers, most of which have a rotating orifice called a spinner, put out 4 to 30 gallons per hour and wet a much larger surface area than do drip emitters. They clog less than drippers because the discharge flow rate is higher and their orifice is larger. However, they still need filters upstream. In sandier soils, where lateral subsurface movement of water is small, a microsprinkler is often the preferred choice. One of the major drawbacks of some microsprinklers is that output from the emitter varies with distance: the amount of water deposited on the outside two-thirds of the wetted pattern may be as little as one-third of the amount deposited on the inside of the wetted pattern. Near the emitter, the soil is often wetted below the root zone.

Fan sprays can overcome the poor output uniformity of microsprinklers by directing fingers of water in various directions. A number of patterns can be obtained, such as butterfly, rectangular, and so on. Microsprinklers and fan sprays can be found in patterns of less than 360°. To prevent disease, the tree trunk must not be wetted; with all of these emitters, a pattern should be selected that keeps the trunks dry.

Many brands and models of emitters are available. Although the quality of low-flow emitters has markedly improved in the last 10 years, there can be problems in their manufacture. Information about specific emitter performance can be obtained from the Center for Irrigation Technology at Fresno State University, http://www.fresnostate.edu/jcast/cit/index.html, and through Cooperative Extension in the UC Davis Department of Land, Air, and Water Resources, http://lawr.ucdavis.edu/cooperative_extension.htm.

Water Supply and Quality

Most low-flow systems require relatively frequent applications of water during peak demand periods. With drip systems, irrigations may be as frequent as once or twice a day. In some irrigation districts, water deliveries may not meet irrigation scheduling demands; in this case, it is necessary to work out an agreement with the agency supplying water. If frequent deliveries are not possible, a pond or tank system should be installed to provide a reservoir, or a supplemental well should be considered.

Water quality can affect plant health as well as the operation of low-flow systems. Organic or mineral sediments that clog emitters are a major water quality concern. These sediments can be filtered, but if major problems with backflushing occur, a continual injection of 2 parts per million (ppm) of chlorine prior to filtration may be required to control organic sediments, or a prefilter may be needed for sand sediments.

With private and municipal well water, chemical precipitation leading to clogging is the major problem. Lime (calcium carbonate) or iron or manganese precipitates are the most common sources of clogging. The presence of white or reddish to blackish deposits is associated with these two types of precipitates. If water contains more than 100 ppm (or 2 milliequivalents per liter, meq/l) bicarbonates and has a pH greater than 7.5, lime precipitation will eventually clog the emitter orifice. Keeping the water pH below 7 will help considerably. Injecting urea sulfuric acid products in the system, or the more dangerous sulfuric or phosphoric acids, decreases this problem. Acid injection rates vary with water quality but typically consist of 1 to 2 gallons of acid per 5,000 gallons of water; fertilizer supply houses may provide a recommendation based on a water sample. Iron and manganese precipitates can be much more difficult to control. Prior to running the water through the system, it is helpful to allow the water to settle out in a pond. Injecting chlorine and installing filters are other methods for controlling precipitates. Running acid water through the system for an hour may dissolve iron precipitates, but it may be necessary to inject acid and let it sit in the lines overnight before flushing. Some pressure-compensating emitters are damaged by water with a pH of 4 or below. Check with the manufacturer before lowering the pH of the water. It may be necessary to run chlorine or calcium hypochlorite through the system at the end of the irrigation if bacterial slimes caused by precipitates persist.

Water quality is also an important consideration for plant growth. The specific salts of boron, chloride, and sodium and the general salinity (total dissolved solids, or TDS, and electrical conductivity, or EC) must be considered. For more information, see chapter 8, "Soil and Water Quality and Amendment."

System Testing and Maintenance

Distribution Uniformity

The efficiency of an irrigation system is indicated by its distribution uniformity (DU). A 100% DU means that every emitter is putting out exactly the same amount of water. If the DU is low, the system must run longer to provide all trees with enough water, but this will cause some trees to get more water than they need, which wastes water and may not be good for the trees. Pressure losses in lines and uneven terrain can make it impossible to achieve 100% DU, but 80% is attainable and 95% is not unheard of. Even in well-designed new systems, clogging and leaks can rapidly reduce distribution uniformity; the way to ensure a high DU is through annual monitoring and regular maintenance.

Measuring DU

Distribution uniformity is obtained by measuring the output of a specified number of emitters. For example, if a block has 100 emitters, lay out an evenly spaced grid across the orchard so that all parts of the orchard can be sampled. Identify a minimum of 12 emitters to sample; the more emitters sampled, the more accurate the DU.

Turn the system on and invert the first emitter to be sampled over a graduated cylinder or measuring cup, capture the water for a specified time (such as 15 seconds), and record the volume of water emitted. After sampling each emitter, arrange the volumes from low to high. Add the values and find the average. Then select the volumes from the one-quarter of the emitters that put out the least, calculate their average, and divide it by the average of all the emitters. Multiply the result by 100 to get the percentage of DU (see the example below). If the DU is less than 80%, it must be corrected (see below). If it is greater than 80%, try to raise it to 90% or better.

Example: Martha runs 12 emitters for 20 seconds each into a graduated cylinder and records the following volumes of water (in ounces) for each: 8.5, 8, 7, 8, 7.5, 6.5, 7, 7.5, 8, 8, 8.5, and 7.5. The 12 values, arranged from low to high and summed, are

$$6.5 + 7 + 7 + 7.5 + 7.5 + 7.5 + 8 + 8 + 8 + 8 + 8.5 + 8.5 \text{ oz} = 92.0 \text{ oz.}$$

The average output is

$$92 \text{ oz} \div 12 \text{ emitters} = 7.7 \text{ oz/emitter.}$$

The low-quarter amounts (the first three numbers in the series), summed and averaged, are

$$6.5 + 7 + 7 \text{ oz} = 20.5 \text{ oz}$$

$$20.5 \text{ oz} \div 3 \text{ emitters} = 6.8 \text{ oz/emitter.}$$

The low-quarter average divided by the average of all the emitters is the DU:

$$(6.8 \div 7.7) \times 100 = 88.3\%.$$

This irrigation system DU of 88% is not bad, but if the orchard is on flat ground, there should be room for improvement.

Correcting low DU

Poor design may cause an irrigation system to have pressure problems that affect the DU. For example, low DU may be caused by pressure differences among individual emitters in the system; it may be necessary to install pressure regulators or pressure-compensating emitters to equalize the pressures. Or, if there is not enough pressure throughout the system to produce a satisfactory DU, it may be necessary to break the system into two or more irrigation blocks with separate valves. If the system has been properly designed, however, low DU is likely to be caused by poor maintenance—the most common cause of poor DU.

Routine Maintenance

Routine maintenance includes checking for leaks, backwashing filters, periodically flushing lines, injecting chlorination or acids, and cleaning or replacing clogged emitters. Coyotes frequently bite polyethylene tubing to get water. Thus, walking the lines in coyote country to inspect for leaks is critical. Most coyote damage occurs during coyote pup season in spring and in the fall, when surrounding hills have dried out. Putting out water for the coyotes can decrease the damage, but it may be necessary to repair lines before every irrigation. Also, lines and emitters may be damaged during and after harvest.

Clogged emitters can often be identified by reduced flows (a small wetted area) and sometimes by the sound they make (jammed spinners, intermittent spray, etc.). Emitters are designed to operate in a given pressure range, and if the pressure falls below that range, the output of the emitter may be significantly reduced. Clogged filters reduce the system pressure and thus reduce application rates and DU. Filters should be backflushed whenever there is a reduction in outflow pressure by 5 pounds per square inch (psi) or more; the frequency of backflushing depends on the water quality. Automatic backwashing filters are available and are relatively inexpensive; they initiate backwashing as soon as a given pressure differential exists.

Clogged emitters may need to be cleaned or replaced. Before replacing an emitter, identify the cause of clogging. If organic slimes or chemical precipitates are the cause, inject acid or chlorine. If earwigs or other insects enter and damage the emitters, it may be necessary to replace the emitters with insect-proof models. Although some emitters are designed to be disassembled and cleaned, nearly all drip emitters are sealed. Most microsprinkler models clog at the orifice in the head and can be cleaned and reinstalled.

Flushing is key to system maintenance. Periodically flushing lateral lines by opening the lines and allowing them to run clear is essential. Filters trap only the larger sediments; smaller sediments can gradually accumulate in laterals and eventually clog the emitters. For more information on microirrigation maintenance, see Schwankl et al. 2008.

Irrigation Scheduling

Irrigation efficiency requires not only uniform irrigation but also the application of the appropriate amount at the correct intervals. Irrigators must know the system's water application rate in inches per day, inches per hour, or gallons per hour, and they must also know when and how much water to apply. Scheduling irrigation, as opposed to applying a fixed amount at a fixed time, strives to determine the water used by trees and then replace that amount (for detailed information on irrigation scheduling, see Goldhamer and Snyder 1989 and Hanson et al. 1999).

The overall use of water in a field, through transpiration from the tree and evaporation from the ground, is called evapotranspiration (ET). Evapotranspiration varies seasonally and from year to year for a given location. The California Department of Water Resources (DWR) has developed a map of the average daily ET for various zones in California (fig. 11.5). These zones are distinctive because the parameters that drive water loss—total sunlight, wind, relative humidity, and temperature—differ in each zone. The Central Valley is hot and cloudless in the summer, whereas the intensity of the marine layer along the coast and the amount of sunshine differ from year to year. In southern California coastal valleys, the average annual irrigation requirement for citrus is about 2 feet of applied water per year (2 ac-ft per ac, or 651,702 gal per ac). This value varies from as little as 18 inches to as much as 3 feet from year to year; spring and fall conditions contribute the most to the variability in ET. In the Central Valley, on the other hand, a mature tree consistently requires the same amount of water every year (about 4 ft).

One of the most important variables in the quantity of water to be applied is the length of the rainfall season and the effectiveness of the rainfall. Effective rainfall is defined as the amount of rainfall that is retained in the root zone of the tree. For example, assume that a tree has a rooting depth of 2 feet and each foot holds 1 inch of available water. If this tree has just been fully irrigated (or if it rained 2 inches yesterday), and it rains 2 inches today, none of today's rain is effective since the soil was already moist before the rain (although the rain was useful, since it leached salts out of the soil). Also, rainfall of less than 0.25 inch is not considered to be effective.

The amount of water to apply at each irrigation depends on the amount of water remaining in the root zone. A loamy soil in which a microsprinkler with a 20-foot diameter throw has wetted the soil 2 feet deep holds about 200 gallons of water at 50% water-holding capacity. Exceeding this amount of water will help leach salts, but exceeding it too far only pushes existing water out of the root zone.

It is best to observe one or two irrigation cycles to determine how long to run the system to achieve a certain depth of infiltration. This can be done with a shovel or, more easily, with a pointed rod or tensiometers. Water moves in a wetting front, and the wetted soil allows a rod to be pushed into the depth of dry soil. Run the system to find out how long it takes water to infiltrate to a depth of 2 to 3 feet. That information will indicate how long to run the system when irrigating. Applying water 2 to 3 feet deep may take several hours. If runoff occurs, turn the system off for a few hours, then turn it on again to obtain the total run time required to wet the required depth. If runoff is severe, use emitters with a smaller flow rate.

Irrigation schedules may be plant based, soil based, or weather based. Many of the technologies used to create these schedules are proven and have been in use for years; others are more experimental and have not been fully tested. In several cases improved electronics and digitalization have added features to older technologies. Growers should be familiar with all types of irrigation scheduling, because a combination of types is often the most effective.

Plant-Based Scheduling

The plant is the ideal subject for measuring water use, as it integrates all the factors driving water loss, soil moisture, and stresses such as soil salinity. To be useful tools in irrigation scheduling, plant-based measuring devices must provide indicators of stress before it harms current or potential yield. Methods for measuring plant water use include the following:

- Pressure chambers (pressure bombs or Scholander pressure chambers) measure plant water tension by applying a comparable air pressure to a leaf or stem. The amount of pressure required to create equilibrium with the plant sap indicates the level of plant water stress.
- Trunk diameter fluctuations (shrinkage or swelling), measured continuously with linear variable displacement transducers (LVDTs), can be used to calculate parameters that are directly related to tree stress (see Goldhamer and Fereres 2001).
- Stem flow gauges can estimate transpiration by placing a heat source on the trunk of the tree and measuring the temperature differential along the trunk.

Figure 11.5 Evapotranspiration zones as determined by the California Department of Water Resources. For a more detailed version of this map, see the CIMIS website, http://wwwcimis.water.ca.gov/cimis/cimiSatEtoZones.jsp. *Source:* California Department of Water Resources.

- Porometers measure the ability of a leaf to transpire; if a leaf is under water stress, it transpires less water.
- Infrared thermometry measures the canopy temperature as affected by the rate of transpiration; as the tree experiences water stress, the leaves get warmer.
- Visual symptoms such as wilting and leaf curling are the cheapest method, but they are the most expensive in the long run as they do not produce reliable measurements.

The pressure chamber is currently the state-of-the-art in measuring citrus tree water stress. While the other techniques can be valuable for scientific use, they have not been frequently adopted in commercial agriculture. Part of the problem is logistical; for example, stem flow gauges do not adapt well to the uneven surface of the citrus tree trunk. Also, porometers and infrared thermometry do not provide enough lead time for a grower to irrigate. Recent research indicates that LVDTs show promise for automating irrigation scheduling.

Soil-Based Scheduling

A rule of thumb is that irrigation should be done when about 50% of the water has been depleted from the soil in the plant's root zone. This 50% figure, however, allows a buffer of water in the soil in case the weather suddenly turns hot and windy. Sandy soils hold less water than clay soils and must be irrigated more frequently. A common misperception is that it takes more water to grow plants in sandy soil than in clay soil; however, the total amount required for the whole year is the same for both soil types. The amount of sunlight, wind, temperature, and humidity control how much water a plant needs, and the soil is only the reservoir.

Determining water content by soil texture

To check the water content in the soil based on the soil texture, dig 8 to 16 inches down into the soil with a trowel, shovel, or soil tube and feel the soil. At about 50% available water:

- coarse soil appears almost dry and forms a ball that does not hold shape
- loamy soil forms a dark ball that is somewhat moldable and can form a weak ribbon when squeezed between the fingers
- clayey soil forms a good, dark ball, makes a ribbon an inch or so long, and slightly sticky

This method, however, gives only an approximate water content; instruments can give more precise readings.

Using tensiometers and other instruments

Irrigation timing can be determined more precisely using a tensiometer. These water-filled tubes with a pressure gauge accurately reflect the amount of energy a plant needs to extract water from the soil. The pressure gauge measures tension values in centibar units (cbars). For citrus, when the gauge reads 30 centibars, it is a good time to irrigate. Tensiometers must be placed in the root zone between the emitter and the tree trunk. Having two tensiometers next to each other can be helpful in deciding when to turn the system on and off. For example, a tensiometer at a depth of 1 foot would indicate when to turn the water on, and a second at 3 feet would indicate when to turn the water off. Prevent tensiometers from being damaged during harvesting and other orchard operations by placing a plastic milk crate or some other structure over each device.

Other devices can also be used to measure soil moisture. Gypsum blocks are very effective; the part in the ground is inexpensive but the reading device costs about $250, so a relatively large acreage is required to spread out the cost of the system. Portable meters rely on an electrical current carried by water in the soil. Even $10 meters can give a rough estimate of the soil water content, but they are not very effective in rocky ground, because their sensitive tips break easily.

Soil-based methods monitor an aspect of soil moisture that, depending on the method, requires a correlation to plant water use. Some methods are well understood and inexpensive, others are expensive, inaccurate, inappropriate, or not well researched. Some methods allow multiple site readings, while others require a device to be left in place. Some measure soil-water directly (e.g., oven-drying), and others measure another parameter, such as electrical conductance. Some methods are affected by salts or soil iron content, and others have limited value in the desired soil moisture range. Some, like tensiometers and gypsum blocks, give a reading from a porous material that comes to equilibrium with soil moisture, while many others use the soil directly as the measured medium—an important distinction, since discontinuities in the soil caused by rocks or gopher holes can affect readings. Also, some of the older techniques have been improved. For example, gravimetric oven-drying can now be done by microwave, considerably speeding up the process, and tensiometers and gypsum blocks can now be found with digital readouts and connections to data loggers that make data easier to manage. Many types of monitoring devices are available; table 11.1 describes their characteristics. As with any tool, the value of these devices increases with use and familiarity. Even though several are stationary devices, by placing them in representative positions in the orchard, they can accurately reflect the entire orchard. Some types of device can be stationary or portable, depending on the model. "Ease of use" in table 11.1 indicates the ease of reading the device as well as the maintenance.

Weather-Based Scheduling

Another scheduling technique that has become popular is the use of weather data that have been converted to a crop water-use value. This value is the estimated amount of water an orchard would use. The value is often referred to as the evapotranspiration (ET) of the crop. Evapotranspiration is the amount of water that is lost from a well-watered crop through the leaves (transpiration) and through evaporation from the surface of the soil. Applying the ET amount at an irrigation keeps trees at their optimal moisture content. This technique is often called the water budget method or checkbook scheduling.

The information required for water budget scheduling is available in many areas of California from newspapers, irrigation districts, and over the California Irrigation Management Information System (CIMIS, http://wwwcimis.water.ca.gov/cimis/welcome.jsp; 800-922-4647). The CIMIS network of over 50 weather stations calculates reference evapotranspiration (ET_o). Reference evapotranspiration values are available from many irrigation districts, CIMIS, and certain weekly journals and magazines. Other sources include the County Flood Control in Ventura County and the resource conservation districts in San Diego County. These values are an estimate of the amount of water lost from a well-watered field of grass, the standard reference for all other crops. ET_o is modified for the specific crop with a crop coefficient (K_c). The formula for converting ET_o to crop ET is

$$ET_o \times K_c = ET_{crop}.$$

For a full-grown citrus orchard, a crop coefficient of 0.65 is used in most of the state, but 0.56 is used in the desert growing areas. When trees are young and intercept little energy to drive water loss, a coefficient of 0.05 works well. As the trees increase in size to where their shade covers about 65% of the soil surface of the orchard, the coefficient is gradually increased each year. With rapidly growing trees, the increase is usually about 10 % each year, until about year 8, when the 65% figure is reached. A correction factor must be incorporated for the irrigation system distribution uniformity. If the orchard is cover cropped for part or all of the year, the period during which the cover is present must be incorporated into the water-use calculation. If a young orchard is covered by a perennial cover crop, a coefficient of 0.65 is used regardless of tree size. A winter annual cover, which uses only rainfall for its growth, does not require a correction in a high-rainfall year, but it may require one in a low-rainfall year.

One of the drawbacks of centralized weather stations is the station values can be quite different from those at your citrus orchard. When using evapotranspiration figures it is always important to back up the estimates with field checks in your orchard. An alternative to using centralized weather stations is establishing your own. An electronic station costs about $5,000 and requires regular maintenance. A simpler weather station can be developed with an evaporation pan or an atmometer (atmosphere meter). An atmometer is a closed system with a ceramic head, much like a tensiometer. As water is drawn out of a reservoir, a sight tube shows how

Table 11.1. Characteristics of selected soil monitoring devices

Method	Cost	Ease of use	Accuracy	Reliability	Salt-affected	Stationary
gypsum block	L	H	H	H	L	yes
tensiometer	L	M	H	M	L	yes
portable tensiometer	M	M	H	M	L	no
solid-state tensiometer	M	H	H	H	L	yes
time domain reflectometer	H	M	H	H	M	both
neutron probe	H	L	H	H	L	yes
feel (soil probe)	L	H	H	H	L	no
gravimetric (oven)	L	M	H	H	L	no
conductance	L	H	M	M	H	both
capacitance	M	H	M	H	M	both

Key:
H = high; M = medium; L = low.

much water has been evaporated. Since the physics of evaporation and transpiration are very similar, the values can easily be used in a water budget. The major drawback to the evaporation pan is the maintenance required to keep birds, coyotes, and bees from causing inaccurate readings. Algae also must be kept out of the pan. The atmometer is more expensive (about $300) than a pan, but it is much easier to maintain.

Regardless of the scheduling technique or equipment used, a thorough evaluation of the irrigation system must be performed so that a known amount of water is being applied. Until the volume and distribution of water are known, it makes little sense to schedule applications.

Calculating Water Requirements

Determining the application rate of low-flow systems can be confusing because irrigation scheduling and water-use information is often presented in inches per day, while discharge from low-flow emitters is in gallons per hour. Inches per day can be converted to gallons per day by the following formula:

Water use (gal/day) = tree spacing (ft^2) × tree water use (in/day) × 0.623 (gal/in-ft^2).

For example, an orchard has fully grown trees at a spacing of 20 by 20 feet (400 ft^2). The tree water use is 0.1 inch per day.

$$\text{Water use} = 400 \times 0.1 \times 0.623$$
$$= 25 \text{ gal/day.}$$

With smaller trees, the area of the canopy should be used instead of the plant spacing. Also, extra operating time must be factored in when distribution uniformity is low. With a DU of 80%, allow 25% more operating time to ensure that all trees receive the minimum amount of water required.

A grower can use a device such as a tensiometer to signal when to initiate irrigations and use the CIMIS values since the last irrigation to determine how much to apply. By using soil moisture depletion, once a threshold is found, the same amount of water can often be applied at an irrigation event. Or, if irrigation is done on the same day of the week, the CIMIS values can be accumulated from the previous irrigation and that amount can be applied, taking into account DU and an amount required for leaching.

Regulated Deficit Irrigation

Irrigation is required in virtually all California citrus orchards to produce top yields of high-quality fruit. This is because citrus ET_c far exceeds the effective rainfall that occurs during the summer. However, increased competition for California's water supply from a growing population and environmental protection suggests that growers will be increasingly accountable for their water use and that future water costs will likely be higher. Thus, knowledge of tree response to water stress is important in irrigation management.

What is water stress? As the leaf stomata, the small openings on the underside of citrus leaves, open in the early morning in response to sunlight, water vapor moves from the interior of the leaf through the stomata and into the atmosphere. This process, known as transpiration, causes the leaves to become slightly deficient in water and creates an energy gradient between the leaves and the shoots, trunk, and roots of the plant. At the root-soil interface, this gradient causes water to be extracted from the soil. The transpiration rate depends primarily on weather conditions. As the soil-water becomes limited, the transpiration rate exceeds the extraction rate. Plant water deficits, or water stress, are the result. Without some type of regulation, the leaves would dehydrate, resulting in damage or death. The stomata provide this regulation; they begin to close in response to water stress, thus maintaining a favorable internal water balance in the plant. However, it comes at a price.

The leaf stomata are the conduits not only for transpiration but also for carbon assimilation. Carbon dioxide diffuses from the atmosphere through the stomata and into the leaf, where photosynthesis converts it into sugars, the building blocks necessary for plant and fruit growth and the fuel that powers important plant processes. In essence, the plant trades water for carbon, since both water and carbon dioxide use the same plumbing system at the leaf surface. Maximum transpiration and thus maximum photosynthesis occur when trees are fully irrigated. Reducing carbon uptake (and thus photosynthesis) negatively affects one or more tree organs—roots, shoots, branches, leaves, or fruit. With citrus, water stress usually results in smaller fruit size at harvest. Thus, we normally recommend that citrus growers avoid water stress by careful irrigation management.

However, recent work (Goldhamer and Salinas 2000) has shown that mature navel orange trees can be stressed during certain periods of the season without negative effects on fruit yield. In fact, carefully imposed water stress can significantly reduce peel creasing and thus improve grower profit. Regulated deficit irrigation (RDI) is the controlled imposition of water stress. The goal of RDI in citrus is to reduce the amount of seasonal applied water while maintaining or possibly improving fruit yield and quality. The success of RDI depends on whether the beneficial aspects of water stress can be achieved without concomitant negative impacts on tree processes and other important fruit yield and quality components.

It has been reported (Goldhamer and Salinas 2000) that creasing can occur when the growth rate of the outer layer of the peel (albedo) is exceeded by the enlargement of edible, internal parts of the fruit (endocarp). This apparently results in the formation of weak sites (cracks) in the outer layer and subsequent creasing. By imposing water stress early in the season with Frost nucellar on Troyer rootstock, the occurrence of creasing, and thus fruit considered "juice," in the harvest fruit was significantly reduced. The RDI regime that was most successful in this research, which occurred in Kern County on a moderately shallow soil, applied water at 25% of potential ET_c from mid-May through mid-July. This resulted in a maximum pressure chamber reading of 19.5 bars (1.95 MPa) during this period. It is likely that water stress during this critical, specific period of fruit development slowed the growth rate of the internal fruit segments more than the peel cell enlargement rate, reducing the formation of weak sites in the peel. Fruit size was lower than that of fully irrigated trees at the end of the RDI period, but it recovered within 6 weeks of the reintroduction of full irrigation in mid- July (fig. 11.6).

While navel orange growers, especially those with a severe creasing problem, can improve their profit using RDI, its use in citrus must be carefully managed. Successful implementation of this technique depends on the timely imposition of tree water stress. The approach used in the research described earlier (Goldhamer and Salinas 2000)—irrigating at a fraction of potential ET_c for a given period of the season—is not directly transferable to areas with different soils, soil profiles, and weather conditions. Weather influences ET_c as well as fruit growth, potentially affecting the beneficial RDI effect observed on creasing. Measurements of tree water status must be made with a pressure chamber to improve the precision of RDI management achievable by irrigating at fractions of potential ET_c. For further information, see Ballester et al. 2011.

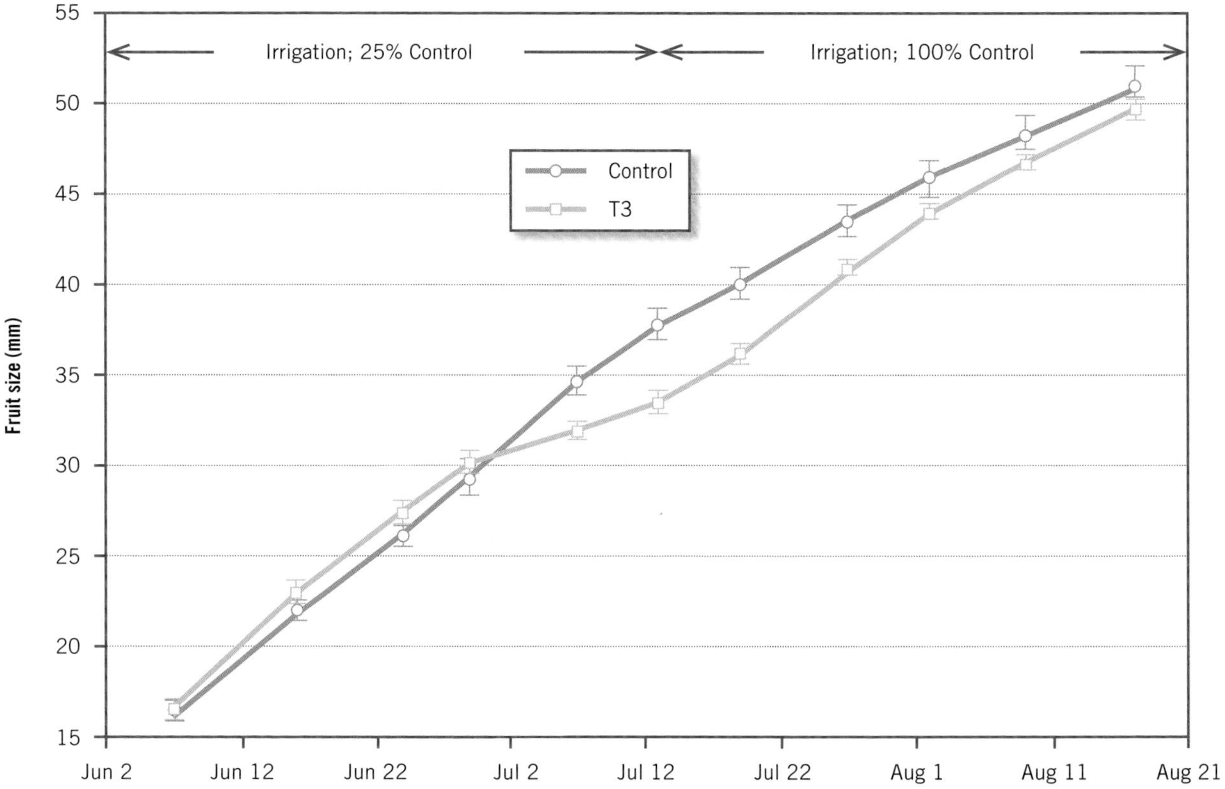

Figure 11.6. Fruit size for regulated deficit irrigation (RDI) regime that applied water at 25% of potential ET_c from mid-May through mid-July (T3) compared with fully irrigated control.

Effect of Irrigation on Disease

High soil moisture content and poor drainage are associated with Phytophthora root rots. This root rot can be caused in winter by *Phytophthora citrophthora* or in summer by *Phytophthora parasitica*. Increasing irrigation intervals, irrigating alternate middles, conversion to drip or microsprinklers from furrow, and installing drain lines have helped in controlling these diseases. On heavier soils, planting on berms or mounds can reduce the likelihood of root rot. The orchard floor should be leveled to avoid water accumulation.

Asphyxiation of roots can occur with excessive water or flooding, and the methods used to help prevent root rot also help with inadequate soil aeration. These methods can also help with lime-induced chlorosis, a soil aeration problem that is exacerbated by a high soil lime content in which trees exhibit an iron deficiency that can be corrected only by improving soil aeration. Correcting this problem is most easily accomplished by adjusting irrigation practices and applying soil acidifiers.

Citrus gummosis is caused by *Phytophthora* species and is attributed to conditions that keep the trunks moist. The incidence can be minimized by avoiding wetting the tree trunk, ensuring good aeration around the trunk, and preventing soil accumulation around the trunk.

Citrus fruit are more prone to oleocellosis when tree moisture status is high. Harvesting fruit shortly after irrigation, early in the morning, or in humid, foggy conditions can cause rupturing of the rind oil glands, leading to rind spotting. Minimizing this problem is accomplished by not harvesting during periods that favor high fruit turgor.

References

Ballester, C., J. Castel, D. S. Intrigliolo, and J. R. Castel. 2011. Response of Navel Lane Late citrus trees to regulated deficit irrigation: Yield components and fruit composition. Irrigation Science, doi: 10.1007/s00271-011-0311-3.

Brown, L. 1972. Irrigation on steep lands. Oakland: University of California Division of Agriculture and University Services Publication 2825.

Center for Irrigation Technology, http://www.fresnostate.edu/jcast/cit/index.html.

Farm Water Quality Management, http://ucanr.edu/sites/farmwaterquality/Fact_Sheets/.

Goldhamer, D. A., and E. Fereres. 2001. Irrigation scheduling protocols using continuously recorded trunk diameter measurements. Irrigation Science 20:115–125.

Goldhamer, D. A, and M. Salinas. 2000. Evaluations of regulated deficit irrigation on mature orange trees grown under high evaporative demand. Proceedings of the 9th Congress of the International Society of Citriculture 1:227–231.

Goldhamer, D. A., and R. L. Snyder. 1989. Irrigation scheduling: A guide for efficient on-farm water management. Oakland: University of California Division of Agriculture and Natural Resources Publication 21454.

Hanson, B., L. Schwankl, and A. Fulton. 1999. Scheduling irrigations: When and how much. Oakland: University of California Division of Agriculture and Natural Resources Publication 3396.

Hanson, B., N. O'Connell, J. Hopmans, J. Simunek, and R. Beede. 2006. Fertigation with microirrigation. Oakland: University of California Division of Agriculture and Natural Resources Publication 21620.

Irrigation Association, http://www.irrigation.org/.

Irrigation Training and Research Center, http://www.itrc.org/.

Schwankl, L., B. Hanson, and T. Prichard. 1998. Microirrigation of trees and vines. Oakland: University of California Division of Agriculture and Natural Resources Publication 3378.

———. 2008. Maintaining microirrigation systems. Oakland: University of California Division of Agriculture and Natural Resources Publication 21637.

Scott, V. H. 1959. Measuring irrigation water. Berkeley: University of California Division of Agricultural Sciences Leaflet 2956.

12 Fertigation

Lawrence J. Schwankl

Fertigation is the injection of fertilizers through an irrigation system. Microirrigation systems are well suited to fertigation because of their application uniformity, their operational frequency, and the ability to easily control water applications. Applying fertilizers through a microirrigation system

- allows fertilizer distribution and water application to be equally uniform
- allows flexibility in timing fertilizer applications
- reduces the labor required to apply fertilizer
- allows less fertilizer to be applied
- lowers costs

Fertilizer Solubility

In order to be injected, fertilizers must be soluble. Fertilizers formulated as a solution can be injected directly into the irrigation system, while dry granular or crystalline fertilizers must be mixed with water to form a solution prior to injection. Fertilizer materials differ widely in water solubility, depending on the physical properties of the fertilizer as well as on irrigation water temperature and pH. Dry fertilizers are tank mixed with water until the granules or crystals dissolve. The solution is then injected into the irrigation system. With solutionizer injection machines, the injected material may be in a slurry of suspended material that goes into solution once it is mixed with the irrigation water.

Fertilizers Commonly Injected

While most fertilizers can be successfully injected, some products may increase the risk of emitter clogging or change the soil pH in the emitter's wetted zone. Table 12.1 summarizes these effects for some commonly injected fertilizers. Consulting with the chemical supply company prior to any injection is always a wise precaution.

Nitrogen Sources

The fertilizer most commonly injected is nitrogen, as many soluble nitrogen sources work well in fertigation. The following are common nitrogen sources, with information on their use in fertigation:

- **Anhydrous ammonia, or aqua ammonia** are nitrogen sources that increase the water pH, which may result in a precipitate if calcium or magnesium, along with comparable levels of bicarbonate, are present in the irrigation water. Volatilization of

Table 12.1. Characteristics of injected fertilizers

Fertilizer	Clogging concern
nitrogen sources	
anhydrous ammonia or aqua ammonia	may increase water pH
urea	minimal concern
ammonium sulfate	minimal concern
calcium nitrate	calcium may combine with bicarbonate in water to form precipitate (calcium carbonate)
phosphate sources	phosphate may combine with calcium or magnesium in water to form precipitate
potassium sources	usually not a clogging problem, but some potassium fertilizers (e.g., potassium sulfate) may not readily go into solution

nitrogen (loss to the atmosphere) may also occur when anhydrous or aqua ammonia are used.

- **Urea** is relatively soluble in irrigation water and is not strongly held by soil particles, so it moves deeper into the soil than ammonia products. Urea is transformed by hydrolysis into ammonium, which is then fixed to soil particles.
- **Ammonium sulfate, ammonium nitrate, and potassium nitrate** are relatively soluble in water and cause only a slight shift in the pH of the soil or water.
- **Calcium nitrate** (CH-9) is 9% nitrogen and 11% calcium. It is relatively soluble in water and causes only a slight shift in the soil or water pH. If the water is high in bicarbonate, however, the calcium content may lead to precipitation of calcium carbonate (lime).
- **Urea ammonium nitrate** (UAN-32) is 16.4% urea nitrogen, 7.4% ammonium nitrate, and 7.4% nitrate-nitrogen. It is a commonly injected fertilizer since it is soluble and has minimal clogging hazard.
- **Ammonium phosphate** can cause soil acidification. If calcium or magnesium levels are high enough in the irrigation water, precipitates may form and clog drip emitters. (For precautions in using ammonium phosphate, see "Phosphate Sources," below.)

Phosphate Sources

Using phosphate fertilizers may cause chemical or physical precipitate clogging. The calcium and magnesium content and the pH of the irrigation water should be considered, since calcium phosphate and magnesium phosphate precipitates may form when the water pH is higher than 7.5. Acidifying the water to a pH of 4.0 to 4.5 with sulfuric acid or phosphoric acid keeps irrigation water pH low and minimizes precipitation problems. Phosphorus is quickly fixed to soil particles and does not move readily into the soil profile, but it has been found to move more easily under microirrigation than under conventional irrigation methods.

Potassium Sources

Injecting potassium fertilizers usually causes few problems, but caution should be taken if potassium fertilizers are mixed with other fertilizers. Potassium, like phosphorus, is fixed by soil particles and does not move readily through the soil profile. Potassium is usually applied in the form of potassium chloride, but for crops sensitive to chloride, potassium sulfate or potassium nitrate may be more appropriate. Potassium sulfate is not very soluble and may not dissolve well in irrigation water.

Injection Devices

Many substances can be injected through irrigation systems, including chlorine, acid, fertilizers, herbicides, micronutrients, nematicides, and fungicides. Of these, fertilizers are most commonly injected. Chlorine or acid injection is used in microirrigation systems to prevent clogging caused by biological growths (algae and bacterial slimes) and chemical precipitation (particularly calcium carbonate). While injecting these chemicals reduces the environmental hazards of their use, fertigation and chemigation can still cause environmental damage, particularly when the injected chemicals move readily with irrigation water through the soil profile. Overirrigation resulting in deep percolation can contaminate ground water when a mobile chemical is injected.

A variety of injection equipment is available, including differential pressure tanks, venturi devices, positive displacement pumps, small centrifugal pumps, and solutionizer machines. It may be possible for a single injection system to inject all the chemicals to be run through the microirrigation system, but it is not unusual to have multiple injection systems. For example, a small centrifugal pump may be used to inject fertilizers, while a positive displacement pump may inject chlorine.

Differential pressure tanks

Differential pressure tanks, often referred to as batch tanks, are the simplest injection devices (fig. 12.1). The inlet of a batch tank is connected to the irrigation system at a point of pressure higher than that of the outlet connection. This pressure differential causes irrigation water to flow through the batch tank containing the chemical to be injected. As the irrigation water flows through the batch tank, some of the chemical goes into solution and passes out of the tank and into the downstream irrigation system. Because the batch tank is connected to the irrigation system, it must be capable of withstanding the operating pressure of the irrigation system.

While relatively inexpensive and simple to use, batch tanks are not ideal, because the chemical mixture in the tank becomes increasingly diluted over the course of the irrigation. The result is a decreasing concentration of injected chemical in the irrigation

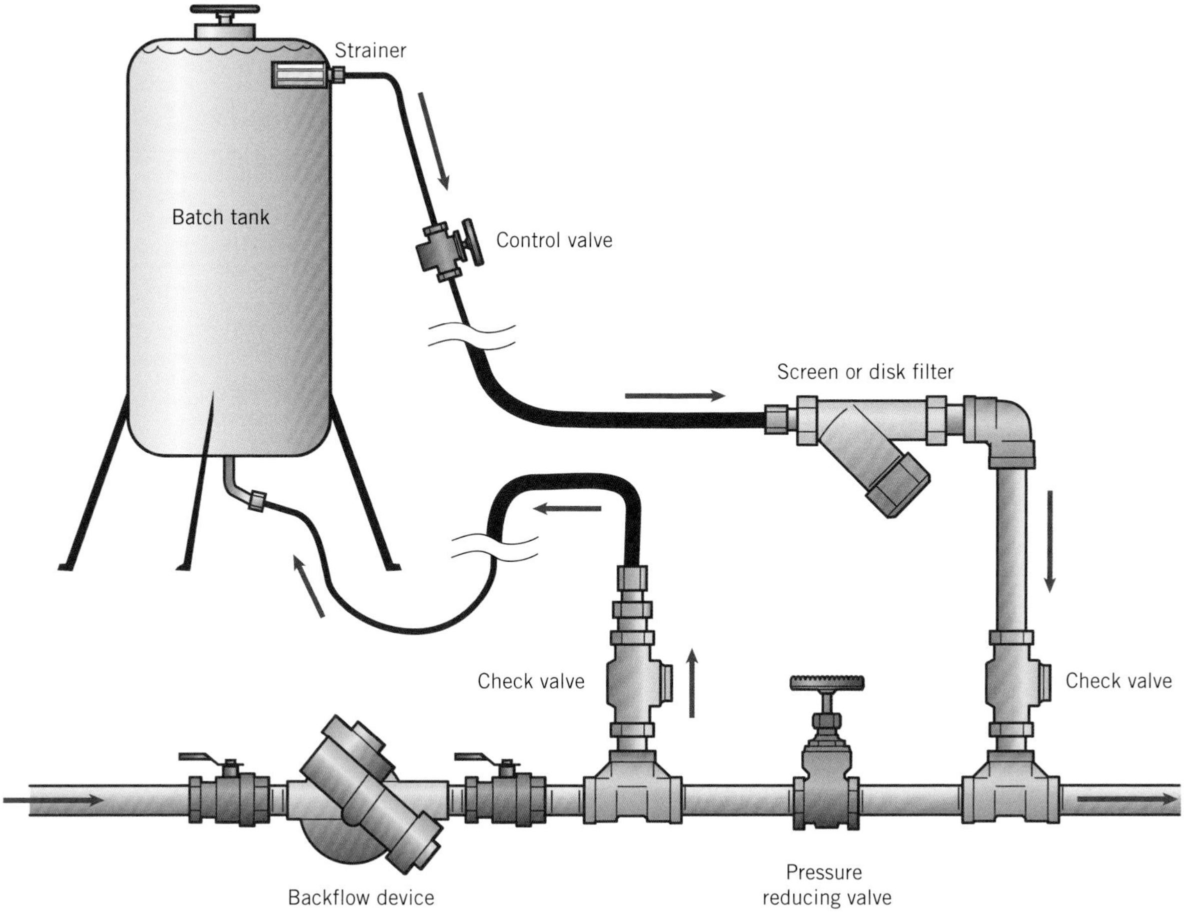

Figure 12.1 Differential pressure, or batch tank, injection system.

water. If a set amount of a chemical, such as fertilizer, is to be injected and the concentration during the injection is not critical, using batch tanks may be appropriate. If the chemical concentration must be kept relatively constant during injection, batch tanks are not appropriate.

Venturi Devices

Venturi devices (fig. 12.2) consist of a constriction in a pipe's flow area that causes a negative pressure, or suction, at the throat of the constriction. These devices are often referred to as Mazzei injectors, using a trade name for a commonly used brand of venturi injector; venturi injectors are also available from other manufacturers.

The venturi injector has a pressure loss of from 10 to 30% due to friction in the venturi. Because of this significant pressure loss, the injector should be installed parallel to the pipeline (fig. 12.3) so that flow through the injector can be turned off using valves when injection is not occurring. When installed this way, the inlet of the venturi injector must be at a pressure 10 to 30% higher than the outlet port or water will not flow through the injector. This is usually achieved by installing the injector across a pressure-reducing valve or a control valve that can be partially closed. The injection rate of a venturi device is determined by the size of the venturi and the pressure differential between inlet and outlet ports. Injection rates as high as 700 gallons per hour are possible with large venturi devices.

Venturi injectors can also be coupled with a small centrifugal pump that draws water from the irrigation system, increases its pressure while moving the water through the venturi, and then returns the water and chemical back into the irrigation system. The small centrifugal pump in the venturi injection system eliminates the need to create a pressure drop for the venturi injector to operate; creating a pressure drop by partially closing a valve or by some other means can be wasteful of energy. A centrifugal pump also can provide a more constant injection rate than plumbing a venturi injector across a pressure drop.

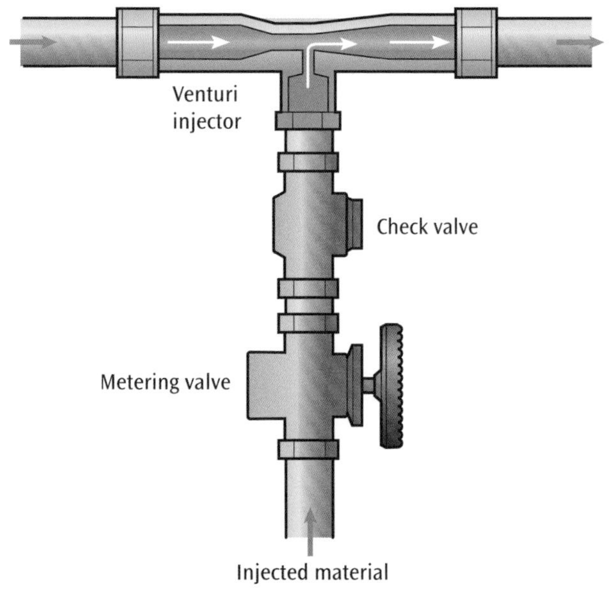

Figure 12.2 Venturi injection device.

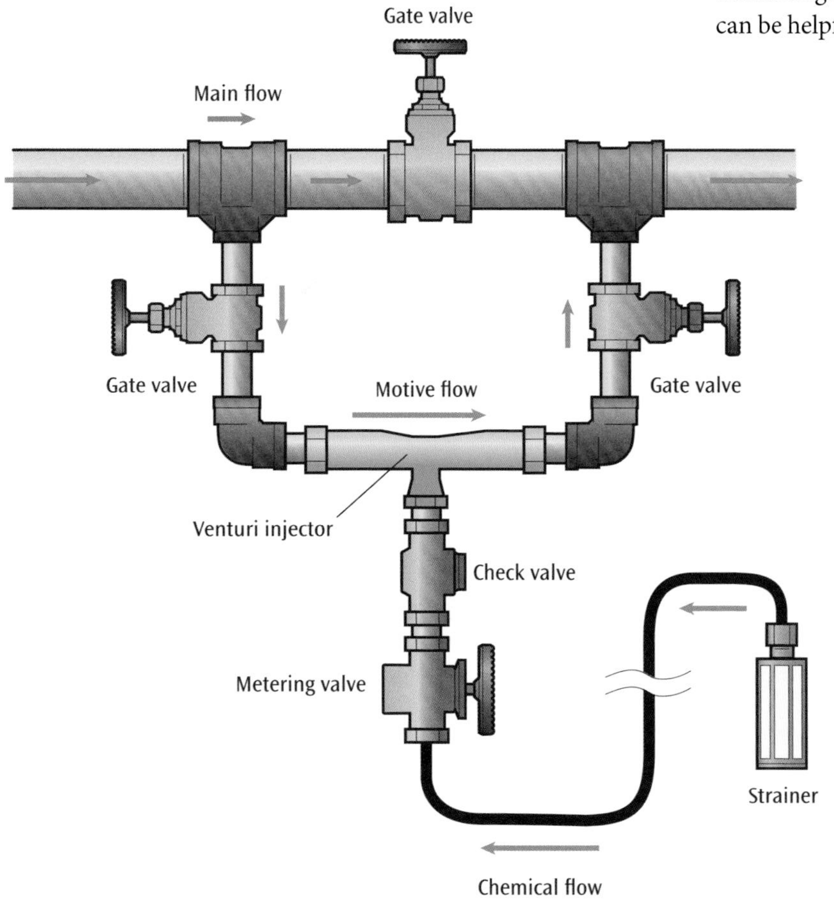

Figure 12.3 Venturi injector installed across a gate valve.

Venturi devices are inexpensive and relatively simple to operate, but they do not inject chemicals at as uniform a concentration as positive displacement pumps. Injecting with venturi devices may be sufficiently accurate for applications such as fertilizer injection, however.

Positive Displacement Pumps

Positive displacement pumps are piston (fig. 12.4) or diaphragm pumps that inject at precise rates. The pumps are powered by electricity or gasoline, or they can be driven by water. The water-driven pumps can be installed in locations that lack electrical power. When a constant and precise injection concentration is needed, positive displacement pumps are preferable. Positive displacement pumps are the most expensive of the injection devices, with costs for electric pumps running $750 or more.

Centrifugal pumps

Centrifugal pumps are frequently used for fertilizer injection. These pumps have a greater flow rate than positive displacement pumps or most venturi injectors, making them appropriate for injecting higher rates of chemical. The centrifugal pumps can be driven by either electricity or gasoline engines. Using a centrifugal pump in conjunction with a flowmeter can be helpful in controlling injection rates.

Solutionizer Machines

Solutionizer machines (fig. 12.5) were developed to inject materials that are not readily soluble. Their most common use is for injecting finely ground gypsum through the irrigation system, but they are also used for injecting fertilizer products such as potassium sulfate. Solutionizer machines inject a slurry of material into the irrigation line, where it then mixes and goes into solution. In microirrigation systems, it is important that these materials be injected upstream of the irrigation system filters to ensure that insoluble materials are filtered out and do not clog the emitters. For example, 95% pure gypsum materials may still contain up to 5% insoluble materials. This would mean that for every 100 pounds of gypsum material injected, 5 pounds of insoluble material may be present. Dry fertilizer materials may also contain significant insoluble material.

Injection Point

The injection point should be located so that the injected fertilizer and the irrigation water can become thoroughly mixed well upstream of any branching of the flow. Because of concerns over fertilizers being flushed out when the microirrigation system filters are backwashed, the injection point should be downstream of the filters in all except the solutionizer injection systems. To ensure that no contaminants are injected into the microirrigation system, a good-quality screen or disk filter should be installed on the line between the chemical tank and the injector. The system should be allowed to fill and come up to full pressure before injection begins. Following injection, the system should be operated to flush the fertilizer from the lines. Leaving residual fertilizer in the line may encourage clogging from chemical precipitates or organic sources such as bacterial slimes.

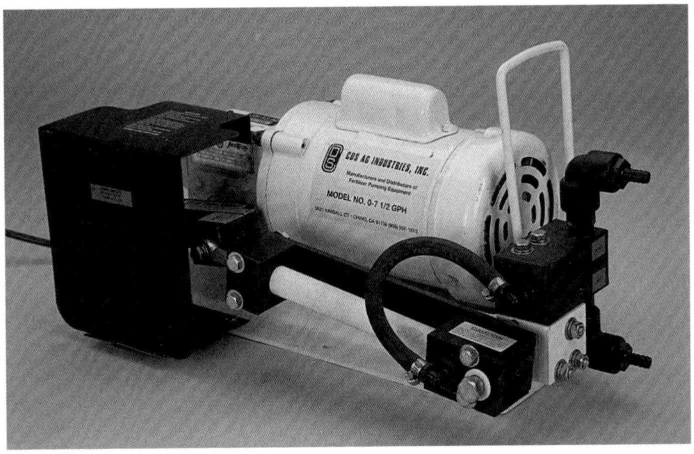

Figure 12.4 Positive displacement injector pump. *Photo:* S. Paisley.

Chemigating Uniformly

Once injection begins, the injected material does not immediately reach the emitters. There is a travel time for water and injected chemical to move through a microirrigation system. Measurements on commercial fruit orchards indicate that this travel time may range from 30 minutes to well over an hour, depending on the microirrigation system design. To ensure that water and chemical applications are equally uniform, take the following steps:

- **Step 1:** Determine the travel time of chemicals to the farthest hydraulic point in the system. This is a one-time determination and can be done by injecting chlorine into the system (also a good maintenance procedure) and tracing its movement through the system by testing the water for chlorine with a pool or spa test kit.
- **Step 2:** The injection period should be at least as long as it takes the injected material to reach the end of the last lateral line as determined in Step 1. A longer injection period is even better.
- **Step 3:** Once injection has stopped, continue the irrigation for as long as it took the injected material to reach the end of the farthest lateral (Step 1). A longer post-injection irrigation period is even better.

Figure 12.5 Solutionizer injection machine. *Photo:* L. Schwankl.

Make sure, especially with injected materials that easily travel with the water through the soil profile (e.g., nitrate materials), that there is no overirrigation that moves water and injected material through the root zone. Such overirrigation could waste the injected material and lead to ground water contamination.

Chemigation Safety

Appropriate care should be exercised when handling all fertilizer materials, and the safety of personnel should be of highest priority. Environmental safety associated with fertigation should also be a priority. Fertigation regulations vary from state to state. In California, chemigation safety regulations apply only to labeled chemicals, not to fertilizers. While there are no California regulations concerning fertigation, local regulations may exist. The same safety equipment required for injection of labeled chemicals is useful for environmental protection when fertigating.

Figure 12.6 is a sample chemigation layout with safety devices for preventing environmental contamination. Numerous other approved layout configurations are possible that incorporate different injectors and other safety devices. The safety devices in figure 12.6 include the following:

- **Chemigation check valve,** located between the water source and the injection point, prevents chemical from moving back to the water source. The check valve has a one-way spring-loaded flap inside that allows water to pass only downstream. It also has an air vent–vacuum relief valve and a low-pressure drain upstream of the one-way flap closure (fig. 12.7). The vacuum relief valve prevents a vacuum from forming, which could draw chemical through the closed check valve. If some chemical does leak past the closed check valve, it will drain out of the low-pressure drain, which is open when the irrigation system is shut down but closes when the irrigation system is pressurized.

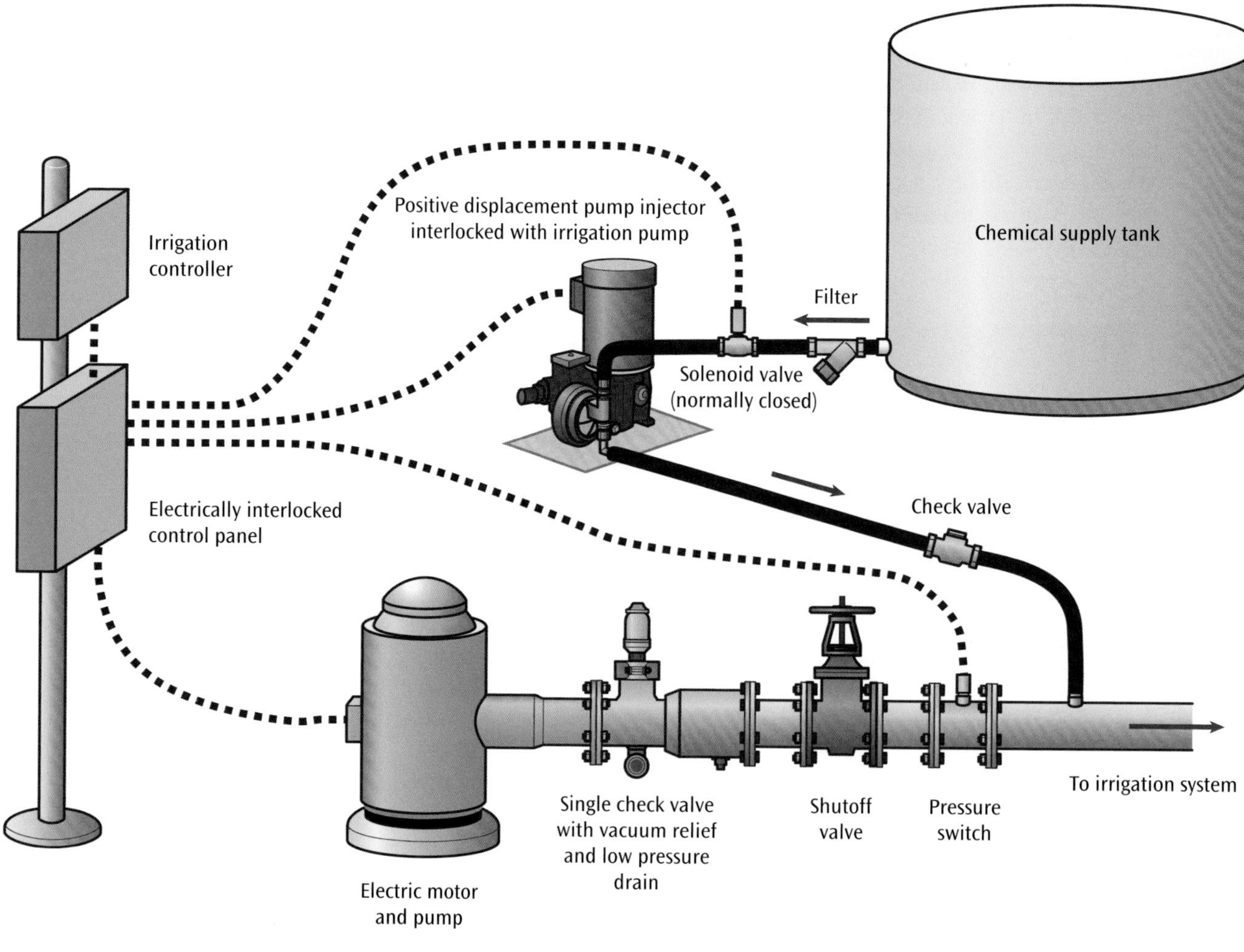

Figure 12.6 Typical layout of chemigation injection system with safety devices installed.

The discharge point of the low-pressure drain should be located so that there is no chance of the drain water entering the well. Even if there is no expectation that chemicals other than fertilizer will be injected through the irrigation system, installation of a chemigation valve is a prudent move. Backflow of fertilizer to a well can result in ground water contamination. At a minimum, leaving room to install a chemigation valve at a later date is wise. Retrofitting an irrigation system lacking space for a chemigation valve can be expensive.

- **Electronic interlock** between the water pump and the fertilizer injector pump prevents the injector from operating if water is not being pumped.
- **Check valve** in the line from the injector to the irrigation system prevents water from flowing back through the injector and overflowing the chemical storage tank.
- **Normally closed solenoid valve** (or a normally closed, hydraulically operated valve) between the chemical tank and the injector keeps chemical in the tank from flowing into the irrigation system when the system is not operating.
- **Pressure switch** in the irrigation system, interlocked to the pump, shuts down the irrigation and injection systems if there is a break in a pipeline or anything else that causes a drop in operating pressure.

References

Hanson, B., N. O'Connell, J. Hopmans, J. Simunek, and R. Beede. 2006. Fertigation with microirrigation. Oakland: University of California Division of Agriculture and Natural Resources Publication 21620.

Figure 12.7 Double chemigation check valve protecting a well from backflow contamination. A double check valve is required in some states and is installed for safety redundancy. *Photo:* UC ANR.

13 Pruning

Craig E. Kallsen and Neil V. O'Connell

The principal reason for pruning citrus is to manage tree growth so that people and equipment are able to enter and work in the orchard to conduct cultural, pest control, and harvesting activities efficiently. Effective pruning requires knowledge of when and how a citrus tree grows. More pruning is being done mechanically in response to increasing labor costs. How pruning is conducted will also have implications for pest and disease control; tree growth, productivity, and fruit quality; and recovery from freeze events.

Growth of the Citrus Tree Canopy

Effective pruning requires knowledge of the citrus tree's growth habit. Commercial citrus cultivars in California have been selected for an upright growth habit. These cultivars are characterized by strong apical dominance within the branches. As long as a branch grows upright, growth of lateral nonterminal shoots is suppressed. However, the weight of leaves and fruit eventually bends the branch over into a more horizontal position. When this occurs, the tip is no longer the most elevated point of the branch. Apical dominance ceases, and the lateral buds at the most elevated part of the branch begin to grow vigorously (fig. 13.1). These new lateral branches become the most upright fruiting branches until they are bent over by fruit, and the cycle repeats. Older wood produced earlier in the cycle becomes shaded by new growth and becomes less productive and eventually dies. Unpruned citrus trees can be characterized as having shells of vigorous, green, fruit-producing foliage covering an ever-expanding core of weak and dead branches. Citrus flowers, and eventually fruit, are typically borne directly on 1-year-old wood or on new, leafy vegetative shoots arising from 1-year-old wood. Sprouts arising from buds on existing scaffold branches will not produce fruit for 2 to 3 years.

Manual and Mechanical Pruning

Pruning refers to the manual or mechanical removal of vegetative or reproductive tissue from the tree. Manual pruning implies the use of human judgment in selecting branches or twigs to be removed or shaped, while mechanical pruning refers to using large, adjustable, circular saw blades mounted on self-propelled machines or sickle bars mounted on tractors. Pruning with large, mechanically operated saws is relatively nonselective in removal of plant tissue. Any tissue in the cutting plane of the saw, living or dead, is pruned.

Manual pruning may involve tearing or pinching back branches and twigs by hand or cutting them with hand clippers or larger shears or loppers. Removal of scaffolds and large branches requires

Figure 13.1 A branch bent by the weight of leaves and fruit shows new midbranch lateral shoot growth as a result of buds being released from apical dominance. *Photo:* C. E. Kallsen

Figure 13.2 Pruning by undercutting dead or dying shaded branches. *Photo:* C. E. Kallsen.

Figure 13.3 Mechanical hedger in action. *Photo:* C. E. Kallsen.

Figure 13.4 Mechanical topper in action. *Photo:* C. E. Kallsen.

handsaws or manually operated electric, gasoline, or pneumatic saws. Generally, manual pruning consists of removing shoots arising from larger branches (thinning). Wood that is typically pruned includes branches crossing from one side of the canopy through the interior of the tree to the other side, intertwined branches, juvenile wood sprouts originating from scaffold branches, and dead or dying branches in the tree interior. As a result of the tendency of the weight of leaves and fruit to pull branches toward the ground, the ends of branches typically become shaded by new growth arising from about the middle of the branch. These shaded, downward-growing branch ends become unproductive and are commonly pruned (undercutting, fig. 13.2).

Mechanical equipment is named for the task that it performs. Hedgers reduce tree width (fig. 13.3), toppers lower tree height (fig. 13.4), and skirters remove vegetation close to the ground. Hedgers are used both along the row and across the rows between trees (cross-hedging). Mechanical pruning is generally limited to nonselective shortening of branches (heading cuts). Both hedging and topping may be conducted at an angle from the vertical or horizontal, respectively. Typically, trees are mechanically hedged with the saws angled outward 5° to 15° from the vertical to increase the amount of light in the lower orchard leaf canopy. A distance of 7 to 8 feet between the outer canopies of trees in adjacent rows is sufficient to permit the passage of equipment for spraying and harvesting operations. Trees are topped either flat or at an angle up to 30° below the horizontal (gable topping, fig. 13.5). Topping at an angle has an advantage over flat topping in that the angled cut throws more of the prunings out of the tree into the middles. To facilitate removal of cut wood and leaves that remain lodged in the tree after cutting, the general order of pruning in the field is first topping, then hedging, followed by manual pruning. Vegetation removed from the trees is usually piled in alternate row middles for mechanical shredding. Large shredders can shred branches up to 4 inches in diameter or larger. Larger branches may need to be disposed of separately.

When and how hedging and topping are conducted in an orchard typically include annually and lightly pruning both sides of the tree and/or topping to preserve 1-year-old wood for the current season's fruit set, or heavily pruning alternate sides of the tree in alternate years. With less-vigorous varieties, intervals between mechanical hedging and topping may be extended. Deeply cutting into the tree canopy can result in excessive removal of fruit-bearing wood and leaves and may result in decreased yield, sunburn of fruit and wood, and weather-related damage to the tree (fig. 13.6). Dependence on maintaining tree height through topping tends to promote a proliferation of branches in the top of the tree, which will eventually severely restrict light penetration into the tree canopy (fig. 13.7).

Figure 13.5 Gable-topped orange trees. *Photo:* C. E. Kallsen.

Citrus Pruning Goals

Pruning citrus is a common and economically justifiable practice among California citrus growers. Reasons for pruning citrus include training a tree to produce and support a commercial fruit crop, making harvest easier, improving worker safety, allowing operation of spray and harvesting equipment, aiding in pest control, and recovering from frost injury. Other common reasons for pruning include improving fruit size and quality and tree rejuvenation.

Training a Young Tree to Produce and Support a Commercial Fruit Crop

Normally, citrus trees are cut to 2 to 2.5 feet tall at the nursery to facilitate transport of the trees, produce new growth that will shade the trunk from heat and cold, and minimize water stress. Generally, for the first few years after planting, young citrus trees are only lightly pruned. Shoots originating from the rootstock (suckers) and low on the scion (water sprouts) are removed. Branches that cross from one side of the tree to the other are typically thinned or removed, but an early objective is to produce an abundance of vegetative growth that will assist in producing new vegetative growth. The tree must produce a canopy of sufficient size before economic fruit production can commence. Attempting to train very young branches as future scaffolds is difficult in citrus due to the growth habit of the tree. As the tree ages and branches originating near the area where the tree was initially cut back in the nursery build strength and girth, good candidates for scaffolds will emerge. The dominance of these proto-scaffolds may be encouraged through thinning cuts (removing entire branches) and undercutting to remove shaded and dying branches. If a particular scaffold is not successful, citrus normally produces many more replacement candidates.

Figure 13.6 View of a severely hedged orange tree showing reduction in yield and slow regrowth of leaf canopy. *Photo:* C. E. Kallsen.

Frequently, young trees will appear to be out of balance because of a few long shoots or branches. Generally, these branches should not be headed back, as the weight of fruit and leaves will pull them into a more normal-looking position. Due to the more brittle wood and willowy vegetative growth habits and alternate-bearing tendencies of some

Figure 13.7 Typical proliferation of new growth at the tree top in the season following early-summer topping. This pattern of growth quickly reduces light penetration into the tree interior. *Photo:* C. E. Kallsen.

mandarin varieties, some growers begin training these trees early to produce a smaller canopy with thicker scaffolds and an open, vase-like shape. This type of training consists of more thinning and heading cuts in the top of the young canopy than would be done for other varieties of citrus.

Placing wraps on the young tree to insulate it from frost and sunburn also usually reduces sprouting of rootstock buds, as does heavy coats of whitewash or white interior latex paint. Once the young tree has produced sufficient leaf canopy to shade the trunk, the sucker growth of many citrus varieties is suppressed.

Pruning a Mature Tree to Produce and Support a Commercial Fruit Crop

Eventually, citrus trees require pruning to allow access to the orchard for cultural and harvest activities. Citrus trees appear to remain most productive if left unpruned or pruned annually, rather than having large amounts of vegetation removed infrequently. Lemons are especially vigorous and produce long, relatively thin branches with little strength but an abundance of fruit. Annual light pruning of lemons, including both thinning and heading cuts, facilitates harvest and produces a tree with less limb breakage. Citrus branches that arise below the graft union should be removed, as they produce the nonmarketable fruit of the rootstock. Annual pruning allows early removal of branches originating low in the tree that eventually grow through the top of the canopy, removal of older shaded branches, and removal of branches that have begun to circle within the canopy in search of light.

Any factor that affects the rate of growth of the tree canopy will affect the need for pruning and the recovery from and response to pruning. Important variables to consider when evaluating citrus pruning experiments that have been published in the scientific literature include planting density, tree age, cultural factors like irrigation and nutrition, tree health and vigor, variety, rootstock-scion interactions, presence of viruses or viroids, and weather (see the references at the end of this chapter). Most pruning experiments have not included values for some of these variables, and thus the results achieved by someone following recommendations based on these experiments may be very different from the results obtained by the experimenter.

Planting Density and Pruning

Planting density affects how a citrus orchard is pruned. Orchards planted with trees in closely spaced rows have smaller-diameter canopies than do those in more widely spaced rows; a minimum of 7 to 8 feet is required between outer canopy branches in adjacent rows to allow clear passage for equipment (see chapter 9, "Establishing the Citrus Orchard").

Likewise, to avoid shading of the lower canopy, closely spaced trees should be shorter than widely spaced trees with similar levels of vigor and productivity. General guidelines used successfully in the past for maintaining tree height for the production of citrus fruit for the table market are that tree height should not exceed 80% of the distance from trunk to trunk in adjacent rows and that the distance between the outer canopy branches of trees in adjacent rows should be roughly 50% of tree height. For example, trees in rows 20 feet apart should have a maximum height of 16 feet, with approximately 8 feet between outer canopy branches of trees in adjacent rows. By following these guidelines, adequate light distribution to the lower tree canopy may be maintained, especially if hedging is done at an angle.

As tree height increases, shading of the lower citrus canopy (skirts) increases. Loss of fruit-bearing branches in the lower canopy is an indication that the trees in the orchard are too tall. In mature orchards, a balance can be achieved between tree height and loss of lower canopy fruitwood that will reduce the need for mechanical skirting, without affecting fruit yield.

Water Use and Pruning

Heavily pruned trees require less water as a result of the smaller canopy and transpiring surface. Light pruning has little effect on water use.

Rootstock and Scion Vigor

Varietal and rootstock differences may greatly affect the growth rate (vigor) of citrus and thus the amount of canopy management that will be required (for more information, see chapter 5, "Rootstocks"). Lisbon lemon and Frost nucellar navel are vigorous, while Satsuma mandarin, old-line Washington navel, and early navel varieties like Thompson Improved and Bonanza are less vigorous. Dwarfing rootstocks like Flying Dragon Trifoliate produce dwarf trees with slow growth rates that require very little pruning. Likewise, many of the Trifoliate rootstocks and Trifoliate hybrids such as C35 grow slowly, and scions of most citrus varieties on these rootstocks have reduced pruning requirements compared with the same varieties on rootstocks with lemon heritage such as Volkameriana, Rough Lemon, and Macrophylla. Most citrus varieties on Sour Orange rootstock are also less vigorous than if grafted onto the lemon rootstocks. Research suggests that certain nonpathogenic viroids such as transmissible small nuclear IIa and IIIb can be introduced into a scion like Washington navel on Trifoliate rootstock to produce smaller trees with greater numbers of large fruit on a per-area basis. While still experimental and research continues, controlling canopy size through the dwarfing effect of nonpathogenic viroidlike entities could reduce pruning requirements and costs.

When to Prune

Citrus may be pruned at any time of the year. Considerations regarding when to prune should include the time of year, presence of mature fruit on the tree, the acreage requiring pruning, and the availability of equipment and labor, as well as other cultural operations affected by the delay resulting from the pruning and shredding. Some young orchards are not pruned until crowding, as defined by mutual shading, begins to affect fruit yield and quality or begins to hinder access to the orchard.

Heavy pruning before or during bloom affects fruit yield less than similar pruning in late spring after the fruit has set. Early-maturing navels, mandarins, tangelos, and lemons in the San Joaquin Valley are typically picked in early fall or early winter, and pruning does not occur until late winter or early spring (late February, March, or early April).

To avoid damage to mature fruit, trees should not be pruned immediately before harvest, and the presence of mature fruit on the tree often precludes an early-spring pruning. Early-maturing Valencia oranges, mid- and late-season navels, and lemons in the San Joaquin Valley are generally pruned after harvest in spring or early summer. Lemons in coastal California and most Valencia orange trees in coastal areas and in the San Joaquin Valley are commonly pruned from late spring to late summer after the final harvest. Late-summer and fall pruning should be avoided in most areas of the San Joaquin Valley and in other cold citrus-growing areas to encourage winter semidormancy, which will further protect the tree from freezes. Pruning in late summer or early fall increases production of freeze-susceptible new growth. Winter pruning should also be avoided in cold locations. A large, thick canopy has been shown to reduce injury to the tree from severe frosts, probably through better heat retention within the tree and improved heat absorption from the ground.

Pruning to Improve Harvest Efficiency and Safety of Fruit Pickers

Pruning can improve the harvest efficiency of hand-picked fruit. Shorter trees with high yields of large fruit generally means lower harvesting costs per picked carton and less risk of injury from manipulating and falling from tall ladders. Research conducted in California in 1962 showed that the speed of harvest improved from 50 field boxes per day per picker for unpruned trees 20 to 22 feet tall to 72 field boxes per day for trees that were hedged and topped at 14.5 feet (Lee 1962). In 2005, labor shortages in California allowed pickers to be more selective in choosing orchards to harvest. Some orchards with trees taller than 15 feet were left unharvested as pickers opted for orchards with shorter trees. When tall trees were picked, growers paid more per bin of fruit, reflecting the increased labor required for harvest.

Pests and Pruning

Pruning can reduce the costs of energy and material in citrus production. Smaller trees require a smaller spray volume for thorough coverage to achieve similar levels of insect or mite control. The upper canopies of tall trees are more difficult to cover adequately than are the tops of shorter trees. Spray penetration into the interior of the canopy, where scale infestations may be more severe, is improved when dead wood and some intertwined scaffold branches have been pruned. (For more information on pests, see chapter 17, "Citrus IPM," as well as chapters 18 to 24 on specific pests.)

Managing the Canopy Environment

Pruning may influence pest populations within the tree by affecting the environment within the tree canopy. Soft scales, such as cottony cushion *(Icerya purchasi)* and citricola scale *(Coccus pseudomagnoliarum)*, proliferate in shady, cool, humid, unpruned tree interiors, whereas an airy canopy provides a habitat more conducive to citrus thrips *(Scirtothrips citri)*, which prefers hot, dry conditions. The relative efficacy of some beneficial insects also appears to be related to microenvironments associated with pruning that vary with species. Research has shown that winter pruning of exterior foliage increased populations of the beneficial mite *E. tularensis* and reduced early-season fruit damage caused by citrus thrips (Grafton-Cardwell and Ouyang 1995). Roof rats *(Rattus rattus)* prefer to live and feed in large, unpruned lemon trees with intertwined scaffolds, perhaps because of the increased structural support for the nest and increased protection from predators that these trees may provide.

Skirting

Skirting is the process of removing branches in the lower tree canopy that contact the ground. Developing fruit and the increased weight associated with branch and leaf growth pull lower branches to the ground as the season progresses. Skirting is typically performed either by hand-pruning or machines. Branches within 2 feet of the ground are removed in the spring before bloom. A tree canopy in contact with the ground is associated with a decrease in fruit quality largely from fruit rots, abrasions, and greater incidences of infestation by pests such as Fuller rose beetle *(Pantomorus cervinus),* bean thrips *(Caliothrips fasciatus),* and the Argentine ant *(Iridomyrmex humilis)* that interfere with biological pest control. Infestations of citrus peel miner *(Marmara gulosa)* can often be found first on fruit touching or in close proximity to the ground. Brown garden snails *(Helix aspera)* are good climbers and are more of a problem in unskirted orchards. By removing the tree skirts, brown garden snails have reduced access to the fruit and may be more effectively controlled by the non-climbing decollate snail *(Rumina decollata).*

Pruning and Disease Control

Pruning, either manual or mechanical, may spread citrus diseases such as psorosis, exocortis, and some viral diseases. Disinfecting hand tools is feasible between trees, but not so for mechanical toppers, hedgers, and skirters. Pruning dead twigs and branches, where fungi reproduce, such as those causing tear staining *(Diaporthe citri* and *Colletotrichum gloeosporioides)* and Botrytis rot *(Botrytis cinerea)* in lemons, may reduce the incidence of some diseases and disorders. Brown rot (*Phytophthora* spp.) often begins to develop on fruit adjacent to the ground from spores splashed from soil or from infected fruit. Pruning tree skirts 2 feet or more above the ground can assist in reducing brown rot disease.

Pruning Freeze-Injured Trees

Severe freeze events damage twigs, branches, and scaffolds, and when severe, may kill trees. Frozen trees should not be pruned after a freeze until the full extent of the damage becomes visible. Typically, after severe freezes, trees initially appear to recover and produce new vegetative and reproductive growth. This new growth may die back, especially once the first high temperatures of late spring or summer occur. Once the extent of the worst damage is apparent, dead wood should be removed to encourage regrowth that can be reformed into a new tree canopy. Pruning cuts should be made into living wood to remove disease-causing organisms that typically colonize frozen wood, such as the fungi *Fusarium solani, Hendersonula torulaoidea,* and *Dothiorella gregaria*. These organisms, if left in the wood, can continue growing and killing trees for years. In dry climates like California, it is not necessary to cover pruning cuts with a protective wax. Trees that lose substantial amounts of canopy should be whitewashed to prevent sunburn.

Tree Rejuvenation

A declining tree may be rejuvenated by being skeletonized (removal of a large portion of the canopy) or buckhorned (cut back to major scaffold limbs). Since oranges are produced on younger wood, this practice greatly reduces fruit yield and quality for 2 years or more. This type of pruning eventually results in a higher ratio of new green leaf tissue to woody structural tissue, which may be partly responsible for the rejuvenating effect. However, results from this practice have been mixed.

Skeletonizing, or buckhorning, trees appears to be useful in trees declining from some root rots such as those caused by some *Phytophthora* species. The purpose of skeletonizing a tree is to bring the aboveground portion of the tree into better balance with a damaged root system. For this type of pruning to be effective, however, the conditions that resulted in the decline of the root system must be first corrected. Typically the soil-water environment is at fault, and improving the drainage and the irrigation system performance or scheduling, along with applying an appropriate fungicide, can improve the chance of the tree surviving. Scaffold limbs and trunks that may have been shaded by the canopy for decades prior to skeletonizing or buckhorning should be whitewashed immediately after pruning regardless of the season to prevent sunburn. Inexpensive interior latex paint mixed 1 part paint to 2 or 3 parts water makes a satisfactory whitewash for this purpose. Whitewashes may also reduce production of suckers and water sprouts. In dry, sunny climates, such as those in the San Joaquin Valley, treating large pruning wounds with tree wax or sealers is unnecessary and may slow healing.

It may not be possible to rejuvenate some declining trees. Late-stage incompatibility is common in older blocks of navels in the San Joaquin Valley, especially with some navel and mandarin varieties on Trifoliate or Trifoliate hybrid rootstocks. Severely diseased trees, such as navel orange trees on Sour Orange rootstock infected with tristeza or trees with stubborn disease, will not respond positively to rejuvenation pruning.

Pruning to Improve Fruit Size and Grade and Its Effect on Yield

In citrus, most of the fruit is produced on the sunlight-exposed outer tree canopy. Theoretically, if a grower wishes to minimize impacts on yield, the outer canopy should be disturbed as little as possible, but this is unavoidable in practice. Removal of intertwined scaffold branches for the purpose of increasing light penetration into the canopy, or a reduction in canopy diameter to avoid the need for annual deep hedging cuts, is usually associated with removal of a significant portion of the outer tree canopy.

Generally, to increase fruit size in the same season in which pruning is done, the pruning should be done by the end of May. Citrus typically bears fruit on wood that is 1 year old or older so the yield, in terms of total weight or total fruit numbers, of severely pruned trees may take several years to recover to prepruning levels. For healthy, vigorous citrus trees that are optimally fertilized and irrigated, experiments suggest that to increase the percentage of large fruit produced per mature tree, at least 20% of the existing tree canopy must be removed. However, the percentage increase in the number of large fruit in the season pruning is done may be accompanied by a decrease in total yield by weight of approximately the same percentage as the percentage of the tree canopy removed. The net result is fewer large fruit per tree. For example, assume that with severe pruning, a mature navel orange tree produces 500 fruit, and 90% of those, or 450, have a diameter greater than 2.8 inches. Assume that a neighboring unpruned tree produces 800 fruit, and only 75%, or 600, have a diameter greater than 2.8 inches. While the pruned tree has a higher percentage of large fruit (90% versus 75%), the total number of large fruit available for harvest is greater on the unpruned tree (450 versus 600). The negative effect on yield may carry into year 2 and later until the canopy of the pruned tree again approximates that of unpruned trees. The total cumulative yield and the number of the fruit in the most valuable size classes on heavily pruned trees may never equal that on unpruned trees, and the expense of heavy pruning may never be recovered.

Orange orchards that typically produce too many fruit to size effectively (usually more than 100,000 fruit per ac) may be densely planted orchards, older orchards of some varieties, and orchards with trees under stress. These orchards may benefit from a severe pruning, resulting in a large reduction in the number of fruit produced per tree but a net increase in the number of fruit in the most valuable size classes. Yields of coastal lemons, especially those planted on Macrophylla rootstock with a vigorous scion like Prior, benefit from light pruning strategies designed to maintain the tree within its allotted space. In unpruned orchards, competition between rapidly growing branches and neighboring trees appears to reduce yields with time. However, several lemon pruning studies have shown that moderate to heavy interior pruning, hedging, or topping resulted in decreased yield when compared with yields of unpruned trees.

Yields of heavily pruned immature citrus trees appear to be less affected by heavy pruning, probably because young trees are primarily vegetative. Immature citrus may have a better ability to compensate for fruitwood removed in heavy pruning. Yields and fruit sizes of pruned and unpruned young trees are typically more similar than what occurs in mature citrus.

Hedging, topping, and skirting (removal of less than 5% of the tree canopy annually or every second year, depending on tree vigor) appear to have little effect on yield or fruit size of mature citrus if conducted in the spring prior to fruit set. Skirting of immature trees reduces fruit yield by reducing the productive area of the tree canopy. Orchard yield will continue to be adversely affected by annual skirting until the continued growth of the trees begins shading the lower canopies, limiting skirt productivity.

Fruit in the Interior of the Canopy

Fruit found inside the tree canopy tend to have less sunburn, more uniform rind characteristics, and less mechanical and citrus thrips scarring. Some growers in California prune citrus to improve light penetration into the canopy with the objective of increasing the percentage of fruit produced in the interior of the tree. However, unless frequent follow-up branch thinning is conducted, little additional fruit is produced in the interior of the tree canopy because of rapid proliferation of vegetative growth. Even large-diameter scaffold branches in citrus have numerous adventitious buds that grow rapidly in response to light (fig. 13.8). Interior canopies of heavily pruned trees are often denser a few months after pruning than they were before pruning, and the bulk of this new vegetative growth is too young to bear fruit. Most fruit located in the interior canopy at harvest is produced close to the outer canopy, and the weight of the developing fruit pulls the fruit-bearing branch down into the inner canopy.

Alternate Bearing

Some growers in California use manual pruning, hedging, and topping to minimize alternate bearing in Minneola tangelo and many mandarin varieties and to reduce crop loads of other citrus varieties in years with a heavy fruit set. Heavy fruit sets, caused by alternate bearing or a cool spring with reduced fruit drop, can result in a large proportion of the fruit being too small to be profitable.

Figure 13.8 A tree showing proliferation of new interior growth in the summer after severe pruning the previous spring. *Photo:* C. E. Kallsen.

References

Burns, R. M., S. B. Boswell, S. F. Wer, and C. D. McCarty. 1975. Comparison of pruning methods of mature Lisbon lemons. Citrograph 60:344, 373–374.

Cameron, S. H., and R. W. Hodgson. 1941. Effect of severity of pruning on top regeneration in citrus. Proceedings of the American Society for Horticultural Science 39:67–72.

Cohen, A., A. Goell, S. Cohen, and R. Ismajovitch. 1988. Effect of leaf distribution in the canopy on the total dry matter production in grapefruit trees. Israel Journal of Botany 37:257–266.

Davies, F. S., and L. G. Albrigo. 1994. Citrus. Crop Production Science in Horticulture Series, vol. 2. Wallingford, UK: CAB International.

Fallahi, E., and M. Kilby. 1997. Rootstock and pruning influences yield and fruit quality of 'Lisbon' lemon. Fruit Varieties Journal 4:242–246.

Francis, H. L., M. Miller, S. Boswell, and C. Colladay. 1975. An economic analysis of three lemon pruning methods. Citrograph 61:12, 24–26, 43.

Golomb, A., D. Reinhartz, and E. Israel. 1988. Management of crowded grapefruit plots by tree-thinning or pruning. Proceedings of the 6th Congress of the International Society of Citriculture 963–968.

Grafton-Cardwell, E. E., and Y. Ouyang. 1995. Manipulation of the predacious mite, *Euseius tularensis* (Acari: Phytoseiidae), with pruning for citrus thrips control. In B. L. Parker et al., eds., Thrips biology and management. New York: Plenum Press. 251–254.

Halma, F. F. 1923. Influence of position on production of laterals by branches. California Citrograph 8:146, 180.

Haney, P. B., J. G. Morse, R. F. Luck, H. Griffiths, E. E. Grafton-Cardwell, and N. V. O'Connell. 1992. Reducing insecticide use and energy costs in citrus pest management. Davis: University of California Statewide Integrated Pest Management Program Publication 15.

Hume, H. H., 1934. The cultivation of citrus fruit. New York: Macmillan.

Johnston, J. C. 1963. Treatment of frost injured trees. California Citrograph 48:106, 116, 117, 119.

Kallsen, C. E. 2005a. Production of commercially valuable sized fruit as a function of navel orange yield. HortTechnology 15:608–612.

———. 2005b. Topping and manual pruning effects on the production of commercially valuable fruit in a midseason navel orange variety. HortTechnology 15:335–341.

Krajewski, A. J., and E. Rabe. 1995. Bud age affects sprouting and flowering in clementine mandarin (*Citrus reticulata* Blanco). HortScience 30:1366–1368.

Lea-Cox, J. D., J. P. Syvertsen, and D. A. Graetz. 2001. Springtime nitrogen uptake, partitioning, and leaching losses from young bearing *Citrus* trees of differing nitrogen status. Journal of the American Society for Horticultural Science 126:242–251.

Lee, B. W. 1962. Hedging and topping of mature Valencia orange trees. California Citrograph 48:42, 52, 53.

Lewis L. N, and C. D. McCarty. 1973. Pruning and girdling citrus. In W. Reuther, et al. eds., The citrus industry. Vol. 3. Rev. ed. Berkeley: University of California Press. 211–229.

Lewis L. N., C. D. McCarty, and P. W. Moore. 1963. Effects of skeletonization on the rejuvenation and production of declining orange trees. Proceedings of the American Society for Horticultural Science 83:295–299.

McCarty, C. D., and L. N. Lewis. 1964. Pruning in relation to tree maintenance. California Citrograph 49:278, 281–284.

McCarty, C. D., S. B. Boswell, R. M. Burns, R. G. Platt, K. W. Opitz, and L. N. Lewis. 1982. Pruning citrus trees. Oakland: University of California Division of Agriculture and Natural Resources Leaflet 2449.

Moore, P. W. 1957. Pruning makes cents. California Citrograph 42:110, 122.

Morales, L. P., and F. S. Davies. 1997. Pruning and skirting affect canopy microclimate, yields and fruit quality of 'Orlando' tangelo. HortScience 35:30–35.

Moreshet, S., Y. Cohen, and M. Fuchs. 1988. Water use and yield of a mature Shamouti orange orchard submitted to root volume restriction and intensive canopy pruning. In R. Goren, K. Mendel, and N. Goren, eds., Citriculture: Proceedings of the Sixth International Citrus Congress, Tel Aviv. Rehovot, Israel: Balaban. 739–746.

Oren, Y. 1988. Pruning clementine mandarin as a method for limiting tree volume and increasing fruit size. Proceedings of the 6th Congress of the International Society of Citriculture 953–956.

Phillips, P. A., N. V. O'Connell, and J. A. Menge. 1990. Citrus skirt pruning: A management technique for Phytophthora brown rot. California Agriculture 44:6–7.

Phillips, R. L. 1980. Hedging and topping practices for Florida citrus. Citrus Industry 61:5–10.

Puffer, R. E. 1963. Girdling and pruning mature navel orange trees. California Citrograph 48:424, 433–434, 436.

Semancik, J. S., and D. J. Gumpf. 2000. Citrus viroids: Detection, relationship to disease, and effects on citrus performance. In 2000 Annual report of the Citrus Research Board. Visalia: Citrus Research Board 31.

Shamel, A. D. 1919. Some results from an experiment with pruning as compared with no pruning of full bearing Washington navel trees. California Citrograph 4:174–175.

———. 1920a. Results of five years' individual tree performance records with pruned and unpruned lemon trees. California Citrograph 5:102, 122–123, 128.

———. 1920b. Results of individual tree performance record studies with pruned and unpruned Marsh grapefruit trees. California Citrograph 5:248, 268–269.

———. 1925. Performance records of pruned and unpruned Washington navel trees for a period of ten years. California Citrograph 10:415, 442–443, 445.

———. 1927. Pruning of citrus trees in California. California Citrograph 13:18–20.

———. 1929. Washington navel pruning and girdling experiments. California Citrograph 14:335, 376, 382.

Shamel, A. D., and C. S. Pomeroy. 1942. Effects of pruning old Washington navel orange trees. Proceedings of the American Society for Horticultural Science 41:71–76.

Stover, E., D. Scotto, C. Wilson, and M. Salyani. 2003. Pesticide spraying in Indian River grapefruit: II. Overview of factors influencing spray efficacy and off target deposition. HortTechnology 13:166–177.

Timmer, L. W., S. M. Garnsey, and J. H. Graham, eds. 2000. Compendium of citrus diseases, 2nd ed. St. Paul, MN: APS Press.

Tucker, D. P. H., T. A. Wheaton, and R. P. Muraro. 1991. Citrus tree spacing and pruning. Gainesville: University of Florida Cooperative Extension Service.

Wheaton, T. A., W. S. Castle, J. D. Whitney, and D. P. H. Tucker. 1999. Horticultural practices for citrus health. In L. W. Timmer and L. W. Duncan, eds. Citrus health management. St. Paul, MN: APS Press. 49–58.

Wheaton, T. A., J. D. Whitney, W. S. Castle, R. P. Muraro, H. W. Browning, and D. P. H. Tucker. 1995. Tree vigor important in citrus tree spacing and topping. Proceedings of the Florida State Horticultural Society 108:63–69.

Whitney, J. D., T. A. Wheaton, W. S. Castle, and D. P. H. Tucker. 1995. Tree height, fruit size, and fruit yield affect manual orange harvesting rates. Proceedings of the Florida State Horticultural Society 108:112–118.

Wilder, J. J. 1943. Some lessons from history. In H. J. Webber and L. D. Batchelor, et al., eds. The citrus industry. Vol. 1. Berkeley: University of California Press; repr. Citrograph 76 (1991): 15, 16, 18, 19, 21, 23.

Zaragoza, S., and E. Alfonso. 1981. Citrus pruning in Spain. Proceedings of the 4th Congress of the International Society of Citriculture 1:172–175.

14 Plant Growth Regulators

Charles W. Coggins Jr. and Carol J. Lovatt

It has long been the goal of growers and researchers alike to be able to manipulate the vegetative and reproductive growth of crop plants. Citrus growers and researchers are no exception. Plant growth regulators, both natural and synthetic, are perhaps the most powerful tools currently available to growers for achieving this goal. There are many examples of the successful use of plant growth regulators to solve citrus production problems, to increase the market value of the fruit, and to improve grower profit. Plant growth regulators have been used successfully as foliar sprays to increase flowering, synchronize bloom, or change the time of flowering to avoid adverse climatic conditions or to shift harvest to when the market is more economically favorable. Foliar-applied plant growth regulators are routinely used to improve fruit set, reduce June drop, or to prevent preharvest drop to increase yield. Plant growth regulator sprays are applied to increase fruit size directly by stimulating fruit growth; they are also applied to increase fruit size indirectly by decreasing fruit number and competition for resources by reducing the number of flowers formed or by promoting flower or fruit abscission. Plant growth regulators have been used as both pre- and postharvest treatments to hasten or slow the ripening process, color development, and maturation of a specific fruit tissue to improve the quality of the product sold in the market. The emerging use of plant growth regulators to overcome the adverse effects of abiotic stresses is proving increasingly successful. Surprisingly, these successes have been achieved with a modest number of commercial plant growth regulators that are members of one of the five classic groups of plant hormones: auxins, cytokinins, gibberellins, abscisic acid, and ethylene. The roles of these plant growth regulators in plant development will be reviewed below. Development and field testing of new commercial plant growth regulators and new uses for those currently registered for use on citrus to determine efficacy and cost-benefit are ongoing.

For a plant growth regulator or combination of plant growth regulators to be successful regarding a particular desired agronomic outcome, the positive responses should be highly reproducible, the negative responses should be either mild or rare, and the overall results should be viable economically. Some of the responses mentioned above and later in this chapter meet these criteria in California and some do not.

Definitions

A hormone is a chemical signal produced in one part of the plant and transported to another part, where it binds to a specific receptor and triggers a response in the target cells or tissues of the plant. Hormones produce profound effects at very low concentrations. Plant growth regulators are synthetic analogs of naturally occurring hormones that are applied exogenously to mimic the effects of hormones (fig. 14.1). Hormones control plant growth and development by influencing cell division (cell number), cell expansion (cell size), and cell differentiation (cell structure and function). Hormones also regulate the plant's response to the environment, which can include the plant's response to environmental stresses. Each hormone has multiple effects depending on its concentration, the target cells or tissues, and the developmental stage of the plant and/or target tissue. Moreover, hormones do not act in isolation. The response to a hormone or plant growth regulator

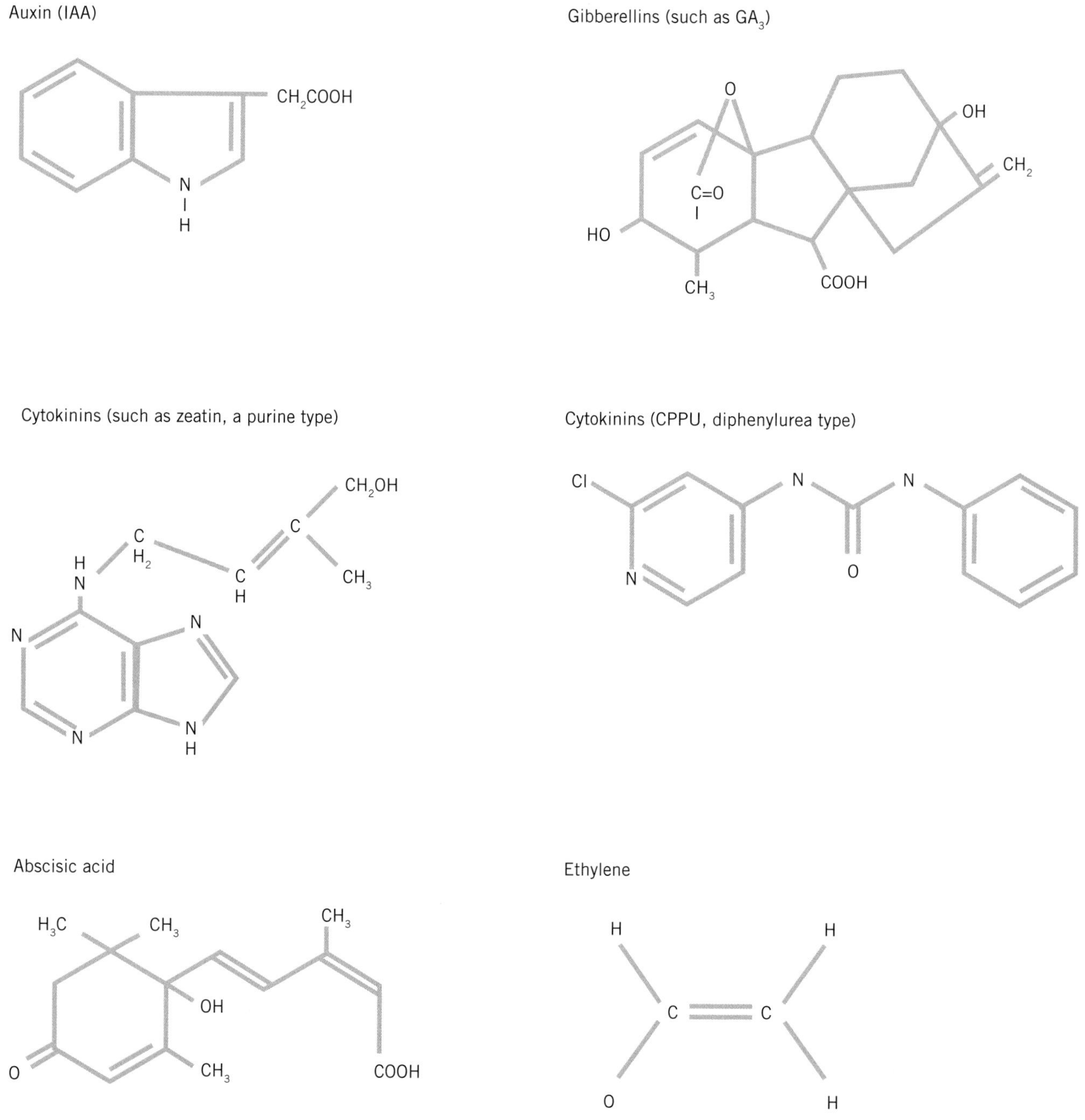

Figure 14.1 Basic molecular structure of the five classic hormones.

depends on its concentration relative to that of other hormones present, the plant's health and nutritional and water status, and on the climate. Each hormone binds to a specific receptor that activates a unique signal transduction pathway, a sequence of reactions that change cell metabolism and ultimately alter the plant's physiology. Due to differences in the presence or absence of receptors and signal transduction pathways in various tissues as they develop, a single hormone can elicit many different responses. In addition, the signal transduction pathway initiated by one hormone intersects (cross-talks) with signal transduction pathways initiated by other hormones, environmental events, or developmental cues, which modify the final physiological response. In this manner, a relatively small number of hormones are able to regulate the physiological processes essential to citrus tree development and crop production.

In the remainder of this chapter no distinction will be made between the terms "hormone" and "plant growth regulator."

Classification and Function

Auxins

Commercial auxins are compounds that elicit responses similar to those initiated by the plant hormone indole-3-acetic acid (IAA). IAA is ubiquitous throughout the plant but is synthesized in actively growing tissues: the embryo in the seed during germination, young leaves during leaf expansion, developing fruit, and the shoot apical meristem during growth. Each spring, IAA stimulates the growth of new conducting cells in the xylem and phloem. IAA synthesized in the shoot apical meristem moves down the shoot to the roots, a process known as polar transport, and promotes shoot and root elongation by causing cells in the zone of cell elongation, located immediately basipetal to the shoot and root apical meristems, respectively, to elongate. The amount of IAA moving down the shoot to the roots is sufficiently high that it inhibits cytokinin biosynthesis in the lateral (axillary) buds along the shoot, also inhibiting the growth of these buds, independent of whether they are vegetative or floral shoot buds. This phenomenon is known as shoot apical dominance. Lord and Eckard (1987) observed this phenomenon for citrus floral shoots, in which the apical flower delayed the growth of the four or five flower buds below it.

IAA is important to fruit set and fruit development; it up-regulates the synthesis of gibberellins, which are also important to fruit development; the export of IAA from developing fruit maintains their sink strength and their capacity to import resources essential for growth. IAA regulates the first steps in parthenocarpy (the development of an ovary into a fruit in the absence of pollination and fertilization) and in stimulated parthenocarpy (the development of the ovary into a fruit in response to pollination but without fertilization), including the up-regulation of gibberellin biosynthesis. Both physiological processes result in seedless fruit. The fruit of numerous citrus cultivars develop parthenocarpically; the Washington navel orange is a well-known example. In addition, stimulated parthenocarpy occurs in this cultivar. Pollination of Washington navel orange flowers with Hamlin orange pollen increases fruit set and fruit size without seed formation.

The export of IAA during fruit development in a heavy crop year can lead to the accumulation of high concentrations of IAA in the buds that develop into the summer vegetative shoots, inhibiting their growth and reducing their contribution to return bloom (Verreynne 2005). When the source of indole-3-acetic acid that inhibits growth is not the shoot apical meristem, the resulting inhibition is termed "correlative inhibition," and it may prove to be an important factor in the effect of crop load on floral intensity the following spring (Verreynne and Lovatt 2009).

Plant tissues have different sensitivities to IAA, so stimulation or inhibition of shoot and root growth and fruit development by IAA or a commercial auxin depends on the concentration applied. The auxin 2,4-dichlorophenoxyacetic acid (2,4-D) provides an excellent example. At a low concentration, 2,4-D effectively increases citrus fruit size, but at a high concentration it is a potent herbicide.

Cytokinins

Commercial cytokinins fall into two groups: modified purines analogous to endogenous cytokinins and diphenylureas that mimic the action of endogenous cytokinins despite having a distinctly different molecule structure. Endogenous cytokinins are synthesized in the roots and transported in the conducting cells of the xylem to the canopy, where they stimulate shoot growth. Cytokinins are also synthesized in leaves. Leaf-produced cytokinins are translocated in the conducting cells of the phloem from the shoots to the roots and stimulate root growth. The synthesis of cytokinins in roots and leaves is carefully regulated by nitrogen availability in each organ.

Cytokinins are best known for stimulating cell division and growth by cell division. They stimulate formation of roots, root growth, seed germination, vegetative and floral shoot growth, fruit set and fruit development, and delay senescence. Cytokinins play an important role in spring budbreak and possibly budbreak at other times in the phenology of the tree. Cytokinins, in addition to indole-3-acetic acid, play a role in shoot apical dominance. IAA inhibits the key regulatory enzyme required for cytokinin synthesis in lateral buds, thereby inhibiting their growth. Evidence for this role comes from pruning, which removes the shoot apex and up-regulates cytokinin synthesis in the lateral buds, releasing them from apical dominance so that new shoots sprout. The loss of the shoot apex, the site of IAA production, results in a greater ratio of cytokinin to IAA in the buds along the shoot and growth of new shoots from the lateral buds. Cytokinins are also important in maintaining the sink strength of fruit and other organs during development.

Gibberellins

It has been estimated that plants synthesize over 120 gibberellins (GAs), although individual plant species have far fewer. It is now understood that GA_1 is the active GA in the majority of plant species.

It is produced in the meristem of apical buds and root apices, in young leaves and fruit, and in the embryo of a seed. Roots also have a very important role in converting GAs from one type to another. Gibberellins produced by or applied to leaves are translocated in the conducting cells of the phloem to the roots and converted to a different GA, which is transported in the conducting cells of the xylem back to the shoots, where it now elicits a response. GAs promote budbreak, seed germination, shoot elongation, leaf expansion, fruit set, and fruit growth. Gibberellins cause cell division in the cells of the node (the point of attachment of a leaf to the shoot) and also cause these cells to elongate, resulting in elongation of the internode (the region of the shoot between nodes) and elongation of the shoot. Gibberellic acid (GA_3), the most commercially important form of GA, differs from GA_1 by the presence of a double bond not found in the GA_1 molecule. Endogenous GA_1 and exogenously applied GA_3 stimulate flowering in some plant species, but not citrus. Foliar-applied GA_3 inhibits the transition from vegetative to reproductive growth in citrus if applied prior to irreversible commitment to flowering. Since it does not affect flowering when applied after irreversible commitment to floral development, GA_3 can be used to pinpoint when buds become committed to floral development (Lord and Eckard 1987).

Abscisic acid

In contrast to the growth-stimulating hormones discussed above, abscisic acid (ABA) slows or inhibits growth and arrests development. Thus, the ratio of ABA to one or more of the growth-promoting hormones is key to controlling physiological processes and growth and development. ABA is a root-produced hormone that is transported to other parts of the plant through the conducting cells of the xylem via the transpiration stream. Root-produced ABA is found in leaves, stems, and fruit, but these tissues also synthesize it. ABA has many effects in plants. The following three are of significant importance to citrus.

The role of abscisic acid in drought stress

When roots experience water deficit, root abscisic acid synthesis increases. Abscisic acid transported to leaves in the conducting cells of the xylem initiates a signal transduction pathway in the guard cells, causing the stomates to close, which preserves the water status of the tree. Leaves receiving inadequate water also produce ABA, which closes their stomates. Thus, tissues experiencing the lack of water produce ABA, the early-warning signal of drought and the elicitor of the plant's response mechanisms to stress. Early in the stress period or under moderate water-deficit stress, ABA-induced stomatal closure reduces transpiration sufficiently to maintain the water status of the tree without significantly affecting tree physiology. However, under severe water-deficit stress, ABA causes prolonged stomatal closure, which reduces photosynthesis and productivity. Lemon cultivars are tolerant to water-deficit stress compared to other cultivars, and *forzatura*, withholding irrigation for two months in the hot summer, is used to force a second bloom in September, upon watering in August, that sets a crop of lemons that are harvested the following summer in time for lemonade season, with no negative effect on the yield of the spring crop (Hake 1995). It has long been speculated that abscisic acid, which is known to down-regulate GA biosynthesis, might regulate flowering in citrus.

The role of abscisic acid in fruit set and development

Our knowledge of the role of ABA in fruit set and development remains incomplete. In sweet oranges, maintaining low amounts of ABA relative to IAA, GA, and/or cytokinins seems to be critical to citrus fruit set and size. Hofman (1988) demonstrated that Valencia fruit borne on leafy inflorescences had higher ratios of cytokinin to ABA than those borne on leafless inflorescences. Under periods of high temperature and water deficit, the ABA concentration increased in fruit borne on both inflorescence types. However, due to the low cytokinin concentration in the fruit borne on leafless inflorescences, the ratio of ABA to cytokinin quickly became greater than 1:1, causing the fruit on the leafless inflorescences to abscise. The higher concentration of cytokinin in the fruit of leafy inflorescences was sufficient to maintain an ABA to cytokinin ratio of less than 1:1 despite the accumulation of ABA, and these fruit persisted on the tree. Erner (1989) hypothesized a similar scenario based on the ratio of ABA to IAA or GA instead of cytokinins. ABA has been shown to increase in concentration as sweet orange fruit mature and to contribute to postharvest disorders related to senescence.

The role of abscisic acid in seed dormancy

ABA plays a role in preventing seed germination. Thus, in order for a seed to germinate, the ABA concentration must decrease or be reduced by washing, cold treatment, or through the accumulation of endogenous gibberellic acid or from its exogenous application.

Ethylene

Ethylene is the only hormone that is a gas. It is known as the ripening hormone and the wound hormone, indicating two of its major roles in plant physiology, but it is also the hormone that causes senescence. Ethylene is synthesized in the tissues of mature fruit, nodes of stems, and aging leaves and flowers. It promotes fruit maturation and senescence in citrus. Citrus fruit do not undergo physiological ripening. For more information, see the section "Ethylene" in chapter 3, "Physiology and Phenology." Depending on the plant species, ethylene promotes or inhibits growth and development of roots, leaves, and flowers. The role of ethylene in these processes in citrus remains unknown. Research has demonstrated a possible role for ethylene in fruit abscission because foliar application of aminoethoxyvinylglycine (AVG), an inhibitor of ethylene biosynthesis, increased fruit retention during both the early drop (fruit set) and June drop periods of Washington navel orange (Gonzalez and Lovatt 2004). Further, this treatment also reduced the incidence of crease, a rind disorder associated with peel senescence, in late-harvested fruit. Ethylene has a role in leaf abscission, and citrus leaves are particularly sensitive to ethylene-induced abscission. Tissues wounded mechanically, chemically, by abiotic stress, or by insect, bacterial, or fungal pathogens produce ethylene, which triggers various responses that protect the plant, including wound repair and leaf abscission.

The UC IPM Citrus Pest Management Guidelines: Plant Growth Regulators

Before implementing a plant growth regulator management strategy and applying treatments, see the Plant Growth Regulator section of the UC IPM Citrus Pest Management Guidelines, http://www.ipm.ucdavis.edu/PMG. This website provides current registered uses, label rates, proper application times, and approved citrus crops. In addition, it includes numerous cautions of which you should be aware when using plant growth regulators to manage your crop, as well as guidelines for ensuring and enhancing the efficacy of plant growth regulator treatments.

Ensuring and Increasing the Efficacy of Plant Growth Regulators

Specific cautions and guidelines designed to ensure and increase the efficacy of plant growth regulator applications is provided under "General Information" in the Plant Growth Regulator section of the UC IPM Citrus Pest Management Guidelines. Careful attention to this information and the additional information discussed below will improve your ability to obtain greater efficacy from your plant growth regulator treatments and to improve your cost to benefit ratio.

Coverage

In general, factors that affect solubilization, penetration, absorption and translocation, or the molecular structure of exogenously applied plant growth regulators affect their efficacy and influence the success of their use. Good spray coverage is extremely important to achieve the desired outcome when using any plant growth regulator. Lower-volume applications (100 gal/ac) are less forgiving of imprecise spraying than are higher-volume applications (500 gal/ac). Using surfactants with plant growth regulators has been shown to greatly improve wetting of the leaf, flower, or fruit surfaces by lowering the surface tension and consequently reducing the contact angle between the liquid and tissue surface and increasing uptake. Complete wetting is achieved when the contact angle is zero. Negative interactions between surfactants and plant growth regulators have been reported and should be taken into account when selecting the surfactant. Because some can cause rind blemishes on citrus fruit and because no clear conclusion can be made about effectiveness of any surfactant over the others, find a suitable surfactant for your citrus crop either by direct performance evaluation with available surfactants under local environmental conditions using your cultivars and strains or by contacting an experienced citrus pest control applicator.

pH

The activity of certain plant growth regulators is affected by the pH of the spray solution. For example, highly alkaline conditions cause decomposition and loss of activity of GA_3 (Coggins 1981). A pH lower than 4 also causes decomposition of GA_3 (Plant Protection Ltd. 1969). Using radioactive GA_3, Greenberg et al. (1984) demonstrated that GA_3 uptake rate at pH 4 was threefold higher than at pH 7. Enhancement of GA_3 activity was independent of the acidifying agent used (Greenberg and Goldschmidt 1988; Greenberg et al. 1992). Results obtained in experiments with Clementine mandarin indicate that Silwet L-77 as well as spray solution acidification increased GA_3 efficacy (BenIsmail et al. 1995). Conversely, Coggins et al. (1992) reported no beneficial effect from acidifying the spray solution under field conditions in California. They reported that inclusion of Silwet L-77 was more promising than lowering the pH. Currently, acidification of the spray solution is practiced in countries such as South Africa and Israel (Gilfillan 1986; Greenberg et al. 1992). Due to pH effects, GA_3 should not be combined with compounds that would increase pH to values greater than 8, such as Bordeaux mixture used to control Phytophthora or whitewash (calcium hydroxide) used to control leafhopper in California (Coggins 1981). Whereas GA_3 appears stable for short periods of time at pH 11, as a general rule the pH should be below 8. Moreover, values below pH 8 may improve GA_3 absorption by plant tissues.

Temperature and Relative Humidity

Temperature and relative humidity are important factors influencing the uptake and effect of plant growth regulator applications. Temperatures should neither be too warm in summer nor too cold in winter. If temperatures are projected to reach approximately 85°F or higher during the day, applications should be made just after daybreak or after the temperature drops in the evening. Uptake is improved with these application times due to higher relative humidity. In addition, the nightly increase in relative humidity in the field may induce renewed uptake of the compound from the dry spray residue and increase its physiological effect, provided that the plant growth regulator is stable on the plant surface. Winter applications should be made midday when temperatures are suitably warm.

Cultural Practices

Cultural practices influence the effectiveness of plant growth regulator sprays. Optimal irrigation, fertilization, and tree health should be the rule in groves that will receive plant growth regulator sprays (see Gilfillan et al. 1973; Wheaton 1981; Gallasch 1984).

Application Time

The physiological state and developmental stage of the target organ or tissues play an important role in the type and degree of the response observed as a result of a plant growth regulator treatment. Timing is also critical to avoid chemical residues in the fruit. Application of 2,4-D to control preharvest drop of mature fruit, for example, should be made at least 7 days prior to harvest in California (Erner and Coggins 1989). Knowing the proper time for the application of each plant growth regulator is critical for best results (see the UC IPM Citrus Pest Management Guidelines, http://www.ipm.ucdavis.edu/PMG).

Plant Growth Regulator Uses Registered for Citrus in California

Currently, 2,4-dichlorophenoxyacetic acid (2,4-D), gibberellic acid (GA_3), and naphthalene acetic acid (NAA) are registered for preharvest use in citrus fruit production in California; 2,4-D and GA_3 are also registered for postharvest use and ethyl 1-naphthaleneacetate is registered for use on nonbearing trees. Though few in number, these plant growth regulators have multiple uses on numerous commercially important citrus cultivars. They solve a wide range of problems and provide significant economic benefits to the grower. This section is designed to remind growers of the plant growth regulator cultural management tools available to them. For additional information, see the UC IPM Citrus Pest Management Guidelines, http://www.ipm.ucdavis.edu/PMG.

Fruit Set

Gibberellic acid (GA_3) is registered to increase fruit set and yield of Clementine mandarin; one or two applications should be made between 50% petal fall and 3 weeks after petal fall. Growers are cautioned that excessive set can occur, which results in small fruit with low commercial value and also increases the potential for a light bloom and crop the following year. The results of a detailed study of the effects of GA_3 on mandarin yield and fruit size are presented in Chao et al. (2011). Several GA_3 concentrations (5, 10, 15, and 25 mg/l) and application times (at 60% bloom, 90% bloom, 75% petal fall, 10 days after 75% petal fall, and July) were tested in numerous combinations from two applications to four applications. Some treatments were initiated early at 60% bloom; others were delayed until 90% bloom. The results of this research demonstrated that the efficacy of GA_3 was crop load dependent and indicate that crop load should be taken into consideration when using GA_3 to increase mandarin yield in order to obtain the maximum yield of commercially valuable size fruit.

Fruit Size

The details have been worked out for using 2,4-D to increase fruit size of navel and Valencia oranges and grapefruit. Whether a grower plans early and wants to apply 2,4-D when the fruit are only 0.20 to 0.25 inch in diameter or only notices the need to increase fruit size later in the season when fruit are already

0.60 to 0.75 inch in diameter, the UC IPM Citrus Pest Management Guidelines provide directions on exactly how much 2,4-D to apply in each case for each cultivar. Included in the directions are the pros and cons of the 2,4-D treatments. For Valencia fruit, there is a potential for increased rind roughness with a fruit sizing application of 2,4-D. In contrast, 2,4-D applied to increase navel orange fruit size has the added benefits of reducing mature fruit drop, delaying granulation, and decreasing rind splitting. Similarly, this treatment reduces mature fruit drop in grapefruit. In addition, 2,4-D can be applied to mandarins and mandarin hybrids to increase fruit size. The application is made 21 to 35 days after 75% petal fall. Growers are cautioned that this application of 2,4-D can cause fruit dryness (less juice) in mandarins, especially Nules Clementine, and in mandarin hybrids in orchards particularly prone to granulation.

Fruit Thinning

Reducing the number of fruit on a heavy bearing citrus tree can increase fruit size and crop value and can reduce the potential for or the severity of alternate bearing in an orchard. Naphthaleneacetic acid (NAA) is registered for this use on citrus. However, fruit thinning with NAA can be tricky because efficacy is highly dependent on climate (Coggins and Hield 1967). The section on fruit thinning with NAA in the UC IPM Citrus Pest Management Guidelines, http://www.ipm.ucdavis.edu/PMG, provides critical information about the interactions among NAA concentration, air temperature, and the degree of fruit abscission that will give growers greater control over the degree of thinning that will occur. "In general, inadequate thinning occurs from the lowest label rate of NAA when maximum daytime temperatures on the day of application and several days thereafter are relatively low (about 85°F). Excessive thinning generally occurs from the highest label rate when maximum daytime temperatures on the day of application and several days thereafter are relatively high (about 100°F). In addition, excessive thinning can occur when NAA is applied to unhealthy or water-deficit-stressed trees."

Delaying Fruit Senescence

One of the best-known uses of a plant growth regulator in citrus production, and perhaps the most economically important one to the citrus industry, is the application of GA_3 to prevent fruit (rind) senescence of navel and Valencia oranges, mandarin hybrids, lemons and limes. Gibbing is the common cultural practice of spraying commercially bearing citrus trees with gibberellic acid (GA_3) prior to color break to delay rind senescence of the fruit in late-harvested orchards. So reliable are the annual economic benefits derived from gibbing citrus fruit for late harvest, it has become a standard horticultural practice in most citrus-producing countries of the world. Not only does this use of GA_3 reduce rind aging and softening, it also has the added benefits of reducing rind staining, water spot, and sticky rind of navels, crease of Valencia oranges, and puffiness of mandarin hybrids. Applications to lemon and lime cultivars are made earlier, when the target crop is one-half to three-quarters of full size and still green, to delay fruit maturity and to keep lime fruit green.

Preharvest Fruit Drop

In most cases, 2,4-D is used to prevent preharvest fruit drop in conjunction with the use of GA_3 to reduce fruit senescence. The addition of 2,4-D in the GA_3 sprays reduces GA_3-induced leaf and fruit drop. Control of preharvest fruit drop with 2,4-D is registered for cultivars of navel, Valencia, grapefruit, lemon, tangelo, and other citrus hybrids. For Valencia and grapefruit trees with mature fruit on the tree past bloom, the application of 2,4-D to control preharvest fruit drop can be timed to also increase fruit size of the young setting fruit. Refer to the UC IPM Citrus Pest Management Guidelines, http://www.ipm.ucdavis.edu/PMG, to determine when to apply 2,4-D to attain maximum preharvest drop of mature fruit while simultaneously increasing fruit size of the new crop or to learn how to coordinate the applications of 2,4-D and GA_3 to best achieve your goals for time of harvest.

Suckers and Water Sprouts

Some cultivars and rootstocks have a high propensity to produce water sprouts and suckers from scaffold limbs, the scion trunk, or rootstock trunk. Benny Boswell and Dean McCarty, at UCR, developed the use of ethyl 1-naphthaleneacetate (NAA) to control sprouting from scaffold limbs, trunks, and rootstocks. NAA is a very cost-effective solution to the problem when it is applied just prior to or during early shoot growth. Used in this manner the treatment can eliminate costly pruning. However, NAA is less effective once the shoots have attained significant growth.

Postharvest Disease and Fruit Quality Control

One of the best treatments for preventing Alternaria rot of lemons is the addition of the plant growth regulator 2,4-D to the water-wax emulsion applied to the fruit or adding it to the final fresh water rinse of the fruit prior to storage. This auxin treatment keeps the buttons from turning black and abscising, which prevents the fungus from entering the fruit and reduces decay. The application of 2,4-D also delays rind aging and softening, further protecting the fruit from decay. Sour rot *(Geotricum candidum)* infection of lemon fruit is reduced by adding GA_3 to the storage wax because GA_3 delays fruit senescence; the treatment also prolongs storage life. For other citrus cultivars, GA_3 included in the storage wax delays aspects of rind senescence and color change and enhances fruit quality and storage life.

Use in Other Citrus-Producing Countries and Florida

This section briefly summarizes plant growth regulators used in other citrus-producing countries and in Florida to solve specific production problems that might be of interest to citrus growers in California. For more detailed information, see the review of the technical and scientific literature published by M. El-Otmani, C. W. Coggins Jr., M. Agustí, and C. J. Lovatt (El-Otmani et al. 2000).

Plant growth regulators are used preharvest in nurseries and groves around the world to promote or prevent vegetative shoot growth, manipulate flowering, modify fruit set and fruit growth, regulate crop load via thinning, reduce labor costs and manipulate the timing of harvest, improve internal fruit quality (percentage juice, total soluble solids, and acidity) and external rind quality (color, firmness, blemish, breakdown), and improve on-tree and postharvest fruit storage. They are also used to increase the rate and percentage of seed germination and are included in media for *in vitro* tissue and organ culture and plant regeneration (a use that will become more important as the need to regenerate transgenic plants increases). Most of these techniques are used to improve the performance of well-established cultivars such as the Valencia orange, Washington navel orange, Satsuma mandarin, Clementine mandarin, and grapefruit.

Several plant growth regulators have been tested as aids in citrus propagation to shorten the time required for seed to germinate, to increase percentage germination, and to enhance the subsequent growth rate of the seedling. For more information, see Bowman 1994; Kalita et al. 2002; and Khan et al. 2002. In the nursery, GA_3 has been used on sour orange seedlings to increase plant height and girth and produce plants ready for budding 8 months earlier than untreated seedlings (Khattab and Guindy 1996). Attempts have also been made to alter the branching habit of nursery trees using foliar-applied auxin plant growth regulators (Abdalla et al. 1978; Hassaballa 1979), but the results have not led to any registered commercial treatments. In Japan, 6-benzyladenine, a cytokinin (a plant growth regulator) is used to accelerate spring budbreak of both vegetative and floral shoots of Wase Satsuma mandarins under greenhouse conditions (I. Iwagaki, pers. comm.). The efficacy of this plant growth regulator to increase spring budbreak and floral intensity of mandarins, especially following a heavy crop, is being tested in California. There is interest in plant growth regulators that inhibit gibberellin biosynthesis (uniconazole, paclobutrazol, and prohexadione-calcium) and in a new product, S-abscisic acid, to reduce vegetative shoot growth as a way to reduce populations of Asian citrus psyllid *(Diaphorina citri)*, the vector of huanglongbing (*Candidatus* Liberibacter). These materials are currently being tested with nursery stock and commercially bearing citrus trees. The ability to regulate flowering is highly desirable, especially an ability to induce precocious flowering in juvenile or immature adult plants (Pillitteri et al. 2004) or increase flowering and prevent an off-crop on an alternate-bearing orchard. Plant growth regulators have been tested for their capacity to enhance flowering and reduce vegetative shoot growth when an off-crop is predicted or to enhance vegetative shoot growth and reduce flower initiation and development when an on-crop is anticipated (see Agustí et al. 1981; Davenport 1983; Guardiola et al. 1982; Lord and Eckard 1987; Stover et al. 2000, 2002, and 2004). In Mexico and Cuba, attempts were made to reduce spring flowering of Washington navel orange with multiple GA_3 applications from November through February in order to enhance summer flowering in May; the inhibitory effect was obtained, but the increase in summer flowering was not consistent (Ortiz et al. 1993). In Mexico, applications of GA_3 to Persian lime in June through August and in October increased flowering and fruit set in December with the promise of a good crop the following year (Vargas 1993).

In addition to 2,4-D, other auxins, such as 2,4,5-trichlorophenoxyacetic acid, 3,5,6-trichloro-2-pyridyloxyacetic acid (3,5,6-TPA), 2,4-dichlorophenoxyproprionic acid, picloram, phenylacetic acid, and naphthaleneacetamide, are used to prevent or induce fruit drop or to increase fruit size by stimulating fruit growth or by causing fruit thinning; the result depends on the concentration used and application time. In trials in California, 3,5,6-TPA showed promise for increasing the yield of commercially valuable large-size mandarin fruit and grower income, but the cost of registering the product for use in California was deemed prohibitive (Chao and Lovatt 2010).

Ethephon was registered in Florida for use as a fruit-loosening agent on lemons, tangerines, and tangerine hybrids (Knapp 1981); it also enhanced fruit color development by increasing carotenoid synthesis and accumulation (Young et al. 1974). However, its registration was discontinued because its efficacy was erratic (Wilson 1983; Wilson et al. 1981). Ethephon is still used in Cuba to facilitate both manual and mechanical harvesting (Pérez et al. 1992). Driven by the high cost of harvesting fruit for juice in Florida, new research into fruit-loosening agents to facilitate mechanical harvesting was undertaken. Methyl jasmonate (a natural plant hormone), methyl-dihydro-jasmonate, and 5-chloro-3-methyl-4-nitro-1H-pyrazole (CMN-pyrazole) successfully induced abscission of mature Valencia oranges and increased the efficiency of mechanical harvest (Hartmond et al. 2000; Kender et al. 2001; Yuan, Hartmond, Grant et al. 2001; Yuan, Hartmond, Kender 2001, 2002; Yuan et al. 2003). The latter material (CMN-pyrazole) was the most effective and registration is being sought through the Environmental Protection Agency (EPA).

Conclusion and Future Prospects

Plant growth regulator use in citriculture is dictated by the immediate objectives and needs of a specific citrus industry. Some production problems (and thus objectives) are common to many countries, and the compounds used to solve these problems are often the same, or similar, but concentrations used and application times may vary. An important goal of researchers has been to find ways to increase or enhance the efficacy of plant growth regulators in order to reduce the quantity applied. Testing of new formulations and adjuvants contributes toward achieving this goal. In addition, a long-term objective of all researchers in this field is to be able to understand the basic mechanisms regulating citrus growth and development in order to identify the specific hormones or other compounds controlling important developmental processes. With this knowledge, key organs or tissues can be targeted at their proper phenological stage to ensure the desired response. The knowledge required to apply some plant growth regulators with maximum efficiency is still lacking. As more is learned about the specific, and possibly unique, roles of endogenous hormones in all aspects of citrus physiology and vegetative and reproductive development, new management strategies based on plant growth regulators will become increasingly available to manipulate key physiological and developmental processes to the benefit of citrus growers and the citrus industry.

References

Abdalla, K. M., A. T. ElWakil, and H. H. ElMasiry. 1978. Nursery treatments with GA in citrus. I. Morphologic response of different rootstocks. Research Bulletin, Faculty of Agriculture, Ain Shams University, no. 936.

Agustí, M., V. Almela, and J. L. Guardiola. 1981. The regulation of fruit cropping in mandarins through the use of growth regulators. Proceedings of the 4th Congress of the International Society of Citriculture 1:216–220.

Ben-Ismail, M. C., E. Feddouli, and M. El-Otmani. 1995. Amelioration de la qualité du fruit du clementinier (*Citrus reticulata* Blanco) par l'applicatión de l'acide gibberellique en prerécolte. In A. Ait-Oubahou and M. El-Otmani, eds., Postharvest physiology, pathology and technologies for horticultural commodities: Recent advances. Agadir, Morocco: Institut Agronomique et Veterinaire Hassan II. 120–123.

Bowman, K. D. 1994. Micropropagation of smooth flat Seville and Yuma citrus rootstocks. Proceedings of the Florida State Horticultural Society 107:15–18.

Chao, C. T., and C. J. Lovatt. 2010. Foliar-applied 3,5,6-trichloro-2-pyridyloxyacetic acid (3,5,6-TPA) increases yield of commercially valuable large-size fruit of 'Fina Sodea' mandarin. Acta Horticulturae 884:433–440.

Chao, C. T., T. Khuong, Y. Zheng, and C. J. Lovatt. 2011. Response of evergreen perennial tree crops to gibberellic acid is crop load-dependent. I: GA_3 increases the yield of commercially valuable 'Nules' Clementine mandarin fruit only in the off-crop year of an alternate bearing orchard. Scientia Horticulturae 130:743–752.

Coggins, C. W., Jr. 1981. The influence of exogenous growth regulators on rind quality and internal quality of citrus fruit. Proceedings of the 4th Congress of the International Society of Citriculture 1:214–216.

Coggins, C. W., Jr., and H. Z. Hield. 1967. Plant growth regulators. In W. Reuther et al., eds., The citrus industry. Vol. II. Rev. ed. Berkeley: University of California Division of Agricultural Sciences. 371–389.

Coggins, C. W., Jr., G. L. Henning, and M. F. Anthony. 1992. Possible methods to increase efficacy of gibberellic acid applied to 'Navel' orange trees. In C. L. Foy, ed., Adjuvants for agrichemicals. Boca Raton: CRC Press. 567–572.

Davenport, T. L. 1983. Daminozide and gibberellin effects on floral induction of *Citrus latifolia*. HortScience 18:947–949.

El-Otmani, M., C. W. Coggins, Jr., M. Agustí, and C. J. Lovatt. 2000. Plant growth regulators in citriculture: World current uses. Critical Reviews in Plant Science 19(5): 395–447.

Erner Y. 1989. Citrus fruit set: Carbohydrate, hormone and leaf mineral relationships. In C. Wright ed., Manipulation of fruiting. London: Butterworth. 233–242.

Erner, Y., and C. W. Coggins, Jr. 1989. Free and bound residues of 2,4-D in 'Marsh' grapefruit and 'Washington' navel orange fruit. Journal of the American Society for Horticultural Science 114:846–850.

Gallasch, P. T. 1984. Practical aspects of the use of ethephon to control alternate cropping of 'Valencia' orange. Proceedings of the 5th Congress of the International Society of Citriculture 1:285–288.

Gilfillan, I. M. 1986. Acidification of gibberellic acid sprays on citrus trees. Citrus Subtropical Fruit Journal 628:8–10.

Gilfillan, I. M., W. Koekemoer, and J. Stevenson. 1973. Extension of the grapefruit harvest season with gibberellic acid. Proceedings of the 1st Symposium of the International Society of Citriculture 3:335–341.

Gonzalez, C. M., and C. J. Lovatt. 2004. Foliar-applied aminoethoxyvinylglycine (AVG) reduces albedo breakdown of late-harvested navel orange fruit: Preliminary results. Proceedings of the 10th Congress of the International Society of Citriculture 3:1062–1065.

Greenberg, J., and E. E. Goldschmidt. 1988. The effectiveness of GA_3 application to citrus fruit. Proceedings of the 6th Congress of the International Society of Citriculture 1:339–342.

Greenberg, J., E. E. Goldschmidt, S. Schechter, S. P. Monselise, and D. Galili. 1984. Improving the uptake of gibberellic acid by citrus fruit and leaves. Proceedings of the Plant Growth Regulation Society of America 1:16–25.

Greenberg, J., Y. Oren, G. Eshel, and E. E. Goldschmidt. 1992. Gibberellin A_3 (GA_3) on 'Minneola' tangelo: Extension of the harvest season and improvement of fruit quality. Proceedings of the 7th Congress of the International Society of Citriculture 1:456–458.

Guardiola, J. L., C. Monerri, and M. Agustí. 1982. The inhibitory effect of gibberellic acid on flowering in *Citrus*. Physiologia Plantarum 55:136–142.

Hake, K. D. 1995. Regulation of flowering in *Citrus limon* by water deficit stress and nitrogen compounds. PhD diss., University of California, Riverside.

Hartmond, U., R. Yuan, J. K. Burns, A. Grant, and W. J. Kender. 2000. Citrus fruit abscission induced by methyljasmonate. Journal of the American Society for Horticultural Science 125(5): 547–552.

Hassaballa, I. A. 1979. Growth and branching angle response of 'Benzahair' lime seedlings to growth regulator sprays. Research Bulletin, Faculty of Agriculture, Ain Shams University, no. 1076.

Hofman, P. J. 1988. Abscisic acid and gibberellins in the fruitlets and leaves of the 'Valencia' orange in relation to fruit growth and retention. Proceedings of the 6th Congress of the International Society of Citriculture 1:355–362.

Kalita, S., B. Singh, P. Rethy, and M. Gogoi. 2002. Effect of plant hormones on seed germination of *Citrus reticulata* 'Blanco.' Journal of Hill Research 15(2): 108–112.

Kender, W. J., U. Hartmond, J. K. Burns, R. Yuan, and L. Pozo. 2001. Methyl jasmonate and CMN-pyrazole applied alone and in combination can cause mature orange abscission. Scientia Horticulturae 88(2): 107–120.

Khan, M. M., U. Muhammad, W. Rashid, and M. A. Ali. 2002. Role of gibberellic acid (GA_3) on citrus seed germination and study of some morphological characteristics. Pakistan Journal of Agricultural Science 39(2): 113–118.

Khattab, M. M., and L. F. Guindy. 1996. Effect of gibberellic acid and Vapor Gard on CO_2 evolved respiration rate and stomatal movement of young sour orange seedlings. Proceedings of the 8th Congress of the International Society of Citriculture 2:1089–1093.

Knapp, J. L. 1981. Florida Citrus Spray Guide. Gainesville: Florida Cooperative Extension Service Circular 393G.

Lord, E. M., and K. J. Eckard. 1987. Shoot development in *Citrus sinensis* L. (Washington navel orange). II. Alteration of developmental fate of flowering shoots after GA_3 treatment. Botanical Gazette 148:17–22.

Ortiz, M. S., G. A. Vargas, and J. R. Espinosa. 1993. Desfasamiento de cosecha en naranja (*Citrus sinensis* (L.) Osbeck) cv. Washington navel en Tenango de Doria, Hidalgo. Memorias del II Simposium International Sobre Sistemas de Producción en Citricos, Programa Interdepartamental de Investigación y Servicio en Citricultura, Chapingo, Mexico. 1:191–194.

Pérez, M., L. Pozo, R. Añón, H. Oliva, C. Noriega, M. Gordillo, M. Aranguren, O. Vento, A. Martínez, M. Castillo, R. del Busto, O. Yánez, and A. Borroto. 1992. Growth regulators on citrus crops under Cuban tropical conditions. Proceedings of the 7th Congress of the International Society of Citriculture 1:503–508.

Pillitteri, L. J., C. J. Lovatt, and L. L. Walling. 2004. Isolation and characterization of a terminal flower homolog and its correlation with juvenility in citrus. Plant Physiology 135: 1540–1551.

Plant Protection, Ltd. 1969. Berelex for Promoting Natural Plant Growth. Information booklet. Fernhurst, Halsmere, Surrey, UK: Imperial Chemical Industries.

Stover, E. W., S. M. Ciliento, and Y. J. Lin. 2002. Opportunities for improving management of Florida citrus using GA_3 applied during bloom induction. Proceedings of the Plant Growth Regulation Society of America 1:63.

Stover, E. W., S. M. Ciliento, and J. J. Salvatore. 2004. Fruit size enhancement of Florida citrus using pruning, NAA, and winter GA_3. Proceedings of the 10th Congress of the International Society of Citriculture 2:488–491.

Stover, E. W., Y. J. Lin, S. M. Ciliento, and M. S. Burton. 2000. GA_3 application timing and rate in fall and winter influences bloom period, flowering intensity, and final crop in several Florida citrus varieties. Proceedings of the 9th Congress of the International Society of Citriculture 1:589.

Vargas, G. A. 1993. Efecto del acido giberéllico y CEPA en el desfasamiento de la floración en limon 'Persa' (*Citrus latifolia* Tan.) en Martiez de la Torre, Ver. Memorias del II Simposium International Sobre Sistemas de Producción en Citricos, Programa Interdepartamental de Investigación y Servicio en Citricultura, Chapingo, Mexico. 195–202.

Verreynne, J. S. 2005. The mechanism and underlying physiology perpetuating alternate bearing in 'Pixie' mandarin (*Citrus reticulata* Blanco). PhD diss., University of California, Riverside.

Verreynne, J. S., and C. J. Lovatt. 2009. Effect of crop load on bud break influences alternate bearing in 'Pixie' mandarin. Journal of the American Society for Horticultural Science 134:1–9.

Wheaton, T. A. 1981. Fruit thinning of Florida mandarins using plant growth regulators. Proceedings of the 4th Congress of the International Society of Citriculture 1:263–268.

Wilson, W. C. 1983. The use of exogenous plant growth regulators on citrus. In L. G. Nickell, ed., Plant growth regulating chemicals. Vol. 1. Boca Raton, FL: CRC Press. 207–232.

Wilson, W. C., G. E. Coppock, and J. A. Attaway. 1981. Growth regulators facilitate harvesting of oranges. Proceedings of the 4th Congress of the International Society of Citriculture 1:278–281.

Young, R. H., O. L. Jahn, and J. J. Smoot. 1974. Coloring and loosening of citrus fruit with ethephon. Proceedings of the Florida State Horticultural Society 87:24–28.

Yuan, R., U. Hartmond, A. Grant, and W. J. Kender. 2001. Physiological factors affecting response of mature 'Valencia' orange fruit to CMN-Pyrazole. I. Effects of young fruit, shoot, and root growth. Journal of the American Society for Horticultural Science 126:414–419.

Yuan, R., U. Hartmond, and W. J. Kender. 2001. Physiological factors affecting response of mature 'Valencia' orange fruit to CMN-Pyrazole. II. Endogenous concentrations of indole-3-acetic acid, abscisic acid, and ethylene. Journal of the American Society for Horticultural Science 126:420–426.

———. 2002. Naphthalene acetic acid and 2,3,5-triiodobenzoic acid affect the response of mature orange fruit to abscission chemicals. HortScience 37:348–352.

Yuan, R., W. J. Kender, and J. K. Burns. 2003. Young fruit and auxin transport inhibitors affect the response of mature 'Valencia' oranges to abscission materials via changing endogenous plant hormones. Journal of the American Society for Horticultural Science 128:302–308.

15 Frost Protection

Neil V. O'Connell and Richard L. Snyder

Citrus: A Subtropical Plant

Cultivated species within the genus *Citrus* are indigenous to the humid tropical regions of China, the southeastern Asian countries, including the western border areas of India and Pakistan, and the islands of the Philippines and Indonesia. Citrus plants were introduced into the West Indies by Columbus. From there they made their way to Florida and eventually to California. In the tropical citrus-producing regions, citrus grows continuously, as warm weather prevails throughout the year. Under subtropical conditions, the trees are controlled by the rhythm of the seasons: they are dormant during the winter and blossom and grow in the spring. In California, citrus is grown under a variety of weather conditions, ranging from a moderate coastal climate to warm inland valleys to hot, dry desert conditions. Threatening temperatures are typically experienced in the inland valleys and the desert on a number of nights during the frost season, from November to mid-February.

Frost versus Freeze

A confusing use of the terms "frost" and "freeze" occurs in the citrus production literature. A freeze occurs when cold air blows in at low levels and replaces the warmer air that was present before the cold air arrived. Freezes are associated with wind speeds greater than 5 miles per hour and little or no temperature inversion at night (an inversion is present when the temperature increases with height above the ground). Frosts occur on clear nights when an orchard experiences a large net loss of heat through radiation, which causes the surface and air temperature to fall below 32°F. Frost events are associated with temperature inversions and temperatures below 32°F near the surface on nights with winds lower than 5 miles per hour.

During sunny daytime conditions, heating of the air, trees, and soil occurs within the orchard. Under clear nighttime skies, more heat is radiated away from an orchard than it receives, so the temperature drops. The temperature falls faster near the radiating surface, causing the temperature to increase with height above the ground (an inversion). At a certain height above the ground, the temperature reaches a maximum, and then it begins to decrease with further height (a lapse condition). The level where the temperature reaches a peak and changes from an inversion to a lapse condition is called the ceiling. A weak inversion occurs when the peak temperature aloft is only slightly higher than that near the surface. With a strong inversion, the temperature increase with height is large.

History of California Freeze Events

Frequency

Radiation frost conditions are common in California, occurring on a number of nights in a typical winter, particularly in inland and desert valleys. Freezes are less common than frost but are far more damaging. On average, major freezes have occurred every 10 to 20 years, with the most recent in 1990 and 1998.

Extent of Damage

Radiation frosts typically cause a range of fruit loss on mature trees, with tree structure damage limited to young trees and to immature growth on larger trees. Freezes not only cause considerable fruit damage but typically kill young trees and severely damage mature trees. Freezes cause severe damage to more susceptible cultivars such as lemon and to trees in a more active or susceptible state following late pruning operations, as well as to weak, nonvigorous trees.

Minimum Nighttime Temperature

When nighttime temperatures drop below a critical threshold temperature, citrus trees and fruit may suffer damage. The longer the temperature is below the critical temperature, the more likely that ice will form and cause damage.

Duration

Damage to trees and fruit is relative to the duration the temperature remains at or below the critical threshold. From studies in Florida (Hendershott 1961), it is known that a growing number of fruit freeze as the temperature remains below the critical temperature (rather than all fruit freezing at once). In these controlled-chamber studies, fruit damage was observed after various time periods at a constant temperature of 25°F (table 15.1). Fruit can supercool to as low as 23°F without experiencing freeze damage, depending on hardening, the presence of ice-nucleating bacteria, and several other factors. Since fruit temperature varies throughout the tree, this partially explains why the freezing of all fruit on a tree takes considerable time at 25°F (Hendershott 1961).

Table 15.1. Percentage of fruit frozen for selected periods at 25°F in controlled chambers

Time (hr)	Damage (%)
0.5–1.0	5
2.0–3.0	35–40
4.0–5.0	55–60
6.0–7.0	65–70
8.5	80

Source: Hendershott 1961.

Figure 15.1 Internal drying from frost damage. *Photo:* D. Rosen.

Nature of Damage

During a freeze event, fruit damage may occur in the peel and in the pulp. Peel damage occurs when moisture on the surface freezes; the damaged area collapses and is then invaded by opportunistic decay organisms followed by premature drop of the fruit. During a freeze, juice moves from the juice vesicles into the peel. After the freeze, this water evaporates, the vesicles dry and collapse, and crystals (hesperidin) may form in 5 to 10 days, giving the fruit an off-flavor (fig. 15.1). On frost nights, if the temperature is low enough, extracellular water in the plant tissue freezes, drawing water out of the cells. If cell desiccation is limited, the water moves back into the cells as temperatures rise the next morning, causing no damage. During this process the tissue takes on a black, water-soaked appearance, but it regains a normal appearance as warming takes place. If the cell desiccation is severe, damage to the cell walls can cause death of cells and the plant tissue. Ice marking on the peel or internal damage in the pulp can result in the loss of the fruit as a packable unit. When frost events occur, packinghouses and regulatory organizations such as the county agricultural commissioner's office initiate an intensive fruit inspection program to examine each lot of fruit harvested for the presence of freeze damage.

Varietal Cold Tolerance

Tolerance to damage varies among citrus cultivars. Results from Texas during the 1962 freeze demonstrated that oranges are the most cold tolerant, followed by tangelos, grapefruit, limes, and lemons. Mandarins showed variable degrees of injury (Young and Olson 1963). These observations were confirmed during freeze episodes in California (Citrus Industry 1973).

Rootstock Influence

Rootstock has an influence on the cold tolerance of the tree. Observations were made as early as 1911 that Trifoliate Orange rootstock imparted frost tolerance to the scion (Citrus Industry 1973). Navel oranges were observed to be more frost hardy on

Trifoliate rootstock than when grown on Sweet Orange rootstock. Rough Lemon seemed to be the most susceptible rootstock, Sweet Orange was thought to be less tender, Sour Orange was considered to be fairly hardy, and Trifoliate was very frost resistant. Results from artificially freezing various rootstock and variety combinations demonstrated that Citrange rootstocks provided less cold hardiness than did Sour Orange rootstock (Citrus Industry 1973). Damage and recovery during the 1963 freeze in Florida demonstrated greater tolerance in Valencia than in grapefruit (Citrus Industry 1973). Cleopatra Mandarin and Sour Orange were the most resistant rootstocks. Cultivars with the least tolerance showed the poorest recovery. In a survey of growers following the 1990 California freeze, the damage was least in trees on Trifoliate rootstock, followed by trees on Sour Orange, then Troyer Citrange, and the greatest damage was observed in trees on Rough Lemon rootstock (Pehrson and O'Connell 1991).

Energy Exchange in the Citrus Orchard

All frost protection measures and systems are designed to prevent heat loss or replace heat lost from orchards. Heat transfer and conversion occurs as a result of one of the following four processes:

- **Conduction:** Heat transfer through a solid body or bodies in physical contact, such as movement of heat through a soil.
- **Convection:** Transfer of sensible heat by movement of a liquid or gas such as air. Sensible heat is energy that one senses or measures with a thermometer.
- **Radiation:** Direct transfer of energy through space without an apparent medium (e.g., sunlight).
- **Latent heat:** Phase changes between water vapor and either liquid or solid water. When vapor condenses as liquid water or deposits as ice, latent energy is converted to sensible heat, and the temperature rises. When liquid water or ice vaporizes (evaporation or sublimation), sensible heat is removed from the air to break the bonds between water molecules, so the temperature drops.

Cold Tolerance and Dormancy

Tolerance to low temperature is related to the level of physiological activity of the tree. When temperatures drop in the fall, the physiological activity level of a tree typically begins to drop as well, and in this lowered state of activity, the tree is less susceptible to potentially damaging temperatures. Daytime temperatures of 60°F with nighttime temperatures of 40°F reduce the activity level of the tree and enhance the tree's cold tolerance (i.e., hardening). This tolerance is lost after a few days of warm weather. Late-season pruning tends to maintain a higher level of physiological activity as the trees enter winter; therefore, pruning activity should be completed well in advance of the frost season. For example, mature Valencia trees topped in October experienced severe splits in 8-inch scaffold branches during the 1990 freeze in the San Joaquin Valley. The application of pesticide oils to trees can exacerbate damage from cold and should not be done shortly before the frost season.

Minimum Temperature Threshold

Critical damage threshold temperatures are related not only to variety but also to the maturity of foliage and fruit, the influence of the rootstock on the scion, and the level of hardening. Mature citrus leaves can tolerate temperatures of 23° to 29°F, and dormant wood can survive 20°F for up to 4 hours. However, immature feather growth at the beginning of a growth flush can be damaged on a night with a low temperature as high as 30°F. Young orchards cool more quickly and experience colder minimum temperatures than do large trees in mature orchards. Therefore, active frost protection methods should be started earlier for young orchards. Fruit that is more mature and higher in soluble solids (sugars) withstands lower temperatures, as do larger fruit and fruit with a thicker peel. Critical damage temperatures for citrus fruit that were developed by the U.S. Weather Bureau Fruit Frost Service are given in table 15.2.

Table 15.2. Fruit temperatures at which freezing begins

Fruit	Temperature (°F)
button lemons (up to 0.5-inch diameter)	29.5–30.5
tree-ripe lemons	29.5–30.5
green lemons (larger than 0.5-inch diameter)	28.5–29.5
green oranges	28.5–29.5
half-ripe oranges, grapefruit, and mandarins	28.0–29.0
ripe oranges, grapefruit, and mandarins	27.0–28.0
buds and blossoms	27.0

Source: Opitz et al. 1979.

Frost Protection Strategies

Passive Frost Protection

Procedures employed 1 or more days in advance of an expected cold event, such as application of water to the orchard floor to maximize heat absorption, are regarded as passive frost protection. The main passive methods include site selection, rootstock selection and harvest period management.

Site selection

Selection of the orchard location plays a significant role in frost protection. Cold air typically drains downhill; therefore, in many frost episodes, orchards on the sides of hills will be warmer than those at lower locations. The coldest temperatures typically occur at the tops and bottoms of hills. Historically, plantings of citrus were made on the slopes of hillsides, particularly with the more susceptible lemon varieties. Air drainage can be interrupted, creating a pool of colder air, often as a result of made obstructions such as berm walls, buildings, and canal banks. Investigation of a possible site for an orchard should include review of records regarding minimum temperatures during frost or freeze episodes and any records on the extent of damage. Temperature recording stations can be situated at the site being considered during the winter to document minimum temperatures and their durations. This information not only provides spatial minimum temperature data but also can be compared with a nearby reference weather recording station to improve frost forecasting.

Rootstock selection

Rootstock selection should consider the potential for frost tolerance imparted to the scion. Rootstock tolerance to freeze is highest in Trifoliate, somewhat less in Carrizo, Troyer, and C35, and poor in Rough Lemon, Volkameriana, and Macrophylla (for more information, see chapter 5, "Rootstocks").

Harvest period

Damaging cold frequently occurs from November to February. If possible, harvesting fruit prior to frost is best. Selection of varieties should consider early harvest in frost-prone areas in addition to other considerations.

Tree spacing

Air movement within the orchard is related to the presence of barriers to circulation. Closely spaced trees that create a wall of foliage can interfere with air circulation. Temperature measurements in a navel orange orchard where alternate trees had been removed in a portion of the orchard had a higher minimum temperature than the remainder of the orchard during wind machine operation on a frost night in Tulare County (Pehrson, pers. comm., 1978).

Orchard floor management

Radiation is typically absorbed in the orchard on a clear day and then lost from the soil, trees, and air at night. The presence of plant material on the orchard floor (particularly if it is dense and tall) interferes with heat absorption and may result in a lower temperature on a cold night.

Vegetation-free versus cover cropping

Removing vegetation from the orchard floor maximizes soil heat absorption during the day and release at night. Recent research in which a fall-seeded cover crop had been planted in a portion of an orchard, with the remaining floor bare ground, demonstrated that the portion with the cover crop tended to be colder (O'Connell and Snyder 1999). Comparing a fall-seeded cover crop with bare soil, the area with the cover tended to be colder with increasing height of the vegetation and was significantly colder than the bare ground in an early spring frost when the cover crop was approximately 20 inches tall (O'Connell and Snyder 1999). Applying preemergent and postemergent herbicides for vegetation management gives better results than cultivation or mowing.

Tillage

Heat absorption and storage are enhanced by a firm, undisturbed soil surface. Fall tillage should be avoided for this reason. If tillage is necessary, it should be done early enough to allow the soil to settle and firm up before the arrival of the frost season. Compressing the soil with a heavy roller after cultivating can improve heat absorption and storage.

Mulching

The presence of plant material on the orchard floor can interfere with heat transfer. A typical cultural practice is for tree prunings to be stacked between the tree rows and shredded in place. Recent research in a Kern County orchard during a frost episode where a thick layer of shredded orchard prunings was present on the orchard floor demonstrated that lower nighttime temperatures occurred in the mulch than where the prunings had been removed (O'Connell and Snyder 1999).

Active Frost Protection

Procedures that are employed during a frost event such as wind machines or application of water are referred to as active frost protection.

Wind machines

Concept

Typically, during radiation frosts, skies are clear during the day, and temperatures fall rapidly at sunset. A layer of cold air forms near the surface, and the temperature increases with height above the ground to well above the orchard (an inversion). Wind machines produce turbulence and mix the warmer air aloft with colder air near the ground, which breaks up the temperature inversion. Typically, one 360° rotation of the tower is accomplished in 4.5 minutes with the propeller revolving at a rate of 590 to 600 revolutions per minute. A properly sized machine can circulate the air in a radius of 300 to 400 feet. It is believed that wind machines blowing in the same direction as they rotate provide more protection. The horsepower of the machine is rated at the propeller and is referred to as brake horsepower (bhp). Generally, the brake horsepower needed for protection ranges from 5 to 8 per acre. The amount of orchard warming required depends on the strength of the inversion and the volume of air moved, which is related to the size of the wind machine. Typically, a 100-horsepower wind machine protects 10 acres. Power sources for wind machines include electricity, gas, diesel, and propane. Machines with lower horsepower may be purchased for parcels smaller than 10 acres. For irregular-shaped parcels, machines may be geared such that the amount of time spent in a portion of the rotation of the machine can be increased.

Operation

Wind machine start-up time is influenced by the rate of temperature drop during the night. Based on work by the U.S. National Weather Service, when to start wind machines depends on the brake horsepower of the machine and the predicted low temperature. If temperatures not lower than 6 to 7 degrees below freezing are predicted, the machines should be started at 32°F. If lower temperatures are predicted, low-horsepower machines (less than 7 bhp/ac) should be started at 35°F and high-horsepower machines (above 7 bhp/ac) at 32°F. Wind machines should be started while the temperature measured at the 5-foot sheltered minimum is above the critical damage temperature and before the sheltered temperature falls much below the temperature measured at 40 feet above the orchard. With a well-designed wind machine operating, the temperature within the orchard varies, but the average is typically increased by about one-third of the difference between the 5-foot and 40-foot temperature readings.

Wind machines are stopped after sunrise when the air temperature outside of the orchard has risen to above 32°F. If there was a severe frost or freeze with temperatures well below the critical damage temperature during the night, the fans should be operated later in the morning to slow down the thawing process. There is evidence that less damage occurs when the fruit is frozen during the night but thaws slowly the next morning (J. F. Thompson, pers. comm.).

Electric machines with thermostats may be preset to start at a given temperature, while machines using other power sources require manual startup. Machines are operated until the air temperature reaches 32°F and the fruit is dry. Because of the risk of damage to machines, they should be shut down if foggy or windy (i.e., speeds greater than 5 mph) conditions develop. Machines should be monitored on an hourly basis during operation. Annual maintenance should be performed by ranch personnel or by outside contractors.

Water as frost protection

Concept

The value of water in frost protection relates to physical properties of water. When water cools, it releases heat as the temperature drops. For each 1°F of temperature drop, 1 gallon of water releases 8.35 British thermal units (Btu) of energy; 1 Btu of energy is the amount of energy required to raise the temperature of 1 pound of water 1°F. The energy is released as radiation or sensible heat, which can be transferred to trees to provide frost protection. The water cooling and energy release process continues until the water freezes, releasing an additional 1,200 Btu of energy during the phase change from liquid to ice. If 1 gallon of water at 65°F falls to a temperature of 32°F, 276 Btu are released. Therefore, when 1 gallon of water cools from 65° to 32°F and freezes into ice, it releases a total of 1,476 Btu (1,200 + 276). This heat release from cooling and freezing is the basis of frost protection with microsprinklers. On a per acre basis, with 100 emitters per acre delivering 10 gallons per emitter per hour, the heat released to cool water from 65° to 32°F and freeze it would equal 1,476,000 Btu per acre per hour (1,476 × 1,000). On a typical radiation night, the net energy loss per hour per acre from an orchard is approximately 414,000 Btu. Not all of the heat released by cooling and freezing water efficiently contributes to frost protection because some energy is lost to convection, evaporation, and radiation to the sky. However, if adequate water is applied to the orchard, the net amount of energy gained is often sufficient to balance the net losses. Because heat is released as the water cools, furrow irrigation beneath the tree skirts can also provide protection. In previous California freezes, water applied at a rate of 50 gallons per minute per acre using water at 65°F was observed to provide 2° to 3°F of protection during a radiation frost.

Operation

The volume of water delivered per unit of time generally varies among irrigation systems. After the 1990 freeze, a survey of growers regarding frost protection practices suggested that the best protection came from continuously operated low-volume systems (Pehrson and O'Connell 1991). Less protection resulted from the use of microsprinklers, and the least came from furrow irrigation systems. Although the volume of water applied is often greatest from a furrow system per unit of time, during a freeze, a layer of sheet ice typically forms part way down the furrow, trapping heat and preventing its release.

Experiences during freezes in Florida suggest that low-volume emitters are capable of providing warming of 1° to 3°F at a height of 4 feet in the tree and 1° to 1.5°F at 8 feet in the tree, with inconsistent warming at 12 feet (Parsons and Bowman 2009). These results were obtained under radiation frost conditions with water delivered at 7,600 gallons per acre per hour. Little benefit was obtained under advective conditions (i.e., wind speeds greater than 5 mph) with mature trees; however, young trees may receive some protection to a height of 18 to 30 inches during windy, low–dew point conditions (see table 15.4).

During operation of microsprinkler or low-volume systems, wetting the fruit or foliage should be avoided. If water is applied to fruit or foliage, it must be continuously applied to these surfaces throughout the freeze event. This is not a good practice for citrus orchards.

Sprinklers should be started and stopped when the wet-bulb temperature is above the critical damage temperature. The wet-bulb temperature is the temperature at which the air becomes saturated if it is cooled by evaporation into the air; it is always between the dew point and dry-bulb (air) temperature (somewhat closer to the air temperature). The air temperature to start the sprinklers is estimated by measuring the dew point temperature, then using table 15.3 to determine the starting air temperature.

Orchard heaters

The use of heaters has been largely discontinued due to the high cost of fuel. In small orchards and in field-planted citrus nurseries, it may be economical to use heaters on cold nights. Emissions from heaters are regulated under air quality standards in California enforced by air quality control districts. A list of approved heaters that meet air quality standards is available from local air quality district offices.

Fuels commonly used in heaters have included oil or liquid propane types. A return stack unit burning fuel at a rate of 3.28 pounds per hour produces a heat output of about 105,000 Btu per hour. About 28% of this total is radiation output, with 10% radiating to the ground, 9% to the trees, and 9% to the sky. For frost protection, radiant energy is more efficient than heated air, and radiation emanates best from a hot solid surface such as the steel smokestack of a heater. The radiation source should be kept as close to the tree as possible without burning it and in a position with an unobstructed view, as radiant energy travels in a straight line. A portion of the combustion is converted to sensible heat as heated air and gases from the flame. This heated air rises and mixes within the inversion. This circulation effect within the orchard is enhanced by the action of a wind machine. Sensible heat in the air is transferred to and absorbed by leaves, fruit, and branches, and the warmed surfaces then conduct heat inside the plant tissue. The number of heaters per acre depends on the heat output, typical inversion strength, and presence of cold spots that require more heat. In general, radiant energy transfer is more efficient than sensible heat transfer, so heating an orchard is more efficient using more heaters with a smaller heat output per unit. The amount of protection for each site can be determined only by the grower's experience with the frost problems on that site.

One method of heater distribution is border heating. On the upwind edge, where the prevailing air drift enters the orchard, use two heaters per tree on the outside and one heater per tree in the first two rows in from the edge. On the downwind side of the orchard, use one heater per tree on the outside. On the remaining two sides of the orchard, use one heater per tree on the outside and on the first row in from the outside. To reduce labor costs, a central fuel supply can be used.

Table 15.3. Minimum start and stop air temperatures (°F) for sprinkler frost protection for a range of wet-bulb and dew point temperatures (°F)

Dew point temperature (°F)	Wet-bulb temperature (°F)										
	22.0	23.0	24.0	25.0	26.0	27.0	28.0	29.0	30.0	31.0	32.0
32											32.0
31										31.0	32.7
30									30.0	31.7	33.3
29								29.0	30.6	32.3	34.0
28							28.0	29.6	31.2	32.9	34.6
27						27.0	28.6	30.2	31.8	33.5	35.2
26					26.0	27.6	29.2	30.8	32.4	34.0	35.7
25				25.0	26.5	28.1	29.7	31.3	32.9	34.6	36.3
24			24.0	25.5	27.1	28.6	30.2	31.8	33.5	35.1	36.8
23		23.0	24.5	26.0	27.6	29.1	30.7	32.3	34.0	35.6	37.3
22	22.0	23.5	25.0	26.5	28.1	29.6	31.2	32.8	34.5	36.1	37.8
21	22.5	24.0	25.5	27.0	28.5	30.1	31.7	33.3	34.9	36.6	38.2
20	22.9	24.4	25.9	27.4	29.0	30.6	32.1	33.7	35.4	37.0	38.7
19	23.4	24.9	26.4	27.9	29.4	31.0	32.6	34.2	35.8	37.5	39.1
18	23.8	25.3	26.8	28.3	29.8	31.4	33.0	34.6	36.2	37.9	39.5
17	24.2	25.7	27.2	28.7	30.2	31.8	33.4	35.0	36.6	38.3	39.9
16	24.6	26.1	27.6	29.1	30.6	32.2	33.8	35.4	37.0	38.7	40.3
15	25.0	26.4	27.9	29.5	31.0	32.6	34.2	35.8	37.4	39.0	40.7

Note: Select a wet-bulb temperature that is at or above the critical damage temperature for your crop and locate the appropriate column. Then choose the row with the correct dew point temperature and read the corresponding air temperature from the table to turn your sprinklers on or off. This table assumes a barometric pressure of 1,013 millibars (29.92 in Hg), which is typical at sea level.

Tree wraps

In newly planted orchards, wraps should be installed around the tree trunks. These wraps serve to exclude light and inhibit unwanted suckers from developing on the trunk; prevent sunburn during the summer; and protect the bark from sprays used in weed control. They may also provide some protection from frost. Wraps are most effective during cold episodes of short duration. During cold episodes in Florida, the following responses to various insulating wrap materials were observed: fiberglass wraps, 3° to 6°F temperature increase; polyurethane, 3° to 6°F increase; rigid polystyrene (thick-walled) foam, 4° to 8°F increase; rigid polystyrene (thin-walled), 0° to 2°F; and closed-cell polyethylene foam 2° to 4°F (Jackson and Parsons 1994). The wrap most commonly used in new plantings in California is a closed-cell polyethylene foam material. No wraps have been tested for cold protection benefit under California frost conditions.

Ice nucleation

Plants are known to supercool when exposed to low temperatures and may reach temperatures considerably below 32°F before ice crystals begin to form in the water within their tissues. Frequently, however, certain bacteria have been found to colonize citrus canopies in the fall that interfere with this supercooling and actually serve as sites for ice crystal formation on the plant surface. These ice nucleation active (INA) bacteria move to the trees from weeds and grasses within or outside the orchard or from noncitrus orchards. The number of these bacteria may be reduced by a common citrus cultural practice: a fall application of copper for preharvest fruit diseases. Copper should be applied at least 72 hours prior to the cold night to allow time for the copper to affect the bacteria.

Frost Forecasts

Weather forecasts of varying detail are generally available from a variety of sources including the Internet, newspapers, radios, television, the national weather service, and private forecasters. These forecasts are generally made in the morning, with an updated forecast in late afternoon. Minimum temperatures for historical key locations (often with a reference weather station) are given for that night. The forecast may include a comment as to whether an inversion exists and the strength of the inversion (the increase in temperature with height from about 5 to 33 or 40 feet, which is commonly several degrees). The forecast may include an estimate of when damaging temperatures will be reached. The dew point temperature may also be given (to obtain the dew point, see fig. 15.2; to estimate the dew point at a given temperature and humidity, see table 15.4). Purchasing a battery-powered hand-held instrument to measure air and dew point temperatures is a wise investment. Many are available on the Internet and from local farm stores. The dew point is the temperature when the air reaches 100% relative humidity if the air is cooled without changing the water vapor content of the air. At the dew point temperature, water vapor will begin to condense as liquid water (dew) or ice (frost) on the surface of the trees and ground. With high dew points, the air temperature tends to drop slowly and steadily at night because water vapor intercepts upward long-wave radiation and partially reradiates the energy back downward, and because heat is released as the dew

Legend:
Slowly add ice cubes to the water in a shiny can to lower the can temperature. Stir the water with a thermometer while adding ice cubes to ensure that the can and water are the same temperature. When condensation appears on the outside of the can, note the dewpoint temperature

Figure 15.2 Obtaining the dew point.

Table 15.4. Dew point temperatures (°F) for a range of air temperatures and relative humidities

Relative humidity (%)	Air temperature (°F)					
	32	**36**	**40**	**44**	**48**	**52**
100	32	36	40	44	48	52
90	29	33	37	41	45	49
80	27	30	34	38	42	46
70	23	27	31	35	39	43
60	20	23	27	31	35	39
50	16	19	23	27	30	34
40	10	14	18	21	25	28
30	4	8	11	15	18	22
20	–4	–1	2	6	9	12
10	–18	–15	–12	–9	–6	–3

Note: Select a relative humidity in the left column and an air temperature from the top row. Then, find the corresponding dew point in the table.

or frost forms. Under low–dew point conditions, a rapid temperature drop during the night is likely to be experienced. For many years the citrus industry has made available district-by-district private weather forecasts. For an additional fee, a grower can obtain even more detailed on-site forecasts during frost nights. The forecasts are disseminated via a telephone recording at a phone number provided at the beginning of the frost season.

Monitoring Orchard Conditions

Several types of thermometers are available for use in frost protection. Temperature can be measured with fixed or mobile thermometers. Generally, each orchard has a minimum registering thermometer mounted in a fruit frost shelter. This unit is read periodically during the night for decisions on wind machine startup and for updates on the current air temperature while wind machines are in operation. The unit also registers the lowest temperature during the night. The following day, the registered low temperature should be recorded and the thermometer reset to make it ready for the following evening. Recording devices are available for producing a record of temperatures over time. Mobile thermometers are available for updating current temperature during the night. These include hand-held units for orchard temperature, units for measuring the internal temperature of the fruit, and units that can be mounted in vehicles for registering ambient air temperatures. Wireless weather stations that remotely feed information to a computer are now commonly used for frost protection, pest management, and other applications. Many are available through the Internet. The weather stations should ideally be placed on the upwind side of the orchard.

References

Hendershott, C. H. 1961. Controlled freezing of orange trees and truit. Florida Citrus Experiment Station Mimeo Series 62:6 (Sep. 19).

Jackson, L. J., and L. R. Parsons. 1994. Cold protection methods. University of Florida Department of Horticultural Sciences IFAS Fact Sheet HS-121. IFAS website, http://indian.ifas.ufl.edu/hort/Cold_Protection_Methods.pdf.

O'Connell, N. V., and R. L. Snyder. 1999. Cover crops, mulch lower night temperatures in citrus. California Agriculture 53(5): 37–40. doi: 10.3733/ca.v053n05p37.

Opitz, K., R. F. Brewer, and R. G. Platt. 1979. Protecting citrus from cold losses. Berkeley: University of California Division of Agricultural Sciences Leaflet 2372.

Parsons, L. R., and B. J. Bowman. 2009. Microsprinkler irrigation for cold protection of Florida citrus. University of Florida Department of Horticultural Sciences IFAS Fact Sheet HS-931. Florida Extension EDIS website, http://edis.ifas.ufl.edu/ch182.

Pehrson, J. E., and N. V. O'Connell. 1991. Survey of damage from the 1990 freeze. University of California Cooperative Extension Tulare County and Lindcove Research and Experiment Station.

Turrell, F. M. 1973. The science and technology of frost protection. In W. Reuther, ed., The citrus industry. Vol. 3. Rev. ed. Berkeley: University of California Press. 407–408.

Young, R. H., and E. O. Olson. 1963. Freeze injury to various citrus varieties in the lower Rio Grande Valley of Texas. Proceedings of the American Society for Horticultural Science 83:333–336.

Part 5 Citrus Pest Management

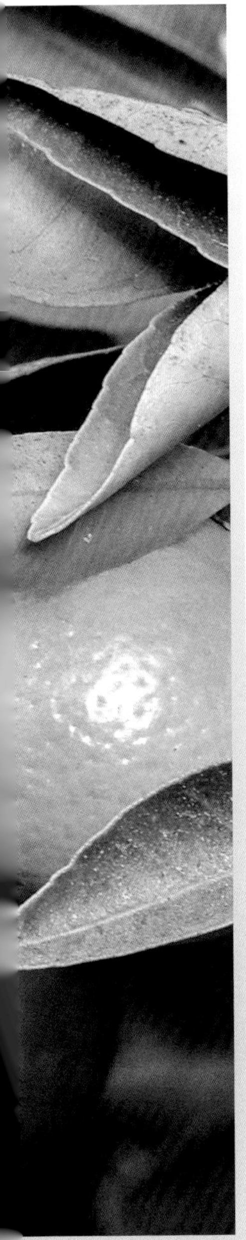

16 Integrated Weed Management

Scott J. Steinmaus

If not managed, weeds are capable of covering the entire citrus orchard floor (fig. 16.1). Weeds are unwanted plants because they

- compete for resources. Competition for light and other resources is most severe on citrus growth in the first 5 years of a new planting. In established orchards, the tree canopy is well above that of weeds so competition for light is rarely a concern. Further, lower branches (skirts) on older trees are maintained very close to the ground so that little light is able to pass through. In both new and established orchards, weed roots encounter surface-applied water and fertilizer before tree roots because weed roots typically occupy shallower soil horizons.
- facilitate other pest problems. Weeds may provide shelter for insects and vertebrate pests and create moisture and temperature conditions around tree trunks that favor pathogens. Weeds also provide habitat for snails, allowing them to move across the orchard floor and into the trees.
- accentuate frost hazard. A vegetated orchard floor experiences cooler ambient temperatures than do bare orchard floors. Temperatures can be about 3° to 5°F cooler, depending on floor vegetation height and atmospheric conditions.
- interfere with cultural operations. Weeds may hinder labor operations such as pruning and harvest. Working a ladder on a vegetated orchard floor can be dangerous; weed species that are a problem for workers with ladders include hairy fleabane (fig. 16.2), horseweed, johnsongrass (fig. 16.3), dallisgrass, and especially vetch. Workers may find some weed species objectionable to work around, including poison oak, puncturevine, spiny cocklebur, stinging nettle, and bristly oxtongue. Weeds may interfere with water flow in furrow-irrigated systems or water spray patterns in sprinkler-irrigated systems.

Weeds do have some beneficial properties: for example, they help stabilize soil in orchards that are on slopes, enable orchard floors to handle traffic better than bare soil under wet conditions, and promote a diverse range of plants, especially perennials, that provide shelter and cover to beneficial organisms.

Weed management in citrus is usually not too difficult because the rooting and aerial portions of citrus trees occupy somewhat different spaces than those of weeds. Most orchards typically have only a few problem weed species. However, the species that are present are usually tolerant of the predominant weed management method. The most commonly encountered weeds are annuals or herbaceous perennials, and each category has its own unique set of strengths.

In the annual weeds that commonly occupy citrus orchards, the seed stage is the most difficult to control, as they possess characteristics that facilitate their dispersal in space and time, such as dispersal by equipment, wind, or irrigation water, combined with dormancy and longevity (see fig. 16.2). Dormant seed will not germinate even if conditions are optimal for germination, and the seed from most weed species can remain viable at least several years, while some can survive for centuries. Herbicides do not kill nongerminated weed seed.

For perennial species, vegetative propagules such as rhizomes, tubers, and stolons are the most difficult organs to kill; they are the primary means by which perennials are spread in space and time (see fig. 16.3). Infrequent soil disturbance favors the perennial weed species that are commonly found in citrus orchards. However, cultivating rhizomes chops them into fragments and spreads them across

Figure 16.1 A weedy orchard floor is usually undesirable but can provide ground cover to prevent erosion. Weeds can compete with citrus trees for resources and interfere with normal orchard operations. *Photo:* D. Rosen.

Figure 16.2 Hairy fleabane *(Conyza bonariensis)* is a summer annual broadleaf common in citrus orchards. Biotypes of hairy fleabane have developed resistance to glyphosate in California citrus orchards. Many species in this family (Asteraceae) spread by wind-blown seed with the aid of a pappus that acts as a parachute in the wind. *Photo:* J. K. Clark.

Figure 16.3 Johnsongrass *(Sorghum halpense)* is a perennial that spreads by seed and an underground modified stem called a rhizome (whitish rootlike structures shown here). *Photo:* J. K. Clark.

the orchard. Each fragment is capable of becoming a new plant. Likewise, tubers of perennial weeds such as yellow nutsedge may be dispersed with soil and equipment movement (fig. 16.4). As with the seed of annual species, these vegetative propagules can be dormant and long-lived, which will contribute to their persistence in time once established. Perennials may also produce viable seed with the same persistent characteristics as those from annual species. Therefore, preventing seed production in all weed species and preventing vegetative propagule formation in perennials are the most prudent weed management objectives for the orchard manager.

Integrated weed management stresses monitoring and cultural control practices, which are essential tools that every orchard manager should implement. When deciding which tool to use in controlling existing weeds, managers may choose from several options, such as chemical control, cover cropping, mechanical control, heating, and biological control.

Monitoring

Monitoring determines which, where, when, and how many weeds are present in an orchard. It involves going into the orchard and surveying the orchard floor for weeds. The objective with each monitoring session is to collect the information needed to complete a scouting report (table 16.1). Information collected for the scouting report will include weed identification, distribution, abundance, and general information about the orchard, all of which are discussed below.

Identification

Weed identification is the first step of an integrated weed management program. Correct identification of the target weed species is necessary to properly select the most effective herbicide or control measure. Also, knowing whether a weed is perennial or annual affects management decisions and the timing of control. For further information about characteristics used to identify annual broadleaf and grass species, sedges, and perennial weeds, see the Weeds section of the UC IPM Citrus Pest Management Guidelines, http://www.ipm.ucdavis.edu/PMG.

Effective control of annual or perennial species germinating from seed requires that they be identified in the seedling stage, beginning at the cotyledon or coleoptile stage through the 4- to 6-leaf (tiller) stage. Larger plants are more likely to recover from control attempts, and they may produce viable seed. Collect many specimens of the same species when identifying seedlings because each species may take on several forms, depending on growing conditions. Look for the average form of all collected specimens. The weed species represented in the seed bank (soil) at a given location will not change drastically with time. This will make identification easier in subsequent years.

- A few dominant species make up 70 to 90% of the seed bank; these are tolerant of the predominant weed control methods because the weed seed bank reflects current and past management practices.
- Several to many species make up 10 to 20% of the seed bank; these are adapted to the geographic region.
- Several species make up less than 1% of the seed bank; these are recalcitrant seed from a previous seed bank or newly introduced species.

To make a positive identification of an unknown weed, consider the characteristics in the following sections, then consult a weed identification reference focused on western U.S. weeds such as *Weeds of California and other Western States* (Ditomaso and Healy 2007, in the references at the end of this chapter) or *Weeds of the West* (see the references). Further assistance can be sought at your local University of California Cooperative Extension office, which may be found online at http://ucanr.org/ce.cfm.

Broadleaf Seedling Characteristics

The characteristics that are used to identify broadleaf annual or perennial seedlings are cotyledon size, shape, and color (fig. 16.5). Additionally, the coloration, texture, and margins of the first few true leaves are used to identify a species. Stem coloration, the presence of spines or thorns, and stem shape may also be used.

Grass Seedling Characteristics

Grass seedlings emerge from a fruit called a caryopsis (seed), leading with a coleoptile above ground and a coleorhiza below ground. The single cotyledon in grasses remains within the caryopsis. Therefore, the morphology of the first few leaves is important, specifically the characteristics associated with the collar region (where the leaf blade joins the stem, fig. 16.6). Characteristics include presence or absence of hairs, fingerlike projections called auricles, and a membranous or hairy collarlike structure called a ligule.

Perennial Characteristics

Newly emerged plants without cotyledons or coleoptiles are perennials that are usually attached to vegetative storage organs such as rhizomes, tubers, stolons, or fleshy roots (see figs. 16.3 and 16.4). The storage organs are usually underground to protect them from harsh winter or summer conditions. Storage organs are modified stems and are usually large and fleshy, with distinct rings called nodes.

Distribution

Monitoring and control efforts must be focused where weeds are located because weeds typically are not distributed uniformly across the orchard floor throughout the entire year. Winter annual species tend to occupy the entire orchard floor during the

Figure 16.4 Yellow nutsedge *(Cyperus esculentus)* is a perennial monocot that is capable of spreading by seed but is most noted for its ability to spread by modified stems called tubers. *Photo:* J. K. Clark.

Table 16.1. Field scouting report used to monitor and survey orchards for weed species identification, extent, and distribution in field

Field Scouting Report

Scout name:____

Field location:____________________________Date:________________

Irrigation (circle): sprinkler furrow rain-fed drip

Soil type (circle): grave sandy loam clay peat

Cover Crop (desired vegetation):________________middles only___entire orchard floor___

Orchard size (acres):__________________________

History:

Insect/pathogen (signs or symptoms):

Neighboring habitats (e.g. creeks, urban, crops, etc.):

Dominant weed species →	Species 1:	Species 2:	Species 3:	Species 4:	Species 5:
Draw field map with landmarks and compass direction (GPS coordinates). Draw distribution of each species in field →					
Percent cover or density (# m^{-2}) →					
Growth stage (seedling, flowering, etc.) →					

Other weed species:

rainy season, while summer annuals grow only in the moist areas around irrigation sources. For dominant species that are uniformly distributed throughout the orchard, surveying 10% of the orchard will suffice, whereas for rare species that grow in isolated clumps, over 90% of the orchard must be surveyed. Weed species composition at a given location does not typically change rapidly even for wind-dispersed species, so monitoring and weed detection become more efficient through experience. Accurately mapping the distribution for each species in the scouting report will ensure that control measures will not be wasted on areas where weeds are not present.

Abundance

Management plans should focus on the most abundant species, as these will be the most problematic. The surveyor should estimate the percentage of cover or density for each species. There are usually only about a half dozen or fewer weed species that become sufficiently abundant to warrant targeted control. Species, especially perennials, that appear for the first time in a scouting report warrant special consideration because they may be eradicated with minimal effort if they are at sufficiently low levels. Unlike species composition, species abundance can change rapidly in a single season when poor weed control allows seed set or establishment of new perennial propagules.

Frequency

Monitor more frequently when weeds are expected to germinate or be at a developmental stage that is susceptible to control. Germination or resprouting between the peak fall and spring periods will occur several days following a soil disturbance, irrigation, or rain; schedule monitoring in anticipation of these events. Determine the proper development stage for control by monitoring once weekly. Monitoring frequency can be reduced to once every 2 to 4 weeks after the fall and spring flushes or if there is little change from one scouting report to the next.

Orchard Information

General information about the orchard, such as soil type, field history, proximity to sensitive habitat, and other pest problems, should be collected during the initial monitoring session (see table 16.1) to aid in devising an integrated pest management plan.

Figure 16.5 Weeds are best identified while in the seedling stage, when they are easier to control. The shape, color, and texture of the leaves and cotyledons are most often used to identify broadleaf seedlings. This is a winter annual, mustard (*Brassica* spp.), showing its first true leaves and cotyledons. *Photo:* J. K. Clark.

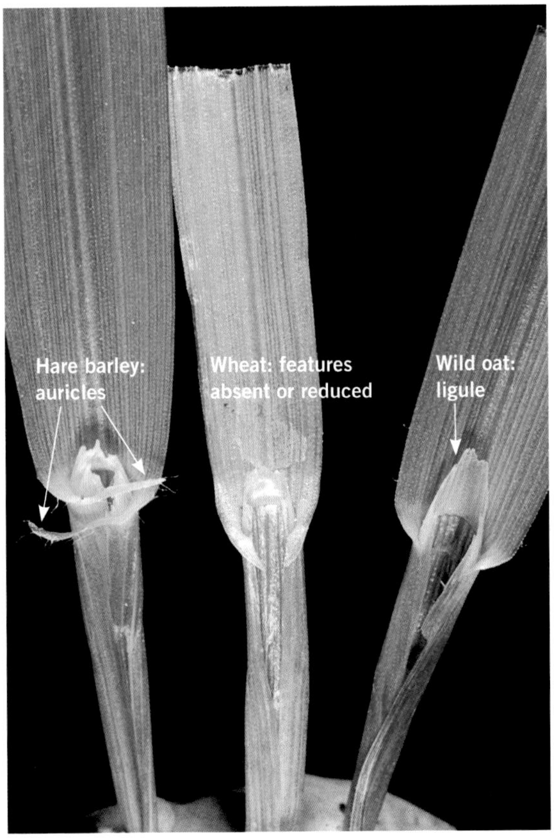

Figure 16.6 The collar region of grasses is most often used to aid in their identification while in the seedling stage. The presence or absence of auricles, ligules, and hairs are the features often assessed. From left to right: hare barley *(Hordeum murinum)*, wheat (*Triticum* spp.), and wild oat *(Avena fatua)*. *Photo:* J. K. Clark.

Cultural Practices

Prevention

The most cost-effective pest management method is to prevent a weed infestation from occurring. Two prudent measures a grower can take to prevent new infestations are practicing good farming hygiene and removing sources of new weed infestations. Farm hygiene includes cleaning soil from equipment as it leaves or enters a field. Removing sources of new infestation includes keeping irrigation ditches clean, since this is a common route by which new infestations begin for species that produce seed that float. Keeping orchard borders weed-free is an important tactic for reducing sources of new weed infestations because wind facilitates immigration of wind-dispersed weeds such as the sunflower family species (see fig. 16.2).

Weeds tend to grow best where there is little or no competition from other plants and little or no management. Any soil that is bare, moist, and sunlit is prime ground for weed growth. Weeds, like most plants, do not grow well in poor conditions, but they grow better in poor conditions than do citrus. Weed species are better able to direct their growth to the organs that acquire the most deficient resource. These adjustments make all resources equally limiting to growth, thus optimizing their performance. Weeds also tend to be luxury consumers of excess resources such as light, water, and nutrients, although these extra resources may not be used for growth. Therefore, the orchard canopy, irrigation, and fertilizer applications should be managed so that conditions are uniformly optimal for citrus growth across the entire orchard before developing a weed management plan.

Irrigation Design

Irrigation systems should put water only where tree roots are located both horizontally and vertically in the soil profile. Weeds are often small, herbaceous species with relatively shallow root systems, and they suffer if long dry periods follow irrigation. Furrow networks that do not allow water to stand for extended periods reduce weed establishment. Deep furrow systems that run along the planting line have "spray shadows" on one side of the furrow when directed herbicide applications are made from the middles. On soils with high clay content or other slow-draining soils, using shorter furrows and lateral furrows that cut across the tree rows reduces the time water stands in furrows. Weed problems are most likely to occur wherever wet conditions exist near the soil surface for extended periods. Moisture also facilitates the breakdown of herbicides due to hydrolysis and increased microbial activity. Wet conditions exist under sprinklers, near misters or drip systems, in furrow bottoms, at furrow ends, and around standpipes. Be prepared to apply postemergent herbicides to weed seedlings that will grow in those areas. Allow sufficient time for herbicide absorption by weeds before rerunning sprinklers or misters.

Fertilizer Application

Do not apply supplemental fertilizer to make up for the nutrients lost to weeds, because the nutrients will likely go to the weeds. Weeds are better competitors for nutrients than are citrus. The shallow root systems of weeds intercept soil-applied fertilizers first. Control weeds before fertilizer applications. Weed response can be especially strong to nitrogen and phosphorous.

Chemical Control

Most weed management systems in citrus orchards rely partially or entirely on herbicides (fig. 16.7). When selecting an herbicide it is imperative that the applicator follow the herbicide label. Weed control efficacy is most often lost as a result of user error. Overapplication or misdirected application can damage trees. Underapplication or misdirected application does not provide acceptable control. Application equipment must be properly calibrated to deliver the recommended rate. Proper timing is achieved by frequently monitoring the orchard to identify when most weeds are at the proper stage for control. Weeds escape herbicidal control by germinating after the application has been made, surviving the application by virtue of their large size, or because they are resistant to or tolerant of the herbicide.

Herbicide Principles

Selectivity

A selective herbicide controls certain plant species but not others that are tolerant due to physical characteristics, inability of herbicide to reach the target site, or rapid metabolism of the herbicide to inactive metabolites. Nonselective herbicides, also called broad-spectrum herbicides, control all treated plant species.

Mobility

Contact herbicides kill on contact and therefore do not move far in the plant. Coverage is very important for contact herbicides because a sufficient amount of plant tissue must be treated and killed, otherwise the plant will recover. Systemic herbicides move from the site of entry through the plant's vascular system to reach the site of action. When using these translocated herbicides, it is important to understand the site of entry, the translocation path, the site of action, and the direction of flow in different plant tissues at different plant life stages. The primary direction of flow in the sugar-conducting phloem tissue is from mature tissues toward roots and young tissues in early to midseason and primarily toward seed or perennial storage organs in the late season. Soil-applied systemic herbicides are absorbed by the roots and move with the transpiration stream in the xylem.

Unless otherwise directed on the herbicide label, do not mix a systemic herbicide with a contact herbicide because the contact herbicide destroys tissue that is needed to carry a systemic herbicide to its site of action. Surface-active agents, also known as surfactants, can be added to herbicide spray solutions to improve coverage, retention, or penetration. Surfactants can enhance the activity of an herbicide by facilitating penetration of the waxy leaf cuticle or by reducing surface tension of spray droplets so that they stick and sit flatter on the leaf surface. Also, the efficacy of certain herbicides such as glyphosate may be limited by hard water with high pH. Water hardness and pH should be adjusted to acceptable levels as specified on the herbicide label using commercially available testing kits and products.

How Weeds Are Killed

The mode of action of an herbicide includes all the physiological, morphological, anatomical, and biochemical processes that results in phytotoxicity. The most common mode of action is photosynthetic inhibition. Herbicides with different chemical structures can have the same mode of action. The mechanism of action is the first vital physiological process or biochemical reaction disrupted by an herbicide. The mechanism of action is more specific than the mode of action. The most common mechanism of action is the destabilization of a specific protein involved in photosynthesis. Rotating the use of herbicides with different mechanisms of action as indicated by the Weed Science Society of America (WSSA) mode of action group number slows the development of herbicide resistance in weeds (table 16.2).

Timing of Application

Preplant incorporated herbicides are applied to the soil before planting and then incorporated by tillage, irrigation, or rainfall. These herbicides control germinating or established weeds. Preemergent herbicides are soil applied before the weeds have emerged from the ground but usually after planting, as specified on the label. Preemergent herbicides are used to control

Figure 16.7 Most citrus orchard floors are kept weed-free with herbicides. *Photo:* S. Steinmaus.

Table 16.2. Mechanisms of action for currently registered herbicides for citrus grown in California

Chemical name	WSSA mode of action group number	Brand name example	Chemical family	Mechanism of action
bromacil	5	Hyvar	uracil	Photosystem 2 electron transport inhibitor by destabilizing D-1 protein.
carfentrazone	14	Shark	triazolone	Protoporphyrinogen oxidase (PROTOX) inhibitor; prevents chlorophyll biosynthesis, leading ultimately to lipid peroxidation.
clethodim	1	Select Max	cyclohexanedione	Acetyl CoA carboxylase (ACCase) inhibitor, stops fatty acid synthesis.
diquat	22	Reglone	bipyridilium	Photosystem 1 electron transport diverter, ultimately resulting in peroxide production and lipid destruction.
diuron	7	Karmex	substituted urea	Photosystem 2 electron transport inhibitor by destabilizing D-1 protein.
EPTC	8	Eptam	thiocarbamate	Unknown; may inhibit fatty acid, protein, and flavenoid biosynthesis.
fluazifop	1	Fusilade	arylophenoxy-propionate	Acetyl CoA carboxylase (ACCase) inhibitor, stops fatty acid synthesis.
flumioxazin	14	Chateau	dicarboximide	Protoporphyrinogen oxidase (PROTOX) inhibitor; prevents chlorophyll biosynthesis, leading ultimately to lipid peroxidation.
glyphosate	9	Roundup	phosphonoglycine	Inhibits aromatic amino acid synthesis by binding to EPSP synthase.
isoxaben	21	Gallery	benzamide	Inhibits cell wall biosynthesis.
MSMA	17	MSMA	organic arsenical	Unknown; results in cell membrane destruction.
napropamide	15	Devrinol	amide	Unknown; inhibits cell division.
norflurazon	12	Solicam	pyridazinone	Blocks carotenoid biosynthesis by inhibiting phytoene desaturase.
oryzalin	3	Surflan	dinitroaniline	Prevents microtubule formation by binding to tubulin, thus stopping cell division.
oxyfluorfen	14	Goal	diphenylether	Protoporphyrinogen oxidase (PROTOX) inhibitor; prevents chlorophyll biosynthesis, leading ultimately to lipid peroxidation.
paraquat	22	Gramaxone	bipyridilium	Photosystem 1 electron transport diverter, ultimately resulting in peroxide production and lipid destruction.
pendimethalin	3	Prowl	dinitroaniline	Prevents microtubule formation by binding to tubulin, thus stopping cell division.
sethoxydim	1	Poast	cyclohexanedione	Acetyl CoA carboxylase (ACCase) inhibitor; stops fatty acid synthesis.
simazine	5	Princep	triazine	Photosystem 2 electron transport inhibitor by destabilizing D-1 protein.
thiazopyr	3	Visor	pyridine	Inhibits cell division by binding to a microtubule-associated protein, not tubulin.
trifluralin	3	Treflan	dinitroaniline	Prevents microtubule formation by binding to tubulin, thus stopping cell division.

Sources: Senseman 2007 and California Department of Pesticide Regulation, http://www.cdpr.ca.gov/docs/registration/regmenu.htm (March 2013). For more information about mixing or rotating herbicdes for herbicide resistance management, see the Herbicide Resistance Action Committee website, www.hracglobal.com.

germinating weed seed or newly emerging seedlings. Postemergent herbicides are applied to the foliage of weeds after they have emerged from the ground to control established weeds. Seasonal timing of applications depends on the dominant weed species and the timing of their susceptibility to the available registered herbicides (see the section "Integrated Weed Management Timeline," below, for further discussion).

Types of Application

A broadcast application is made to the entire orchard floor, whereas a strip application treats a strip 4 to 8 feet wide along the tree line. Strip applications are best suited when using expensive, selective herbicides to kill weeds in the planting line. Strip applications will hinder the development of herbicide resistance in weeds because only a portion of the weed population is treated with this type of herbicide application. Herbicide residuals in the soil are typically shorter with strip applications because only a fraction of the soil will be sprayed. Control of weeds in the middles should use a less expensive and less selective management regime, such as mowed vegetative ground cover. Directed applications are made from spray nozzles that are angled or hooded so that only the low-growing weeds are treated, not the tree foliage. Young trees should have a protective collar around their base to protect them from herbicides. Mature trees typically have a thick bark that cannot be penetrated by herbicides. However, citrus skirts (lower branches) hang very close to the ground, and caution should be exercised to prevent misdirected spray or drift onto them. Applying oxyfluorfen to soil in young orchards should be done with caution, as this compound has the potential for codistillation or "lift-off" with water vapor and could cause off-target crop injury.

Herbicides can be injected into the irrigation system (chemigation). For example, norflurazon can be chemigated in established citrus for control of many common annual weed species such as sprangletop, pigweeds, purslane, and puncturevine. Photosensitive herbicides such as oryzalin and trifluralin may be also applied and incorporated by this method. Labels provide special instructions for this form of application and how to prevent the contamination of the water source.

A recent development in precision spray technology uses a visual, or intelligent, spray system. The soil surface is illuminated with an infrared light source. The reflected light has a unique signature, depending on whether the soil surface is bare or vegetated. The sensor is calibrated to open a solenoid that controls an herbicide spray nozzle when positioned over green vegetation. When positioned over bare ground, the solenoid defaults to the off position. This apparatus can be mounted on an all-terrain vehicle or tractor. Herbicide costs with this precision system will be the same as those of a conventional broadcast or strip application when weed cover is 100%. However, most orchards have substantially less than 100% weed cover in a given year, especially after years of good weed control. In these cases, the precision system will save the same percentage in herbicide costs as the percentage of orchard floor that is bare. Therefore, if weed control has effectively reduced weed cover to 5%, then the visual spray system will save 95% of the herbicide and its associated costs compared with a conventional broadcast application in which 100% of the orchard floor is treated.

Fate of Herbicides in the Environment

Any herbicide fate that does not involve killing the weed represents lost herbicide efficacy. Herbicides cannot realistically be applied with 100% efficiency. Therefore, recognizing and adjusting for the fates discussed below allows growers to improve herbicide efficacy. Herbicide labels provide all information about conditions known to affect herbicide efficacy. They also provide instructions on how to deal with some of the following fates. It is imperative that the herbicide applicator follows herbicide label instructions to ensure proper selectivity and to minimize off-target impacts.

Microbial decomposition

The most important fate of organic herbicides is enzymatic breakdown by soil microbes. Most herbicides are organic, as they contain carbon. Any soil condition that favors microbial growth such as moisture and moderate temperatures favors microbial breakdown of the herbicide. Unfortunately, these are the conditions that are usually desirable for healthy soils. Herbicide rates may have to be increased when applying a soil-active herbicide to soils with high microbial activity. This information will be on the herbicide label if it is pertinent.

Adsorption

The charged binding of compounds such as herbicides in fine-textured soils with high levels of organic matter is called adsorption. Adsorption prevents the herbicide from reaching plant roots. Each herbicide has a unique potential to adsorb to soil particles; the potential is stronger in clay soils and those with high levels of organic matter. Sand is coarse textured and has less surface area and thus fewer binding sites for herbicides than does clay. Labels often recommend the use of higher rates if the herbicide is prone to adsorption, especially on fine-textured soils such as those with high clay content.

Leaching

The downward (or sometimes horizontal) movement of chemicals through the soil profile is called leaching. Leaching may contaminate ground water. The leaching potential of an herbicide is often inversely proportional to its adsorption potential; if the herbicide is held by soil, it will not leach. Leaching, which is especially problematic in coarse, sandy soils, is a function of chemical properties of the herbicide and water volume percolating through the soil profile. Simazine (Princep), diuron (Karmex), bromacil (Hyvar) and norflurazon (Solicam) are herbicides registered for use in California citrus that are prone to movement in sandy soils and thus have use limitations specified on their labels. Applying these herbicides in late winter may avoid heavy rainfall that would otherwise facilitate their leaching. The California Department of Pesticide Regulation (DPR) has stipulated that the Ground Water Protection Area (GWPA) designation replace the Pesticide Management Zone (PMZ) system to identify areas susceptible to ground water contamination based on soil type and depth to ground water (for more information, see chapter 27, "Surface and Ground Water Quality"). Each county agricultural commissioner's office has information on the designated GWPA in their county. Maps of GWPAs within each county may be found online at the DPR website, http://www.cdpr.ca.gov/docs/emon/grndwtr/gwpamaps.htm. Special conditions must be met before the four aforementioned herbicides may be applied in a GWPA, as specified in the herbicide label.

Photochemical decomposition

Any herbicide that has a strong color to it, such as the bright orange trifluralin and oryzalin, will be prone to decomposition when exposed to sunlight. These herbicides should be incorporated into the soil immediately after application either by rain, sprinkler irrigation, or by mechanical soil incorporation.

Drift

Off-target movement of herbicides is called drift and can be reduced by decreasing nozzle pressure, lowering the spray boom, increasing nozzle size, and by spraying when humidity is high and temperature and wind are low.

Volatilization

Volatilization occurs when a substance changes from the solid or liquid state to the gaseous state. It is a function of the vapor pressure of the compound and temperature. Higher temperatures and higher vapor pressures represent a higher potential to volatilize. To reduce losses attributable to volatilization, choose low-volatility formulations of an herbicide when given the choice and spray when temperatures are low, such as in the morning.

Herbicide Resistance Management

The most obvious reason to minimize herbicide use involves nontarget movement, such as by drift or leaching. Public concern over agricultural sources of water and air pollution has made it imperative that pesticides remain on target. A second reason to minimize herbicide use involves the development of herbicide-resistant weed populations. The continuous use of herbicides with the same mechanism of action can lead to the development of resistance to any other herbicide with that same mechanism of action. Aside from user error, herbicide resistance is the most significant reason for poor herbicide efficacy. There are too many examples of herbicide resistance to list here, but more information on resistance can be found on the International Survey of Herbicide Resistant Weeds at the Weed Science Society website, http://weedscience.com, and the Herbicide Resistance Action Committee (HRAC) website, http://www.hracglobal.com. All chemical groups of herbicides have examples of weed species that are resistant to them. Recent examples are the development of resistance to glyphosate (Roundup) in California. After 10 or more years of continuous use of glyphosate, weedy ryegrass *(Lolium rigidum)*, horseweed *(Conyza canadensis)*, and hairy fleabane *(Conyza bonariensis)* populations that were once susceptible to glyphosate now tolerate recommended rates of that herbicide (see fig. 16.2). Minimizing herbicide use for resistance management takes the form of using alternative methods such as tillage, solarization, flaming, and biological control as discussed above. Clearly, the best resistance management approach is to rotate mechanisms of herbicide action and to rotate to alternative methods (see table 16.2). The Weed Science Society of America has developed a numeric coding for each mode of action, which is now appearing on some herbicide labels and can be easily used to avoid using herbicides within the same group number repeatedly. The generally recognized steps orchard managers should take to suppress the development of herbicide-resistant weed populations is to

1. Use integrated weed management practices where several methods are utilized simultaneously. This includes using good orchard principles that enhance tree competitiveness and cleaning equipment between sites.
2. When using chemical methods, use multiple herbicide modes of action with overlapping

weed control spectrums either in rotation, sequences, or mixtures.

3. Use the full recommended herbicide rate and proper application timing for the weed species that are most difficult to control in the orchard.
4. Monitor fields after herbicide application to ensure control has been achieved. Avoid allowing weeds to reproduce by seed or to proliferate vegetatively.
5. Establish orchards in weed-free fields (see the "Integrated Weed Management Timeline" section in this chapter).

If postemergent herbicides are registered for problematic weeds in a given locale, they should be applied only after monitoring has identified that those weeds are present in a given year. Assessing the efficacy of preemergent herbicides is difficult because a pest control manager will not know whether they were necessary in the first place. Based on estimates from many cases of herbicide resistance, rotating to different mechanisms of action should be done no less frequently than every 3 to 5 years to minimize the selective force that ultimately leads to an herbicide-resistant weed population. Herbicide-resistant seed from wind-dispersed species such as hairy fleabane *(Conyza bonariensis)* and horseweed *(C. canadensis)* can move into an orchard from surrounding areas (see fig. 16.2). Monitoring and historical records of weed populations and herbicide efficacy will help alert orchard managers to such problems. Wind-dispersed seed are usually small and are effectively controlled by burial with mechanical cultivation.

Alternative Nonchemical Methods

Cover Cropping

Most citrus orchards in California do not use cover crops except in areas prone to soil erosion, such as on steep slopes or poorly structured soils. Many of the adverse effects associated with weeds apply to growing cover crops. Cover crops increase the threat of frost by about 3° to 5°F and require water and fertilizer in addition to that required by the trees. More water is lost by the evapotranspiration from a carpet of weeds than from a pool of water, and much more than is lost from bare soil with a dry crust or dead weeds on top. However, cover crops can improve soil structure for better water penetration and nutrient retention and can stabilize soil on slopes. A cover crop can simply be the resident vegetation maintained by frequent mowing, or it can involve sowing a particular species to obtain the desired effects (fig.

Figure 16.8 One alternative to a completely bare orchard floor is to maintain a mowed vegetated strip down the middles and strip-apply herbicides down the tree line. *Photo:* S. Steinmaus.

16.8). Leguminous species are frequently chosen as a cover crop plant because they increase the nitrogen content of soils, while small grain cover crops draw down nitrogen levels. Legumes break down rather quickly in the spring after they die back or are mowed, whereas the straw from grain crops will persist on the soil surface throughout the summer if not removed or plowed under. Often, the strategy is to plant a cover crop that is less competitive or is easier to control than the weeds that would otherwise occupy that space.

Cover crops, when used, are grown in established orchards and rotated every few years to avoid pathogen buildup. There is no one cover crop species that will fit all situations and provide all the possible benefits. Cover crop species should be chosen based on site requirements, soil conditions, and availability of water and equipment to manage the cover crop. See *Cover Crops for California Agriculture* (Miller 1989) and the University of California Sustainable Agriculture Research and Education Program (SAREP) website, http://www.sarep.ucdavis.edu, for more information.

Vegetation on the orchard floor lowers ambient temperatures by a few degrees. Growers can estimate the risk of frost for any given location by logging on to the UC IPM website and selecting the Weather Data and Products page. Bare orchard floors experience temperatures that are roughly 3°F warmer than those recorded at CIMIS stations because the latter are typically measured over an irrigated turf ground cover.

Tillage

Soil tillage is a valuable tool for weed control in many agroecosystems, including citrus production. However, it is used primarily during site preparation rather than in established citrus orchards. Repeated plowing and disking brings up perennial roots and rhizomes that will desiccate and die in the summer heat if no water is applied or no rain falls for several months.

Additionally, aside from soil preparation before planting and reforming furrows in furrow-irrigated orchards, tillage is not often used in established citrus orchards for the following reasons:

- Fine feeder roots that absorb nutrients and water reside in the top 2 feet of the soil and are damaged during tillage.
- Damaged roots allow openings for pathogen entry.
- It is difficult to control weeds in the tree planting line.
- Tillage brings new weed seed from the deep seed bank to the surface.
- Tillage creates favorable conditions for weed seed germination (e.g., good soil contact, scarification).
- Tillage can facilitate the spread of fragmented perennial species such as field bindweed (fig. 16.9).
- Soil compaction, creating a plow sole, interferes with water and root penetration.
- Dust raised during tillage can contribute to other pest problems and reduce air quality.
- Tillage promotes soil erosion, especially in orchards on slopes or in areas prone to wind.
- The cost of tillage tends to be greater than systems that rely on herbicides.

Even though cover crops and tillage have numerous drawbacks, they can be viable alternatives to herbicides. As with cover cropping, growers must balance the benefits with the drawbacks of tillage when developing a weed management plan.

Figure 16.9 Field bindweed *(Convolvulus arvensis)* is a perennial broadleaf species that spreads by seed and perenniating parts such as rhizomes and plant fragments. Tillage can facilitate its spread throughout an orchard. *Photo:* J. K. Clark.

Heat

Two methods involving heat can be used to manage small weed seedlings and seed. These methods work best on small-seeded weed species such as those typically dispersed by wind. Common purslane *(Portulaca oleracea)* is a good indicator species (fig. 16.10): if purslane is controlled, most species will be suppressed. Plants and seed, as with most biological organisms, are more sensitive to wet heat than to dry heat. Steam injection systems take advantage of this fact, but they are still being investigated and are not commonly used commercially. Soil is a poor conductor for heat because of the air spaces interspersed among soil particles. Therefore, heat does not travel very far in soil from where it is applied. Consequently, surface seed are killed by the methods discussed below, but seed below about 2 inches are not. Most weed seed germinate from within 1 inch of the soil surface, so the methods are viable if the soil remains undisturbed after the application of heat. The main drawbacks with methods that rely on heat are the cost of fuel to generate the heat, lack of effective control on larger weeds, and increased fire hazard. However, as with tillage and cover crops, they provide alternatives to herbicides.

Flame

Torches fed by propane or specially designed liquid petroleum gas (LPG) mounted on a tractor-drawn rig may be used to kill small weeds. Flaming causes the cells in a plant to swell until the membrane ruptures, causing cell death. The size and succulence of the target plant and the duration of the flaming determine the killing potential. Best results are achieved with small plants (2 in tall or shorter) that are flamed for one to several seconds. Selective flaming can take advantage of the size difference that usually exists between the citrus trees and weeds. Flaming works best on broadleaf species; grasses tend to recover. Caution must be exercised while using these methods so as to not burn tree trunks or trunk collars, or to ignite dry vegetation.

Solarization

Clear 3-mil plastic laid over wet soil and sealed along all edges generates sufficient heat to kill the seed of many weed species lying at or near the soil surface in most citrus-growing regions in California (see fig. 16.10). If used, solarization should be done in the summer prior to planting, as this method is impractical in an established orchard. Heat is generated as a result of the greenhouse effect, in which sunlight passes through plastic that is nearly transparent to short-wave radiation but not to long-wave radiation. Radiation that is reflected off or reradiated from the soil has long wavelengths and is trapped underneath the plastic. The wet soil aids in heat absorption and conductance of heat into the upper soil layers. Moisture that evaporates from the wet soil contributes to a high relative humidity, accentuating killing power. Temperatures in excess of 170°F should kill most weed seed, pathogens, and nematodes. Higher temperatures may be achieved under double layers of clear plastic or bubble plastic, which could be especially beneficial in cool areas. Weed seed mortality is determined by the maximum temperature and the duration of exposure to high temperatures. Several hours at 150°F may be as lethal as half an hour at 170°F (see Elmore 1997 for more information).

Figure 16.10 Common purslane *(Portulaca oleraceae)* is a common summer broadleaf annual weed. It is a good indicator plant for the effectiveness of soil solarization, in that if common purslane seed is killed, then most weed seeds will be killed. *Photo:* J. K. Clark.

Biological Control

Biological control is the use of natural organisms to reduce pest populations to a level that is no longer economically damaging. It is not commonly used for weeds in citrus production, but it can provide an alternative to herbicides. The augmentative approach is the only form of biological control that may provide acceptable levels of control for certain weed species in citrus. Augmentation involves rearing and releasing large numbers of biocontrol agents to attack or infect a target weed. With the augmentation approach, the agent frequently cannot sustain itself from year to year in the environment, so must be reapplied when warranted by monitoring.

The classical approach to biological control involves the release of a non-native organism to control a non-native weed, but it usually does not provide the level of weed control required by most citrus producers. This approach is more suited to large areas of low-value land such as rangelands. The conservation approach involves the conservation or protection of native parasites, predators, or diseases of weeds so they can better control the weed. Allowing perennial members of the Apiaceae (carrot) family to grow on the perimeters of the orchard is an example of this form of control, but additional forms of control must be used in most citrus situations. The broad-spectrum approach involves some form of grazing, but it is not typically done in citrus because of the damage to the low skirts of citrus trees that is likely to occur. However, geese and ducks may effectively control weed seedlings.

Special Considerations for Controlling Annuals

As a general rule, herbicides do not kill ungerminated seed because the seed coat allows very little exchange with the outside environment unless dormancy has been broken and the seed has become imbibed. Immediately following germination, new seedlings are susceptible to soil-applied preemergent herbicides as they absorb water and nutrients through the coleoptillar nodes in grasses and through the new roots or hypocotyls in broadleaf species. The seed contains sufficient reserves to support the initial growth of the seedling but rarely enough to push growth beyond the first set of true leaves or tillers without photosynthetic contributions. Annuals will be most sensitive to postemergent control immediately after the switch from relying on seed reserves to photosynthesis, which corresponds to about the 2 to 6 true leaf stage. Relative exchange rates for gases, nutrients, and water are maximal at this time, and little effort is required to kill the plants (see figs. 16.2, 16.5, and 16.10 for examples).

Special Considerations for Controlling Perennials

Efficient control can be obtained by understanding the direction of carbohydrate movement in perennials. In the early spring, carbohydrates from underground vegetative storage organs such as rhizomes move to support the growth of new shoots (fig. 16.11). By late spring or early summer, carbohydrate reserves are at their lowest and new shoots have not yet begun to refill those reserves. At this time, repeated removal of shoots using contact herbicides, mowing, flaming, scraping, or other suitable tillage operation can significantly impact these species. After shoot tissues are removed, the belowground storage organ must commit more reserves to support new shoot growth. Eventually, this process, called carbohydrate starvation, leads to depletion of reserves and ultimately death in many perennial species. For species, however, such as field bindweed (fig. 16.9), tillage may remove competition, thus favoring its establishment. Field bindweed has required more than 16 successive cultivations to eventually control it by carbohydrate starvation. The alternative strategy of applying a systemic herbicide late in the season is usually required for this species. Rhizomes of several important citrus weed species such as johnsongrass and bermudagrass may be controlled before planting the orchard if they are tilled up and allowed to desiccate for several weeks when conditions are hot and dry. If repeated shoot removal cannot be done through the entire season or if conditions are moist, a systemic herbicide should be applied in the late summer just as carbohydrates from photosynthesis begin translocation to the vegetative storage organs. Burial of vegetative propagules such as tubers and rhizomes is not prudent, since these structures possess substantial reserves (see figs. 16.3 and 16.4 for examples). Further, rhizomes, stolons, and tubers are not readily controlled with direct application of herbicides.

Integrated Weed Management Timeline

Preplant

A field scouting timeline should be prepared for a prospective orchard site well before planting (fig. 16.12). Unless the history for the site is known, a report should be completed for an entire year to determine the species composition of the weed seed bank. Important winter annual weed species in California citrus will be most abundant in the late winter, while summer annuals are most abundant in early summer. Perennials can be growing at any time during the year, but they are usually most vigorous in spring and summer. In the first few years, the entire orchard should be scouted. Less ground may be scouted as the location of dominant weed populations becomes known, since the established weed populations do not normally move or change in numbers, especially if there is no soil disturbance. Recommendations for preemergent herbicides are made on the basis of the previous year's scouting reports.

Once the identity, abundance, and distribution of weed species are known, a weed management program can be devised. Developing a competitive orchard leads to maximum yield and suppression of weeds. Before planting,

- match the proper citrus variety with a specific location
- prepare the site so it provides uniformly optimal conditions for citrus
- repeatedly purge and monitor the weed seed bank

The uppermost layer of soil should be as free of weed seed as possible before planting. To achieve this, plow and disk the soil, then irrigate it and till it again after weed seedlings emerge; repeat the process until weed growth is minimized. Three or four cycles will have a major impact on most annual weed species with small seed, which typically have little dormancy. Instead of cultivating on the last cycle, use a nonselective herbicide, flaming, solarization, or any other method that does not disturb the soil surface.

Selecting an herbicide must be based on the weed species present in the orchard. Herbicides must be registered for use in California citrus for the weeds to be controlled. For more information, see table 16.2 and the Weeds section of the UC IPM Citrus Pest Management Guidelines at http://www.ipm.ucdavis.edu/PMG. The rhizomes and roots of perennials may require tillage with V-sweeps with the front point slanted down in the summer prior to planting to bring vegetative propagules such as rhizomes and tubers to the soil surface. Follow this operation with a long, dry, fallow period (3 to 4

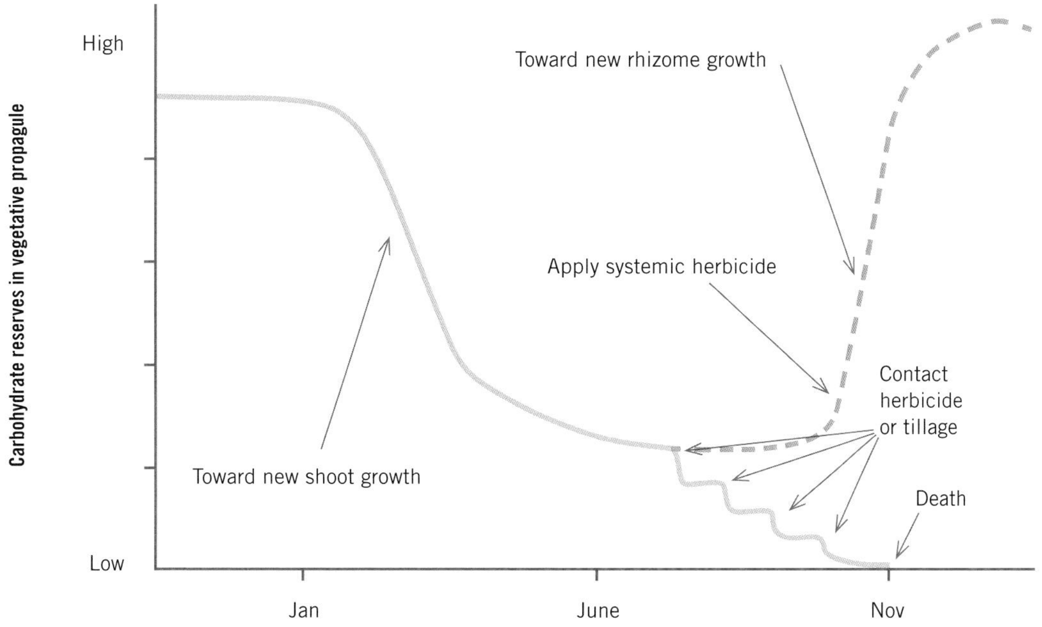

Figure 16.11 Timeline for controlling perennial weed species by carbohydrate starvation of the vegetative propagule. The goal is to deplete the overwintering structure, usually a modified stem such as a rhizome, stolon, or tuber, of carbohydrates so the plant cannot survive the winter. If carbohydrate starvation is not attempted using tillage, the modified stem will be replenished by the photosynthesis of the current year's growth (dotted line). In this situation, a systemic herbicide may be applied late in the season so that it will be moved into the rhizome and kill it.

weeks) to desiccate the propagules (see figs. 16.3 and 16.4 for examples).

Special Considerations for Newly Planted Orchards

Weed control is critical in the first few years because trees are most susceptible to competition and other stresses during this time. Wrap plastic collars around young trees to protect them from herbicide drift and other damage. The use of collars, however, does not guarantee that trees will be completely protected from drift or misdirected herbicide applications.

Tightly woven cloth mulch that is resistant to ultraviolet light can be laid down around the base of new trees to provide a weed-free barrier between the orchard floor and the trees. The cloth must be of a sufficiently tight weave to block light that would otherwise support weed growth, yet breathable enough to avoid anaerobic conditions. Using trunk collars and extending cloth barriers 4 feet from the tree should provide adequate protection from herbicide sprays used to treat the remainder of the orchard floor. Maintaining a spray buffer around the tree can be important when trees have not developed a deep root system. Preemergent herbicides that are mobile in sandy soils with low levels of organic matter, such as simazine, norflurazon, and diuron, can move and possibly damage the trees if a buffer does not exist.

Avoid the use of cover crops or any other vegetation in the first several years. Growing alternative crops between the tree lines during the nonbearing years of young orchards has been practiced, but these crops may detract from citrus performance. Use the previous year's scouting report to determine which preemergent herbicide to apply, as the previous year's weeds will be the subsequent year's weed floras.

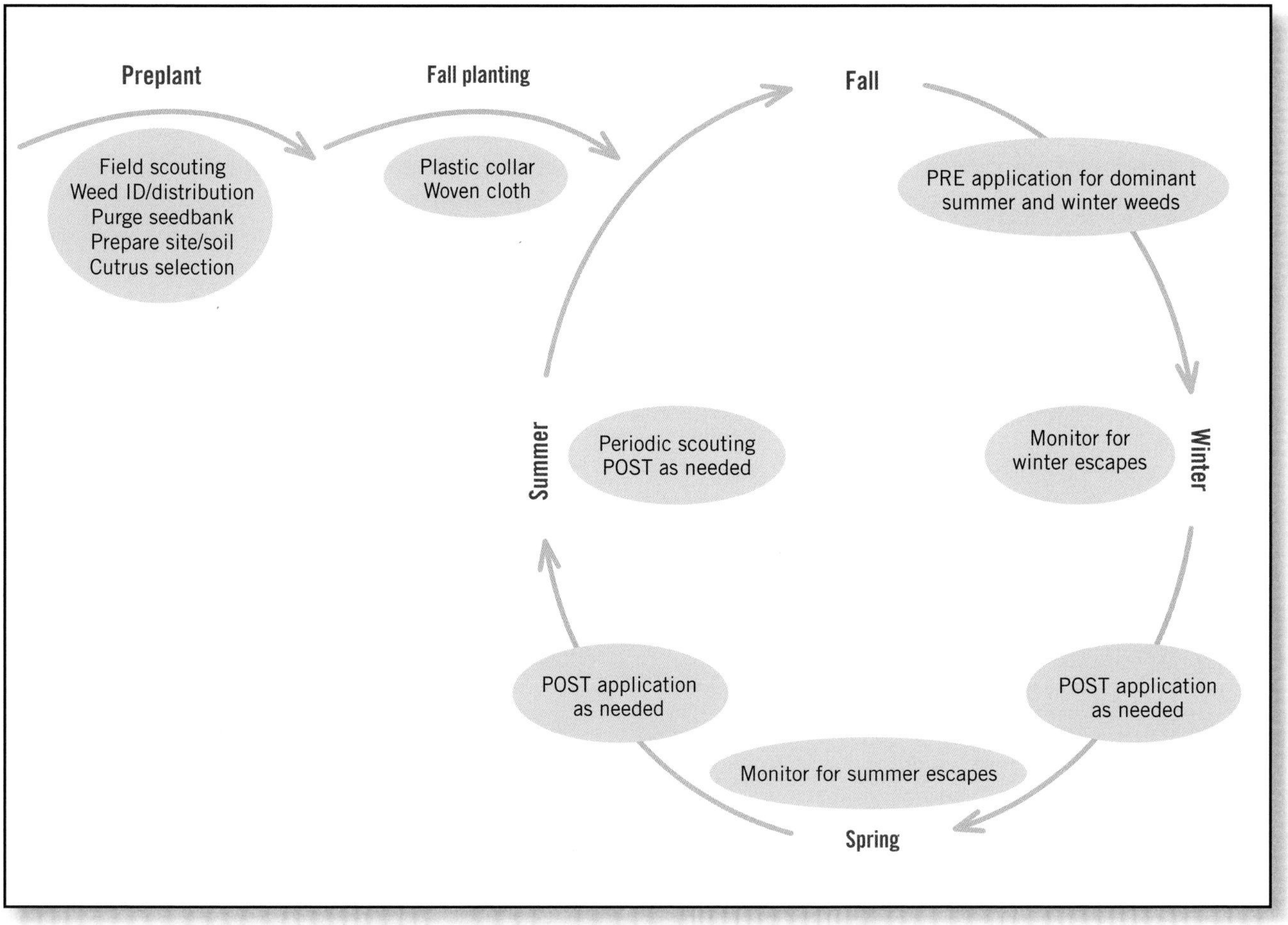

Figure 16.12 Integrated weed management timeline, beginning with initial scouting before planting through a full annual cycle, which should be repeated following citrus establishment. PRE indicates an herbicide application before the weeds emerge (preemergent), while POST indicates an herbicide application after the weeds emerge (postemergent).

Established Orchards

Monitor in established orchards when winter and summer annual weed species emerge from the soil, typically after the first substantial rain in the late fall and following rain in the late winter, respectively. These reports aid in developing recommendations for postemergent herbicide applications in the current year. The reports also allow the grower to follow any changes in weed populations, aiding in recommendations for preemergent applications in the subsequent year. Frequent monitoring will determine when weeds are at the proper stage for optimal control. Weeds often escape herbicide treatments by germinating after the application of the herbicide. Applicators should not wait for every weed to be at the proper stage for control; treatment time should occur when most, not necessarily all, of the population is at the proper stage for control.

The most common form of orchard floor management for citrus is to maintain a completely clean floor using herbicides (see fig. 16.7). Though not as common, strips 4 to 8 feet wide along the planting line are kept clean with herbicides, and the middles are vegetated with a sowed cover crop or resident vegetation. (fig. 16.8) Cover crops are shallow-incorporated into the soil or mowed periodically, beginning in the spring. Mowing can be mechanical or by applying a low rate of glyphosate. Mowing 8 to 10 times throughout the year may be required to keep vegetation at the desired height. Allow time for desirable species to flower and set seed for the subsequent year's cover crop. Orchards that are furrow irrigated or basin flooded with berms require refurrowing or scraping to reestablish irrigation efficiency. This operation also effectively controls weeds in these areas. Use evapotranspiration (ET) demand computations to determine the optimum irrigation scheduling for citrus (see chapter 11, "Irrigation"). Long dry periods after irrigation will help to suppress most shallow-rooted annuals. Choose herbicides that control the weed species detected during monitoring in scouting reports. One or more preemergent herbicides are typically applied from December to March, when a majority of the typical weed species are germinating and beginning growth. Early spring and late winter is when the temperature and moisture are beginning to be conducive to weed growth. A single postemergent herbicide application or tillage is made in late spring through early summer to control newly emerged summer annuals. Several of the preemergent herbicides (e.g., simazine, diuron, oryzalin, and trifluralin) have sufficiently long soil residual activity to provide control of summer annuals. If perennial species are present, more postemergent applications may be required late in the season, when their storage organs are more susceptible to systemic herbicides (see fig. 16.11). Remember to rotate or mix mechanisms of action for herbicides (see table 16.2) every few years or find alternative methods for control. Returning to a mode of action that was used for several consecutive years should be delayed as long as possible, up to 10 years, to most effectively prevent the development of herbicide-resistant weed populations. However, the length of this delay depends on the specific mode of action and other considerations (see the Herbicide Resistance Action Committee (HRAC) website, http://www.hracglobal.com, for more information). Monitoring reveals changes in the weed flora over time, so the management program must change to prevent a dominant flora from developing. Use untreated soil around the root zone when replanting dead or damaged trees in an established orchard if herbicides with long soil residual activity were recently used. Be prepared for a new flush of weeds wherever the soil is disturbed and moistened.

References

California Department of Pesticide Regulation (DPR). Ground Water Protection Program website, http://www.cdpr.ca.gov/docs/emon/grndwtr/.

DiTomaso, J. M. 2005. Grass and grass-like weeds of California. CD. Davis: University of California Cooperative Extension Weed Research and Information Center; California Invasive Plant Council website, http://www.cal-ipc.org/resources/booksandcds/grasses.php.

———. 2006. Broadleaf weeds of California. CD. Davis: University of California Cooperative Extension Weed Research and Information Center; California Invasive Plant Council website, http://www.cal-ipc.org/resources/booksandcds/broadleaf.php.

DiTomaso, J. M., and E. A. Healy. 2007. Weeds of California and other western states. Oakland: University of California Agriculture and Natural Resources Publication 3488.

Elmore, C. E. 1997. Soil solarization: A nonpesticidal method for controlling diseases, nematodes, and weeds. Oakland: University of California Agriculture and Natural Resources Publication 21377.

Herbicide Resistance Action Committee (HRAC) website, http://www.hracglobal.com.

Miller, P. 1989. Cover crops for California agriculture. Oakland: University of California Agriculture and Natural Resources Publication 21471.

Senseman, S. A., ed. 2007. Herbicide handbook. 9th ed. Weed Science Society of America. Lawrence, KS.

University of California Integrated Pest Management Program. UC IPM citrus pest management guidelines. UC IPM website, http://www.ipm.ucdavis.edu/PMG.

University of California Sustainable Agriculture Research and Education Program (SAREP) website, http://www.sarep.ucdavis.edu.

Watson, A. K., ed. 1993. Biological control of weeds handbook. 1993. Champaign, IL: Weed Science Society of America Monograph 7.

WeedScience.com. International survey of herbicide-resistant weeds. WeedScience website, http://weedscience.com.

Weeds of the West. 2002. 9th ed. D. A. Ball, T. D. Cudney, et al., eds. Laramie: Western Society of Weed Science, in cooperation with the Western United States Land Grant Universities Cooperative Extension Services and the University of Wyoming.

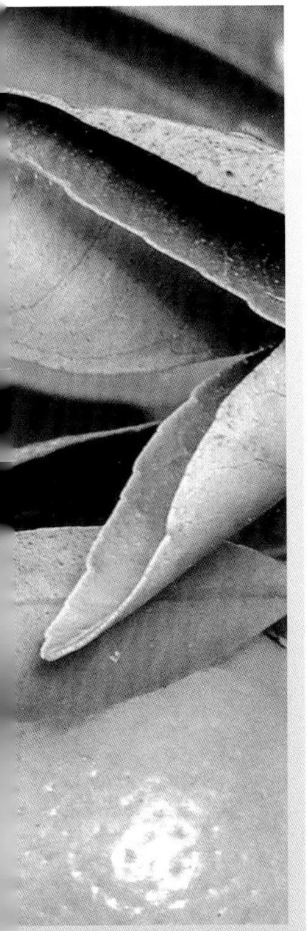

17 Citrus IPM

Elizabeth E. Grafton-Cardwell, David H. Headrick, Peggy A. Mauk, and Joseph G. Morse

Definition and Goals of Citrus IPM

General IPM versus Biologically Based IPM

Integrated pest management (IPM) is the long-term control of pests through the use of natural enemies (parasitoids and predators), cultural controls (pruning, reduction in dust, maintenance of tree health, resistant varieties, timing of harvest and fertilization to reduce pest impact, etc.), and selective pesticides (pesticides that allow natural enemies to survive). Pesticide use is reserved for pest populations that have exceeded their economic thresholds. The vast majority of California citrus growers currently are conducting a form of IPM because they or their pest control advisers (PCAs) sample for pests to determine whether a pesticide treatment is necessary, limiting pesticide use. For some pests, natural enemies are also sampled, because if they are found to be at sufficient levels relative to the pest density, a treatment may not be needed.

Growers practicing IPM look at the organisms in the citrus crop as a system composed of interrelated pests and natural enemies in which control tactics for one species may affect the others. For example, a treatment for worm pests (the larvae of various Lepidoptera) with a carbamate such as methomyl (Lannate) in the springtime would greatly reduce levels of predatory mites needed for citrus red mite control later in the season. The IPM approach is a long-term perspective of the pest complex as a whole that avoids disruption of natural enemies as much as possible.

Growers practicing IPM avoid the use of broad-spectrum pesticides and use the lowest effective rates of these pesticides whenever possible, including pyrethroids, organophosphates, carbamates, and neonicotinoids. The most recent registrations of new insecticides have tended to be more selective than these previously used groups of insecticides. Thus, the same insecticides are used by growers who intentionally practice IPM and those who apply pesticides without regard to natural enemies or long-term control. The term "biologically based IPM" was coined for growers who release and preserve natural enemies as their first line of defense and use pesticides as a last resort, emphasizing the use of selective pesticides.

Long-Term Reasons for Adopting IPM

In the San Joaquin Valley citrus-growing region, where the majority of citrus is now grown, several issues suggest that a biologically based IPM approach is superior in the long-term. First, key pests have developed resistance to a number of insecticides used for their control. Citrus thrips developed resistance to DDT within a few years of its first use in the 1940s. Resistance to organophosphate and carbamate insecticides was widespread by the late 1980s, and resistance to pyrethroids developed in a number of San Joaquin Valley citrus thrips populations during the late 1990s. San Joaquin Valley California red scale populations developed resistance to organophosphate and carbamate insecticides in the 1990s, and citricola scale populations developed resistance to organophosphates in the 2000s. Relying on natural enemies to the maximum extent possible reduces problems with pesticide resistance because it reduces the frequency with which insecticides are applied. Second, the general public and regulatory agencies are demanding fewer pesticide residues in food, air, and ground water. The Food Quality Protection Act (FQPA) requires a reduction in certain types of pesticides, and volatile organic compounds

(VOCs) generated by emulsifiable concentrate (EC) formulations of pesticides must be reduced. A pest management program that is based primarily on natural enemies and cultural controls addresses these regulatory concerns.

Factors Affecting the Stability of IPM

Biologically based IPM is a delicate balance of pests and natural enemies coupled with very careful use of insecticides, especially selective insecticides, that have minimal impact on natural enemies. A number of factors can tip the balance in favor of the pests. Weather is a primary factor. For example, during the heat of summer, predatory vedalia beetles do not grow or reproduce very well. If these predators have not completed the job of dramatically lowering cottony cushion scale populations in the spring, scale populations will resurge during the summer months. Another example is that during especially hot San Joaquin Valley summers, California red scale grows faster and survives better than the parasitoids and predators that attack it, delaying effective biological control until later in the fall.

Another factor influencing the stability of IPM is the introduction of invasive pest species and the pesticides needed to control them. When a new pest species enters a region, it is often relatively free of natural enemies for several years. The invasive pest outbreak may require broad-spectrum insecticide applications to keep the pest below economically damaging levels. If a broad-spectrum insecticide is the only type that is effective against the invasive pest, it may disrupt natural enemies needed to control other pests. A good example of this situation is the glassy-winged sharpshooter management program (fig. 17.1). The glassy-winged sharpshooter uses citrus as an oviposition host, and populations can reach high densities on that crop. It then moves from citrus to weeds, where it picks up the bacterium *Xylella fastidiosa*. From there it can move into neighboring grapes and through its feeding on weeds and grapes introduce the bacterium that causes Pierce's disease in grapes. While the glassy-winged sharpshooter rarely causes significant direct damage to citrus, its densities in citrus must be kept low to help prevent Pierce's disease transmission in nearby grapes. The most effective insecticides that reduce glassy-winged sharpshooter numbers in citrus are broad-spectrum pyrethroid and neonicotinoid insecticides. The gradual increase in frequency of these applications and the broad-spectrum nature of the insecticides used can disrupt the stability of the citrus IPM program. In a similar fashion, when the Asian citrus psyllid establishes in the San Joaquin Valley, it is anticipated that changes in pesticide choices will shift in response to a desire to manage it, especially if the devastating bacterial disease huanglongbing that it vectors becomes widespread in California. The control program is likely to rely on broad-spectrum pesticides that will disrupt the citrus IPM program.

Figure 17.1 Glassy-winged sharpshooter uses citrus as an oviposition host. *Photo:* E. E. Grafton-Cardwell.

Considerations before Starting a Biologically Based IPM Program

Regional Differences in the Success of Biological Control

Using natural enemies to manage citrus production has historically been easier in the southern interior and coastal areas of California than in the San Joaquin Valley. This is largely because the climate is milder in the southern and interior areas. A well-known example that highlights the differences between the two regions is California red scale and its key parasitoid, *Aphytis melinus*. In the San Joaquin Valley, cold winters eliminate the smaller instars of California red scale. In this region, the spring season begins with the scale population in a fairly uniform stage of development, primarily adult females and males. Once the adults begin to produce crawlers and die off, the only scale stages available (1st instar nymphs) are too small for the parasitoids to use as hosts. Thus, the parasitoids struggle to find enough scales in the right stage to deposit their eggs on. As the scale stages progress to 2nd and 3rd instars during May, the situation for the parasitoids begins to improve. However, in June or July, the extreme heat of summer slows adult parasitoid activity and causes the growth and development of the scale to speed up and sometimes outpace the development of parasitoid populations. Thus, *Aphytis* populations are unable to gain control of California red scale until fall (September to October) in the San Joaquin Valley (fig. 17.2). In southern California, where temperatures are milder, all stages are available for parasitism during most periods of the year, and parasitoid population development is more stable.

Isolated areas, such as the Fillmore Protective District of Ventura County, often show more success with their IPM program. Growers in the Fillmore District cooperated for many years to produce natural enemies such as *Metaphycus helvolus* for black scale and *Aphytis* for California red scale for release over large areas. They also took a regional approach to insecticide use, reducing the impact of insecticide drift from neighboring orchards and the impacts on natural enemies when they move between orchards.

Influence of Citrus Variety

Most pests cause similar damage to different varieties of citrus. However, there are several exceptions. Citrus peelminer shows a preference for the fruit of grapefruit, pummelo, and smooth-skinned early navels (Atwood, T.I., and Fukumoto) (fig. 17.3). Mandarins that flush frequently provide an ideal habitat for aphids. Foliage mining by citrus leafminer can reduce yields of lime and multicropping lemons (based on data from Florida), but generally, grapefruit, mandarins, and oranges are less affected. Cottony cushion scale populations build more easily on citrus varieties with dense foliage such as mandarins, Fukumoto navels, and grapefruit. Citrus thrips tend to damage the rind of navel orange varieties more easily than Valencias. Citrus bud mite

Figure 17.2 *Aphytis* wasps have difficulty gaining control of California red scale in the San Joaquin Valley because of extremes of heat and cold. *Photo:* E. E. Grafton-Cardwell.

Figure 17.3 Citrus peelminer shows a preference for the fruit of grapefruit, pummelos, and smooth-skinned early navels. *Photo:* E. E. Grafton-Cardwell.

are especially damaging to lemons grown on the coast. Careful choice of citrus variety or cultivar can help reduce the incidence of some of these pests. Pests may need to be managed somewhat differently on different varieties (e.g., the thresholds for citrus thrips treatments are lower on navels than on Valencias).

Tree Health and Canopy Management

A healthy, well-pruned orchard is often less susceptible to insect damage than an unhealthy orchard. As a first line of defense against insect pests, growers should plant rootstocks and scions proven to be adapted to the local climate, evaluate the irrigation system to make sure that water is applied uniformly, and conduct leaf analyses to ensure that trees are effectively fertilized. Pest mites reproduce faster on water-stressed trees. A number of pests, such as cottony cushion scale, citricola scale, earwigs, black scale, citrus peelminer, and mealybugs, prefer high humidity and the cool conditions inside a dense tree canopy. Parasitoids of California red scale are more effective in parasitizing scale on fruit than on wood and can search more effectively if the tree canopy is open. Thus, choosing citrus varieties that do not grow densely (or opening the canopy through internal pruning) can reduce pest numbers.

Neighboring Crops

In most cases, citrus pests are fairly specific to citrus, and neighboring crops do not affect their densities. Examples of pests that are fairly specific to citrus include California red scale, citricola scale, cottony cushion scale, citrus thrips, citrus leafminer, and citrus red mite. Occasionally, neighboring crops produce pests that can attack citrus. For example, citrus peelminer can arrive from willows in riparian areas or may move en masse into citrus when grapes, cotton, or beans begin to senesce. The aphid that vectors citrus tristeza virus (CTV), *Aphis gossypii,* can move into citrus from melons, cotton, or pomegranate orchards (fig. 17.4). Bean thrips rarely reproduce on citrus but build up on other crops and weeds in and around citrus. Bean thrips are a problem because adults find the navel on navel oranges to be an excellent overwintering site, and they can become an issue when the fruit is exported. Sometimes, citrus pests create problems for other crops. Examples of pests that build up in numbers in citrus and then move into neighboring crops include citrus cutworm and forktailed bush katydids, which attack stone fruit, and glassy-winged sharpshooter, which spreads Pierce's disease in grapes.

The negative impact of neighboring crops on citrus is not always the pest that comes from that crop but the pesticides that are applied to it. For example, *Aphytis* is very sensitive to sulfur, and when a citrus orchard is located next to a crop that receives frequent sulfur applications, such as grape vineyards, the *Aphytis* simply do not prosper. Because of the neighboring crop and pesticide effect, small blocks with many neighbors may be more difficult to manage with natural enemies and require more intense monitoring than larger blocks (> 20 ac). A regional approach to pest management using natural enemies is more effective than a patchwork quilt of very different practices.

Figure 17.4 Aphids that vector citrus tristeza virus can move into citrus from neighboring pomegranate orchards. *Photo:* E. E. Grafton-Cardwell.

Avoiding Broad-Spectrum Pesticide Residues

One of the most critical aspects of successful biological control is minimizing the effect of pesticides on natural enemies through careful choice of chemical, rate, timing, and coverage of the tree. All pesticides potentially have an impact on one or more natural enemies; however, some pesticide effects are far more severe than others. Pesticides that kill a broad range of insects, including both pests and natural enemies, are called broad-spectrum pesticides. Pesticides that tend to allow natural enemies to survive are called selective, or soft, pesticides. Table 17.1 lists a number of insecticides and their effect on natural enemies. Insecticides in the organophosphate, carbamate, neonicotinoid, and pyrethroid groups are broad-spectrum pesticides.

Some broad-spectrum pesticides have greater impacts than others due to their longevity on citrus foliage, fruit, or wood at levels that impact natural enemies. Many of the pyrethroids and organophosphates are quite persistent, with toxic residues affecting natural enemies for one to several months. Other quite toxic products do not persist very long at all; for example, naled (Dibrom) kills much of what is present at the time of application but persists for only several days.

Insecticides may be toxic to only a portion of a natural enemy population. For example, oil smothers all insects and mites that it comes in contact with (both pests and beneficials), but natural enemies that emerge or arrive later are relatively unaffected because oil residues do not persist very long. Insecticides may be highly toxic to one group of natural enemies and relatively safe for another. The insect growth regulators Esteem and Applaud are relatively nontoxic to wasp parasitoids needed for California red scale and citricola scale control, but they are highly toxic to vedalia beetles needed for cottony cushion scale control (fig. 17.5A, 17.5B). Many miticides are toxic to predatory mites that are important for citrus thrips and citrus red mite control, but they have relatively little impact on insect predators and parasitoids. Finally, many populations of predatory mites and beetles have developed resistance to organophosphates because of decades of exposure and can tolerate occasional treatments. In contrast, the parasitic wasps seem to be more sensitive, even when they have developed some level of resistance.

A broad-spectrum organophosphate insecticide such as Lorsban (chlorpyrifos) can be made more selective by using a very low rate. Rates of a few ounces of Lorsban used for katydid control are relatively harmless to *Aphytis* wasps. Keep in mind, however, that even if a natural enemy has tolerance or resistance to an insecticide, if you spray that insecticide repeatedly, the parasitoid population eventually succumbs. Thus, if possible, the recommended use of Lorsban is limited to one application per year in the biologically based citrus IPM program.

The general rule of thumb is to avoid insecticide use or to use the most selective insecticide early in the season to maximize early-season growth of natural enemy populations. Later in the season, if pest populations continue to build and threaten to cause economic damage, a less selective insecticide can be used, if needed.

If you are starting an *Aphytis melinus* wasp release program and the orchard has been treated with a pesticide in the previous or current year, determine whether residues exist on the leaves that are harmful to the wasps. A simple test to determine

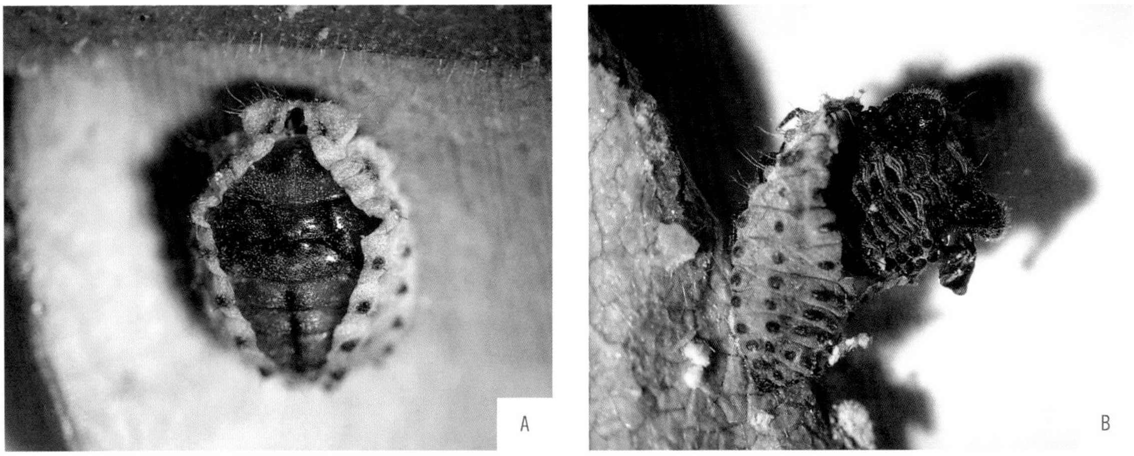

Figure 17.5 A healthy vedalia beetle pupa (A); a vedalia beetle pupa whose development was disrupted and altered by the insect growth regulator pyriproxyfen (B). *Photos:* E. E. Grafton-Cardwell.

Table 17.1. Summary of the level of selectivity of various pesticides used in citrus pest management

Pesticide group	Chemicals and trade names	Soft on most natural enemies	Toxic to predatory mites	Toxic to beetles	Toxic to parasitoids
organophosphates	azinphos-methyl (Guthion) dimethoate (Cygon) naled (Dibrom) malathion, phosmet (Imidan) chlorpyrifos (Lorsban advanced) methidathion (Supracide)	none	populations with resistance can survive organophosphate treatments	populations with resistance can survive organophosphate treatments	populations with resistance can survive very low rates of these insecticides
carbamates	formetanate (Carzol) methomyl (Lannate) carbaryl (Sevin)	none	populations with resistance can survive carbamate treatments	populations with resistance can survive carbamate treatments	populations with resistance can survive very low rates of these insecticides
neonicotinoids	imidacloprid (Provado, Admire Pro, and generics) acetamiprid (Assail) thiamethoxam (Platinum, Actara)	none	neonicotinoids suppress populations whether applied as a foliar or a systemic	neonicotinoids suppress populations whether applied as a foliar or a systemic	neonicotinoids suppress populations whether applied as a foliar or a systemic
pyrethroids	beta-cyfluthrin (Baythroid XLR) fenpropathrin (Danitol) zeta-cypermethrin (Mustang)	none	pyrethroids are highly toxic	pyrethroids are highly toxic	pyrethroids are highly toxic
pyrethrins	pyrethrin (Pyganic, Evergreen)	Pyganic, Evergreen	relatively nontoxic	relatively nontoxic	relatively nontoxic
insect growth regulators	methoxyfenozide (Intrepid) buprofezin (Applaud) pyriproxyfen (Esteem) diflubenzuron (Micromite)	Intrepid	relatively nontoxic	Applaud, Esteem, Micromite	relatively nontoxic
avermectins and spinosyns	abamectin (Agri-Mek and generics) spinosad (Success) spinetoram (Delegate)	Success	Agri-Mek and Delegate are somewhat toxic	relatively nontoxic	Delegate is somewhat toxic
Bacillus thuringiensis	Bt (Dipel, Javelin, MPIII)	Dipel, Javelin, MPIII	relatively nontoxic	relatively nontoxic	relatively nontoxic
botanicals	sabadilla (Veratran D) azadirachtin (Neemix)	Veratran D, Neemix	relatively nontoxic	relatively nontoxic	relatively nontoxic
diamides	cyantraniliprole (Altacor)	Altacor	relatively nontoxic	relatively nontoxic	relatively nontoxic
miticides	propargite (Omite) fenbutatin oxide (Vendex) pyridaben (Nexter) acequinocyl (Kanemite) hexythiazox (Onager) fenpyroximate (FujiMite) bifenazate (Acramite)	Onager, Vendex	FujiMite, Kanemite, Nexter, Omite, Fujimite, Acramite	relatively nontoxic	relatively nontoxic
tetronic and tetramic acid inhibitors	spirodiclofen (Envidor) spirotetramat (Movento)	Movento, Envidor	Movento, Envidor	relatively nontoxic	relatively nontoxic
miscellaneous insecticides	petroleum oil cryolite (Kryocide, Prokil Cryolite) sulfur	Kryocide, oil (≤ 0.5%)	> 1% if outside coverage or when using thorough coverage	relatively nontoxic	sulfur

Note: For current information on registered pesticides, see the UC IPM Citrus Pest Management Guidelines, http://www.ipm.ucdavis.edu/PMG.

whether toxic residues are present involves placing foliage from the orchard into a 1-gallon glass jar (fig. 17.6). Streak a very small amount of honey on the inside of the glass jar, then place living adult *Aphytis* wasps in the jar and cover the top with a cloth or paper towel that is secured in place. In addition, prepare a control jar with untreated foliage so you can make sure the *Aphytis* are healthy. Look at the *Aphytis* after 24 hours to see whether there is a high level of kill. If more than 25% of the *Aphytis* are dead in the jar with the treated foliage, the residues have not worn off enough to initiate *Aphytis* releases. If mortality is high, repeat the test in a few weeks. If survival has increased and 75 to 100% of the *Aphytis* are alive when exposed to the treated foliage, *Aphytis* releases can begin. If the *Aphytis* are dead in the control jar, they were not healthy to begin with, and you must redo the test with fresh *Aphytis*.

Insecticide Strategies That Minimize Impacts

Various strategies can minimize the impact insecticides have on natural enemies. At the start of the season, avoid using the broad-spectrum insecticides (most organophosphates, carbamates, pyrethroids, and neonicotinoids), because natural enemies are coming out of dormancy and building their populations. Broad-spectrum insecticides can stop that population growth.

Broad-spectrum insecticides can also disrupt natural enemies at the end of the summer. *Aphytis* wasps have their greatest impact during September and October. An Assail treatment to disinfest citrus of glassy-winged sharpshooter just prior to harvest in November or December is not as disruptive as an Assail treatment applied in spring or summer. In late fall, most natural enemies are in a more protected state and are less active, and they are thus less exposed to pesticide residues. Winter rains help rinse the pesticide residues off the trees, and natural enemies are usually not strongly affected when they become active early in spring.

If a broad-spectrum insecticide is needed, there are ways to make it more selective, such as applying it using a low rate, treating only the part of the orchard with high pest densities, and limiting broad-spectrum pesticide applications to once per year. Often, natural enemies recover after a month or two when a broad-spectrum insecticide is used. However, they do not survive multiple applications well.

Strategies That Maximize the Effects of Natural Enemies

Dust, manure, gypsum, Surround (kaolin clay), and whitewash reduce the effectiveness of parasitoids (fig. 17.7). That is the major reason why the dusty edges of citrus orchards often have the most California red scale. A layer of dust or chemical prevents parasitoids from finding their hosts and can cause the parasitoid to abandon the plant and search for cleaner plants. Growers should water dusty roads throughout the season and wait to apply whitewash until late fall, when *Aphytis* has finished parasitizing scale.

As previously described, growers should prune to promote an open tree canopy to allow parasitoids access to scale insects and reduce the humidity that promotes a number of pests. Growers should also prune the tree skirts so that the branches do not touch the ground to reduce damage due to brown garden snail and Fuller rose beetle.

Some growers have found that planting a cover crop (annual grasses, vetch, and a bell bean combination) or allowing natural vegetation to grow can improve biological control of pests by increasing humidity and reducing the orchard floor temperature in summer. Organic growers who allow a heavy cover of weeds or plant a cover crop rarely have a citrus thrips problem. This may be because in

Figure 17.6 Place treated and untreated foliage in 1-gallon glass jars. *Photo:* E. E. Grafton-Cardwell.

Figure 17.7 Kaolin clay, dust, and lime can make it difficult for natural enemies to get to their prey. *Photo:* E. E. Grafton-Cardwell.

organic situations, the soil where citrus thrips pupate has higher levels of organisms that prey on the pupae. California red scale is often easier to control using *Aphytis* releases in older trees because the large canopy provides a cooler environment for the wasps; this effect is further improved by addition of a cover crop.

Ants protect honeydew-producing pests such as whiteflies, many scale insects, and aphids from predators and parasitoids. They can also interfere with biological control of pests that do not produce honeydew, such as California red scale. Good ant control is often critical to achieve effective biological control in citrus and reduces the need for insecticide treatments. The ant species that most severely disrupt biological control are the Argentine and native gray ants. A new technology that is under development is reducing these species of ants with liquid sugar mixed with a toxicant in a bait station (fig. 17.8). The bait stations are distributed throughout the orchard. The sugar is the attractant, and the toxicant is added at a concentration that is not detected by the ant (too high of a rate of the toxicant results in avoidance). The ant takes the liquid into its crop and feeds it to other workers and immature ants in the nest, thereby providing greater control of ants than surface sprays with organophosphate insecticides.

San Joaquin Valley Citrus IPM: Pests That Have Effective Biological Control

A number of citrus pests have effective natural enemies that can regulate pest populations below economically injurious levels. These pests tend to become problematic when their natural enemies are disrupted by broad-spectrum pesticides, by dust, or when environmental conditions are not favorable. Listed below are several examples of pests that can be managed with natural enemies and the occasional use of a selective insecticide. Photos of these insects and mites can be found in *Integrated Pest Management for Citrus* (Dreistadt 2012), and treatment recommendations are provided in the UC IPM Citrus Pest Management Guidelines, http://www.ipm.ucdavis.edu/PMG.

Figure 17.8 Liquid baits containing sugar mixed with a toxicant are under development for control of ants. *Photo:* E. E. Grafton-Cardwell.

California red scale *(Aonidiella aurantii).* California red scale is attacked by a number of natural enemies, including lacewings, predatory beetles, *Comperiella bifasciata,* and *Aphytis melinus* wasps. In the San Joaquin Valley, citrus growers can purchase *Aphytis* wasps from insectaries and release them at rates of 5,000 to 6,000 per acre every 2 weeks from March through October, for a total of 50,000 to 100,000 per acre. In the San Joaquin Valley, these releases help supplement natural populations of *Aphytis,* especially in the early season. High-pressure scale washers in packinghouses can remove live and dead scales that remain at the end of the season. Occasional treatments (every few years) using selective insecticides can reduce scale populations without severely affecting natural enemies.

Brown soft scale *(Coccus hesperidum).* Brown soft scale has multiple overlapping generations. Unless protected by ants, these soft scales are easily controlled by parasitic wasps (various *Metaphycus* spp.). The only control needed for these pests is ant control. The ants farm the scale honeydew and protect the scale from parasitoids and predators.

Citrus red mite *(Panonychus citri).* Citrus red mite responds to tree stress (mite levels increase with water-stressed and dusty trees) and to weather (mild winters or cool weather in early summer increase the survival of mites). Generally, the predatory mite *Euseius tularensis* keeps citrus red mite numbers below an economic threshold of eight adult female mites per leaf during April through June. In June or July, when the temperature normally rises above 90°F, a virus spreads through citrus red mite populations, causing them to collapse. Thus, growers rarely treat for citrus red mite unless densities are above the threshold during February and March when the predatory mites and virus are not active. Most miticides are toxic to both pest mites and beneficial mites. Thus, miticide treatments reduce *E. tularensis* densities needed during April and May for citrus thrips control. Oil treatments of 1 to 2% in the early spring or the addition of 0.5% oil to a citrus thrips treatment at petal fall are common methods of reducing citrus red mite levels without severely affecting predatory mites.

Cottony cushion scale *(Icerya purchasi)*. Vedalia beetles have been controlling cottony cushion scale for more than 100 years in California. They are extremely effective because they find small scale colonies quickly, reproduce rapidly, and are specific to cottony cushion scale. Their control of cottony cushion scale is superior to the standard organophosphate and carbamate insecticide treatments because vedalia beetles more completely control the cottony cushion scale populations and because they do not disrupt natural enemies needed for control of other pests. Vedalia beetles must arrive in the orchards naturally, and sometimes they do not arrive early enough for full control of cottony cushion scale. In some areas of California, a parasitic tachinid fly is also important in control of cottony cushion scale.

Citrus cutworm *(Egira xylomyges)*. A number of parasitic wasps attack the eggs and larvae of the citrus cutworm. If the parasitoids are not fully effective, cutworms are fairly easily controlled with the very selective *Bacillus thuringiensis* (Bt) pesticide treatment. The Bt insecticides are specific to lepidopteran caterpillar larvae and do not harm parasitoid and predator species.

Brown garden snail *(Helix aspera)*. The brown garden snail can feed on ripe fruit and flush growth, but the pest typically survives on the trunks of citrus trees and near sprinkler emitters. In situations where brown garden snails persist, growers are encouraged to use predatory decollate snails for control in counties where their use is legal. These predators take 5 to 10 years from the time they are introduced to gain control of the brown garden snail population.

Citrus leafminer *(Phyllocnistis citrella)*. The citrus leafminer attacks new flush growth of citrus. This species arrived in California from Mexico in the year 2000 and gradually moved northward, and it now infests most of California. It is expected that naturally occurring parasitoids will build up and maintain leafminer populations below economic levels in mature citrus. Therefore, insecticides are unnecessary on mature trees, as they can tolerate flush damage. One possible exception, which is currently being researched, is lemons in coastal California that have multiple crops and may suffer yield loss due to season-long citrus leafminer damage. Pesticide applications are warranted for nursery trees and newly planted trees if infestations are observed.

Citrus peelminer *(Marmara gulosa)*. A new strain of this pest arrived in the San Joaquin Valley in 1998 and has since caused more damage than was previously experienced in other citrus-growing areas. This species mines the rind of green fruit, primarily grapefruit, pummelos, and susceptible navels (Fukumoto, T.I., Atwood). Rind damage in the San Joaquin Valley ranges from 3 to 50% of the fruit, depending on the citrus variety and peelminer population development in a particular year. Insecticides are not very effective on this species because it attacks interior low-hanging fruit that are difficult to reach with insecticide sprays. Use of insecticides can be counterproductive due to elimination of natural enemy populations. Citrus peelminer populations have declined since the arrival of citrus leafminer, which shares many of the same natural enemy parasites.

San Joaquin Valley: Pests That Have Inadequate Biological Control

A number of pests do not have adequate biological control and often require pesticide intervention.

Citrus thrips *(Scritothrips citri)*. The major natural enemies of citrus thrips in California are thought to be predatory mites in the genus *Euseius*, although additional research is needed to determine what impact other predators have on this pest. *Euseius* will eat pollen, various mite species, dead and live small insects, and even a bit of leaf sap. Because they are not specific to citrus thrips, they do not always maintain citrus thrips populations below an economic threshold. Growers have a tendency to use insecticides to control citrus thrips because of the lack of complete control by predators and a low tolerance for citrus thrips damage of fruit.

Citrus thrips damage the fruit in the early spring just after petal fall. In biologically based IPM programs, PCAs monitor citrus thrips nymphs on fruit twice a week, and if thrips densities are below the economic threshold, they do not treat with insecticides. Biologically based IPM growers treat for citrus thrips less often than do growers following other types of management programs. This has indirect benefits above the cost savings by avoiding pesticide resistance (the more often you spray, the faster resistance develops) and by avoiding disruption of natural enemies needed for control of other pests. Biologically based IPM growers use oil in combination with moderate to highly selective pesticides containing spinosyns, avermectins, or sabadilla plus sugar for citrus thrips control. Additional broad-spectrum insecticides available include organophosphates, carbamates, and pyrethroids. However, a number of populations of citrus thrips have developed resistance to these broad-spectrum insecticides, which are toxic to natural enemies.

Forktailed bush katydid *(Scuddaria furcata).* Katydids damage fruit quickly by taking large bites from the surface of the fruit during the first few weeks after petal fall. Most of their natural enemies attack the katydid's egg stage. If the egg is not parasitized and the katydid emerges, the young nymph can rapidly scar young fruit. Insecticides that contain the stomach poison cryolite or the insect growth regulator diflubenzuron (Micromite) can be used to control katydids, but these insecticides are slow acting. Growers must apply these pesticides before petal fall (before young fruit are present) because these insecticides will not act quickly enough to prevent fruit damage. Katydids are extremely sensitive to organophosphate and pyrethroid insecticides; when organophosphates were commonly used for citrus thrips and California red scale, katydids were rarely seen. Since growers have shifted to softer insecticides, katydids have become a chronic problem in citrus. Growers often tank-mix a low rate of organophosphate or pyrethroid with the citrus thrips treatment to simultaneously manage citrus thrips and katydid populations.

Citricola scale *(Coccus pseudomagnoliarum).* Citricola scale has only one generation per year, and it grows and molts very slowly. During the summer, the population consists of adult females and their hatching eggs. During the early fall, the population consists of young nymphs that are too small for the successful development of parasitic wasps (*Metaphycus* and *Coccophagus* spp.). In the San Joaquin Valley, nearby alternate hosts such as black scale and brown soft scale are often lacking, so the parasitoids have a period of time (July to September) when there are no suitable hosts. Biological control of citricola scale does not begin to develop until late fall and spring, when the scales molt to larger instars. In the San Joaquin Valley, the fall months are cool and the parasitoids develop slowly. Meanwhile, the citricola population that escapes parasitism produces tremendous amounts of honeydew in the springtime on which sooty mold grows. Growers must treat citricola scale in the previous fall to avoid the sooty mold and yield problems that occur the following spring.

When the commonly applied pesticides for California red scale were organophosphate and carbamate insecticides, citricola scale was easily controlled at the same time. Now that the common insecticides for California red scale are insect growth regulators, citricola scale has become a frequent and serious problem because insect growth regulators do not control this pest very well. Heavy densities of citricola scale can dramatically reduce the yield of citrus trees. Growers generally use broad-spectrum insecticides to control this pest.

Biologically Based IPM in the San Joaquin Valley

This section provides general information on how to manage pests using biologically based IPM tactics (modified from Haney et al. 1992). For more detailed information on sampling methods, consult the UC IPM Citrus Pest Management Guidelines, www.ipm.ucdavis.edu/PMG/. Several of the most serious citrus pests are most active and damaging in the spring just after petal fall, from the end of April to early May (fig. 17.9). Citrus thrips, katydids, and cutworms scar fruit as it is forming. Scale insects may reduce yield (citricola scale or cottony cushion scale) or tree vigor and can cosmetically damage the fruit (especially California red scale). Viewing the system as a whole helps to keep in mind the effects of insecticide treatments on natural enemies needed for management of all of the pests.

A typical biologically based IPM program begins with releases of *Aphytis* wasps for California red scale control in early March and continues those releases every 2 weeks (50,000 to 100,000 per ac over the season) through October (fig. 17.10). In late February, red scale pheromone traps are hung at a rate of 1 per 2.5 acres to determine when male scales are first flying. This biofix of first male flight plus the accumulation of degree-days through the year provide information about when the males fly and crawlers emerge during each of the subsequent generations. California red scale is monitored by the use of pheromone traps and by inspection of fruit from July through October to determine whether biological control is functioning sufficiently. A difficulty in assessing the success of biological control of California red scale lies in the fact that treatments to reduce scale numbers are best applied in July, yet the success of *Aphytis* control of scale cannot be observed until August or September. If *Aphytis* parasitism increases during monthly inspections of scale-infested fruit from August to October, the majority of scale will be killed and high-pressure washing can eliminate the remainder.

During March, scouting for cottony cushion scale takes place. If populations are found and vedalia beetles are not present, beetles are collected from other orchards and released. Only small numbers (25 beetles per 10 ac) are needed, and they can generally control a cottony cushion scale population in 6 weeks

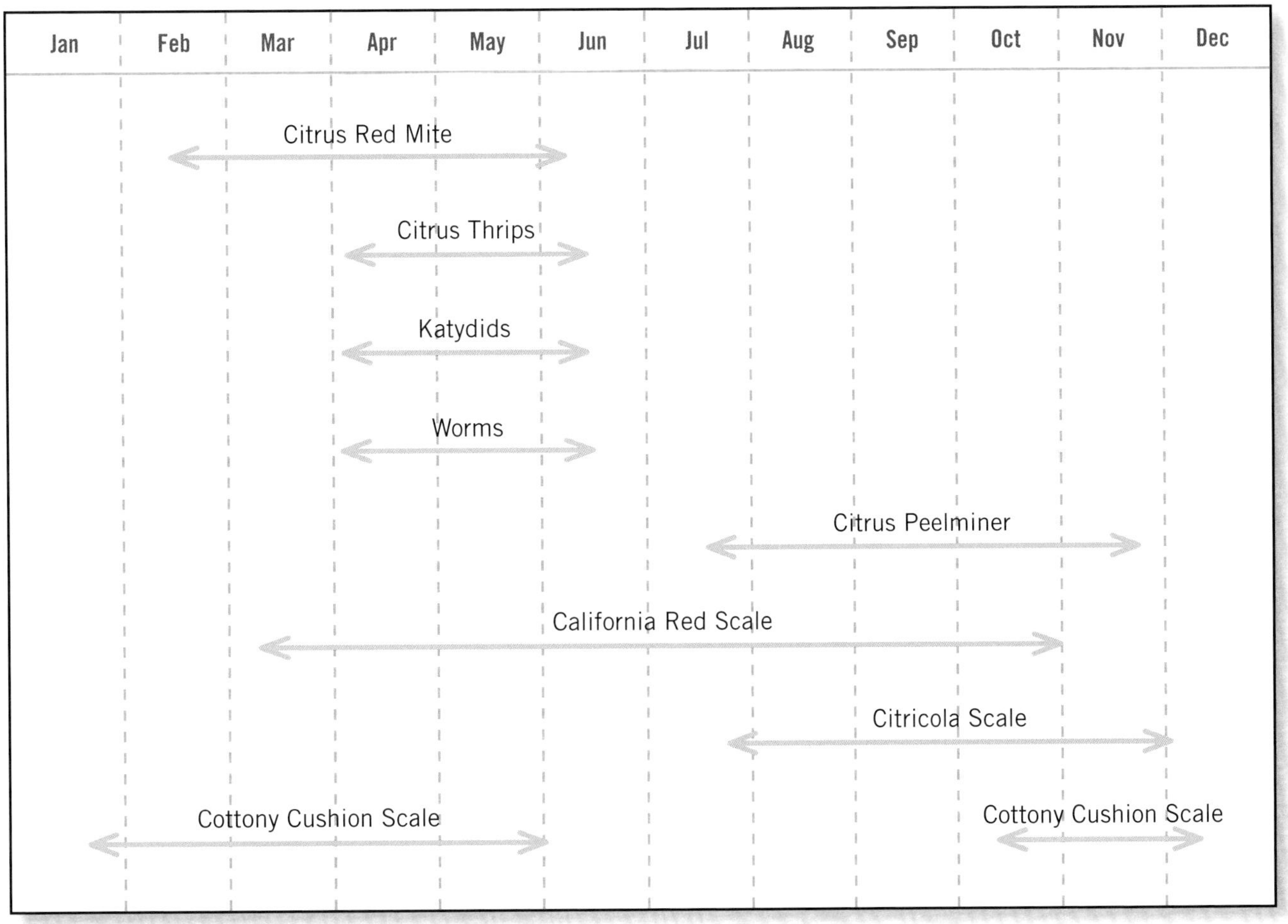

Figure 17.9. Monitoring periods for selected San Joaquin Valley citrus pests.

Figure 17.10 *Aphytis* wasps are released every 2 weeks from early March through October. *Photo:* E. E. Grafton-Cardwell.

(fig. 17.11). It is critical to establish the vedalia beetle early enough in the season so that the cottony cushion scale is eliminated before June, when the insect growth regulators may be applied for California red scale control. Vedalia beetle is very sensitive to these insecticides and will not prosper once these insecticides are applied in nearby orchards. Vedalia beetle is also very sensitive to high temperature and will not reproduce during the hot months of summer, which again emphasizes the importance of establishing them in infested orchards in early spring.

From February through May, sample leaves for citrus red mite and predatory mites. If more than eight adult mites per leaf are found, then treatment may be needed (fig. 17.12). Generally, high levels of citrus red mites result in high densities of predatory mites, which aid in controlling citrus thrips later in the season. Thus, moderate levels of citrus red mite are considered advantageous for biological control. If populations are extremely high, an acaricide may be necessary to gain control of the population. Growers frequently use oil treatments, alone or in combination with other insecticides, to suppress citrus red mite populations.

Figure 17.11 Vedalia beetle adults and other stages can be moved from one orchard to another. *Photo:* J. K. Clark.

Figure 17.12 Citrus red mites are sampled on leaves. *Photo:* E. E. Grafton-Cardwell.

Figure 17.13 Young citrus fruit is examined for citrus thrips nymphs. *Photo:* E. E. Grafton-Cardwell.

At petal fall, the twice weekly sampling for citrus thrips begins (fig. 17.13). Biologically based IPM growers ignore adult thrips (which cause little spring fruit damage) and use selective insecticides based on the levels of immature thrips attacking the fruit. For navel oranges, the treatment threshold is 5% of the fruit infested with thrips nymphs when predatory mites are low (< 0.5 per leaf) and 10% of the fruit when predatory mites are high (≥ 0.5 per leaf). For Valencia oranges, the threshold is higher because the fruit is more difficult for thrips to damage: 10% of the fruit when predatory mites are low (< 0.5 per leaf) and 20% of the fruit when predatory mites are high (≥ 0.5 per leaf). In years with few citrus thrips (spring rains or cool weather during bloom), biologically based IPM growers can often avoid treating for citrus thrips because densities remain below these thresholds. In years with many citrus thrips (warm, dry springs) or when the citrus thrips hatch is prolonged, biologically based IPM growers may have to treat several times, a level which is similar to pesticide based growers.

Katydids can also be a problem at petal fall. They can be very damaging because they often move around the tree and take one bite out of each piece of fruit. It is difficult to define an economic threshold for this pest. If one or more katydids are found during a 3-minute time search on 20 trees and leaf or fruit damage is occurring, action needs to be taken quickly (fig. 17.14). Biologically based IPM growers use slow-acting insecticides such as cryolite or Micromite before petal fall to reduce katydids. If action is not taken until after petal fall, when the risk of fruit damage is high, growers add a low rate of an organophosphate or pyrethroid to their citrus thrips spray to reduce this pest while preserving natural enemies.

During August and September, sampling for citricola scale is accomplished by taking leaf samples from the northeast quadrant of 25 trees in each of four rows and determining whether scales are alive or dead (fig. 17.15). If the number of nymphs exceeds the economic threshold of 0.5 per leaf, a treatment is needed. Treatment is needed in the fall because the population will increase the following spring and lower yield and cause sooty mold to grow on the honeydew it produces, reducing the quality of the fruit. Research indicates that about 40% of citricola scale populations have resistance to the organophosphate Lorsban, the most commonly used insecticide for citricola scale control. The alternative insecticides suppress scale for only 1 year. The lack of effective, long-lasting, selective insecticides and poor biological control make citricola scale the most difficult pest to manage with IPM tactics.

Citrus peelminer is a common problem for a limited number of citrus varieties in the San Joaquin Valley (pummelos, grapefruit, and the navels Fukumoto, Atwood, and T.I.). Biologically based IPM growers treat two times for this pest, but they can achieve at best a 50% reduction in mined fruit due to this pest.

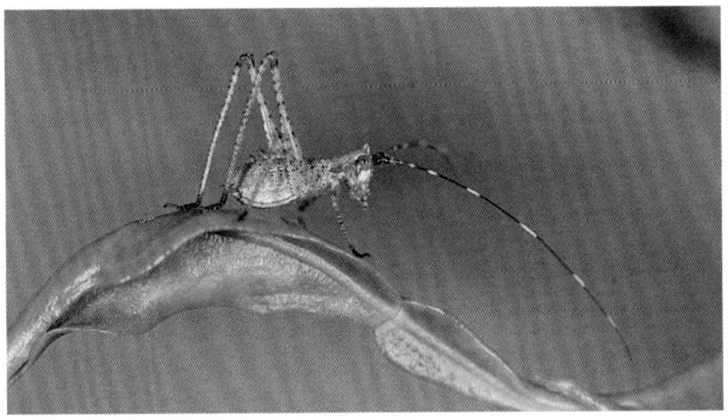

Figure 17.14 Katydid nymphs can damage fruit quickly. *Photo:* E. E. Grafton-Cardwell.

IPM in the Southern Interior Region

Commercial California citrus production began in southern California when William Wolfskill planted an orchard in the village of Los Angeles in 1841. By 1871, with completion of the transcontinental railroads, citrus production became the second California gold rush. Shipments of the Washington navel orange to the East Coast during winter resulted in substantial income to many southern California growers.

The ease of growing citrus changed, however, with the introduction of a number of invasive pests, including California red scale, yellow scale, black scale, citricola scale, citrophilus mealybug, and cottony cushion scale. The cottony cushion scale seriously threatened citrus production in the 1880s, killing a large number of citrus trees. The industry was saved by the importation of the vedalia beetle and *Cryptochaetum* fly in the late 1880s. The successful control of cottony cushion scale by the beetle initiated the use of biological control as a scientific discipline. Chemical control was absent during this early period because the insecticides that were available were marginally effective or severely affected the trees. Biological control remained the principal control tactic, and it suppressed the citrophilus mealybug, citricola scale, red scale, and yellow scale and substantially ameliorated black scale and several mealybug pests. It is this legacy of successful natural enemies that provides the foundation for integrated pest management in southern California citrus.

Initial chemical control efforts were rudimentary and were often a last-ditch effort out of desperation. They involved oils that were often phytotoxic and hydrogen cyanide fumigation. Interestingly, two citrus pests, California red scale and citricola scale, developed resistance to hydrogen cyanide fumigation, an omen of future outcomes of the overuse of pesticides. However, the introduction of DDT in the later half of the 1940s finally provided

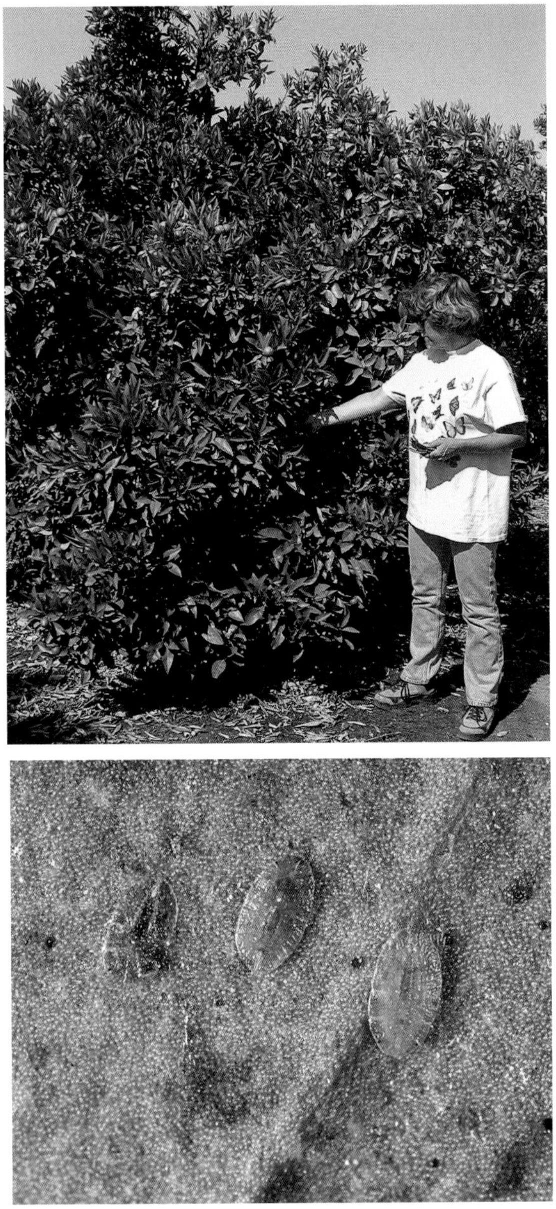

Figure 17.15 Citricola scale are sampled by removing shaded leaves from the northeast corner of the tree (A). *Photo:* E. E. Grafton-Cardwell. Dead, dying, and healthy citricola scale nymphs (B). *Photo:* J. K. Clark.

outstanding suppression of several severe citrus pests, but at a price. It also induced severe outbreaks of several other citrus pests and indigenous plant-feeding species that had previously been sporadic pests at best. After about 60 years of outstanding suppression, cottony cushion scale erupted as a pest again because of DDT's effects on the vedalia beetle.

The development of citrus protective districts evolved as another legacy left by early citrus pest control efforts. In the early 1910s, the first protective district was established to battle the citrophilus mealybug and later as a means of cost sharing among involved growers for the expense of using hydrogen cyanide fumigation for scale control. One such district, the Fillmore Protective District, developed unique methods of black scale parasitoid mass production. The parasitoid *Metaphycus helvolus* was not fully effective in suppressing black scale without augmenting its numbers in citrus (fig. 17.16). Later, the Fillmore Protective District adopted, adapted, and expanded the *Aphytis melinus* mass-production program, developed by Paul DeBach after *A. melinus* was imported and introduced into California in the mid-1950s for the control of California red scale. The pest management practices the protective districts pioneered, along with the use of pest management advisors, laid the groundwork in developing an integrated pest management program for southern California citrus. This stimulated the emergence of several independent pest control advisers and commercial insectaries for *Aphytis* production that fostered biologically based pest management in many southern California locations. Studies conducted in the 1980s showed that the pest management costs borne by the grower members of the Fillmore Protective District were one-fifth of those borne by nonmembers in that citrus-producing region.

The southern interior citrus-growing region includes western Riverside and San Bernardino Counties and inland portions of San Diego, Orange, and Los Angeles Counties. The interior district tends to be warmer and drier in the summer and colder in the winter than the coast. Citrus production in southern interior California (a mixture of citrus varieties) has fewer pest problems than the San Joaquin Valley because its milder climate promotes natural enemies. Currently, 13 pest species, many of them homopterous pests such as whiteflies and scales, are adequately controlled by introduced natural enemies. Another 8 pest species are controlled by native natural enemies. However, the current use of broad-spectrum insecticides for invasive pests disrupts the efficacy of some natural enemies. Urban development in southern California has displaced much of the citrus production. What citrus remains faces significant problems along the urban-agriculture interface. Typical production practices, especially pesticide use, have been under fire from the public and regulatory agencies. Air and water quality issues have emerged, many of which negatively impact citrus production. Add to this the invasive pest species that frequently initially infest southern California citrus plantings (e.g., glassy-winged sharpshooter, citrus leafminer, Diaprepes root weevil, Asian citrus psyllid, and a number of fruit fly species), and the result is that today's growers face challenges that strain IPM practices. The detection of pests such as fruit flies often leads to state or market-imposed quarantines that severely limit marketing and sales. Thus, growers face significant challenges to their viability.

Figure 17.16 Metaphycus wasps reared by the former Fillmore Insectary and released to control black scale. *Photo:* J. K. Clark.

Major Pests in Southern Interior Citrus

Citrus thrips *(Scirtothrips citri).* Citrus thrips are considered a key pest in southern California, especially in the hot, dry regions. Monitoring and treatment thresholds are the same as those described for the San Joaquin Valley. The *Euseius* predator common in this region is *E. stipulatus.* Feeding on the fruit by citrus flat mite can make citrus thrips scarring worse in southern California (fig. 17.17). If a treatment is warranted, the use of more selective insecticides is recommended. Citrus thrips develop resistance to insecticides quickly; thus, judicious use of newly registered materials and resistance management techniques are strongly advised because the development of new thrips insecticides by agrichemical manufacturers has not been a high priority.

California red scale *(Aonidiella aurantii)***.** In most situations, California red scale is controlled by *Aphytis* and other natural enemies such as the ladybird beetle *(Rhyzobius lophanthae)* and the parasitoid *Comperiella bifasciata*. In southern California, if *Aphytis* releases are needed, growers release 10,000 to 40,000 per acre over the year.

Black scale *(Saissetia oleae)*. Historically, black scale was one of the worst citrus pests in southern California, but it is now under substantial control by a number of natural enemies, including *Metaphychus helvolus*, which was introduced from South Africa in 1937. Nine primary and six secondary parasitoid species are known to attack black scale in California; hence, it is now a less severe problem, with outbreaks occurring in a particular orchard about every 10 years. Why black scale periodically reaches economic levels is not well understood, but some of the factors influencing its status include the presence of ants, dust, and pesticides in addition to cyclical weather.

Citrus mealybug *(Planococcus citri)*. Mealybugs appear to be a greater problem on grapefruit than on other citrus varieties (fig. 17.18). Ants interfere with natural enemy activity and should be controlled.

Argentine ant *(Linepithema humilis)*. The Argentine ant is one of the most important pests in the southern California region because it disrupts biological control of a number of pests, including those that produce honeydew (mealybugs, whiteflies, and soft scales) and those that do not produce honeydew (armored scales). Ants prevent parasitoids and predators from gaining access to the pests, leading to pest population buildup (fig. 17.19). Natural enemy species imported for biological control that did not evolve with ants are especially affected. Currently registered solid ant baits are effective only against fire ants, as these species are attracted to and can feed on the soy oil–corncob grit formulation. Argentine ants feed on liquid, sugar-based, baited toxicants.

Glassy-winged sharpshooter *(Homalodisca vitripennis)*. Glassy-winged sharpshooter can spread strains of the bacterium *Xylella fastidiosa*, which cause Pierce's disease of grapes, almond leaf scorch, oleander leaf scorch, plum leaf scorch, and phony peach disease. A form of *X. fastidiosa*, citrus variegated chlorosis, affects citrus, but it is not yet found in California. Glassy-winged sharpshooter populations appear to be declining in southern California, probably because of improved biological control. Insecticide treatments for glassy-winged sharpshooter are required in many areas of southern California to reduce overall population densities near ornamental nurseries and other affected crops. Neonicotinoid insecticide treatments for glassy-winged sharpshooter may disrupt natural enemies of

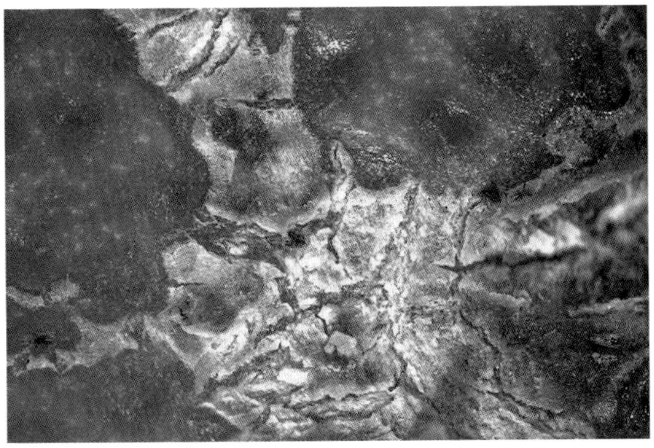

Figure 17.17 Tiny reddish flat mites can feed in and expand scars of fruit. *Photo:* E. E. Grafton-Cardwell.

Figure 17.18 Citrus mealybug infesting the stem end of citrus fruit. *Photo:* E. E. Grafton-Cardwell.

Figure 17.19 Argentine ants protect honeydew-producing pests such as soft scales, mealybugs, and whiteflies. *Photo:* J. K. Clark.

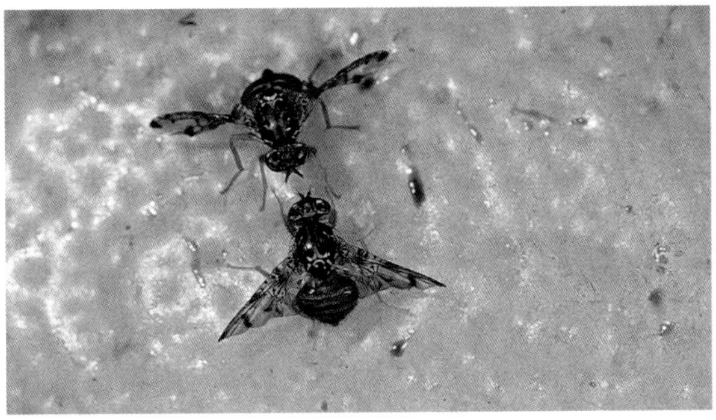

Figure 17.20 Adult Mediterranean fruit flies are periodically detected and eradicated in California. *Photo:* J. K. Clark.

other pest species, causing secondary pest outbreaks. Research is currently under way to better elucidate these interactions.

Exotic fruit flies, various species. Non-native fruit fly species such as Mexican fruit fly and Mediterranean fruit fly are periodically detected in California, most frequently in southern California (fig. 17.20). Verified trap catches of fruit fly species initiate a sequence of events involving federal, state, and local agencies to isolate, treat, and eradicate potential infestations. Frequently, these efforts are accompanied by quarantines on shipping fruit from the infested areas until the flies are eradicated. This creates a heavy financial burden for the growers involved. All current eradication methods, such as the sterile insect technique and baited toxicants (spinosad, GF-120), are compatible with the citrus IPM program.

IPM in the Coastal Region

California's coastal region, from Santa Barbara County south to the San Diego–Mexico border, has a mild climate influenced by marine air. Lemons are the major variety of citrus grown, although significant orange and grapefruit acreage is also found there. Coastal citrus production has had fewer pest problems than production in the San Joaquin Valley because the mild climate allows better year-round biological control. This situation is changing somewhat due to the increased use of broad-spectrum insecticides for glassy-winged sharpshooter and Asian citrus psyllid. Additionally, management practices in adjacent crops, especially those that use insect growth regulators, have had some negative impacts on IPM practices in citrus. Compounding these problems is the ever-increasing interface between citrus production and urban development and the cost of water. Typical production practices in these zones are increasingly under fire from the public and regulatory agencies, and growers are having to scale back or discontinue certain cultural and pest management practices.

Major Pests of Coastal Citrus

Citrus bud mite *(Eriophyes sheldoni).* Bud mite does not have adequate biological control, and there is renewed debate over whether pesticide treatments for this pest are economically justified for lemons. The distortions of leaves and fruit that the mite cause are easy to detect, but they represent damage that occurred as early as the previous season (fig. 17.21). Thus, applications of pesticides are difficult to time and may not be thorough enough to reach these tiny concealed mites. Research has clearly shown that lemon trees differentially abort blooms and fruit damaged by bud mite, reducing the economic impact of this pest. The debate focuses on whether the cost of treatment is recovered in higher yields or less fruit damage. One of the consequences of one possible treatment, high rates of narrow-range oil, is a phytotoxic effect that reduces yield.

Silver mite on lemon *(Phyllocoptruta oleivora). P. oleivora* is also known as citrus rust mite on other citrus varieties (fig. 17.22). There are no effective biologically based options for silver mite. Silver mite is a sporadic pest that only occasionally requires treatment.

Citrus broad mite *(Polyphagotarsonemus latus).* Broad mites are occasional pests of coastal lemons from late July through early October. Controlling ants helps reduce mite numbers.

Citrus thrips *(Scirtothrips citri).* Control of citrus thrips is occasionally economically justified in coastal regions. More consistent and accurate monitoring may reveal that thrips infrequently reach economic levels in certain varieties such as oranges and grapefruit. The July to August lemon fruit set may occasionally require treatment. If a treatment is warranted, using more-selective insecticides is recommended.

Glassy-winged sharpshooter *(Homalodisca vitripennis).* The coastal region maintains some of the most highly valued vineyards in the world. Citrus plantings contribute to the growth of glassy-winged sharpshooter populations, but the impact of glassy-winged sharpshooter populations on citrus yield and fruit quality has not yet been determined. Treatments for glassy-winged sharpshooter in citrus are aimed at reducing overall populations to reduce the vectoring of Pierce's disease within and between weeds and vineyards. These treatments tend to be broad-spectrum neonicotinoid pesticides that often have unintended consequences in citrus, such as disruption of natural enemies for other pest species and secondary pest outbreaks.

Black scale *(Saissetia oleae).* Similar to control in the southern interior areas, black scale has substantial biological control in the coastal area and is only a sporadic problem in some orchards. Why black scale reaches economic levels is not well understood, but some of the factors include presence of ants, dust levels, pesticide use, and weather cycles. New insect growth regulators (IGRs) may provide a selective tool for use in years when populations do not respond to biological control.

California red scale *(Aonidiella aurantii).* This species is generally under good biological control in coastal regions because the mild climate allows for continuous scale population development. This, in turn, provides enough preferred stages for oviposition by *Aphytis melinus* for its continuous growth and development throughout the year. Biological control of red scale is not as effective on young trees, on lemons, and where *A. melinus* is disrupted by ants, dust, or chemicals. In some areas, augmentative releases of *A. melinus* at 10,000 to 40,000 per acre per year have been shown to provide substantial benefits in aiding natural populations in controlling more aggressive California red scale populations (fig. 17.23). On the coast, other parasitoids and predators, especially coccinellid beetles and *Encarsia perniciosi,* play an important role in biological control of California red scale.

Argentine ant *(Linepithema humilis).* Ant control is essential for successful management of homopterous pests. Ants prevent parasitoids and predators from attacking honeydew-producing pests, such as whiteflies, and non-honeydew-producing pests, such as California red scale. Natural enemy species imported for biological control that did not evolve alongside ants are especially affected. The result is population spikes of mealybugs, scales, and whiteflies that normally are under good control. Currently registered ant baits are effective only against fire ants, as these species prefer the soy oil–corncob grit formulation. Argentine ants feed on liquid, sugar-based, baited toxicants, and these are still under development for citrus.

IPM in the Desert Region

Desert citriculture is located primarily in the Coachella, Palo Verde, and Imperial Valleys, areas characterized by low humidity and extremely high temperatures in the summer and fall, with large daytime-to-nighttime temperature fluctuations. The types of citrus varieties grown are limited by the climate. This region produces the majority of the state's grapefruit, but other varieties such as mandarins are also grown.

Deserts generally have low species diversity; as a result there are typically fewer pest species, as only a few are able to tolerate the extreme conditions. However, those that can survive often thrive due to citrus being an evergreen, irrigated, and healthy host plant throughout the year. The pest complex in the desert is unique and frequently changes, with some pest species predominating for several years and then subsiding in importance. Some of our recent invasive species were initially discovered in this region. Thus, these growers are often the first to deal with new pests and the disruptions that these pests bring to established IPM programs.

Major Pests of Desert Citrus

Citrus thrips *(Scirtothrips citri).* Citrus thrips are considered a key pest in the desert region. Monitoring for citrus thrips nymphs on susceptible fruit twice a week will determine population densities relative to economic thresholds (fig. 17.24). Thresholds vary from 5 to 20% infested fruit, depending on the citrus variety and the presence of natural enemies. If treatment is warranted, the use of more-selective insecticides is recommended. Citrus thrips are known to develop resistance to insecticides quickly, so judicious use of newly registered materials and resistance management techniques, such as rotation of pesticides, are advised.

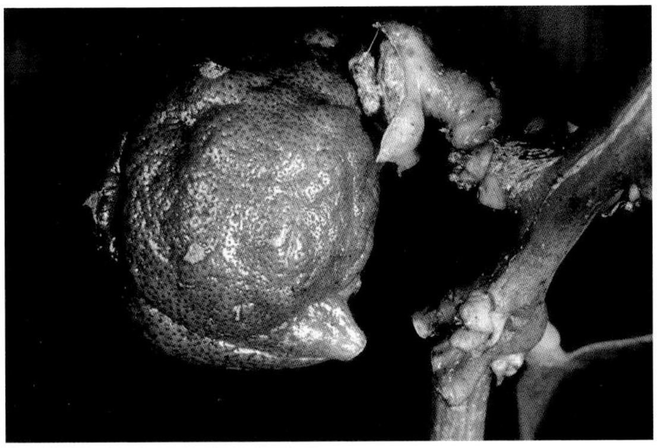

Figure 17.21 Bud mites damage lemons as they are developing, causing fruit to be misshapen. *Photo:* J. K. Clark.

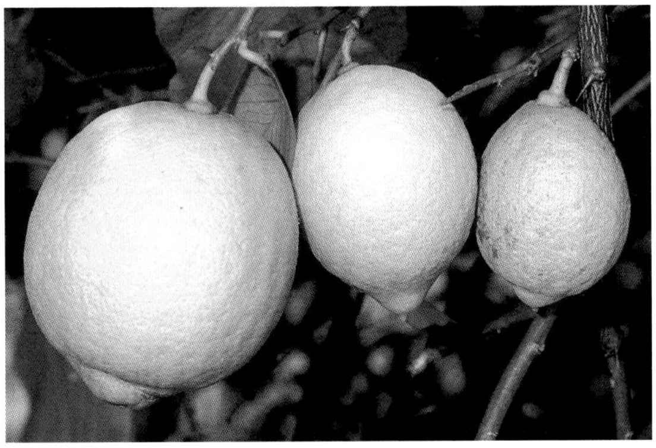

Figure 17.22 Silver mite damage of lemons. *Photo:* J. K. Clark.

Figure 17.23 Releases of *Aphytis* wasps assist control of California red scale. *Photo:* E. E. Grafton-Cardwell.

Mites. Citrus red mite *(Panonychus citri)*, citrus flat mite (*Brevipalpus* spp.), Yuma spider mite *(Eotetranychus yumensis)*, and Texas citrus mite *(Eutetranychus banksi)* feed on citrus leaves and occasionally infest fruit. Extensive leaf feeding causes defoliation, resulting in exposed fruit and subsequent sunburn of the rind. Citrus flat mite can be an economically important pest on mandarins (tangerines). The flat mite is not very common, but it is heat tolerant and populations can persist through the summer. The Yuma spider mite is a pest on young grapefruit and lemon, especially where dust is a problem. Most mite species have natural predators that maintain populations below economic levels. Upsets occur with the use of certain broad-spectrum pesticides.

Citrus leafminer *(Phyllocnistis citrella)*. This species was first discovered in 2000 near the California-Mexico border (Imperial Valley), and within 2 years it had invaded the Coachella Valley. Populations have been low and difficult to detect in the Imperial Valley, but they are persisting on flush growth in the Coachella Valley, especially in orchards with aggressive irrigation practices. Because of the presence of a related native species, citrus peelminer, a number of effective natural enemies are present that are adopting the leafminer as a host. Pesticides are unnecessary on mature trees. On nursery trees or newly planted trees, pesticide applications, especially Admire, may be warranted if mining damages leaves (fig. 17.25). It is expected that naturally occurring biological control will help maintain leafminer populations below economic levels.

Citrus peelminer *(Marmara gulosa)*. This species mines the rind of green fruit, primarily grapefruit and grapefruit hybrid varieties. Because the damage is direct to the fruit, even small populations can be economically significant. The population in the desert appears to be a strain of this species that is native to California. It has an extensive group of eulophid parasitoids that attack it and in most cases maintain populations below economic levels, although high peelminer populations appear sporadically in some orchards. Pesticides are not very effective on this species, reducing populations only by 50%. Broad-spectrum insecticides can be counterproductive due to elimination of natural enemy populations.

California red scale *(Aonidiella aurantii)*. This species is under eradication in some areas of the Coachella Valley. California red scale does not fare well in the extreme heat of this region.

Brown garden snail *(Helix aspera)*. The brown garden snail can feed on ripe fruit and flush growth, but the pest typically survives on the trunks of citrus trees and near sprinkler emitters. Although populations can build to substantial numbers where irrigation practices are aggressive, control is not generally necessary, as the snails rarely move into the canopy and feed on fruit. In situations where brown garden snails persist, growers are encouraged to prune tree skirts and establish decollate snails for eventual control (fig. 17.26).

Woolly whitefly *(Aleurothrixus floccosus)*. This pest is a fairly recent introduction to the desert growing region. Populations can build to extremely high levels, especially on grapefruit leaves (fig. 17.27). Woolly whitefly produces copious amounts of honeydew, creating problems with sooty mold and cultural practices, especially for harvesting crews. Natural enemies were introduced into California in the 1960s, moved into Coachella in the 1980s and 1990s, but have not been consistently effective. Currently, treatments for glassy-winged sharpshooter appear to have secondarily provided control of woolly whitefly. Whitefly species are especially amenable to biological control, and there should be additional efforts in finding desert-adapted species of parasitoids to control this pest once glassy-winged sharpshooter treatments are reduced.

Figure 17.24 Citrus thrips, *Scirtothrips citri*. *Photo:* J. K. Clark.

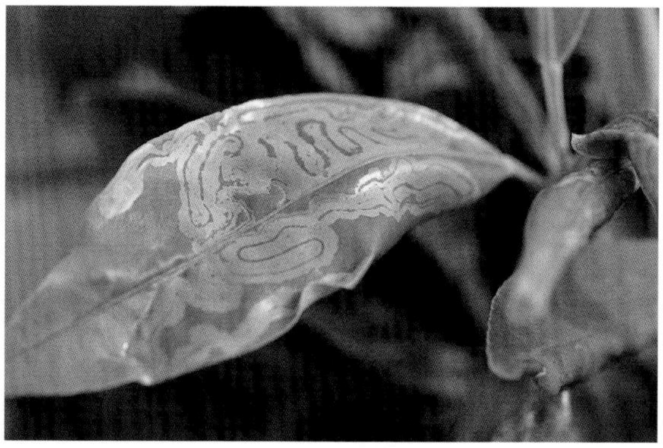

Figure 17.25 Citrus leafminer can slow the growth of young trees. *Photo:* E. E. Grafton-Cardwell.

Figure 17.26 Decollate snail consuming a brown garden snail. *Photo:* J. K. Clark.

Figure 17.27 Woolly whitefly infesting the underside of a leaf. *Photo:* J. K. Clark.

Invasive Pests That Disrupt Citrus IPM

Currently, the greatest difficulty in conducting a biologically based IPM program is having IPM tools available when new (invasive) pests enter the region. As the world's human population grows and movement of plant material increases between U.S. states and foreign countries, the rate of introduction of invasive pests and diseases has rapidly increased. In the past 10 years, the imported fire ant, glassy-winged sharpshooter, new strains of citrus peelminer, citrus leafminer, Diaprepes root weevil *(Diaprepes abbreviatus)*, several species of fruit flies, and Asian citrus psyllid *(Diaphorina citri)* have arrived and disrupted citrus IPM in one or more regions of the state. Additional species of pests are threatening to enter California from Florida, Texas, Hawaii, and Mexico. This section provides three examples of the difficulties of managing citrus IPM with the entrance of new pests and the need for strict quarantine and eradication programs.

Glassy-winged sharpshooter. Glassy-winged sharpshooter entered southern California in the late 1980s, probably on shipments of nursery ornamentals. It has spread throughout the southern half of California (from the Mexico border up through Fresno County). In 2000, grape growers in the Temecula Valley realized that it was acting as a vector and spreading the bacterium that causes Pierce's disease. There is currently no cure for Pierce's disease, which causes infected vines to die within 1 to 3 years. The glassy-winged sharpshooter uses citrus as one of its preferred oviposition hosts. The citrus industry is asked to reduce glassy-winged sharpshooter numbers in citrus so that they do not invade neighboring grapes and spread Pierce's disease in and among the weeds and vines (citrus is not a source of the disease, only the vector). Citrus growers apply broad-spectrum insecticides for glassy-winged sharpshooter during the field season, which can negatively affect natural enemies needed for other pests of citrus. Areawide treatment programs for citrus, implemented when populations start to increase on yellow sticky traps, have reduced glassy-winged sharpshooter densities to low levels statewide, lessening the need for treatments in citrus. It is expected that long-term strategies such as host plant resistance will eventually eliminate the need for these treatments.

Diaprepes root weevil. The Diaprepes root weevil (fig. 17.28) is native to the Caribbean and infests a wide variety of plants in Florida. It was found infesting ornamental plants in urban areas of Los Angeles and Orange Counties and urban areas and

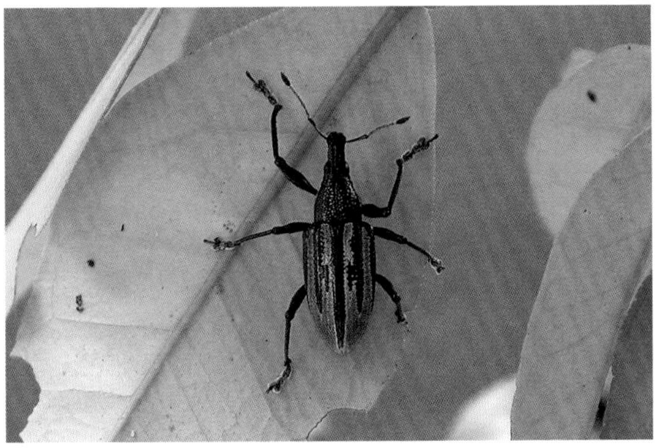

Figure 17.28 Diaprepes root weevil adults feed on leaves. *Photo:* J. K. Clark.

lemon orchards in San Diego County in 2005 and 2006. In spite of compliance agreements requiring insecticide treatment of containerized plants prior to shipment from Florida nurseries to California, it is likely that this pest arrived via the larval stage in the soil of containerized ornamentals or palm trees that were subsequently planted in urban landscaping. An eradication effort by the California Department of Food and Agriculture was under way for several years but was discontinued due to fiscal constraints and expansion of the detection area. Egg parasitoids were later introduced from Florida to assist in population suppression. Diaprepes larvae feed on the roots of plants and can cause enough damage to the root system of citrus to cause the tree to decline and die (fig. 17.29). In Florida, citrus growers spend as much as $400 per acre to treat for Diaprepes larvae in the soil, as well as for Phytophthora disease, which is exacerbated by Diaprepes root feeding. Many of the insecticides used to control the aboveground and belowground stages of this pest are broad-spectrum, which will disrupt the IPM program if this pest gains a foothold in the major citrus-growing regions of the state. In California, nurseries are required to treat the soil of potted plants to prevent nursery infestations and potential movement of Diaprepes to new areas. Because it is a fairly slow-growing pest, with the larval stage spending 10 to 18 months underground, and slow to disperse, efforts to control it biologically with fungi, nematodes, and parasitoids are hoped to keep it in check.

Asian citrus psyllid. The Asian citrus psyllid is one of the most catastrophic invasive pests to arrive in California because it is a vector of a deadly disease. This insect was first found in California in 2008 and has spread through most of southern California. By itself, it is an irritating pest because it injects a toxin when it feeds on leaves and stems (fig. 17.30), which causes the leaves to turn back or twist and curl. This damage could be mitigated with insecticides or natural enemies. However, Asian citrus psyllid is also a vector of the devastating bacterial disease known as huanglongbing (HLB), or citrus greening. HLB causes leaves to yellow, the fruit to be reduced in size and develop asymmetrically, and the juice to become bitter. The tree eventually dies. There is no cure, and infected trees must be removed. A single infected tree was found in central Los Angeles in 2012 because a homeowner grafted diseased budwood to a citrus tree. The tree was destroyed, but the disease may have spread to other trees in the area. HLB is also spreading from Florida through Mexico toward California in infected psyllids. The invasion of Asian citrus psyllid in California citrus is still new, but it has already begun to dramatically affect citrus IPM, because frequent applications of broad-spectrum insecticides are the best way to reduce the psyllid populations and slow the spread and severity of the disease. A parasitic wasp, *Tamarixia radiata,* is being released to help control the psyllid, but it does not control the psyllid quickly or effectively enough to prevent disease spread. Treatment strategies have been developed to use pesticides when they are most effective (winter dormant period and during leaf flushing) and to minimize broad-spectrum pesticide use in order to preserve natural enemies and reduce the risk of selecting for pesticide resistance. However, all of these treatment strategies will at best slow the rate of spread of the psyllid and disease and buy time for researchers to develop a long-term strategy to protect citrus, such as a modified psyllid, a spray that prevents the psyllid from picking up the disease, or a disease-resistant plant. The psyllid is small and difficult to detect. It can spread rapidly among the millions of urban and commercial citrus trees in California, and once it acquires the bacterium, it carries it for the rest of its life. This pest and disease complex is a very severe challenge to the California citrus pest management program.

Figure 17.29 Diaprepes root weevil larvae damage the root system of citrus. *Photo:* E. E. Grafton-Cardwell.

Figure 17.30 Asian citrus psyllid adults feeding on leaves. *Photo:* M. E. Rogers, University of Florida.

Conclusions

Control of citrus pests with natural enemies and careful use of selective insecticides is certainly attainable in California citrus. It has been used by a number of growers during the past decades. In some years, the program provides an immediate cost savings, because when pest pressures are low, careful sampling demonstrates that pesticide treatments are not necessary. Factors such as extremes of weather, invasive pests, pesticide drift, and variability in the effectiveness of biological control can make a biologically based program difficult. However, creating a healthy environment for natural enemies and reducing pesticide use will have many long-term benefits, including delaying pesticide resistance and reducing pesticide residues in the environment. Additional research to improve the practicality and implementation of this approach is worthwhile.

Citrus IPM Online Resources

Citrograph, http://www.citrusresearch.org/citrograph (known prior to 1968 as *California Citrograph*). *Citrograph* magazine is published by the Citrus Research Board and features articles on citriculture, including pest management.

UC ANR Online Learning Site, Citrus IPM Modules, http://class.ucanr.org/. Six modules teach the biology and management of citrus pests such as California red scale, citricola scale, citrus red mite, katydid, cottony cushion scale, and citrus peelminer.

UC Citrus Entomology Laboratory at the Kearney Agricultural Center, www.ucanr.org/sites/KACCitrusEntomology. Committed to providing the citrus growers of the San Joaquin Valley with timely pest management information.

UC IPM website, http://www.ipm.ucdavis.edu/. Provides up-to-date pest management information for major crops grown in California. For each crop, Pest Management Guidelines describe the pests, how to monitor for them, and degree-day units to predict major events, and also provides economic thresholds, information on natural enemies, and pesticide recommendations. The year-round citrus guidelines give a seasonal context for management practices.

References

Bernal, J. S., R. F. Luck, J. G. Morse, and M. S. Drury. 2001. Seasonal and scale size relationships between citricola scale (Homoptera: Coccidae) and its parasitoid complex (Hymenoptera: Chalcidoidea) on San Joaquin Valley citrus. Biological Control 20:210–221.

Bethke, J., J. Morse, F. Byrne, E. Grafton-Cardwell, K. Godfrey, and M. Hoddle. 2012. Asian citrus psyllid pesticide testing begins at the Chula Vista Insectary. Citrograph 3(3): 36–40.

Dreistadt, S. 2012. Integrated pest management for citrus. 3rd ed. Oakland: University of California Agriculture and Natural Resources Publication 3303.

Forster, L. D., R. F. Luck, and E. E. Grafton-Cardwell. 1995. Life stages of California red scale and its parasitoids. Oakland: University of California Division of Agriculture and Natural Resources Publication 21529.

Graebner, L., D. S. Moreno, and J. L. Baritelle. 1984. The Fillmore citrus protective district: A success story in integrated pest management. Bulletin of the Entomological Society of America 30:27–33.

Grafton-Cardwell, B. 2012. What are the University of California sources for citrus integrated pest management information? Citrograph 3(2): 44–46.

Grafton-Cardwell, B., and J. Stewart. 2012. Proper monitoring and management of California red scale in the San Joaquin Valley. Citrograph 3(2): 12–14, 16, 18–21.

Grafton-Cardwell, B., J. Morse, and B. Taylor. 2011. Asian citrus psyllid treatment strategies for California-Arizona. Citrograph 2(5): 5–10.

Grafton-Cardwell, B., C. Reagan, S. Scott, Y. Ouyang, and G. Montez. 2010. Citricola scale, the scourge of the San Joaquin Valley. Citrograph 1(4): 18–21.

Grafton-Cardwell, E. E. 2002. Stages of the cottony cushion scale *(Icerya purchasi)* and its natural enemy, the vedalia beetle *(Rodolia cardinalis)*. Oakland: University of California Agriculture and Natural Resources Publication 8051. http://anrcatalog.ucdavis.edu/pdf/8051.pdf.

Grafton-Cardwell, E. E., and C. Reagan. 2003. Surround use in citrus increases California red scale. UC Plant Protection Quarterly 13(3): 5–7.

Grafton-Cardwell, E. E., and J. Stewart-Leslie. 1998. *Aphytis* Release Programs. Citrograph 83(11): 3, 18–19.

Grafton-Cardwell, E. E., K. E. Godfrey, D. H. Headrick, P. A. Mauk, and J. E. Pena. 2008. University of California Agriculture and Natural Resources Publication 8321. http://anrcatalog.ucdavis.edu/pdf/8321.pdf.

Grafton-Cardwell, E. E., K. Godfrey, J. Pena, C. McCoy, and R. Luck. 2004. Diaprepes root weevil. University of California Division of Agriculture and Natural Resources Publication 8131. http://anrcatalog.ucdavis.edu/pdf/8131.pdf.

Grafton-Cardwell, E. E., K. Godfrey, M. Rogers, C. Childers, and P. Stansly. 2006. Asian citrus psyllid. University of California Division of Agriculture and Natural Resources Publication 8205. http://anrcatalog.ucdavis.edu/pdf/8205.pdf.

Grafton-Cardwell, E. E., N.V. O'Connell, C. E. Kallsen, and J. G. Morse. 2003. Photographic guide to citrus fruit scarring. Oakland: University of California Division of Agriculture and Natural Resources Publication 8090.

Grafton-Cardwell, E. E., Y. Ouyang, R. Striggow, and S. Vehrs. 2001. Armored scale insecticide resistance challenges San Joaquin Valley citrus growers. California Agriculture 55(5): 20–25.

Haney, P. B., J. G. Morse, R. F. Luck, H. Griffiths, E. E. Grafton-Cardwell, and N. V. O'Connell. 1992. Reducing insecticide use and energy costs in citrus pest management. Davis: University of California Statewide IPM Program Publication 15.

Kapranas, A., J. G. Morse, P. Pacheco, L. D. Forster, and R. F. Luck. 2007. Survey of brown soft scale *Coccus heseridum* L. parasitoids in southern California citrus. Biological Control 42:288–299.

Khan, I., and J. G. Morse. 1998. Citrus thrips (Thysanoptera: Thripidae) resistance monitoring in California. Journal of Economic Entomology 91:361–366.

Lampson, L. J., and J. G. Morse. 1992. A survey of black scale, *Saissetia oleae* [Hom.: Coccidae] parasitoids [Hym.: Chalcidoidea] in southern California. Entomophaga 37:373–390.

Luck, R. F., J. G. Morse, P. B. Haney, H. J. Griffiths, J. M. Barcinas, T. J. Roberts, E. E. Grafton-Cardwell, and N. V. O'Connell. 1992. Citrus IPM: It Works! California Grower 16(4): 25–27.

Morse, J., and B. Grafton-Cardwell. 2012a. Management of citrus thrips to reduce the evolution of resistance. Citrograph 3(2): 22, 24, 26–30.

———. 2012b. The evolution of biologically-based integrated pest management in California citrus: History and perspective. Citrograph 3(2): 32–34, 36–38, 40–43.

Schweizer, H., J. G. Morse, R. F. Luck, and L. D. Forster. 2002. Augmentative releases of a parasitoid *(Metaphycus* sp. nr. *flavus)* against citricola scale *(Coccus pseudomagnoliarum)* on oranges in the San Joaquin Valley of California. Biological Control 24:153–166.

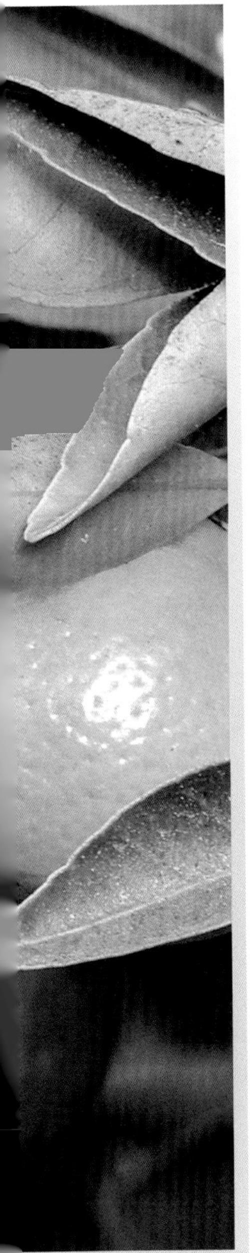

18 Managing Pesticide Resistance in Insects, Mites, Weeds, and Fungi

Joseph G. Morse, Elizabeth E. Grafton-Cardwell, James E. Adaskaveg, Helga Förster, and Joseph M. DiTomaso

Commonalities in Resistance Principles

Pesticide resistance is a major problem when managing arthropods, weeds, and fungi affecting citrus in California. Because of resistance, previously effective treatments can be rendered ineffective, resulting in substantial crop damage and economic loss if chemicals are not used judiciously and either in rotation or in combination. This chapter discusses resistance management for citrus pests and pathogens and finds that the concepts and methods useful in managing resistance are similar for these organisms. To avoid pesticide resistance, it is important that growers know what class of chemistry each pesticide represents (see tables 18.1, 18.2, and 18.4) so they can rotate among treatments from different chemical classes or use different chemistries in mixtures (particularly useful for fungi). These practices minimize the overuse of any one chemical class and the resultant increased selection pressure on a particular pest or pathogen species.

Resistance is an evolutionary process in insect, mite, weed, and fungal populations that are exposed to pesticides. Several different mechanisms allow resistant pests and pathogens to survive a pesticide treatment. These include

- increased metabolism of the pesticide (more rapid breakdown often due to enhanced enzyme activity or overproduction of protective enzymes)
- an altered target site (the site of pesticide action is modified by the organism so that the pesticide has less of an impact)
- decreased penetration, enhanced excretion, or sequestration away from the site of activity (different mechanisms that result in less pesticide reaching the target site)
- behavioral resistance (avoidance of the pesticide or treated foliage by insects or mites)

For resistance to be heritable (i.e., transferred from one pest or pathogen generation to the next), it must have a genetic basis. We often discuss this in terms of resistance genes, the gene or group of genes that allow some members of a population to survive exposure to the pesticide. Resistance is typically due to one or more mutations to pre-existing genes. These genes are selected for when pesticides are applied because they allow pests or pathogens with these genes to survive and reproduce with a greater frequency than do members of the population that do not have the mutated genes.

Resistance management strategies can be classified as preventative or corrective (or both). Regardless of whether one is considering insects, mites, weeds, or fungi, preventative resistance management strategies involve reducing how often and how frequently pesticides in a particular class of chemistry (i.e., pesticides with a similar mode of action) are used against the pest or pathogen population. Maximizing nonchemical methods of managing the pest or pathogen population, rotating between different classes of chemistry, and using effective application methods when a treatment is needed (so that later retreatment is not needed or is not needed as quickly) are important in resistance management regardless of the type of pest or pathogen that is being controlled. Methods of implementing corrective resistance management are similar in nature to preventative methods, but they also include withholding or reducing the use of any pesticide in the same class of chemistry as the class in which resistance has been observed.

Another common principle of resistance management is that resistance is most easily managed when detected early or if it is proactively prevented from occurring. Once levels of resistance are present at a moderate degree in a pest or pathogen population, that particular class of chemistry may be of little value for an extended period of time. In the case of weeds and fungi, if resistance is detected early, it may be possible to eradicate members of the resistant population (e.g., by rogueing out resistant weeds in part of a field or using sanitation to eradicate resistant fungal populations in a packinghouse). This is less successful with insects or mites given the speed with which they may disperse. Resistance is more stable in a population after resistance has been present for a period of time and the same pesticide class continues to be used. Once resistance is observed, withholding use of that class of chemistry for an extended period of time often allows resistance to revert, so that products from that chemistry are again effective. This will occur if the resistant population is less fit or at a disadvantage compared with the wild-type population. Unfortunately, resistance often does not revert completely, and this is one reason why preventative resistance management makes the most sense. Also, in some cases, resistant populations are of similar fitness as susceptible populations, which is often the case with weeds. In this case, even in the absence of pesticide use, the frequency of resistant weeds in the population is unlikely to decline.

Pesticide Resistance in Arthropod (Insect and Mite) Pests of Citrus

Since the introduction of modern insecticides in the mid-1940s, pesticides have been relied on for control of key pests of citrus such as California red scale, soft scales, citrus thrips, Argentine ant, and various plant-feeding mites. Pesticides are not needed in every situation (e.g., biological control of California red scale is usually adequate in southern California), but when economically damaging populations are present, growers often turn to chemical control. Pesticides are generally effective and are relatively inexpensive, and in most cases, the negative impacts of their use are not readily apparent. As with any management technique, however, one should carefully weigh the benefits and cost of using a pesticide. In addition to the direct cost of its use, two additional costs should be considered when deciding on the need for a pesticide application: the impact of the pesticide on natural enemies important in the natural maintenance of various pest species at subeconomic levels, and the impact of the pesticide on the development of pesticide resistance in both the target pest and nontarget pest species. This chapter deals mainly with the pesticide resistance issue and its impact on the target pest.

In California citrus, three arthropods in particular, California red scale, citrus thrips, and citrus red mite, have a history of developing resistance to pesticides. California red scale was one of the first insects on any commodity reported to develop pesticide resistance when hydrogen cyanide resistance appeared in the 1930s (Quayle 1938). From the first development of parathion in the 1940s until the 1990s, citrus growers depended on organophosphate (OP) and carbamate insecticides for insect control. In the 1990s, resistance of California red scale to the OPs chlorpyrifos (Lorsban) and methidathion (Supracide) and to the carbamate insecticide carbaryl (Sevin) appeared in the San Joaquin Valley (Grafton-Cardwell et al. 2001). The high cost of multiple OP or carbamate sprays for California red scale control caused citrus growers to switch to use of the insect growth regulators pyriproxyfen (Esteem) and buprofezin (Applaud), oil treatments, and releases of *Aphytis melinus* wasps (Morse et al. 2007).

Citrus thrips has developed resistance to any chemical used repeatedly for its control, including tartaric emetic (1940), DDT (1946), dieldrin (1954), malathion (1961), dimethoate (Cygon) (1980), formetanate (Carzol) (1986), cyfluthrin (Baythroid) (1996), and spinetoram (Delegate) (2011) (Morse and Brawner 1986; Immaraju et al. 1989; Khan and Morse 1998). Citrus red mite populations have developed resistance to several organophosphates. Although it took a longer period of time, various populations of citrus red mite have developed resistance to the older group of acaricides, including dicofol (Kelthane), fenbutatin oxide (Vendex), and propargite (Omite), especially in citrus nursery situations.

Key Concept: Localized versus Widespread Resistance

Pesticide resistance in a pest population often appears first in a particular citrus-growing region where use of that pesticide has been frequent or continuous; resistance may spread to other areas, depending in part on whether pesticide use with that class of chemistry remains frequent and the degree to which resistant pests move from one area to another. As a general principle, if resistance is detected early and pesticides in other classes of chemistry are used in rotation, resistance may decline or not spread to other areas.

What is it about California red scale, citrus thrips, and citrus red mite that led to the appearance of pesticide resistance? Part of the answer is that these three pests have been treated regularly with pesticides because of relatively low thresholds for the damage they cause. For example, a citrus thrips treatment threshold often used on San Joaquin Valley navel oranges is 10% or more of the fruit infested with larvae for the first 6 weeks after petal fall; growers usually attempt to reduce California red scale to levels that will result in fewer than 10 scales per fruit at harvest. Citrus red mite initially had a damage threshold of one to two adult females per leaf, but research raised the treatment threshold for this species in the San Joaquin Valley to eight adult females per leaf, greatly reducing the number of treatments applied for red mite in recent years. Second, only a few pesticides were available for control of these species from the 1950s to the 1980s, and this limited the ability of growers to rotate pesticides between different classes of chemistry (for California red scale and citrus thrips, treatments were primarily organophosphate and carbamate insecticides). Repeated exposure to a single class of chemistry is well known to be one of the most important factors in selecting for pesticide resistance. Third, mites and thrips are notorious among arthropods for their inherent ability to develop resistance to pesticides. One reason for this is that they have relatively short generation times, leading to two or more generations being selected from a single pesticide treatment if the pesticide is persistent. A fourth problem is that a pest may be treated only once during the year directly, but it is present on citrus for most of the year. Thus, treatments applied for other pest species contribute to the selection for pesticide resistance, in addition to the treatments applied specifically for that pest. California red scale has a longer generation time than either citrus red mite or citrus thrips, but springtime organophosphate or carbamate treatments for orangeworms (butterflies and moths [Order Lepidoptera] on citrus are collectively referred to as orangeworms) or citrus thrips likely contributed to the development of resistance to these pesticides in California red scale (in addition to the treatments directly applied for California red scale in summer). In addition to this problem of multiple applications of pesticides, California red scale does not travel very far (dispersal is mainly by crawlers being moved on flying insects, by being blown in the wind, or by unintentional movement by pickers or on bins or equipment), and thus few scales escape treatment and very few susceptible male scales migrate into a treated orchard to mate with resistant females and dilute resistance.

Methods of Slowing the Evolution of Arthropod Resistance

Many articles have been written on the factors influencing how quickly pesticide resistance evolves in arthropod pest species (e.g., Tabashnik and Croft 1982; Roush and McKenzie 1987; Groeters and Tabashnik 2000). In a practical sense, there are two basic ways a grower can reduce the rate at which pesticide resistance develops in a particular arthropod pest: reduce the number of pesticide treatments per year to as few as possible with a particular class of chemistry, and rotate between different pesticide classes (Immaraju Morse, and Hobza 1989). It is important to remember that pesticides applied for other pests also select for resistance in the target pest. Therefore, each chemical class should be used as infrequently as possible during the year. Limiting pesticide treatments as much as possible may sound rather simplistic, but in most cases, the major genes responsible for resistance carry a negative fitness early in the selection process. That is, populations of insects or mites carrying those genes grow more slowly or produce fewer progeny than do their susceptible counterparts in the absence of pesticide exposure. The result is that if pesticide treatments are rare, the resistant insects or mites reproduce poorly and resistance in the populations declines (it is selected against). With citrus thrips, for example, a spinetoram (Delegate) spray in spring might lead to thrips showing a slight loss of susceptibility to spinetoram in a particular orchard. During the remainder of the year, however, if one avoids using this class of chemistry (Group 5, spinosyns; see table 18.1), thrips carrying the gene(s) conferring resistance will not survive as well as susceptible thrips, and the levels of resistance to spinetoram will decline in the population. Ideally, one might use spinetoram only once every second year in order to maintain thrips susceptibility to the spinosyn group of pesticides. With few pesticides available for citrus thrips management and growers often treating every year, this level of moderation is often difficult to achieve.

Notice that in table 18.1, carbamate (Group 1A) and organophosphate (1B) insecticides have the same mode of action. Thus, resistance in thrips or scales to one insecticide in this group leads to partial or full cross-resistance to all of the insecticides in that group. For many years, organophosphates and carbamates were the main insecticides available for control of citrus pests, and it is not surprising that citrus thrips and California red scale developed high levels of resistance to these pesticides. Currently, we have more diversity in pesticide chemistry than in the past, and there is more flexibility

Table 18.1. Insecticides and acaricides used for pests of California citrus, classified by mode of action

Group	Primary target site	Chemical subgroup or exemplifying active ingredient	Pesticides in the group
1A	acetylcholinesterase inhibitors	carbamates	carbaryl (Sevin) formetanate hydrochloride (Carzol)methomyl (Lannate)
1B		organophosphates	azinphos-methyl (Guthion) chlorpyrifos (Lorsban) dimethoate (Cygon) malathion methidathion (Supracide) naled (Dibrom)
3	sodium channel modulators	pyrethroids, pyrethrins, DDT	cyfluthrin (Baythroid) fenpropathrin (Danitol) pyrethrins and pyrethrum zeta-cypermethrin (Mustang)
4	nicotinic acetylcholine receptor agonists or antagonists	neonicotinoids	acetamiprid (Assail) imidacloprid (Admire [systemic]; Provado [foliar]; Macho, Couraze, Nuprid, Alias, and others) thiamethoxam (Platinum [systemic]; Actara [foliar])
5	nicotinic acetylcholine receptor allosteric activators	spinosyns	spinetoram (Delegate) spinosad (Success, Entrust)
6	chloride channel activators	av, ermectins, milbemectins	abamectin (Agri-Mek, Zoro,Clinch, and others)
7C	juvenile hormone mimics	pyriproxyfen	pyriproxyfen (Esteem)
10A	mite growth inhibitors (unknown target protein)	hexythiazox, clofentezine	etoxazole (Zeal) hexythiazox (Onager, Savey)
11B2	microbial disruptors of insect midgut membranes	microbial pesticides, Bt ssp. *kurstaki*	*Bacillus thuringiensis* products (Dipel, Javelin, MVP)
12	inhibitors of mitochondrial ATP synthase	organotin miticides, propargite	fenbutatin oxide (Vendex) propargite (Omite)
15	inhibitors of chitin biosynthesis, type 0, Lepidopteran	benzoylureas	diflubenzuron (Micromite, Dimilin)
16	inhibitors of chitin biosynthesis, type 1, Hemipteran	buprofezin	buprofezin (Applaud)
20B	mitochondrial complex III electron transport inhibitors (coupling site II)	acequinocyl	acequinocyl (Kanemite)
21	mitochondrial complex I electron transport inhibitors	METI acaricides, rotenone	pyridaben (Nexter) fenproximate (FujiMite)
23	inhibitors of lipid synthesis, gowth regulation	tetronic acid derivatives	spirodiclofen (Envidor) spirotetramat (Movento)
25	mitochondrial complex II electron transport inhibitors	bifenazate	bifenazate (Acramite)
28	ryanodine receptor modulators	diamide	chlorantraniliprole (Altacor)
—	unclassified	azadiractin	azadiractin (various Neem products)
—	unclassified	cryolite	cryolite (Kryocide)
—	unclassified	dicofo	dicofol (Kelthane)
—	unclassified	oils	various petroleum oils (415, 440, 455, 470 narrow-range oils)
—	unclassified	sulfur	various sulfur formulations

in rotating among chemical classes and avoiding cross-resistance.

Insects and mites can also develop resistance to multiple classes of insecticides (multiple resistance). For example, some populations of citrus thrips are resistant to both Groups 1A and 1B (carbamates and organophosphates) and Group 3 (pyrethroids). While some insecticides fall into separate modes of action in table 18.1, there are concerns that through cross-resistance or multiple resistance, insects can develop resistance to a new pesticide group that it has not previously been exposed to. For example, citrus thrips may develop cross-resistance to abamectin (Agri-Mek and a number of generic abamectins) and spinosad (Success, Entrust) due to the use of similar genes for resistance, even though these insecticides are from different mode of action groups. Studies are needed to detect resistance at its first occurrence in order to make resistance management recommendations and avoid cross-resistance scenarios.

A problem in implementing arthropod resistance management is that it is not always possible to rotate among different classes of chemistry as much as is desired for the purposes of resistance management. For example, in 2012, abamectin, cyfluthrin (Baythroid), dimethoate (Cygon), fenpropathrin (Danitol), formetanate (Carzol), sabadilla (Veratran D), spinosad, spinetoram (Delegate), and spirotetramat (Movento) were registered for citrus thrips control and recommended by the University of California (see the UC IPM Citrus Pest Management Guidelines, http://www.ipm.ucdavis.edu/PMG). However, because there is cross-resistance between dimethoate and formetanate, between cyfluthrin and fenpropathrin, and between spinosad and spinetoram and possibly also abamectin (this latter case is yet unclear), this reduces the number of different insecticide groups to five. If a citrus thrips population has resistance to organophosphates, carbamates, and pyrethroids, a not uncommon situation in San Joaquin Valley citrus, the number of insecticide groups that can be rotated is reduced to three: spinosad-spinetoram-abamectin, sabadilla, and spirotetramat. If spinetoram is the most effective, most cost effective, and most selective insecticide available for citrus thrips control, growers tend to use it year in and year out for thrips control. This is just the sort of pesticide selection pressure that will eventually select for resistance.

In 2012, insecticides registered for California red scale control and recommended by the University of California included buprofezin (Applaud), carbaryl (Sevin), chlorpyrifos (Lorsban), imidacloprid (Admire), methidathion (Supracide), narrow-range oils, pyriproxyfen (Esteem), and spirotetramat (Movento). We know that there is cross-resistance between chlorpyrifos, methidathion, and carbaryl, and many red scale populations have resistance to this group of insecticides, reducing the actual number of scale treatments available to five (buprofezin, imidacloprid, oil, pyriproxyfen, and spirotetramat). Because of the high efficacy of pyriproxyfen, it is the preferred insecticide treatment, and growers tend to use only it for every California red scale treatment, selecting for resistance to this insecticide, in spite of the fact that a number of other insecticides are available. For California red scale, an even better way to delay insecticide resistance is to use natural enemies such as augmentative releases of *Aphytis* wasps for biological control rather than using insecticides.

Key Concept: Reducing Selection Pressure Is Key to Good Resistance Management

Any practice that reduces frequent use of a particular class of pesticide chemistry is important in slowing the rate at which resistance evolves. Maximizing the use of biological control agents, monitoring and using treatments only when economic thresholds are exceeded, and rotating between different classes of pesticide chemistry are some of the better methods that can be used to manage resistance.

Recently, we have found that San Joaquin Valley populations of citricola scale have begun to develop resistance to chlorpyrifos. Citricola scale was previously controlled by high rates (6 to 12 lb a.i. per ac) of this insecticide that were applied for California red scale control, but they also reduced citricola scale populations below the economic threshold for 3 to 5 years. Alternative insecticides, such as acetamiprid (Assail) and buprofezin (Applaud), suppress citricola scale populations for only 1 year. Thus, chlorpyrifos resistance causes growers to treat for citricola scale more frequently because chlorpyrifos and the alternative insecticides are less effective. Development of resistance to chlorpyrifos started in citrus thrips in the 1980s, in California red scale in the 1990s, and in citricola scale in the 2000s. The timing of the development of resistance is due in large part to the fact that the more generations an insect has per year, the faster it develops resistance. Citrus thrips have six to eight generations per year, California red scales have four generations per year, and citricola scale has only one generation per year.

Resistance develops in a particular citrus orchard mainly as a result of the treatments that are applied in that orchard, especially for pests such as California red scale that show limited dispersal. Even for a relatively mobile species such as citrus thrips, research suggests that pesticide-use patterns in surrounding orchards are much less important than treatments applied in the particular orchard of interest.

In summary, to practice arthropod resistance management, growers should try to limit pesticide treatments to those that are absolutely needed and rotate pesticides among the classes of chemistry that are available. For example, in years when relatively low citrus thrips populations are present, one might consider not treating at all or using a somewhat less effective pesticide such as sabadilla (Veratran D) because it has a chemistry that is different from the other control options available for this pest. One might also try to use one insecticide from the spinosad, abamectin, and spinetoram (Delegate) group only once every 2 years. At present, new pesticides useful in control of citrus thrips are not being introduced very quickly; thus, if resistance develops to the insecticides we currently have registered, effective citrus thrips treatments could become problematic.

In a similar fashion, one might consider trying to reduce the frequency of pyriproxyfen use for California red scale control to once every second or third year. In alternate years, *Aphytis melinus* releases alone or in combination with selective pesticides such as buprofezin, spirotetramat, or oil would reduce resistance to pyriproxyfen.

Pesticide resistance management makes economic sense: if resistance develops in a particular orchard, relatively few new pesticides will be available to provide effective control. Resources such as the Insecticide Resistance Action Committee website, http://www.irac-online.org/eClassification/, and the UC IPM Citrus Pest Management Guidelines, http://ucipm.ucdavis.edu/PMG/, are very helpful for determining the mode of action of each insecticide and developing a resistance management plan.

Pesticide Resistance in Weeds of Citrus

Weed resistance to herbicides is a more recent phenomenon than resistance in arthropods and fungi. The first report of herbicide resistance occurred with the discovery of triazine-resistant common groundsel *(Senecio vulgaris)* in western Washington in the late 1960s. Since that time, over 315 weed biotypes representing 183 species in 59 countries have evolved resistance to herbicides, and resistant weeds infest more than 280,000 fields and 7.4 million acres of cropland worldwide. Today, at least 64 weeds have developed resistance to herbicides in the United States (Heap 2013; see the WeedScience.com International Survey of Herbicide Resistant Weeds at http://weedscience.com, and also chapter 16, "Integrated Weed Management").

Prior to 1989, only one weed species in California, common groundsel, had developed resistance to herbicides. Today, however, 18 weed species have been reported to be herbicide resistant in the state, and several others anecdotally appear to have developed resistance. Of these cases, 9 weed species have developed resistance to herbicides used in rice production. Many of these rice weeds have developed resistance to the sulfonylurea herbicide bensulfuron (Londax). A roadside survey conducted in 1995 and 1996 found that resistance to sulfonylurea herbicides was very common in Russian thistle *(Salsola tragus)* (Prather et al. 2000). Most recently, a rigid ryegrass *(Lolium rigidum)* biotype exhibited resistance to glyphosate (Roundup) in a northern California orchard (Simarmata et al. 2005). Rigid ryegrass is not common in California, but it readily hybridizes with both Italian ryegrass *(Lolium multiflorum)* and perennial ryegrass *(Lolium perenne)* (Sherwood and Jasieniuk 2009). It appears that resistant hybrids of *Lolium* have spread to roadsides in many areas of the state. Other common weed species in citrus are now beginning to show resistance to glyphosate, in particular hairy fleabane *(Conyza bonariensis)* and horseweed *(Conyza canadensis)* (Heap 2013).

Mechanisms of Weed Resistance

The mechanisms by which weed species have developed resistance to herbicides can include alterations in the site of action, increased metabolism of the herbicide, overproduction of protective enzymes, and perhaps changes in herbicide transport processes that increase the plant's ability to sequester the compound away from the site of action. The most common method is either through an altered site of action or enhanced metabolism. In the case of altered site of action, resistance is usually based on a mutation or alteration in the target protein. The application of the herbicide selects for plants with this alteration, as they are the ones that escape injury. Rapid metabolism of herbicides to nontoxic metabolites is perhaps the most frequent mechanism for selectivity differences (tolerance) among many plant species, particularly nontarget crops. However, enhanced metabolism accounts for far fewer cases of herbicide resistance than do alterations in the site of action. Weed biotypes with enhanced metabolism have a much lower level of resistance than do weeds expressing resistance through site-of-action changes.

Factors That Lead to Herbicide Resistance

Herbicide resistance has evolved independently several times among taxonomically unrelated species. A number of factors can lead to, or accelerate, the development of herbicide resistance. Most of these factors increase selection pressure for the particular genome that confers resistance. These factors can be classified into three groups: chemical properties, weed characteristics, and cultural practices.

The chemical properties of some herbicides can lead to more rapid development of resistance than is seen in other herbicidal compounds. In general, herbicides with a single site of action are more prone to develop resistance among weeds (table 18.2). This is particularly true for herbicides with a high risk of resistance, including the ALS (acetolactate synthase) inhibitors such as the sulfonylurea or imidazolinone herbicides, as well as the ACCase (acetyl CoA carboxylase) inhibitors, which include the graminicides (cyclohexanediones and aryloxyphenoxypropionate). No ALS inhibitors are currently registered for use in California citrus. However, the same principle does not hold true for glyphosate, although its activity is also on a specific enzyme, EPSP synthase. Only a relative few cases of herbicide resistance have been reported for glyphosate worldwide, despite its widespread use (see table 18.2). In addition, herbicides that are detoxified by weeds by a common metabolic pathway (i.e., cytochrome P_{450} monooxygenase or glutathione-*S*-transferase) are more likely to develop resistance among weed species than would herbicides that do not metabolize rapidly in plants or are degraded by a less common pathway (Preston 2004). Finally, broad-spectrum herbicides are more likely to lead to resistance in weeds because they have a wide range of weed species from which to select.

Key Concept: Reduce Overreliance on a Single Herbicide Mode of Action

The single most important factor contributing to herbicide resistance in weeds is the continuous use of a single herbicide or group of herbicides with the same mode of action. Weeds most often develop resistance in response to repeated and exclusive exposure, which renders the herbicide ineffective over time.

Some weed species are much more likely to develop resistance to herbicides than are other species. Of the most common weed species developing resistance to herbicides worldwide, the top 10 all have an annual growth habit, are widely distributed throughout the world, and proliferate in a variety of climatic and environmental conditions (table 18.3). This suggests that genetically diverse species are most likely to contain a gene that confers resistance to a particular herbicide. Other weed features associated with resistance include high seed production, relatively rapid turnover of the seed bank, several reproductive generations per growing season, and extreme susceptibility to a particular herbicide.

Of equal importance in the development of herbicide resistance in weeds are the cultural practices used in a weed management system. Any practice that increases the selective pressure on a weed population will increase the likelihood of resistance development. For example, reliance only on herbicides for weed control, a shift away from crop rotation to monocultures, or continuous or repeated use of a single herbicide or several herbicides with a similar mode of action can increase the probability of selecting for herbicide-resistant weeds. Because a limited number of herbicides are available in citrus and rotations are not practical, weeds in citrus are more sensitive to the development of herbicide resistance, and particular attention should be given to the type of herbicides applied. A weed developing resistance to one herbicide typically has cross-resistance to many, if not all, herbicides with the same mode of action. For example, a weed that has developed resistance to clethodim (Select Max, Prism) through a site of action change will often be resistant to sethoxydim (Poast) and fluazifop-p-butyl (Fusilade) as well.

Management and prevention of herbicide-resistant weeds

The management of established resistant biotypes and a reduction in the occurrence of resistance could be more effectively accomplished with a greater reliance on integrated weed management approaches. These strategies can incorporate other cultural or mechanical control tools into a weed management plan for citrus, including cultivation, mowing, drip irrigation systems that minimize weed infestations, and biological control (when feasible). In addition, the use of herbicide combinations with overlapping spectrums, or rotation to herbicides with a different mode of action, will slow the development of herbicide-resistant weeds. Table 18.2 lists the herbicides registered for use in citrus in California and includes the number of herbicide-resistant weed species reported for that herbicide worldwide, nationally, and statewide. While some chemical families can lead to many cases of herbicide resistance in weeds, others do not appear to readily select for resistant weeds.

Table 18.2. Herbicide families and compounds used in California citrus and the number of weed species developing resistance worldwide, within the United States, and in California

Chemical family	Herbicide	Mode of action	Number of weed species developing resistance		
			Worldwide	United States	CA
acetamides	napropamide (Devrinol)	cell division inhibitor, specific site of action unknown	0	0	0
benzamides	isoxaben (Gallery)	cellulose inhibitor, specific site of action unknown	0	0	0
bipyridilium	paraquat (Gramoxone)	electron acceptor from photosystem I in photosynthesis	26	5	1
cyclohexanediones and aryloxyphenoxy-propionate	clethodim (Prism, Select Max) fluazifop-p-butyl (Fusilade) sethoxydim (Poast)	inhibits specific enzyme ACCase	39	15	3
dinitroaniline	oryzalin (Surflan) pendimethalin (Prowl) trifluralin (Treflan, Trilin)	binds to specific protein tubulin important in cell division	11	6	0
diphenylethers	oxyfluorfen (Goal)	inhibits activity of specific enzyme protoporphyrinogen oxidase (PPO)	6	2	0
glycine	glyphosate (Roundup, others)	inhibits specific enzyme EPSP synthase	24	14	6
organoarsenicals	MSMA (MSMA)	general metabolic disruptor, specific site of action unknown	1	1	0
pyridazinones	norflurazon (Solicam)	inhibition of carotenoid biosynthesis at specific enzyme phytoene desaturase	0	0	0
pyridines	thiazopyr (Visor)	blocks cell division by inhibiting microtubule assembly	0	0	0
thiocarbamates	EPTC (Eptam)	inhibits lipid synthesis, probably no specific enzyme site	0	0	0
triazines	simazine (Princep)	electron transport inhibitor of photosystem II in photosynthesis, acts on specific D1 enzyme	69	26	1
uracils and ureas	bromacil (Hyvar) diuron (Karmex, Direx)	electron transport inhibitor of photosystem II in photosynthesis, acts on specific D1 enzyme	23	8	1

A critical aspect in the prevention and management of herbicide-resistant weeds is to recognize resistant populations early in their establishment. It has been reported that resistant weed populations go undetected until growers observe about a 30% failure in the control of a particular species (Prather et al. 2000). By that time, the resistant-weed seed bank will have accumulated to a level that makes eradication nearly impossible and management a long-term process. Thus, monitoring the initial evolution of resistance by recognizing patterns of weed escapes typical to resistant plants is a critical aspect to management and prevention.

Early Recognition of Resistance

Growers should be alert to the appearance of a patch of a particular weed species that survives what should have been an effective treatment. The earlier resistance is dealt with, by rotating to other chemistries or roguing suspected resistant weeds, the easier it is to contain the situation effectively.

Rarely does the pattern of development of herbicide-resistant weeds occur as a complete failure of control on one or several species within the entire

Table 18.3. Top 10 weed species most susceptible to the development of herbicide resistance, the number of chemical families they are reported to be resistant to, and countries with resistance; in order of total number of resistant cases

Common name	Scientific name	Family	Chemical families with resistance	Countries with resistance	Total reported cases
horseweed	*Conyza canadensis*	Asteraceae	5	15	52
common lambsquarters	*Chenopodium album*	Chenopodiaceae	4	18	44
rigid ryegrass	*Lolium rigidum*	Poaceae	11	12	42
wild oat	*Avena fatua*	Poaceae	6	14	44
redroot pigweed	*Amaranthus retroflexus*	Amaranthaceae	4	15	42
kochia	*Kochia scoparia*	Chenopodiaceae	4	3	42
barnyardgrass	*Echinochloa crus-galli*	Poaceae	10	19	35
smooth pigweed	*Amaranthus hydridus*	Amaranthaceae	3	8	23
goosegrass	*Eleusine indica*	Poaceae	7	6	22
green foxtail	*Setaria viridis*	Poaceae	4	5	15

field, or as a strip that would indicate an application skip. Even isolated escapes are seldom associated with resistant weeds, although it is certainly possible. There are numerous other reasons why individual plants can escape injury following herbicide treatment. These include uneven application rates or application skips, subsequent emergence after a postemergent application, degradation of a preemergent herbicide, or presence of an insensitive stage of development at the time of treatment. The key element to preventing herbicide-resistant weeds from becoming well established is to monitor the site soon after application to determine whether the distribution patterns of surviving weeds represent the development of herbicide resistance. The most typical pattern associated with the development and expansion of herbicide-resistant weeds is a small patch of a single species that has escaped treatment when the same species is controlled in other nearby patches. These patches generally expand over time. They should be isolated and controlled by hand-weeding, by using another herbicide, or by using a mechanical method to prevent plants from producing seed.

Once an herbicide-resistant weed becomes well established in an orchard, agricultural crop, or noncrop area, the efficacy of this herbicide, as well as other herbicides with a similar mode of action, will be dramatically limited. Because the registration of new products for weed control in orchards is uncommon, it is important to manage with weed resistance in mind to prevent the loss of current weed control options. The development of herbicide-resistant weeds continues to increase exponentially worldwide, nationally, and within California. Prevention or management against the development of herbicide-resistant weeds requires a multiyear program. Ideally, such a strategy should be initiated before resistance develops and not afterward.

Pesticide Resistance in Fungal Pathogens of Citrus

Fungicide resistance in fungal pathogen populations of citrus in California is mostly a problem in managing soilborne root and trunk diseases and decays of fruit after harvest. Historically, foliar fungicide treatments have been needed only infrequently for crop protection in California due to low rainfall and subsequently low disease pressure in most production areas. Furthermore, most foliar fungal diseases of citrus have been successfully managed with inorganic compounds such as copper or zinc that work through multiple sites of action, making them less prone to selecting resistant populations of fungal pathogens.

In California, fungicides with a single-site mode of action are commonly used for managing soilborne diseases caused by *Phytophthora* species and on harvested fruit before storage or before shipment of the fruit to market. Fungicide resistance has become a major problem in the postharvest management of decays caused by *Penicillium* species, which are among the most important postharvest decays of citrus crops worldwide. Fungicides with a single-site mode of action are used to control these diseases because they can be highly effective. Unfortunately, they are more likely to lead to resistance in heterogeneous species or in organisms with high reproductive rates than are multisite fungicides because the increased potential in variability of the single target site may allow selection of less-sensitive individuals.

Future economic considerations in a highly competitive market may also lead to the development of resistance to foliar fungicides, especially when the same fungicide products or classes (see the Fungicide Resistance Action Committee website, http://www.frac.info/frac/index.htm) are used in the field and in the packinghouse for the same pathogen. An example of this is azoxystrobin, which is registered on citrus for use before and after harvest for postharvest disease management. Ideally, fungicides of different classes than those registered for preharvest uses should be registered for postharvest use. This principle was applied with the development and registration of fludioxonil and pyrimethanil, which were limited to management of postharvest decays of most citrus crops and selected other commodities (Adaskaveg et al. 2003; Kanetis et al. 2008). On lemons, however, pyrimethanil and a premixture of fludioxonil and cyprodinil are registered as preharvest treatments.

Although increased use of preharvest fungicides could further exacerbate postharvest fungicide resistance, the development of new preharvest and postharvest treatments and an improved knowledge of the causes of resistance development will likely lead to the implementation of better resistance management strategies in the future.

Types of Fungicide Resistance and Factors That Contribute to Resistance

Fungicide resistance is a genetically inherited character that allows the fungus to withstand a chemical that previously inhibited its growth (Kendall and Hollomon 1998). Selection and mutation processes in the heterogeneous pathogen population can lead to an increase in resistant individuals under the selection pressure of fungicide usage and eventually, with continuous exposure to a particular fungicide, to a population that is mostly resistant (Hewitt 1998). This reduced sensitivity results in the loss of fungicide efficacy and eventually leads to crop loss. The shift in pathogen population may be temporary or permanent, depending on the fitness of the new population as compared with the original one. Understanding the basics of resistance development and the characteristics of resistant pathogen populations is critical for developing an antiresistance strategy during fungicide use. Resistance in fungal populations has been described as either qualitative (monogenic) or quantitative (polygenic) (Kendall and Hollomon 1998), and these two types are often correlated with single-site or multisite modes of action, respectively (figs. 18.1 and 18.2). In qualitative resistance, a single mutation or a small number of mutations in major genes results in the sudden shift from a sensitive to a resistant population. Populations of pathogens with qualitative resistance generally remain parasitically fit and are stable in the absence of the fungicide. Consequently, the efficacy of the fungicide is lost indefinitely. Examples of this type of resistance are found with the methyl benzimidazole carbamate (MBC) fungicides (e.g., thiabendazole, or TBZ) for control of Penicillium decays of citrus (Eckert 1988).

In quantitative resistance, numerous mutations result in changes that contribute toward the development of a heterogeneous resistant population that is comprised of individuals with different degrees of sensitivity. In this type of resistance, there is no sudden shift but rather a gradual change or selection toward a resistant population with the continued use of the fungicide. Because the multiple changes generally make a resistant population less fit than the sensitive population, the population will revert to sensitivity over time in the absence of the selection pressure (i.e., the fungicide). This type of resistance is typical for the demethylation-inhibiting (DMI) fungicides (e.g., imazalil) used on citrus for control of Penicillium decays (Holmes and Eckert 1999). Pathogen populations that are resistant to imazalil are therefore less common in a packinghouse at the beginning of a new harvest season (fall or early winter) than late in the season (late spring or summer).

Resistance against fungicides that have a single-site mode of action, which includes most of the newer materials (table 18.4), is more likely to develop than against materials with a multisite mode of action that disrupt several vital growth processes. Thus, the resistance potential for materials with a single-site mode of action is much higher. According to their potential to develop resistance, fungicides have been grouped into those that show a low, moderate, or high risk of promoting resistance. In addition, the type of pathogen can determine the resistance potential. For example, high reproduction potential, genetic diversity, and inherent resistance frequency are factors that determine resistance potential in a population. Pathogens with a high reproduction potential, such as *Penicillium* species, are more likely to develop resistance than are pathogens that reproduce more slowly. Generally, fungicide resistance is specific to a chemical class or

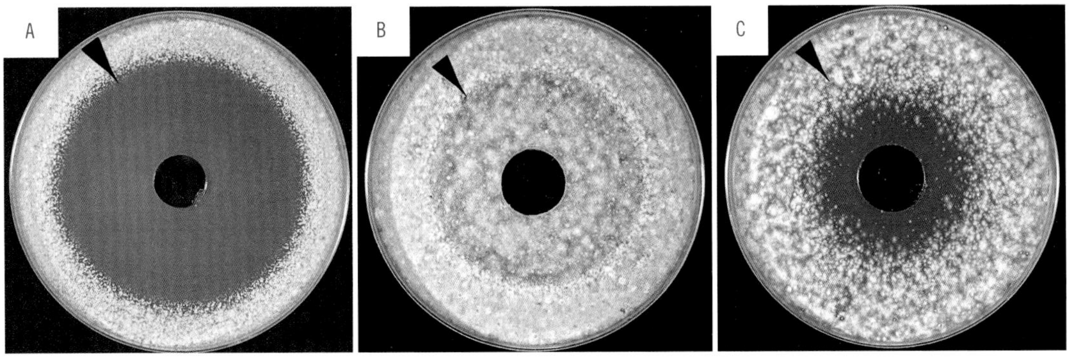

Figures 18.1 A, B, and C

Visualization of types of resistance using the spiral gradient dilution technology that generates agar plates with a fungicide concentration gradient (high concentrations are toward the center and low concentrations are toward the edge of the plate). Examples of packinghouse air samplings of *Penicillium digitatum* populations are shown using plates amended with (A) *fludioxonil* (no resistance), (B) *thiabendazole* (TBZ: qualitative resistance), or (C) *imazalil* (quantitative resistance). With up to the 95% effective concentration (EC_{95}) of each fungicide (black arrows), a high density of fungal colonies is found after 5 days on incubation. No colonies at concentrations greater than EC_{95} are present in the absence of resistance (A). For qualitative resistance, two rather uniform growth patterns are observed across the plate: growth at concentrations lower (sensitive and resistant isolates grow in this region) and higher (only resistant isolates grow in this region) than the EC_{95} values. This indicates the presence of two populations distinct in their fungicide sensitivity (B). For quantitative resistance, there is a gradual decrease in colony density from EC_{95} to higher concentrations toward the center of the plate, indicating a range of sensitivities within the population resulting from a step-wise accumulation of resistance genes (C). *Source:* Adapted from Kanetis et al. 2010.

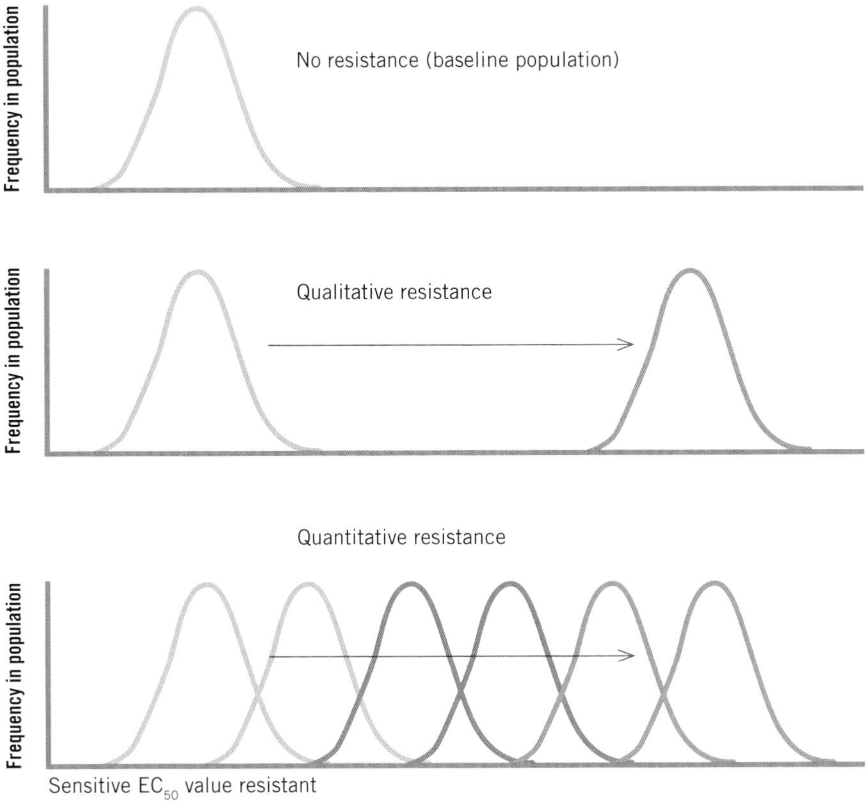

Figure 18.2 Frequency distribution of EC_{50} values in fungal populations with no resistance, with qualitative resistance (e.g., MBC fungicides), and with quantitative resistance (e.g., DMI fungicides). Only one population with a distinct baseline range of sensitivities is observed in a sensitive population (no resistance). For qualitative resistance, a shift in fungicide sensitivity is observed by the presence of two distinct populations: a sensitive baseline population and a resistant population. For quantitative resistance, there is a gradual shift to increased EC_{50} values, resulting in a range of sensitivities within the population due to a step-wise accumulation of resistance genes. For both qualitative and quantitative resistance, frequencies of resistant isolates as compared with sensitive isolates can vary widely (i.e., the heights of each distribution may be different). *Source:* Adapted from Brent 1995.

Table 18.4. Fungicides for managing fungal diseases of California citrus classified by class and mode of action

Products registered on citrus	Use	Chemical class	FRAC code	Primary target site	Potential for resistance*
thiabendazole (TBZ)	postharvest	methyl benzimidazole carbamates (MBC)	1	beta-tubulin assembly	high (*Penicillium* spp.)
imazalil (Freshgard, Imazalil, Deccocil, etc.)	postharvest	demethylation inhibitors; imidazoles (DMI)	3	C-14 demethylation in sterol biosynthesis	medium (*Penicillium* spp.)
fenbuconazole	preharvest	demethylation inhibitors; triazoles	3	C-14 demethylation in sterol biosynthesis	medium
propiconazole	postharvest	demethylation inhibitors; triazoles	3	C-14 demethylation in sterol biosynthesis	medium (*Penicillium* spp.)
mefenoxam‡ (Ridomil Gold)	preharvest	phenylamides	4	RNA polymerase I	high (*Phytophthora* spp.)
pyrimethanil (Scala§, Penbotec)	pre- and postharvest	anilinopyrimidines	9	methionine biosynthesis (proposed)	high (*Penicillium* and *Botrytis* spp.)
azoxystrobin (Abound, Diploma)	pre- and postharvest	quinone outside inhibitors (QoI fungicides)	11	complex III in fungal respiration	high
fludioxonil (Graduate)	postharvest	phenylpyrroles	12	MAP protein kinase in osmotic signal transduction (speculative)	medium
fosetyl-Al (Aliette), phosphorus acid‡ (Phosgard, Prophyt, etc.)	pre- and postharvest	phosphonates	33	unknown	low
sodium ortho-phenylphenate	postharvest	phenols	M	nonspecific DNA damage	medium (*Penicillium* spp.)
ferbam†	preharvest	dithiocarbamates	M3	nonspecific	low
chorothalonil†	preharvest	chloronitriles	M5	nonspecific	low
copper and zinc‡	preharvest	inorganics	M9	nonspecific, multisite	low

Notes:

For fungicide modes of action, see the FRAC website, http://www.frac.info/frac/index.htm. For current label information, see sources such as CDMS (http://www.cdms.net/LabelsMsds/LMDefault.as) and Agrian (http://www.agrian.com/labelcenter/results.cfm).

*Fungicide resistance has been identified in California for pathogens in parentheses.

†Proposed for registration in the United States and California as of 2012. Note that ferbam is only federally registered and is not available in California.

‡Phosphonate and phenylamide fungicides are for management of fungal-like organisms in the kingdom Stramenopila; whereas copper fungicides are for management of fungi and fungal-like organisms.

§For lemons only.

mode of action and reflects a change in the target site. If a fungal isolate is resistant to one chemical within a given class, it generally is also resistant to other chemicals in the same class (cross-resistance). For example, benomyl-resistant populations of fungi are cross-resistant to thiophanate methyl, thiabendazole, and carbendazim. Still, numerous examples are also known where a fungal isolate is resistant to members of more than one fungicide class (multiple resistance). This has been interpreted as changes in each fungicide's target site rather than changes in uptake or detoxification of the pesticide, as is often the case for insecticide resistance. Still, nonspecific detoxification systems such as ABC membrane transporter systems that "pump out" the fungicide toxicant from the cytoplasm of the fungus have been reported in *P. digitatum*. Multiple resistance to unrelated compounds has been reported for *P. digitatum* to biphenyl, benomyl, and 2-aminobutane (Dave et al. 1980) and to imazalil, thiabendazole, and ortho-phenylphenol (Holmes and Eckert 1999).

Based on these concepts, resistance to a particular fungicide is most likely to develop if a single-site mode of action fungicide is applied multiple times to a large population of pathogen propagules. In a California citrus packinghouse situation, lemons are especially prone to the development of resistant *Penicillium* species populations because packinghouses receive shipments of fruit almost year-round, fruit are commonly being stored for up to 4 months before marketing (during which time resistant individuals can be selected and can multiply), and fruit are being treated before and after the storage period, often with the same fungicide. Additional factors that favor the development of resistance include suboptimal application methods due to poorly adjusted equipment and the use of improper fungicide rates that deposit low levels of fungicide residue on the fruit. The low or sublethal rates lead to greater survival, and, subsequently, a larger population is exposed to selection.

Three fungicides, ortho-phenylphenol, thiabendazole, and imazalil, have been extensively used to control postharvest Penicillium decays of citrus (see table 18.4). Each fungicide was introduced separately after resistance to the previously registered materials had already occurred: ortho-phenylphenol in the 1930s, thiabendazole in the 1970s, and imazalil in the 1980s. Several new fungicides have been registered for postharvest use on citrus, including azoxystrobin, fludioxonil, pyrimethanil, and propiconazole (see table 18.4) (Kanetis et al. 2007; Adaskaveg and Förster 2010; McKay et al. 2012). Although the potential for resistance development exists for these fungicides as well, their almost simultaneous introduction offers a unique opportunity to keep the development of resistant pathogen populations to a minimum while still using the older materials by employing antiresistance (resistance management) strategies.

Antiresistance Strategies: Fungicide Stewardship

In fungicide stewardship, a pathogen population's resistance potential to a given product is also a determining factor in fungicide usage (Kanetis et al. 2010). The goal of antiresistance strategies is to modify decay management programs so that the risk of a pathogen population developing fungicide resistance is minimized.

Proper sanitation of fruit and equipment as an antiresistance strategy reduces the amount of pathogen propagules exposed to the fungicide. Thus, a smaller population is being exposed to selection for resistance. Additionally, a routine monitoring program, where fungicide sensitivities of the current pathogen population are compared with those of a population that was sampled before the use of a new fungicide (baseline sensitivity), should be implemented for the early detection of fungicide-resistant isolates. Baseline sensitivities for azoxystrobin, fludioxonil, and pyrimethanil have been established for isolates of *P. digitatum* and *P. italicum* in California, and new high-throughput monitoring methods have been developed (Förster et al. 2004; Kanetis et al. 2010). With a resistance-monitoring program, fungicide usage in rotations and mixtures can be adjusted according to the composition of the pathogen population.

To help delay the development of resistant populations, a new chemical should be gradually introduced into the current chemical program, and it should never be used exclusively. As indicated above, continuous exposure of any pathogen to a fungicide may result in the rapid selection of resistance to the fungicide and shifts in population structure. Fungicides with a single-site mode of action, like most of the pre- and postharvest fungicides registered on citrus for foliar diseases, in addition to the preharvest fungicide mefenoxam for soilborne diseases (see table 18.4), should not be used alone on a continuous basis but instead in rotations or mixtures with fungicides having different modes of action.

Fungicide Rotation Is Essential

Continuous exposure of any pathogen to a single-site MOA fungicide may result in the rapid selection of resistance to the fungicide and shifts in population structure. Thus, rotations and mixtures or rotations of mixtures of different MOA fungicides are essential antiresistance strategies.

If fungicide mixtures are employed as a strategy for resistance management, they must be used from the introduction of the new fungicide or before resistance is detected (table 18.5). This is because a pathogen population that is already resistant to one of the mixture components will be selected for resistance to the second fungicide, and thus, the mixture will not remain effective. Models describing the development of resistance to benomyl in target populations have indicated, however, that once resistance is established in a population, mixtures with another material will reduce the rate of additional selection of the resistant population (Delp 1979). Mixtures should be used in each application and at effective rates for each component in the mixture. Furthermore, each material should have a similarly high efficacy and performance against target populations; otherwise selection of resistant populations may still occur.

Table 18.5. Fungicide premixtures for managing fungal diseases of California citrus classified by class and mode of action

Products registered on citrus	Use	Chemical class	FRAC code	Potential for resistance
difenoconazole-azoxystrobin (Quadris Top)	preharvest	demethylation inhibitors-triazoles; QoI fungicides	3, 11	medium
cyprodinil-fludioxonil (Switch)*	preharvest	anilinopyrimidines; phenylpyrroles	9, 12	medium
azoxystrobin-fludioxonil (Graduate A+)	postharvest	QoI fungicides; phenylpyrroles	11, 12	medium
boscalid-pyraclostrobin (Pristine)	preharvest	succinate dehydrogenase inhibitors (SDHI fungicides); QoI fungicides	7, 11	medium
imazalil-pyrimethanil (Philabuster)	postharvest	demethylation inhibitors-triazole; anilinopyrimidines	3, 9	high

Notes:

For fungicide modes of action, see the FRAC website, http://www.frac.info/frac/index.htm. For current label information, see sources such as CDMS, http://www.cdms.net/LabelsMsds/LMDefault.as, and Agrian, http://www.agrian.com/labelcenter/results.cfm.

*For lemons and limes only.

In summary, antiresistance strategies should include the following integrated fruit health and pest management approaches:

- Optimize the health of the fruit and reduce decay:
  - ▹ Minimize injuries during harvest and handling.
  - ▹ Store fruit at optimal temperature and air circulation to delay senescence.
  - ▹ Limit maximum storage times.
  - ▹ Use plant growth regulators to delay senescence.
- Reduce pathogen numbers:
  - ▹ Sanitize fruit and equipment with multisite, broad-spectrum toxicants.
  - ▹ Regulate air quality (spore load) using filtration and directed airflow in the packinghouse.
  - ▹ Chemically disinfest and physically remove culled fruit from the packinghouse.
- Use labeled fungicide rates and optimal application methods to increase performance and minimize decay:
  - ▹ Integrate alkaline salts (e.g., bicarbonates) with fungicide programs.
  - ▹ Use heated fungicide solutions (about 122°F).
  - ▹ Use compatible sanitizers in recirculating fungicide treatments.
- Use rotations or mixtures of different chemical classes (modes of action) of fungicides
  - ▹ for different lots of fruit
  - ▹ over regular time intervals
  - ▹ in spatially stratified storage rooms.
- Minimize the total number of applications of any chemical class to only one application per fruit lot.
- Monitor pathogen populations routinely for shifts in fungicide sensitivity and change the fungicide program according to the composition of the population.
- Separate repacking of fruit that may have decay and potentially resistant strains of the pathogen from the packinghouse where they were originally packed.

Use this simple acronym to follow the "RULES" for fungicide resistance management:

- **R**otate between different fungicide classes or use mixtures prior to the development of resistance.
- **U**se labeled rates (do not cut rates, as this imposes sublethal selection pressure on fungal populations).
- **L**imit the total number of fungicide applications of any one class ideally to one and no more than two per season.
- **E**ducate yourself about fungicide activity, mode of action, and class (the FRAC system).
- **S**tart a fungicide management program when possible with multisite mode of action fungicides that includes washing equipment before harvest and fruit immediately after harvest with sanitizers, as well as applying integrated treatments such as sodium bicarbonate.

References

Adaskaveg, J. E., and H. Förster. 2010. New developments in postharvest fungicide registrations for edible horticultural crops and use strategies in the United States. In D. Prusky and M. L. Gullino, eds., Post-harvest pathology; plant pathology in the 21st century: Contributions to the 9th International Congress. Dordrecht: Springer Verlag. 107–117.

Adaskaveg, J. E., H. Förster, and N. F. Sommer. 2003. Principles of postharvest pathology and management of decays of edible horticultural crops. In A. Kader, ed., Postharvest technology of horticultural crops. 4th ed. Oakland: University of California Agriculture and Natural Resources Publication 3311. 163–195.

Brent K. J. 1995. Fungicide resistance in crop pathogens: How can it be managed? FRAC Monograph No. 1. Brussels, Belgium: GIFAP.

Dave, B. A., H. J. Kaplan, and J. F. Petrie. 1980. The isolation of *Penicillium digitatum* Sacc. strains tolerant to 2-AB, SOPP, TBZ, and benomyl. Proceedings of the Florida State Horticultural Society 93:344–347.

Delp, C. J. 1979. Resistance to plant disease control agents–How to cope with it. 9th International Congress of Plant Protection. Washington, D.C.

Eckert, J. W. 1988. Dynamics of benzimidazole-resistant *Penicillia* in the development of postharvest decays of citrus and pome fruit. In C. J. Delp, ed., Fungicide resistance in North America. St. Paul, MN: APS Press. 31–35.

Förster, H., L. Kanetis, and J. E. Adaskaveg. 2004. Spiral gradient dilution, a rapid method for determining growth responses and EC_{50} values in fungus-fungicide interactions. Phytopathology 94:163–170.

FRAC (Fungicide Resistance Action Committee) website, http://frac-online.org/.

Grafton-Cardwell, E., Y. Ouyang, R. Striggow, and S. Vehrs. 2001. Armored scale insecticide resistance challenges San Joaquin Valley citrus growers. California Agriculture 55(5): 20–25.

Groeters, F. R., and B. E. Tabashnik. 2000. Roles of selection intensity, major genes, and minor genes in evolution of insecticide resistance. Journal of Economic Entomology 93:1580–1587.

Heap, I. 2013. The international survey of herbicide resistant weeds. WeedScience website, www.weedscience.com.

Hewitt, H. G. 1998. Fungicides in crop protection. Cambridge, UK: CAB International.

Holmes, G. J., and J. W. Eckert. 1999. Sensitivity of *Penicillium digitatum* and *P. italicum* to postharvest citrus fungicides in California. Phytopathology 89:716–721.

Immaraju, J. A., J. G. Morse, and R. F. Hobza. 1989. Field evaluation of insecticide rotation and mixtures as strategies for citrus thrips (Thysanoptera: Thripidae) resistance management in California. Journal of Economic Entomology 83:306–314.

Immaraju, J. A., J. G. Morse, and D. J. Kersten. 1989. Citrus thrips (Thysanoptera: Thripidae) pesticide resistance in the Coachella and San Joaquin Valleys of California. Journal of Economic Entomology 82:374–380.

IRAC (Insecticide Resistance Action Committee) website, http://irac-online.org/.

Kanetis, L., H. Förster, and J. E. Adaskaveg. 2007. Comparative efficacy of the new postharvest fungicides azoxystrobin, fludioxonil, and pyrimethanil for managing citrus green mold. Plant Disease 91:1502–1511.

———. 2008. Baseline sensitivities for new postharvest fungicides against *Penicillium* spp. on citrus and multiple resistance evaluations in *P. digitatum*. Plant Disease 92:301–310.

———. 2010. Determination of natural resistance frequencies in *Penicillium digitatum* using a new air-sampling method and characterization of fludioxonil- and pyrimethanil-resistant isolates. Phytopathology 100:738–746.

Kendall, S. J., and D. W. Hollomon. 1998. Fungicide resistance. In D. Hutson and J. Miyamoto, eds., Fungicidal activity: Chemical and biological approaches to plant protection. New York: Wiley. 87–108.

Khan, I., and J. G. Morse. 1998. Citrus thrips (Thysanoptera: Thripidae) resistance monitoring in California. Journal of Economic Entomology 91:361–366.

McKay, A. H., H. Förster, and J. E. Adaskaveg. 2012. Efficacy and application strategies for propiconazole as a new postharvest fungicide for managing sour rot and green mold of citrus fruit. Plant Disease 96:235–242.

Morse, J. G., and O. L. Brawner. 1986. Toxicity of pesticides to *Scirtothrips citri* (Thysanoptera: Thripidae) and implications to resistance management. Journal of Economic Entomology 79:565–570.

Morse, J. G., R. F. Luck, and E. E. Grafton-Cardwell. 2007. The evolution of biologically-based integrated pest management on citrus in California. UC Plant Protection Quarterly 17(1): 1–11.

Prather, T. S., J. M. DiTomaso, and J. S. Holt. 2000. Herbicide resistance: Definitions and management strategies. Oakland: University of California Agriculture and Natural Resources Publication 8012. UC ANR catalog website, http://anrcatalog.ucdavis.edu/Weeds/8012.aspx.

Preston, C. 2004. Herbicide resistance in weeds endowed by enhanced detoxification: Complications for management. Weed Science 52:448–453.

Quayle, H. J. 1938. The development of resistance to hydrocyanic acid in certain scale insects. Hilgardia 11:183–210.

Roush, R. T., and J. A. McKenzie. 1987. Ecological genetics of insecticide and acaracide resistance. Annual Review of Entomology 32:361–380.

Sherwood, A. M., and M. Jasieniuk. 2009. Molecular identification of weedy glyphosate-resistant *Lolium* (Poaceae) in California. Weed Research 49:354–364.

Simarmata, M., S. Bughrara, and D. Penner. 2005. Inheritance of glyphosate resistance in rigid ryegrass *(Lolium rigidum)* from California. Weed Science 53:615–619.

Tabashnik, B. E., and B. A. Croft. 1982. Managing pesticide resistance in crop-arthropod complexes: Interactions between biological and operational factors. Environmental Entomology 11:1137–1144.

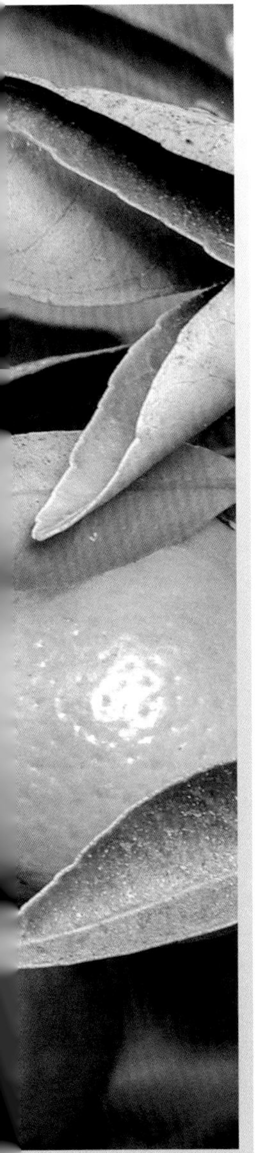

19 Invasive Pests: Insects

David H. Headrick

Invasive species are organisms that enter in new areas where they have not previously occurred and whose introduction causes economic and environmental harm or harm to human health. Not all insect species that arrive in California cause problems; many never establish and disappear. Some may establish but are not considered harmful to human endeavors or the environment, and thus are not considered pests. Some species arrive and become pests to varying degrees. Many times we do not anticipate these invasions, nor can we predict whether a species will become a serious pest. But certain species are known to be dangerous to humans or agriculture, such as the Mediterranean fruit fly; their arrivals are anticipated, and they become the targets of aggressive eradication programs.

Although the California citrus industry is focused on invasive species that are currently causing economic harm, the industry and its partners in the research community know the importance of looking ahead at species that are not yet here but may cause economic and environmental harm. This chapter provides basic information about potential invasive insects that do not occur in California but represent a known, significant threat to citrus production in other parts of the world. For detailed discussions and photographs of invasive insects currently causing harm, see chapter 17, "Citrus IPM," and for plant pathogens and quarantine regulations, see chapter 24, "Invasive Pests: Exotic Plant Pathogens."

The process of invasion—the arrival or spread of a harmful pest species—generally occurs in this sequence: arrival, establishment (successfully developing through multiple successive generations), range expansion and proliferation (reaching densities that can cause economic damage), and eventual naturalization (becoming a recognized part of the landscape). This process usually takes many years. For example, glassy-winged sharpshooter (*Homalodisca vitripennis* (Germar)) was first noted in the late 1980s in Ventura and Irvine, but it began an aggressive expansion and became a cause of concern in the 1990s and continues to expand its range today. Invasion of California by non-native insect species is taking place with greater frequency than ever before, and some of these species are causing tremendous harm to citrus production and the economy.

The concept of an invasion by a non-native insect species becomes confusing when interpreting "non-native" solely on the basis of the state's geopolitical borders. Insects do not pay attention to the lines on a map; therefore, it is possible to have an invasion spread within the state. This concept is the basis of many of our quarantine regulations that are implemented to slow the movement of invasive species from infested to uninfested parts of the state. At this level, the citrus industry and regulatory agencies typically work along county borders.

Invasive species are a concern for agriculture because of their fiscal impacts on all sectors of the economy. There are annual costs for state and federally funded exclusion and detection programs. Following detection, implementation of quarantine regulations to aid eradication or slow the spread can result in costs that include loss of product and revenue. There are also costs for eventual regulatory and phytosanitary compliance and changes to production practices. Eradication programs require a rapid buildup of personnel, facilities, and materials but also include lengthy environmental impact studies, public relations campaigns, and costs of inevitable litigation. If an invasive pest becomes established, we are burdened with the long-term annual costs of control in the affected cropping system, including national and international quarantine restrictions that affect marketing and price increases for the domestic consumer.

The California citrus industry has been assiduously dealing with invasive insects since its inception. In the 1880s and early 1900s, cottony cushion scale (*Iceyra purchasi* Maskell) and citrophilus mealybug (*Pseudococcus calceolariae* (Maskell)) virtually eliminated the citrus industry in southern California. Many invasive insects of importance to the citrus industry were controlled using classical biological control (the importation and release of natural enemies from the invasive pest's native range). Today, responses to invasive species are far more complicated and require multiple management tactics involving costly mitigation technologies (e.g., monitoring and sampling programs, pesticides, and pheromones) and implementation of research-derived management programs.

Currently, the citrus industry in California is dealing with several invasive species, including Asian citrus psyllid (*Diaphorina citri* Kuwayama); glassy-winged sharpshooter (*Homalodisca vitripennis* (Germar)); Diaprepes root weevil (*Diaprepes abbreviatus* L.); and citrus leafminer (*Phyllocnistis citrella* (Stainton)). One of the main concerns about invasive insects is their effect on daily production practices, typically in the form of quarantine restrictions and regulations and additional broad-spectrum pesticide applications until other controls can be developed. Invasive species are highly disruptive to established integrated pest management programs and can influence management decisions for years, such as the pest management responses to glassy-winged sharpshooter and the Asian citrus psyllid (see chapter 17, "Citrus IPM").

Invasive species such as Asian citrus psyllid and citrus leafminer are citrus specialists—they feed on little else—and thus may be perceived as a problem only for citrus growers. The risks other invasive species pose can be far reaching; they are directly related to the large number of host plant species attacked and so are not isolated to the citrus industry. Mediterranean fruit fly is known from approximately 200 different fruit and vegetables, most of which are grown in California. According to a University of California report, if Mediterranean fruit fly were to become established, California could expect to lose $538 million in agricultural output, $259 million in total income, $283 million in gross state product, and 7,900 jobs. Similarly, the costs associated with the establishment of Oriental fruit fly in California could range up to $176 million in crop losses, additional pesticide use, and quarantine requirements. The costs of detection and eradication programs may seem expensive at $16 million per year just for Mediterranean fruit fly, but they are trivial compared with the complete spectrum of losses if these generalist pests became established in California.

Researchers have compiled a list of over 100 of the most damaging insect species of citrus from around the world that pose a threat to California citrus production based on similarities in climate, citrus varieties, relatedness to other pest insects, and lack of effective management tools. Although pest status in one area of the world does not necessarily equate to pest status in California, it does raise our awareness of the efforts needed to exclude these species. Table 19.1 presents a representative number of species from a variety of insect orders to demonstrate the broad range of potential pests and how they may impact current management practices; a detailed examination of six of these species is the focus of the rest of this chapter.

These insect species are not known to occur in California but are representative of some of the species that are considered a significant risk for invasion. Some of them do occur in other places in the United States or North America, making exclusion more challenging. The descriptions provided include scientific and common names; the U.S. distribution, if present; identification information; the life cycle in areas where it occurs; the native distribution where it occurs outside of the United States and the host plant species used for development; and information on the types of damage caused and control methods.

African Citrus Psyllid, *Trioza erytreae* (del Guercio)

Other names. Two-spotted citrus psyllid.

Current U.S. distribution. Not known to be in the United States.

Identification. Small to minute insects with egg, nymph, and adult stages (fig. 19.1). Eggs are cylindrical, tapering to a point, and orange; they are laid on the margins of tender leaves of flush growth.

Figure 19.1 Adult African citrus psyllid. *Photo:* P. Stephen.

Nymphs are typically sessile feeders, compressed onto a leaf surface; the piercing-sucking mouthparts are on the underside of the body and not visible from above. The body is oval and flattened; individual nymphs vary from yellow to green to dark gray, with two small red eyespots visible on the head and two larger dark spots on the lateral margins of the lower abdomen; there is a distinct marginal fringe of white, waxy filaments. Individuals are usually clustered into colonies on the underside of young leaves. Feeding results in distortion of the leaf tissues under the nymph body into a cup-shaped depression or "open gall." Adults are dark gray to brown; females (0.09 in long) are slightly larger than males; the abdomen ends in a sharp point in females (ovipositor), and male abdomens are blunt. Wings are held together over the abdomen when at rest. When feeding, adults have a distinctive head-down stance. African citrus psyllid differs in appearance from the Asian citrus psyllid (established in southern California) in that the African citrus psyllid adult has clear wings and a darker body. Also, African citrus psyllid nymphs form cup-shaped feeding depressions in leaves, whereas the Asian species does not; the Asian citrus psyllid nymphs often produce long, curled, waxy honeydew tubules.

Life cycle. Populations in the native range tend to favor cool, moist conditions at over 1,600 feet elevation, where citrus flushes are prolonged. African citrus psyllid is sensitive to hot, dry weather, with the eggs and 1st instar nymphs being the most vulnerable. Adult longevity ranges from 17 to 50 days. Mating is reported to occur two to four times per day, with oviposition immediately following. Females remain fertile for up to 2 weeks in the absence of males. Females can produce 2,000 eggs over their lifetime, with daily maximum egg production during the middle of their adult life. Eggs are inserted into the leaf epidermis. Egg eclosion (hatch) takes place 6 to 15 days after oviposition, and nymphal development takes 17 to 43 days through 5 instars. Growth and development of all feeding stages can be prolonged in the absence of young foliage.

Distribution and host plants. *Trioza erytreae* is native to southern Africa and completes its life cycle on host plant species limited to the Rutaceae (citrus family): wild hosts are *Clausena anisata* (Willd.) Hook.f. ex Benth. and *Vepris lanceolata* (Lam.) G.Don; cultivated hosts are in the genus *Citrus*, especially lemons (*C. limon* (L.) Burm. F) and limes (*C. aurantiifolia* Swingle). This species has been found in Cameroon, Comoros, Ethiopia, Kenya, Madagascar, Malawi, Mauritius, Réunion, Rwanda, South Africa, St. Helena, Sudan, Swaziland, Tanzania, Uganda, Zaire, Zambia, and Zimbabwe. Outside of Africa, *T. erytreae* has been recorded in Saudi Arabia and Yemen, and outbreaks on Madeira (Portugal) and the Canary Islands (Spain) have also been reported.

Damage and control. *T. erytreae* nymphal feeding severely galls tender leaves, resulting in stunted and deformed flush growth; leaves and surrounding surfaces become dusted with nymphal and adult fecal pellets. This species also vectors the bacteria that causes citrus greening disease (see chapter 22, "Bacterial Diseases"). Insecticides are used to reduce *T. erytreae* in commercial orchards, but areas such as Cameroon report little efficacy against high populations. In Réunion, effective biological control of *T. erytreae* has been achieved with the introduction of *Tamarixia dryi* (Waterston) from South Africa. In South Africa, numerous natural enemies occur but are considered ineffective in population control in commercial settings. The economic impacts of this species are estimated to be similar to that of Asian citrus psyllid for growers in the United States, adding about $500 per acre per year in direct costs of pesticide treatments.

Brown Citrus Aphid, *Toxoptera citricida* (Kirkaldy)

Current U.S. distribution. Florida (1995).

Identification. Brown citrus aphid is a relatively large-bodied aphid compared with other citrus-feeding aphid species. Aphids occur in colonies of mixed stages on flush growth of citrus (fig. 19.2); adults occur as both winged (alatae) and wingless forms (apterae); nymphs are always wingless. Wingless adults are shiny black, and nymphs are dark reddish brown. Adult winged forms are distinguished by their three dark-colored basal antennal segments. In California, the common occurrence of other dark-colored aphid species on flush growth could make field identification of brown citrus aphid difficult.

Figure 19.2 Mixed stages of brown citrus aphid. *Photo:* W. Hunter.

Life cycle. This species of aphid typically has active colonies throughout the year; although an overwintering egg stage is known to occur in temperate areas, overwintering eggs are not typically the source of new colonies. Females live for 28 to 48 days, producing 5 to 6 nymphs per day and up to 80 nymphs per lifetime. In Florida, the growth of the colony and the production of alates that fly to colonize new plants depend on the presence of tender flush, which lasts only 3 to 4 weeks. The decline of flush growth can limit and even stop further colony development.

Distribution and host plants. Brown citrus aphid is found throughout Southeast Asia, where it is assumed to have originated. It has spread to sub-Saharan Africa, including Ghana, as well as Somalia and Tunisia. It also occurs in Australia, New Zealand, the Pacific Islands, South and Central America, the Caribbean, and, most recently, Florida. It is not yet established in California; incoming plant material is screened to prevent its introduction. The host plant range of brown citrus aphid is mainly restricted to *Citrus* species.

Damage and control. Brown citrus aphid infests the stems and new leaves of citrus trees. Heavy populations build up on young shoots, weakening trees and stunting growth. It is an efficient vector of citrus tristeza virus (CTV), which causes the disease tristeza (see chapter 23, "Virus and Viroid Diseases"). Virus transmission is 6 to 25 times more efficient than in other aphid species found in California.

There is a low threshold for the presence of brown citrus aphid in citrus orchards because just one infected insect can spread CTV and control options are therefore limited. Cultural control practices for CTV primarily rely on reduction of virus inoculum through tree removal and production of disease-free nursery stock. Biological control agents of brown citrus aphid have not been successful, and no single natural enemy has shown potential for consistent population reduction to the point of reducing disease spread. Chemical control is expensive and disruptive and has not been shown to effectively reduce the transmission of CTV. Removal of CTV-infected trees is the only effective control, as there is no known cure for this disease.

Fruit Flies

Current U.S. distribution. Several species of tropical fruit flies (family Tephritidae) can be serious pests of citrus in Hawaii and Florida. Some of the most well known examples include the Caribbean fruit fly (*Anastrepha suspensa* (Loew)), Mexican fruit fly (*Anastrepha ludens* (Loew)), Mediterranean fruit fly (*Ceratitis capitata* (Wiedemann)), and Oriental fruit fly (*Bactrocera dorsalis* (Hendel)). Less well known species such as guava fruit fly (*Bactrocera correcta* (Bezzi)), melon fly (*Bactrocera cucurbitae* (Coquillett)), and Queensland fruit fly (*Bactrocera tryoni* (Froggatt)) are becoming more frequent invaders into California than they have been in the past.

Identification. Tephritid fruit flies are typically large, with conspicuously patterned wings (fig. 19.3). Wing patterns vary from spots to stripes, depending on the species, and are used in taxonomic identifications. Adult bodies are often distinctly colored; some have contrasting stripes that make them appear somewhat wasplike. Larvae resemble typical fly maggots: white, with a tapered head that has dark mouth hooks and a bluntly rounded tail end.

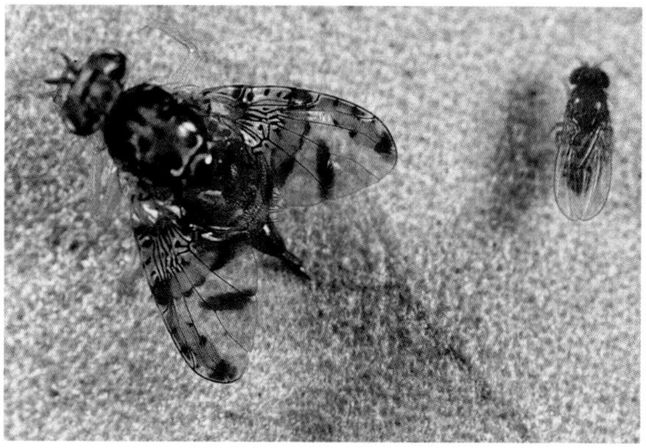

Figure 19.3 Patterned wings of the Mediterranean fruit fly. *Photo:* J. K. Clark.

Life cycle. Fruit-feeding species have multiple generations per year as long as suitable plant tissues are available for larval development. Diapause can occur in the egg or pupal stages during weather extremes. The life cycle of most fruit fly species takes about 1 month, but it varies with temperature. Adults are typically long-lived, ranging from a few weeks to 6 months or more under favorable conditions. Adults become sexually mature within a few days after emergence, after which they mate and lay eggs. Fruit fly females commonly mate more than once. Adult female fruit flies insert their eggs in living plant tissues, usually several eggs at a time. Female fruit flies can lay up to several hundred eggs over their lifetime. There are 3 larval instars; larval development may take 2 to 3 weeks, depending on the environmental conditions and the type of host plant. The larvae of fruit-infesting fruit fly species exit the fruit after completing their larval development and drop to the ground to pupate in the soil or leaf litter. Pupation lasts from 1 to 3 weeks, after which adults emerge from the puparium in the soil or leaf litter and begin the life cycle again.

Distribution and host plants. The host range for many economically important fruit fly species is extensive. For example, the Mediterranean fruit fly has been recorded from over 200 species of fruit and vegetables. Mexican fruit fly prefers citrus, but it also is known to successfully breed in nearly 30 other species of tropical, subtropical, and deciduous fruit, including apple, avocado, cherimoya, pear, and persimmon.

Damage and control. Fruit fly larvae feed in the citrus fruit, rendering it unsuitable for consumption. Losses to growers are incurred directly by a loss of fruit due to infestation and indirectly due to quarantine regulations restricting movement of fruit. California has a comprehensive and highly effective detection program to quickly find and eradicate fruit flies. The detection program targets a wide range of fruit fly species using baited traps on a grid pattern. If detected, a series of state and federal protocols are enacted to delimit the infestation, establish quarantines, and begin eradication. Fruit flies are susceptible to the sterile insect technique, which allows for complete population eradication. In other states where pest fruit flies are established, management tactics include foliar sprays, poisoned baits (some species-specific), soil drenching, fruit stripping, and sterile insect release programs.

Golden Headed Weevil, *Compsus auricephalus* (Say)

Current U.S. distribution. Parts of the U.S. Gulf Coast; Mexico to Panama.

Identification. A distinctive weevil with a short snout; adults whitish, with light green markings (scales) on the elytra and body; head golden to yellow (fig. 19.4). There is considerable variation in the color and size of individuals in this species. Larvae are typical legless, C-shaped, white or translucent grubs that feed on roots in soil.

Life cycle. No information available; adults are collected in June and July in southern Texas.

Figure 19.4 Golden headed weevil. *Photo:* A. Sergeev.

Distribution and host plants. Eleven species of *Compsus* are known in the Neotropics, with only *C. auricephalus* extending to the United States, where it has been recorded in Arkansas, Colorado, Georgia, Louisiana, Mississippi, and Texas. It is also known in Costa Rica, Guatemala, Mexico, Nicaragua, and Panama. *Compsus auricephalus* was reported to be common on citrus in Texas after a 1983 freeze event. Older collection records report feeding on the roots of weeds in Texas. Newer records have recorded this species on mesquite *(Prosopis glandulosa* Torr. var. *glandulosa)* flowers and pods in Texas and on the genus *Zea* in Nicaragua.

Damage and control. Significant damage is reported in Texas; *Compsus* is under an eradication program there, but the current status is unknown. The weevils are reported to cause a rapid tree decline starting with leaf wilt, yellowing, and defoliation, followed by tree death in 4 to 5 weeks. Larval feeding takes place on main and large lateral roots, resulting in deep grooves or channeling that subsequently leads to attack by Phytophthora root rot pathogens; the combination of weevil feeding and pathogens appears to be the cause of tree death. There are no current control recommendations. The larvae of this species are soil dwelling and are difficult and costly to control.

Spined Citrus Bug, *Biprorulus bibax* (Breddin)

Current U.S. distribution. Not found in the United States.

Identification. Eggs are initially white, then become mottled with black and red as they develop. Early-instar nymphs are marked with black, green, yellow, white, and orange. Later-instar nymphs are mainly green, with black markings. Adults are green, 0.6 to 0.8 inch long, and have a pair of prominent dark spines on the "shoulder" of the thorax just behind the head (fig. 19.5).

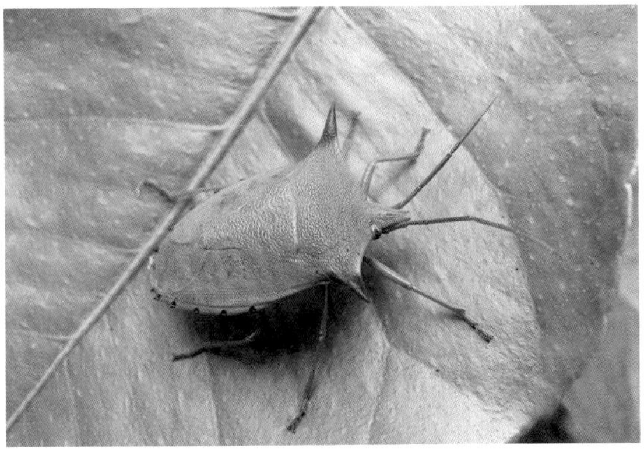

Figure 19.5 Spined citrus bug. *Photo:* D. Papacek.

Life cycle. An egg stage, 5 nymphal instars, and an adult. Oviposition occurs from October to April in Australia, with two peaks in November and February to March. Eggs are laid on leaves, fruit, or twigs in clusters of 4 to 36 eggs, as is common with other species in this family. The number of eggs per clutch is generally higher during the early part of the season. The 1st instar nymphs congregate on empty eggshells and remain together until ready to molt to the next instar, after which they disperse. Nymphal instars are observed throughout most of the year, except in winter. Adults have been recorded throughout the year, with the greatest number in winter, when they tend to aggregate on native host plants. Adult females live for up to 18 months; in Australia, there are three generations per year in temperate areas and four generations in subtropical or tropical areas. A total of 455 degree-days (DD) are required for development from egg to adult, with a lower developmental threshold of 57.7°F. Adult males produce an aggregation pheromone that is attractive to both genders, resulting in clusters of up to 50 individuals on their native host plants near citrus plantings. Adults then disperse in spring and begin oviposition on nearby citrus (lemon) production areas. The pheromone is commercially available in the United States.

Distribution and host plants. The native host plant species is the desert lime *(Eremocitrus glauca)*. In the 1980s, this species became a damaging pest of citrus in temperate areas of Australia such as New South Wales, Victoria, and South Australia; it also occurs in Queensland.

Damage and control. Adults and post–1st instar nymphs feed on all stages of citrus fruit. Lemons are preferred over other citrus types in Australia. The insects use their piercing-sucking mouthparts to pierce the rind and feed on fluids from the interior of the fruit, which dries and stains the fruit segments brown and causes gumming on the skin and premature fruit drop. Management in Australia includes cultural and biological controls. Use of the commercially available aggregation pheromone allows for removal of overwintering adults. Biological control involves conserving already-present natural enemies. Chemical control measures are recommended when only 10% of trees are infested with adults or nymphs.

Sri Lanka Weevil, *Myllocerus undecimpustulatus undatus* Marshall

Current U.S. distribution. Florida (2000).

Identification. *Myllocerus undecimpustulatus undatus* is a short-snouted weevil subspecies from Sri Lanka; many related species are known on the Indian subcontinent. The adults are a dull whitish color, with a mottled pattern of darker spots on the body and a somewhat yellowish head; there is much variation in spot patterning among individuals (fig. 19.6). The femora on all legs have a subapical expansion near the attachment point with the tibia, similar to the boll weevil (*Anthonomus grandis* Boheman). The front and middle leg expansions have distinct bidentate projections or spines, and the hind femora have three spines. The elytra (forewings) have strongly angled humeri (the shoulders near the point of attachment) and project beyond the width of the prothorax. No weevils in California have these characteristics.

Figure 19.6 Sri Lanka weevil. *Photo:* L. Buss.

Life cycle. Adult females insert eggs singly into soil or leaf litter and may lay up to 360 eggs over a 3-day period. Eggs hatch in 3 to 5 days, and larvae burrow through the soil to feed on roots of host plants for 1 to 2 months. Pupation takes place in the soil or leaf litter and lasts approximately 1 week. Adults emerge from pupation and live from 10 to 150 days; there can be several generations per year, depending on temperature and availability of resources.

Table 19.1. Common insect pests of citrus outside the United States

Insect order Common name: Area of origin	Scientific name	Feeding damage
Blattodea		
Asian cockroach: Japan	*Blatella asahinai* Mizukubo	foliage, flush terminals
Coleoptera		
golden headed weevil: Neotropics	*Compsus auricephalus* (Say)	foliage and roots
little leaf notcher: Sri Lanka	*Artipus floridanus* (Horn)	foliage and roots
southern root weevil: Mexico	*Pachnaeus litus* (Germar)	foliage and roots
Sri Lanka weevil: southern India, Sri Lanka, Pakistan	*Myllocerus undecimpustulatus undatus* Marshall	foliage and roots
Diptera		
Chinese citrus fly: Bhutan, China	*Bactrocera minax* (Enderlein)	fruit
citrus gall midge: South America, Neotropics	*Prodiplosis longifila* (Gagne)	galls, flowers
Hemiptera		
African citrus psyllid: Africa	*Trioza erytreae* (del Guercio)	foliage, flush terminals
brown citrus aphid: widespread distribution	*Toxoptera citricida* (Kirkaldy)	foliage, flush terminals
spined citrus bug: Australia	*Biprorulus bibax* (Breddin)	fruit
passionvine mealybug: Neotropics	*Planococcus minor* (Maskell)	phloem
cottony citrus scale: Australia	*Pulvinaria polygonata* (Cockerell)	phloem
citrus snow scale: Asia	*Unaspis citri* (Comstock)	phloem
Isoptera		
eastern subterranean termite: eastern U.S.	*Reticulitermes flavipes* (Kollar)	roots
Lepidoptera		
summer fruit tortrix: Europe and Asia	*Adoxophyes orana* (Fischer von Röslerstamm)	foliage
fruit piercing moth: Asia, Australia	*Eudocima fullonia* (Clerck)	foliage and fruit
false codling moth: Africa	*Thaumatotibia leucotreta* (Meyrick)	fruit
Thysanoptera		
citrus thrips: South Africa	*Scirtothrips aurantii* Faure	fruit
chili thrips: Southeast Asia, India	*Scirtothrips dorsalis* (Hood)	foliage, flowers
plague thrips: Australia	*Thrips imaginis* (Bagnall)	foliage, flowers

Distribution and host plants. This subspecies is known to originate from Sri Lanka. The genus has over 300 species that occur over a large biogeographic range, including India, the rest of southern Asia, Indonesia, China, Japan, Australia, and Africa, as well as Europe, northern Asia, and the Middle East. No species from this genus is known to originate from North America.

This pest has a wide host range. In Florida, it has at least 103 host plant species, including natives, ornamentals, and fruit crops such as acacia, avocado, black olive, bottlebrush, carambola, citrus, crepe myrtle, ficus, live oak, longan, loquat, lychee, mango, muscadine grape, orange jasmine, papaya, palms, passion fruit, pigeon plum, plumbago, pygmy date palm, and Surinam cherry. The large number of host plant species increases the probability of this weevil being transported and redistributed in potted ornamental and nursery stock plants to other parts of the United States and California.

Damage and control. Adults feed aggressively on leaves of host plant species; they chew the leaves from the margins, creating deep notches in the leaf blade and sometimes leaving only the main veins untouched. Feeding takes place on all leaf stages, but new flush is highly vulnerable. Trees eventually decline and can be stunted from continual loss of leaf tissues. Larvae feed on roots; the host plant range may not fully match that of the adult, and work is ongoing to verify the larval host range in Florida. Control options are limited for growers in Florida. There are no specific biological control agents, and only a limited number of effective cultural control methods are available. These include destruction of abandoned orchards and cleanup of host plants in vacant lots. Disking between rows in orchards can reduce larval and pupal populations but may destroy roots.

Chemical control has limited efficacy. Several common pesticides are registered for adult control on citrus, but adults are known to disperse from pesticide applications. Egg, larval, and pupal stages occur in the soil and are protected from pesticide applications.

References

Dreistadt, S. H. 2012. Integrated pest management for citrus. 3rd ed. Oakland: University of California Agriculture and Natural Resources Publication 3303.

Elton, C. S. 2000. The ecology of invasion by animals and plants. Chicago: University of Chicago Press.

Heywood, V. H. 1989. Patterns, extents and modes of invasions by terrestrial plants. In J. A. Drake et al., eds., Biological invasions: A global perspective. New York: Wiley. 31–60.

National Research Council. 2002. Predicting invasions of nonindigenous plants and plant pests. Washington, DC: National Academies Press.

Perrault, A., M. Bennett, S. Burgiel, A. Delach, and C. Muffett. 2003. Invasive species, agriculture and trade: Case studies from the NAFTA context. Proceedings of the 2nd North American Symposium on Assessing the Environmental Effects of Trade. Mexico City. Montreal: North American Commission for Environmental Cooperation.

Pimentel, D., ed. 2002. Biological lnvasions: Economic and environmental costs of alien plant, animal, and microbe species. Boca Raton: CRC Press.

Sawyer, R. 1996. To make a spotless orange: Biological control in California. Ames: Iowa State University Press.

Van Driesche, J., and R. Van Driesche. 2000. Nature out of place: Biological invasions in the global age. Washington, DC: Island Press.

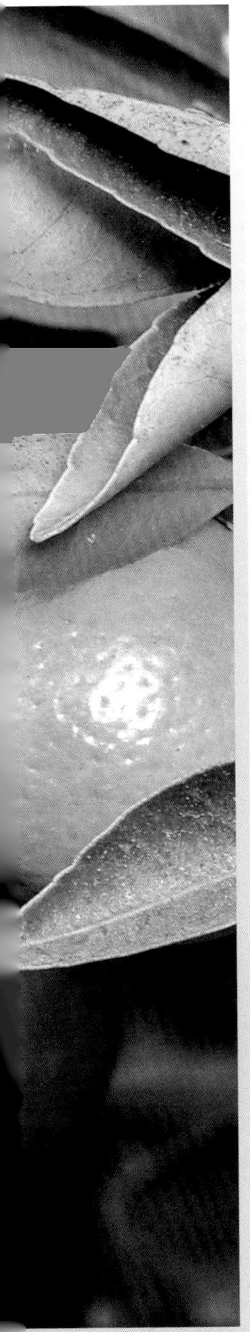

20 Nematodes

J. Ole Becker

Nematodes are aquatic invertebrate roundworms that are mostly threadlike, unsegmented, and typically about 0.016 to 0.040 inch long. They are the most abundant multicellular animals on earth. Nematodes occur in almost every terrestrial and aquatic environment. In spite of their ubiquitous presence, nematodes are inconspicuous because of their typically minute size, nearly transparent body, and hidden habitats. Most nematodes live in soil, either on land or in sediment layers of oceans, lakes, or rivers. These organisms play an important role in the cycling of soil nutrients by feeding on microorganisms such as bacteria, fungi, and algae. Other beneficial nematodes include those that function as biological control agents against certain insect larvae or snails. Some nematode species are specialized parasites of animals or plants. Nematodes are the causal agents for a number of severe diseases in humans, pets, and livestock. Plant-parasitic nematodes constitute only a small fraction of the total number of nematode species, but their impact on U.S. agriculture results in an estimated crop production loss of at least $11 billion annually.

Plant-parasitic nematodes are easily distinguished from other nematode species by the presence of a stylet, a syringelike structure adapted for feeding on host tissues. They push the stylet into plant cells and withdraw the nutritious contents with the pumping action of an internal muscular bulb. Damage to the root varies, depending on the nematode species and the response by the plant host. Ectoparasitic nematodes feed, molt, and reproduce on the outside of their hosts. Short stylet-bearing nematodes, such as stubby root nematodes, affect only the surface layers of the epidermis. Nematodes with long stylets, such as sting nematodes, may inflict more severe damage by damaging the growing point of a root. Endoparasites such as root knot nematodes spend most of their life inside the root tissues. These parasites are among the most damaging because their reproductive capability can result in high population densities. The roots not only provide them with nutrients, they also serve as protective housing against threats such as predators, pathogens, drought, or pesticides. The typical life cycle of a plant-parasitic nematode includes the egg, four juvenile stages, and the adult. The first-stage (J1) and second-stage (J2) juveniles develop in the egg, from which the latter one hatches. The minimum and maximum soil temperature for egg hatch, development, and feeding activity of citrus parasitic nematodes is generally from 68° to 86°F.

The Citrus Nematode

Although more than 200 species of plant-parasitic nematodes have been reported in association with roots of citrus trees, only a few have been shown to be of economic importance. The citrus nematode *(Tylenchulus semipenetrans)* is the only major nematode pest in California citrus. It was first discovered in California in 1912, but it was another four decades before its importance on citrus tree health was realized. The worldwide distribution of the nematode suggests that it has been spread by infested planting material.

Under optimal conditions, the citrus nematode completes its life cycle in about 6 to 8 weeks. Its lifestyle is commonly referred to as semi-endoparasitic. Freshly hatched second-stage juveniles (figs. 20.1 and 20.2) seek out young feeder roots and start feeding on epidermal cells. They eventually penetrate into tissues deep in the cortex of young feeder roots. When females mature, the posterior of their bodies enlarges and extends out from the root surface (figs. 20.3 and 20.4). Each female produces approximately 100 eggs surrounded by a gel that provides protection against desiccation as well as predators and pathogens.

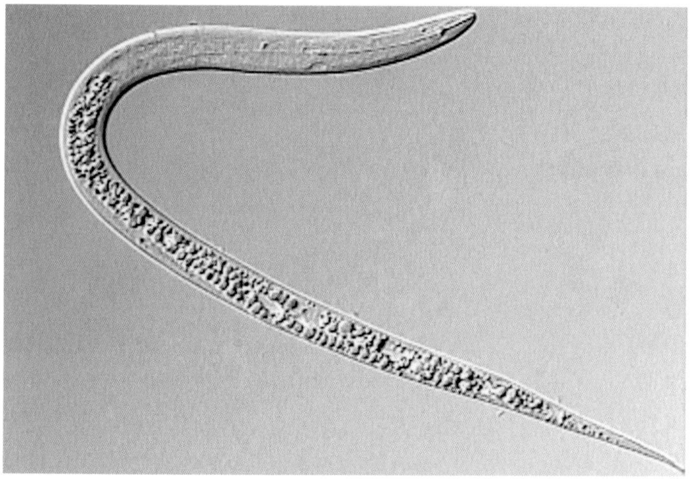

Figure 20.1 Second-stage juvenile (J2) of *Tylenchulus semipenetrans.* *Photo:* P. De Ley.

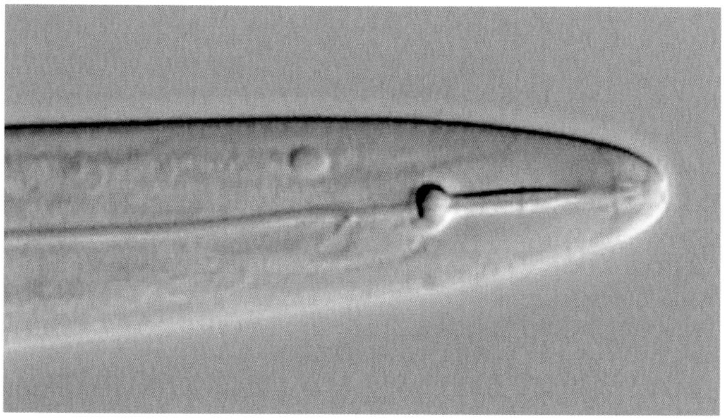

Figure 20.2 Head region of the citrus nematode (J2) with mouth stylet. *Photo:* P. De Ley.

Both males and females undergo four molts, but males remain outside the root and are very short-lived. Reproduction of the citrus nematode occurs both sexually and asexually. The highest nematode populations are typically found in late spring and late autumn following citrus root flushes. Infection, growth, and reproduction occur from 68° to 86°F. Citrus nematodes can survive for several months to a year or more in moist, cool soil. However, in the absence of a host, population levels decrease dramatically in a few months as body food reserves are rapidly depleted, especially in warm, dry, aerated soils.

Two physiological races of *T. semipenetrans,* Citrus and Poncirus, occur in California. The Citrus race is dominant and is parasitic on *Citrus* species, grape, olive, and persimmon. Trifoliate oranges and hybrids with *Citrus* species are generally tolerant to resistant to this race, although the degree of resistance to infection and damage varies. The Poncirus race has been reported in California, but its distribution is not known. It reproduces well on citrus, trifoliate oranges, and grapes, but not on olive.

Damage cause by *T. semipenetrans* is referred to as citrus slow decline, which appropriately describes the gradual deterioration of the tree (fig. 20. 5). Early symptoms above ground are reductions in leaf and fruit size, followed by a decline in tree appearance. Yellowing, leaf curling, and dieback are consequences of insufficient root development and feeder root decay. Parasitized trees also become more sensitive to other biotic or abiotic stresses such as fungal pathogens, insect pests, waterlogging, drought, salinity, and cold. Feeder roots heavily infested with mature females often appear thicker and darker than those without nematodes (fig. 20. 6). The roots are encrusted with soil particles sticking to the gel surrounding the nematode eggs.

Ideally, citrus management strategies should be based on nematode-free planting material, nematode-resistant rootstocks, nematode-free planting sites, and avoidance of postplanting infestation by nematode-infested irrigation or runoff water. Such preplant nematode management and sanitation measures are the most important means for optimizing young tree growth and securing long-term productivity.

Resistance Breeding

Breeding for nematode-resistant rootstocks must be an ongoing effort as new biotypes of nematodes evolve and disseminate. *Poncirus trifoliata,* a spiny, hardy citrus relative native to northern China, has been used as a main source of resistance. The resistance is based on various mechanisms, including wound tissue formation in tolerant cultivars that walls off the feeding nematodes and hypersensitive responses to larvae feeding in resistant cultivars. Some rootstock hybrids combine disease resistance against *Phytophthora* species and citrus tristeza virus (CTV) with good resistance against most populations of the citrus nematode (see chapter 5, "Rootstocks"). The good performance of Trifoliate Orange and its hybrids in the presence of the Citrus race of the citrus nematode and *Phytophthora* species fungi can provide several decades of relief from soilborne disease problems in California.

Cultural Management

Cultural management of nematode pests is a long-term approach with the goal of reducing citrus nematode population levels. The economic threshold for these pests can be quite variable, depending on the size and overall health of the tree. Minimizing other forms of stress such as disease pressure (especially by *Phytophthora* spp.) or water- and nutrient-related deficiencies helps trees to tolerate low levels of nematode parasitism. Although nematode-destroying organisms occur abundantly in the citrus rhizosphere, little is known about their contribution to nematode population decline. Planting of nematode-free rootstocks and avoiding spread of citrus nematodes by irrigation or runoff water are the key components for preventing the introduction of nematodes into a new grove or reinfestation of previously fumigated soil, since nematodes do not move very far by themselves.

Replanting of old citrus groves requires special precautions because citrus nematodes can survive host-free periods of many months or even years if large citrus roots continue to live in moist soil. Dry, aerated, warm soil promotes the decline of citrus nematode populations. Failure of citrus replants to grow as well as the previous plantings or as those grown in soil never before planted with citrus is referred to as the citrus replant problem. Most studies indicate that the buildup of citrus nematodes and deleterious root pathogens such as *Phytophthora* species and *Fusarium* species during prolonged cropping to citrus is the primary cause for the citrus replant problem.

Chemical Control

In the absence of obvious diagnostic symptoms of nematode damage, such as root galling caused by root knot nematodes, nematode disease problems can easily be attributed to other causes. Consequently, nematicides historically played an important role in proving that plant-parasitic nematodes can cause economic losses in citrus production. However, massive use of nematicides with very slow biodegradation led to ground water contamination and consequent closure of drinking water wells in California and the socioeconomic consequences of that period are still a burden for California's taxpayers. Preplant fumigation of old citrus groves with a high citrus nematode infestation is still an option to avoid early damage to young susceptible trees. However, no treatment can eliminate citrus nematodes that survive at depths up to 8 feet. Occasionally, nondesired effects of soil fumigation have been observed such as stunting or malnutrition due to the elimination of myccorhizae. Citrus nematode management with postplant-applied nematicides has resulted in variable and inconsistent return of investments. Most studies indicate that yield or quality improvements might occur only after 2 or more years of nematicide treatments.

Figure 20.3 Stained posteria ends of citrus nematode females and their eggs on a citrus feeder root. *Photo:* J. Ole Becker.

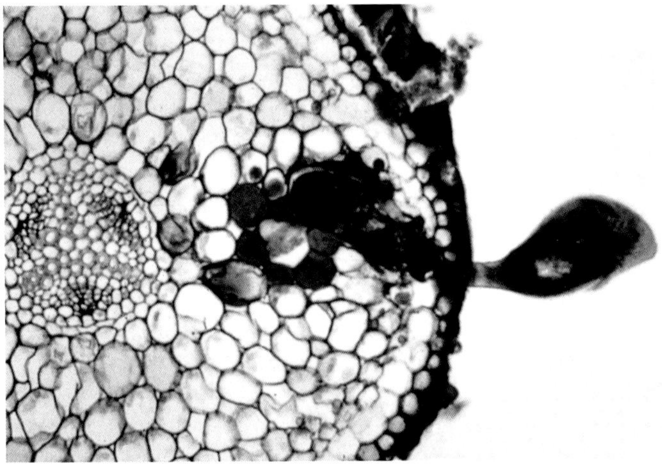

Figure 20.4 Mature female of the citrus nematode feeding on cortex cells in citrus root while its posterior end is exposed to the rhizosphere. *Photo:* S. Van Gundy.

Information about the presence or absence of citrus nematodes in a new or replanted grove is required to make a rational decision about the timing of planting, potential cultivation, or soil fumigation, as well as choice of rootstock. Postplant use of a nematicide should also be based on a previous determination of citrus nematode population levels. Proper sampling and extraction as well as interpretation of the results are therefore important key components to nematode management. Growers should contact soil analytical laboratories to inquire about their services and fees. Collection of soil samples should be based on an understanding of the soil and cropping history. Nematodes are not uniformly distributed in grove soils. In order to sample an orchard adequately, the land should be divided into representative sampling blocks of 2 to 4 acres that are distinguished by soil texture, tree appearance, cropping history, and so on. A sample from each block should consist of 20 or more soil and feeder root subsamples taken from the top 1 foot beneath the tree canopy using a soil auger or a shovel. The collected subsamples should then be thoroughly mixed in a bag or bucket and about 1 quart of soil with some roots poured into a plastic bag for nematode extraction. Each bag should be clearly labeled. The samples should not be exposed to direct sunlight and should be stored in a cooler. Samples must be processed for nematode enumeration as quickly as possible to obtain reliable data. For the citrus nematode, the most accurate method of reporting population levels is to determine the number of females per gram of root. Current guidelines for chemical management of citrus nematodes define action thresholds at about 1,000 juveniles per 100 cubic centimeters of soil and 1,000 females per gram of feeder roots (see the UC IPM Citrus Pest Management Guidelines, Citrus Nematodes, http://www.ipm.ucdavis.edu/PMG).

Figure 20.5 Advanced stage of slow decline caused by citrus nematodes *(T. semipenetrans)*. *Photo:* J. O. Becker.

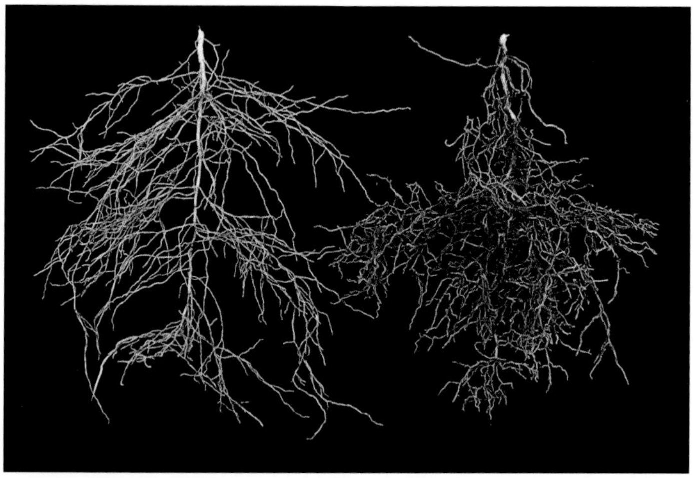

Figure 20.6 Citrus feeder roots not infected (left) and infected (right) with citrus nematodes. *Photo:* J. O. Becker.

Other Plant-Parasitic Nematodes

The sheath nematode *(Hemicycliophora arenaria)* occurs on citrus in the inland deserts of southern California. Despite its wide host range, this nematode is currently considered of minor economic importance. Various other nematodes that are serious pests in other citrus-growing areas are currently not a problem in California citrus. These are primarily the burrowing nematode *(Radopholus similis)*, which causes spreading decline in Florida; several lesion nematodes (*Pratylenchus* spp.), worldwide; the sting nematode *(Belonolaimus longicaudatus)*, in the southeastern United States; and root knot nematodes (*Meloidogyne* spp.), in Asia. The reason that these nematodes are not established in California's citrus is partially due to geographical isolation by mountains, deserts, and the Pacific Ocean. The foresight of those who developed quarantine inspections along California's borders and other entry points such as harbors and airports has helped limit the introduction of exotic or invasive nematode species. Another major asset to avoid the dissemination of nematode species is the regulatory requirement of certified disease- and pest-free citrus nursery rootstocks (see chapter 5, "Rootstocks").

References

Barker, K. R., G. A. Pederson, and G. L. Windham, eds. 1998. Plant and nematode interactions. Madison, WI: American Society of Agronomy; Crop Science Society of America; Soil Science Society of America.

Dreistadt, S. H. 2012. Integrated pest management for citrus. 3rd ed. Oakland: University of California Agriculture and Natural Resources Publication 3303.

Duncan, L. W. 1999. Nematode diseases of citrus. In L. W. Timmer and L. W. Duncan, eds., Citrus health management. St. Paul, MN: APS Press. 136–148.

University of California Statewide Integrated Pest Management Program. UC IPM citrus pest management guidelines. UC IPM website, http://www.ipm.ucdavis.edu/PMG.

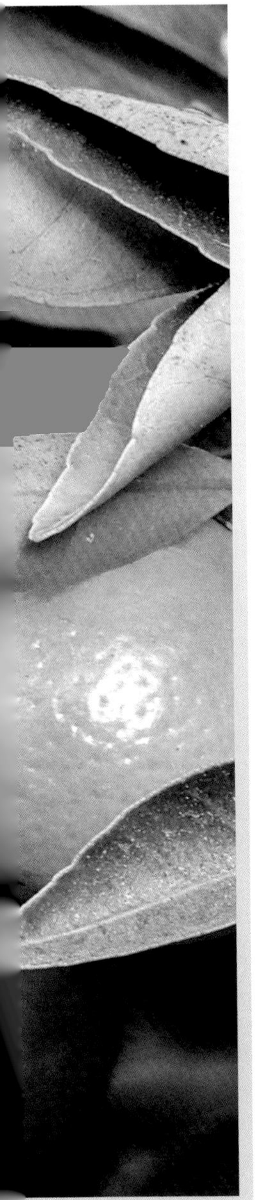

21 Fungal Diseases

James E. Adaskaveg, Helga Förster, and Peggy A. Mauk

Citrus is susceptible to numerous diseases that reduce the yield and quality of the crop, weaken trees, and even cause tree death. Many of the diseases can be controlled readily with a basic knowledge of citrus propagation, horticulture, and preventive practices. Diseases of citrus can be biotic or abiotic in nature. Infectious microorganisms, such as viruses, fungi, and bacteria, cause biotic diseases; adverse environments or genetic disorders cause abiotic diseases. Environmental stresses can also predispose citrus trees to diseases or enhance the severity of a specific disease.

The major biotic diseases of citrus established in California are caused by viruses and viroids, a phytoplasma, a bacterium, and numerous fungal and fungal-like species, such as organisms in the Stramenopila-Oomycota. Viruses, viroids, and phytoplasmas can infect all portions of citrus trees and can be effectively managed through careful propagation techniques that prevent plant infections. These diseases are discussed in chapter 7, "The California Citrus Clonal Protection Program." Fungal diseases of roots and trunks are of the greatest concern in California citrus production. Other fungal and bacterial diseases of foliage and fruit can occur in most growing regions of the state and can cause serious losses, but they are generally less of a problem in the semiarid climate of citrus-producing areas in California, except for certain postharvest decays. The damage caused by many diseases can depend on the variety of the rootstock and scion, as well as the environment, including soil conditions that ultimately affect the nutrient and water status of the plant. Thus, the incidence and severity of a disease depend on the virulence of the pathogen, the genetic susceptibility and the growth stage of the host, and the specific microclimate such as temperature and wetness or relative humidity.

Many of the diseases have similar primary symptoms that affect leaves, stems, trunks, or roots and result in the general decline of trees. Symptoms can include light green or necrotic foliage, poor growth flushes of shoots and roots, leaf drop, branch dieback, and trunk cankers. These symptoms may eventually affect the overall growth habit, size of tree, and most important, crop yield.

Nursery Diseases

Damping-Off Diseases

Damping-off is a worldwide problem that can affect recently germinated seedlings of any citrus cultivar. Pre- and postemergent damping-off diseases in citrus seedbeds can be caused by a number of fungal pathogens, including *Phytophthora* and *Rhizoctonia*, as well as occasionally *Thielaviopsis* and *Pythium* species. Under warm conditions, typical of greenhouses, and with excessive moisture, damping-off can be a serious problem.

Symptoms. Seedlings with damping-off can be killed rapidly. During preemergence, the fungi infect and destroy germinating seed before the seedlings reach the soil surface, resulting in a sparse stand of plants. The fungi can also infect the seedling stem at the soil level as a postemergent disease, resulting in seedlings toppling over. Cankers can develop on the maturing seedling stems, but once true leaves are formed, the plants become more resistant.

Economics. These diseases are considered minor in California, but some nurseries can experience devastating losses.

Control. The use of noninfested soil or soil that has been pasteurized or chemically fumigated is the standard approach. Other control strategies include avoiding nitrogen overfertilization, wet soils, high humidity, excessive shading, and crowding of seedlings. Soil pH should be adjusted to a range of 4 to 5.5, and soil temperatures should be maintained from 86° to 90°F to avoid these diseases. Temperatures can be lowered to 75° to 80°F once seedlings are 5 inches high. Hot water seed treatments can be effective against damping-off pathogens such as *Phytophthora* and *Pythium* species, but most citrus seed will not tolerate the necessary temperature of 125°F.

Black Root Rot

Black root rot is caused by the fungus *Thielaviopsis basicola* (Berk. & Broome) Ferraris and is mainly a disease of nurseries and greenhouses. The disease does not cause losses in the field. The potential for black root rot to severely damage the California citrus industry is very low.

Symptoms. The fungus infects the feeder roots of citrus plants, causing distinct black or brown root lesions. Under severe conditions, the fungus can destroy the root cortex, leaving the stringy vascular cylinder. Plants with severe black root rot display interveinal chlorosis, probably as a result of nutrient deficiency. Black root rot can cause severe stunting and chlorosis of nursery-grown citrus seedlings, but it is not considered a problem in field-grown trees.

Economics. In California, the disease is considered minor.

Disease cycle and epidemiology. Chlamydospores (thick-walled, single-celled survival spores) may be introduced to container-grown citrus plants on peat or other organic matter. Chlamydospores are very resistant to chemical or fumigation control. Once in the containers, the fungus infects roots and grows throughout the soil. It produces phialospores (thin-walled, single-celled dissemination spores) on the surface of the soil, which may be splashed by water or become airborne and infest the entire greenhouse. Cool temperatures and high soil moisture prevalent under winter conditions favor black root rot. The pathogen population is reduced during the summer, but it increases again the following winter.

Control. Citrus species vary in their susceptibility to *T. basicola,* in order of very susceptible to least susceptible: Cleopatra Mandarin, West Indian lime and Carrizo Citrange, Troyer Citrange, Homosassa Sweet Orange, Siamelo, and New Mexican Trifoliate Orange. Potting mixes must be kept free of *T. basicola*. Peat and organic matter are often sources of the fungus. Adequate drainage and aeration should be maintained in all potting mixes. Bark and other coarse materials will provide better drainage. Benches, pots, and utensils must be sterilized before reusing. Avoid high pH, salt, low light, and cool temperatures, which reduce the vigor of citrus seedlings.

Root Diseases

Armillaria Root Rot

The fungus *Armillaria mellea* (Vahl) P. Kumm. causes this root disease of citrus in California. The fungus is also known as oak root fungus. Research has shown that the previously recognized species *A. mellea* is actually a complex of several species, with differences in virulence, host specificity, and ecology. Evidence exists, however, that only one of these species is pathogenic to citrus, and this is the one that retains the original name *A. mellea.* In citrus-producing areas in the southern United States, *Armillaria tabescens* (Scop.) Emel is also known to cause Armillaria root rot. Many woody plants, including oaks, are susceptible. High soil moisture is a favorable environmental factor, and the disease is widespread along streambeds and flood plains throughout the state where oak trees and other hosts grow.

Symptoms. *Armillaria mellea* infects, kills, and rots the major support roots and trunks of citrus trees. As a result, the tree undergoes a slow decline, with the foliage becoming chlorotic, thinning, and eventually dying. Often the disease first appears on one side of the tree before the entire tree declines in subsequent years. Positive identification of the disease is based on the presence of clusters of the distinctive honey-colored, white-gilled mushrooms that develop at the base of the tree after rains during late fall and early winter and the presence of white, fan-shaped mycelial mats between the bark and rotting wood. Infected tissue emits a strong odor of mushrooms. Often, black shoestring-like rhizomorphs (rootlike aggregations of hyphae) that can penetrate healthy roots are also present on the surface of decaying roots in moist soil. In the orchard, a pattern of ever-enlarging circles of diseased trees reflects this method of infection. These circles are called infection centers.

Economics. The disease can be fairly serious in orchards planted on newly cleared land. It can decimate entire orchards, and the cost can be high for individual growers, but the disease tends to be localized, and the damage to the entire citrus industry is minor.

Disease cycle and epidemiology. Armillaria root rot is spread by rhizomorphs that grow from root-to-root contacts in the soil. The fungus survives only in association with infected roots and does not live independently in the soil. The basidiospores (sexual spores or meiospores) of the fungus that are produced on the mushroom stage present relatively little threat to healthy citrus trees. The fungus must build up inoculum in buried wood or roots before it can successfully attack a healthy tree. Therefore, orchards planted in fields that previously contained native tree vegetation or orchards previously infected with *A. mellea* are at the greatest risk. Weakened trees or trees planted close to old stumps are also more likely to become infected by *A. mellea*. The disease is very slow in its development because it spreads by growth of rhizomorphs. Small clusters of diseased trees form infection centers that can enlarge and coalesce over several years to destroy entire orchards.

Control. Diseased trees and roots should be removed to prevent spread of the fungus by root-to-root contact. All roots greater than 0.5 inch in diameter and any wood or old stumps should be taken out of the planting sites before establishing citrus orchards. Several healthy trees surrounding disease centers should also be removed because the disease is invariably distributed well beyond trees showing visible symptoms. Maintain health and vigor in the orchard because strong, rapidly growing citrus trees are quite resistant to *Armillaria*. Leave the land fallow for more than 1 year before replanting in an *Armillaria*-infested area. Soil fumigants may be used effectively in soil infested with small, shallow roots, but they are often not effective because the pathogen persists in roots buried deep in the soil for many years. Thus, eradication is seldom achieved. Removing the soil from around the crown of the tree and allowing the bark to dry out can temporarily extend tree life. This often delays girdling of the trunk by the advancing rhizomorphs until the mycelium in the wood grows from the roots into the trunk.

Dry Root Rot

Dry root rot is caused by the soilborne fungus *Fusarium solani* (Mart.) Sacc., which invades the root system. The fungus is considered a facultative pathogen—that is, a saprophyte that feeds on dead organic matter but under certain circumstances becomes a pathogen and causes disease. It is ubiquitous on leaves, bark, and roots of citrus but usually does not instigate disease.

Symptoms. Citrus trees infected with dry root rot show a moist, dark decay in the bark of major roots or the root crown, which later has a cracked appearance. The wood below this dead bark is hard and dry, with a reddish to brownish gray or purple stain. Occasionally, on infected larger roots, only the inner bark is decayed, and the wood is not discolored. This is in contrast to Phytophthora root rot, which mostly occurs on the feeder roots. Furthermore, dry rot lesions do not ooze gum as do Phytophthora trunk rot lesions. Trees infected with dry rot may lack vigor and have reduced canopies. Some of the leaves may have an abnormal, curled appearance. A period of 2 to 3 years may pass from the time of infection until noticeable wilt. Trees affected with dry root rot may suddenly wilt and die, with the leaves drying in place after a period of prolonged summer heat.

Economics. Dry root rot normally affects only a few trees in a citrus orchard. Sometimes, however, large numbers of trees are killed, resulting in economic losses for individual growers.

Disease cycle and epidemiology. *Fusarium solani* is naturally associated with decaying leaves and sloughed bark of citrus trees; however, trees must be stressed in order to succumb to infection. Lemons appear to be especially susceptible to dry root rot because of their propensity to overproduce and starve their root systems for carbohydrates. Stress factors that are known to initiate dry root rot infections are gophers, root weevils, rootstock-scion incompatibilities, tristeza, sieve tube necrosis, overwatering, poor drainage, excess fertilizer, and Phytophthora root rot, as well as injuries from cultivation, herbicide usage, and nematodes.

Control. There are no effective control measures known for dry root rot, and all citrus rootstocks are susceptible to the disease. The best control is to correct stress factors, such as excess fertilizer, gophers, Phytophthora root rot, poor drainage, and overwatering, that initiate the disease. In poorly drained soils, removing the soil from the root crowns and planting trees on soil mounds or berms reduces damage from this disease.

Phytophthora Root Rot, Foot Rot, and Gummosis

Phytophthora root rot is a destructive Stramenopila-Oomycota disease of citrus occurring in all growing regions of California. Phytophthora root rot can be caused by a number of soilborne *Phytophthora* species. On citrus, *Phytophthora parasitica* Dastur (syn. *P. nicotianae* Breda de Haan), *P. citrophthora* (R.E. Sm. & E.H. Sm.) Leonian, *P. hibernalis* Carne, *P. syringae* (Kleb.) Kleb., and *P. palmivora* (E.J. Butler) E.J. Butler have all been reported. In California, *P. parasitica* and *P. citrophthora* are the primary root rot pathogens, whereas *P. parasitica* also causes foot rot. Young, fibrous roots of citrus are more susceptible to root rot than are older roots. Furthermore, the disease varies seasonally and annually. Activity of *P. parasitica* coincides with active root growth periods (flushes) in the late spring and late summer, whereas *P. citrophthora* is mainly active during the winter. Both species can also cause gummosis, but in California *P. citrophthora* is more common. Brown rot of fruit is also caused by *Phytophthora* species (see "Foliar and Fruit Diseases," below).

Symptoms. Phytophthora root rot causes a slow tree decline that results in foliage yellowing, leaf drop, and twig dieback. In severe cases, when a significant portion of the root system is damaged, large limbs may succumb. On feeder roots, portions of the outer cortical tissue disintegrate, leaving only the white, threadlike stele (central part of the root) protruding from the decaying outer tissue. The pathogen can also infect the cambial layer under the bark of the trunk or large roots. The tree reacts to the cambial layer damage by producing copious amounts of amber to brown gum. Foot rot symptoms develop when infected root crowns gum at the soil line. Gummosis symptoms develop when the bark of trunks is infected higher in the tree and gum is produced. Citrus gum is water soluble and disappears after heavy rains. Badly infected trees have small, pale green leaves with yellow veins, as is typical of mechanically girdled trees. If the Phytophthora lesion encircles the tree and girdles it, the tree will die. If the lesion ceases expansion or the pathogen dies, callus tissue will grow into the dead areas of the bark in an attempt to heal the wound. Lesions do not extend below the bud union of trees on resistant rootstocks. On trees with susceptible rootstocks, the lesion may extend downward into roots and upward into branches and may also occur underground, with canopy symptoms developing without any visible sign of Phytophthora lesions.

Economics. Phytophthora diseases are widespread, with nearly all orchards in California harboring at least one of the pathogens. The root disease is particularly insidious because mild symptoms are not always noticeable, yet yield is reduced. Annual losses from root rot, fruit brown rot, and branch and trunk gummosis can be substantial in the United States. These losses occur even though most citrus in the United States is planted on resistant rootstocks.

Disease cycle and epidemiology. Depending on the species, *Phytophthora* spp. survive as chlamydospores, oospores, and/or hyphae (somatic filaments of fungi) in root debris in the soil (fig. 21.1). Under proper conditions, chlamydospores germinate to produce sporangia (saclike structures containing zoospores) that contain many zoospores (motile asexual spores or mitospores). Irrigation or rainfall stimulates the release of zoospores. Because the zoospores are motile in water, saturated conditions are necessary in order for the zoospores to swim to roots. Root exudates (chemicals given off by healthy roots) attract zoospores. The zoospores infect feeder roots, and a new generation of chlamydospores and sporangia is produced that restarts the disease cycle (see fig. 21.1). A new generation of zoospores is released from sporangia after each rain or irrigation. Thus, during the growing season, many cycles of sporangium production may occur on infected roots, resulting in enormous numbers of zoospores during wet periods (note: zoospores do not persist). Gummosis and foot rot occur when zoospores attack and penetrate through the bark. In the case of gummosis, zoospores must be splashed up onto the trunk. Disease cycles may be initiated at different times of the year for different *Phytophthora* species. As mentioned above, *P. citrophthora* is usually active in the cooler winter, while *P. parasitica* is active during the warmer summer months. Salt exacerbates *Phytophthora* root damage as does the feeding of the root weevil (*Diaprepes* spp.)

Control. The first line of defense against Phytophthora root rot is the use of resistant rootstocks. Many are nearly immune to gummosis but are only tolerant of root rot. Rootstocks most resistant to *Phytophthora* species include Ponderosa Lemon, Swingle Citrumelo, Rubidoux Trifoliate, African Shaddock, C32 Citrumelo, C35 Citrumelo, and Schaub Rough Lemon. Table 21.1 indicates the relative susceptibility of scions and rootstocks of *Citrus* species. Irrigation management is an excellent way to control Phytophthora root rot. The disease is greatly affected by the amount of rainfall and irrigation water because the zoospores depend on water for movement. Flood or furrow irrigation can increase the damage caused by *Phytophthora* species. Irrigation with minisprinklers and allowing the surface soil to dry between irrigations reduces root rot. In areas chronically affected by Phytophthora root rot, two irrigation lines may be used alternately to allow the soil to dry completely on one side of the tree while the other side is receiving water. Control of root rot may also be achieved with the use of systemic fungicides applied as foliar or soil treatments. These fungicides may also be sprayed or painted on gummosis lesions to reduce the severity of gummosis. These fungicides will not be economical unless *Phytophthora* populations are greater than 15 propagules per gram of rhizosphere soil, susceptible rootstocks are present, or the orchard has chronic drainage problems. Soil sampling procedures for the enumeration of populations of *Phytophthora* species are described in figure 21.2 and table 21.2. Fungicides, together with nematicides or a soil fumigant, should also be used to protect replants during the first 2 years of growth. Mulching with organic matter, especially animal waste products, may be beneficial for water conservation and plant nutrition, but this practice may increase populations of *Phytophthora* species.

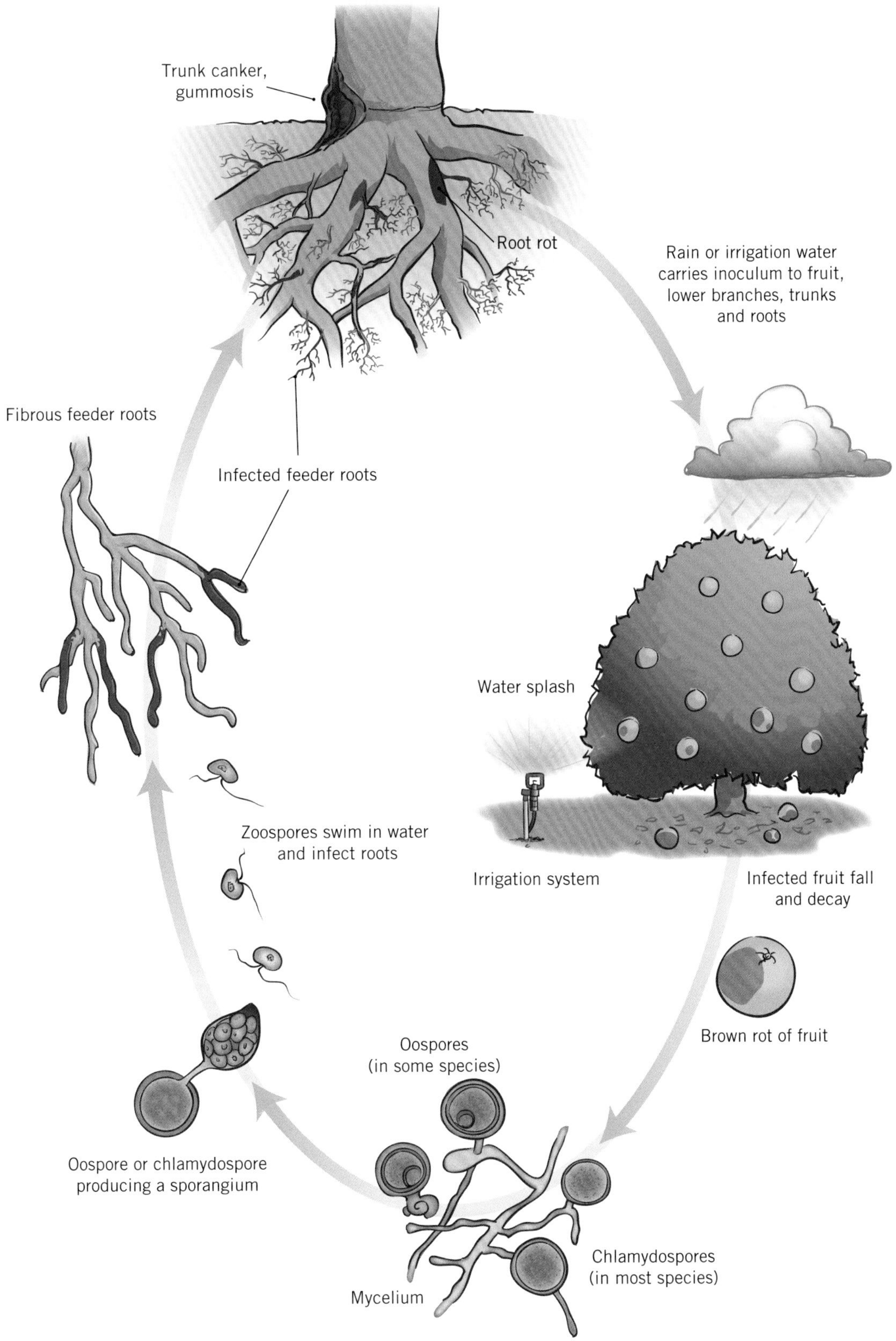

Figure 21.1 Disease cycle of *Phytophthora* species causing root rots, trunk cankers, gummosis, and brown rot of fruit on citrus.

Table 21.1. Susceptibility to Phytophthora root rot in common scion and rootstock cultivars of citrus

Propagation material	Selection name	Susceptibility
scions	grapefruit	very susceptible
	lemons	very susceptible
	limes	very susceptible
	mandarins, tangerines	susceptible
	sweet oranges	very susceptible
rootstocks	Alemow	moderately resistant
	C32 Citrumelo	resistant
	C35 Citrumelo	resistant
	Carrizo Citrange	tolerant
	Cleopatra Mandarin	tolerant
	Ponderosa Lemon	resistant
	Rangpur	tolerant
	Rough Lemon*	tolerant
	Rubidoux Trifoliate	resistant
	Schaub Rough Lemon	resistant
	Sour Orange	moderately resistant
	Swingle Citrumelo	resistant
	Trifoliate Orange	resistant
	Volkamer Lemon	tolerant

Note: *Most selections of Rough Lemon are tolerant, but some are susceptible.

Other Root Diseases

Rosellinia Root Rot

Rosellinia root rot caused by *Rosellinia necatrix* Berl. ex Prill. can be a problem in high-rainfall areas on recently cleared land. The fungus attacks a wide range of woody hosts, and once established on dead stumps or logs, it can spread to adjacent living trees. The result is an infection center similar to that caused by Armillaria root rot.

Symptoms. Citrus trees infected with *R. necatrix* become chlorotic and eventually lose their leaves and die. This is the result of severe root loss or major girdling of the trunk. *Rosellinia necatrix* produces white, fan-shaped mycelial growth under the bark of infected roots. Roots of diseased trees eventually become covered with the dark brown to black mycelium of the fungus, and abundant perithecia, the sexual fruiting bodies of the fungus, may be produced.

Economics. The disease is considered minor in California, but in some countries it is increasing in importance.

Disease cycle and epidemiology. *Rosellinia necatrix* spreads from infested dead stumps, roots, or logs to nearby living trees on newly planted citrus land. The infection site will expand, affecting additional healthy trees. Because grape is a good host for this pathogen, citrus planted in areas where grapes were previously grown is at risk. The spread is largely by mycelial (the body of the fungus made up of hyphae, or cellular threads) growth and thus is usually slow. The sexually produced ascospores (meiospores) can infect dead citrus roots or stumps, and the fungus then spreads to healthy trees.

Control. Remove stumps and roots from the land before planting. Use care when planting into areas where grapes were previously grown. Trench around infected trees or stumps to prevent spread of the disease. The fungus is intolerant of drought, so infested areas should be thoroughly drained and dried to reduce the spread of the pathogen. Chemical soil fumigation and trenching around infected areas reduce the disease incidence.

Phymatotrichum Root Rot

This destructive disease, also called Texas root rot or ozonium root rot, is caused by the fungus *Phymatotrichopsis omnivora* (Duggar) Hennebert. It is more common in other citrus-producing areas of the southwestern United States, but it has been reported from isolated locations in southeastern California. The fungal pathogen is indigenous to heavy, alkaline soils and has a wide host range, with over 2,000 species of dicotyledonous plants; monocotyledonous plants are not affected.

Symptoms and ecology. Symptoms of the disease are yellowing or bronzing of leaves, wilting, subsequent leaf drop, and tree death. Infected mature trees can survive for several years. Death of grapefruit, lime, orange, and mandarin seedlings has occurred in nurseries. In summer, following a rain or irrigation, whitish yellow, cottony, flat spore mats 3 to 9 inches in diameter develop on the soil surface. The fungus produces yellow to brown rhizomorphic mycelial strands on the surface of infected roots. The fungus overwinters by producing sclerotia (firm masses of hyphae for fungal survival) in the soil near host roots, and these germinate best at a temperature of approximately 80°F. The fungus survives well at soil depths of 8 to 18 inches.

Economics. In California, the disease is of minor importance.

Control. Sour Orange and Trifoliate Orange rootstocks are the least susceptible. Susceptible crops, such as cotton and alfalfa, can be planted as biological indicators for the presence of the pathogen.

Macrophomina Root Rot

This is a minor root rot disease that has been reported in California, Arizona, and other

1. Randomly select 20 to 40 locations within a 10-acre orchard block with mild or moderate tree decline presumably from Phytophthora root rot.

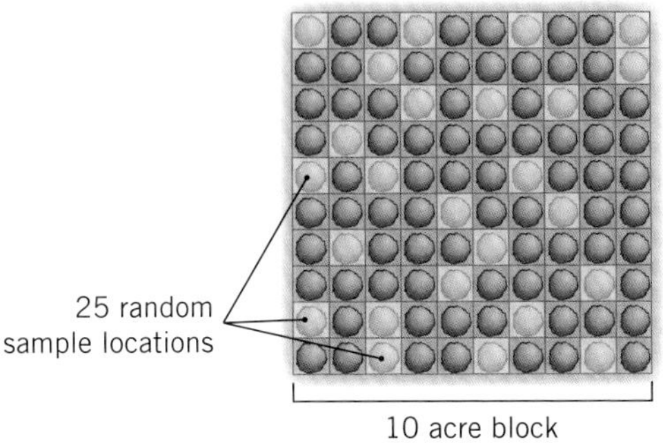

2. Individual tree samples are collected within the tree 'dripline' or near the irrigation emitter where roots are concentrated.

3. Composite samples in a plastic bag:
   · Seal bag to retain soil moisture
   · Do not refrigerate
   · Do not overheat
   · Ship to lab within 24 to 48 hr

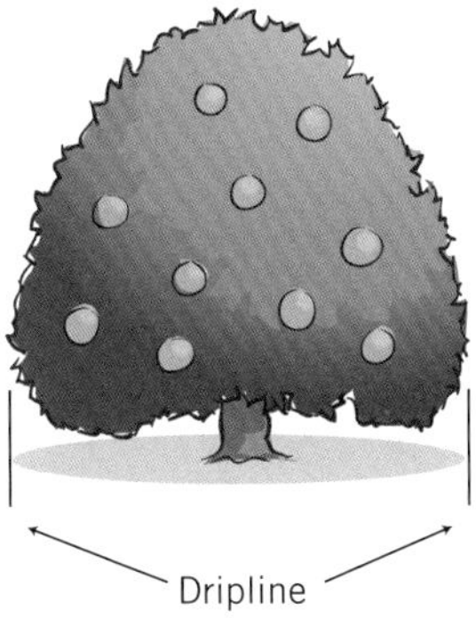

4. Mix soil in sampling bag and pour into a cup with drain holes. Add water and allow the sample to incubate for 24 to 48 hr at 68 to 77°F.

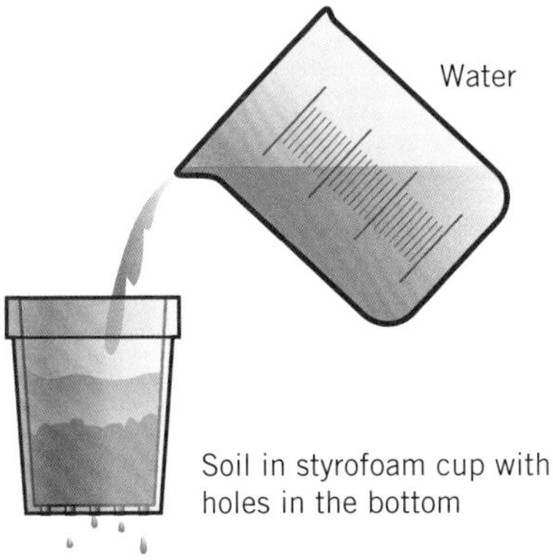

5. Approximately 10 g of soil is removed and mixed with 90 ml of sterile water.

6. Five to ten 1-ml samples are transferred using a pipette and each sample is then spread onto a Petri plate containing PARPH selective medium (see Definition box).

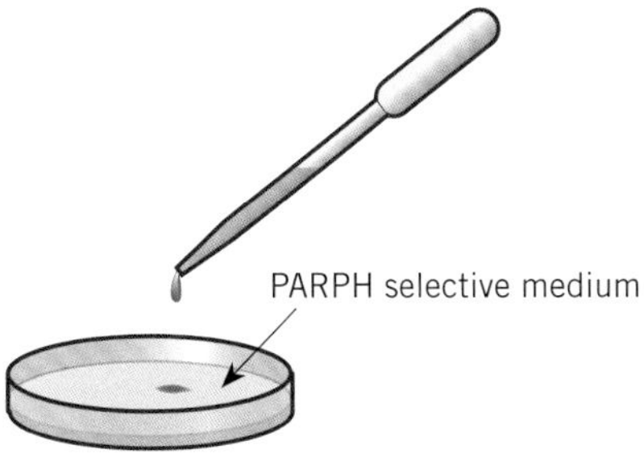

7. Plates are incubated for 48 to 72 hr at 68 to 77°F. Note: incubation temperatures are dependent on sampling season and suspected Phytophthora spp.

8. Colonies per plate are counted and totaled to express propagules/g soil.

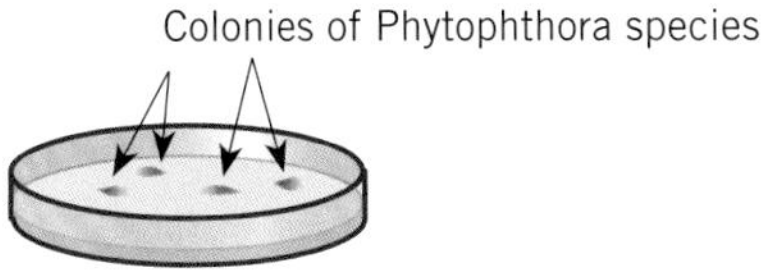

9. Populations in excess of 10 to 20 propagules/g soil are considered high and may cause damage to citrus roots depending on the disease history, soil type, topography, irrigation methodology, and rootstock cultivar.

Figure 21.2 Sampling soils to detect and monitor populations of *Phytophthora* species in mature citrus orchards.

citrus-producing areas of the world. The disease is caused by the fungus *Macrophomina phaseoli* (Maubl.) Ashby and develops on heavily watered citrus. The fungal pathogen has a wide geographical distribution and has been reported to cause root and lower stem infections of over 500 species of dicotyledonous and monocotyledonous plants.

Symptoms and ecology. Citrus on Rough Lemon rootstocks develop soft, watery lesions on the bark of main lateral roots. In general, drying lesions cause darkening of the wood and the cambium (cambial side facing the bark).

Economics. In California, the disease is of minor importance.

Control. Drying out the crown and root system, careful irrigation, and proper drainage to prevent overwatering are effective strategies.

Trunk and Branch Gummosis and Canker Diseases

Diaporthe and Botryosphaeria Gummosis and Canker Diseases (formerly Diplodia, Dothiorella, and Phomopsis Gummosis)

Branch cankers and diebacks are often caused by several Ascomycota fungi that invade bark and wood of citrus and cause host gumming, hence the disease name. Historically, the three main gummosis diseases have been based on the anamorphic names of the pathogens. Thus, *Dothiorella gregaria, Diplodia natalensis* (both with *Botryosphaeria* teleomorphs), and *Phomopsis citri* (*Diaporthe* teleomorph) have been reported to cause Dothiorella, Diplodia, and Phomopsis gummosis, respectively. This nomenclature, however, is outdated due to recent taxonomic

Table 21.2 PARPH-V8 selective medium for the isolation of *Phytophthora* species (see fig. 21.2)

Ingredient	Amount per		Concentration (a.i.) (ppm)
	1.0 liter	0.5 liter	
Basal medium			
clarified V8 concentrate*	50 ml	25 ml	
distilled water	950 ml	475 ml	
agar	15 g	7.5 g	
Amendments			
Delvocid (50% pimaricin)	10 mg = 0.01 g	5 mg	5
ampicillin (sodium salt)	250 mg = 0.25 g	125 mg	~ 250
rifampicin	10 mg = 0.01 g	5 mg	~ 10
Terraclor (75% PCNB)†	66.7 mg = 0.0667 g	33.4 mg	50
hymexazol†	50 mg = 0.05 g	25 mg	50
Tachigaren (70% hymexazol)†	71.4 mg = 0.071 g	35.7 mg	50

Source: Ferguson and Jeffers 1999.

Notes:

*Clarified V8 concentrate = buffered V8 Juice (1.0 g $CaCO_3$ per 100 ml V8 Juice) clarified by centrifugation at 7,000 rpm for 10 minutes. Freeze supernatant for later use.

†PCNB is particularly useful to inhibit soilborne fungi on soil dilution plates. Hymexazol inhibits most *Pythium* species while allowing most *Phytophthora* species to grow.

Directions:

1. Add ingredients for basal medium to a 2-liter flask and mix thoroughly.
2. Autoclave for 20 minutes at 250°F and 15 psi.
3. Add amendments to 5 ml of sterile distilled water, vortex to mix (or sonicate), and add to cooled medium.
4. Pour plates (approximately 15 ml per plate) and allow to polymerize.
5. For storage, refrigerate plates in plastic bags in the dark.
6. Plates are best used within several weeks, but they will keep for months.

changes for some of the pathogens and because additional pathogens have been described for the diseases. Thus, among the pathogens, *Diaporthe citri* and species within the Botryosphaeriales have been implicated on citrus (table 21.3). Species within this latter taxonomic group include *Botryosphaeria rhodina, B. dothidea,* and *B. ribis.* Other species described that are associated with this disease complex include *B. australis, B. iberica, B. lutea, B. parva, B. stevensii, Spencermartinsia viticola,* and *Dothiorella gregaria* (an anamorphic species in the Botryosphaeriaceae) (see table 21.3). All species have a wide host range. Some of these pathogens may also cause postharvest fruit rots (i.e., stem-end rots) of citrus (see the section "Postharvest Decays," below). *Diaporthe citri* (anamorph *Phomopsis citri*) may cause a preharvest disease of fruit, foliage, and twigs known as melanose. This disease is most common in high-rainfall citrus production areas and is rarely observed in drier areas such as California (see the section "Foliar and Fruit Diseases," below).

Symptoms. Twig and branch dieback, as well as limb and trunk cankers, are associated with this disease complex. Cankers develop as the inner bark tissue dies, and a tan-colored zone of gumming may form beyond the invasion margin. In gummosis caused by *B. rhodina,* bark damage is initially confined to small patches of mature or weakened tissues. The wood, however, may be extensively colonized and have a dark grayish discoloration. Ultimately, extensive areas of bark and large branches may be killed. Unlike gummosis caused by other species of *Botryosphaeria,* such as *B. dothidea* and *B. ribis* (including *D. gregaria*), the invaded inner bark and wood surface infected with *B. rhodina* become chocolate-brown and disintegrate, leaving elongated

Table 21.3. Fungal pathogens involved in Diaporthe and Botryosphaeria gummosis and canker diseases of citrus

Disease	Taxonomic phylum and order	Teleomorph	Anamorph	Comment
Diaporthe gummosis	Ascomycota: Diaporthales	*Diaporthe citri* (Faw.) Wolf 1926	*Phomopsis citri* Faw. 1912	historical name: Phomopsis gummosis
Botryosphaeria gummosis	Ascomycota: Botryosphaeriales	*Botryosphaeria australis* Slippers, Crous & M.J. Wingf. 2004	*Neofusicoccum australe* (Slippers, Crous & M.J. Wingfield) Crous, Slippers & A.J.L. Phillips 2006	
		B. dothidea (Moug. Ex Fr.) Ces. & De Not. 1863	*Fusicoccum aesculi* Corda 1829	similar disease symptoms to Dothiorella gummosis
		B. iberica A.J.L. Phillips, J. Luque & A. Alves 2005	*Dothiorella iberica* A.J.L. Phillips, J. Luque & A. Alves 2005	
		B. lutea A.J.L. Phillips 2002	*Neofusicoccum luteum* (Pennycook & Samuels) Crous, Slippers & A.J.L. Phillips 2006	
		B. parva Pennycook & Samuels 1985	*N. parvum* (Pennycook & Samuels) Crous, Slippers & A.J.L. Phillips 2006	
		B. rhodina (Berk. & M.A. Curtis) Arx 1970	*Diplodia natalensis* Pole-Evans 1911; important synonyms: *Botryodiplodia theobromae* Pat. 1892; *Lasiodiplodia theobromae* (Pat.) Griffon & Maubl. 1909	historical name: Diplodia gummosis
		B. ribis Grossenb. & Duggar 1911	*Neofusicoccum ribis* (Slippers, Crous & M.J. Wingf.) Crous, Slippers & A.J.L. Phillips 2006; *Dothiorella ribis* (Grossenb. & Duggar) Sacc. 1884	historical name: Dothiorella gummosis
		B. stevensii Shoemaker 1962	*Diplodia mutila* (Fries) Montagne 1834	
		Botryosphaeria sp.	*N. mangiferae* (Syd. & P. Syd.) Crous, Slippers & A.J.L. Phillips 2006	
		Spencermartinsia viticola (A.J.L. Phillips & J. Luque) A.J.L. Phillips, A. Alves & Crous 2008	*Dothiorella viticola* A.J.L. Phillips & J. Luque 2005	
		unknown	*Dothiorella gregaria* Sacc. 1881	historical name: Dothiorella gummosis

Note: Nomenclatural rules for fungi changed as of January 2013. Currently, only one fungal name is allowed, and the earliest name published has priority.

Sources: Adesmoye et al. 2011; Graham and Menge 2000; Klotz 1978.

cavities or grooves in the bark next to the cambium that may ooze gum. Hence, this is often called pocket gum. Gummosis caused by *B. rhodina* is much more extensive than that caused by *D. citri* and some other *Botryosphaeria* species where lesions are similar and much more self-limiting. Overall, disease symptoms for the numerous *Botryosphaeria* species reported have not been described in detail. Black pycnidia (asexual fruiting structures that produce mitospores) of the fungi may be present on bark and wood.

Economics. These diseases are seldom severe enough to cause major losses in California and are generally considered to be of minor importance. Phytophthora gummosis is of greater importance because entire trees may be killed if not treated.

Disease cycle and epidemiology. The causal fungi are considered to be weak pathogens that require stress factors ranging from mechanical injuries (pruning damage) to environmental conditions (e.g., wind damage, sunburn, frost or cold damage, drought, or flooding) before infecting and causing damage to citrus. These fungi normally inhabit dead twigs and branches in the canopy. Fungal spores are wind or rain disseminated and may infect healthy twigs, wood, or bark of susceptible trees.

Control. Pruning of dying branches well below the dead tissue helps manage the disease because the inoculum for infection originates in dead twigs. The best control measure is keeping trees healthy and vigorously growing. In California, the disease is seldom severe enough to warrant fungicide treatment.

Sclerotinia Twig Blight

Sclerotinia twig blight is a minor disease of citrus. It is caused by *Sclerotinia sclerotiorum* (Lib.) de Bary, which may damage roots and trunks and also cause a fruit rot (see the section "Other Postharvest Fungal Diseases"). The fungus may invade twigs from infected blossoms or fruit or any location on the roots, trunk, or branches; branches and twigs are usually girdled and killed. At first, the lesions are soft and exude gum. Later, the bark becomes gray or buff and is characteristically shredded into long, fibrous strips. Symptoms are similar to those caused by *Botrytis cinerea* Pers., but the lesions are larger, with a more shredded appearance. *S. sclerotiorum* has a very wide host range, including many fruits, vegetables, and flowers, but it is considered to be only a weak pathogen of citrus. Infected flower petals, fruit, dead bark, or other organic substrates in direct contact with healthy bark are usually required for infection. Most infections result from sexual spores produced in apothecia (sexual fruiting bodies, or ascomata) on nearby susceptible host crops, cover crops, or organic material around the base of the tree. Removal of this material or control of the disease in nearby crops is usually all that is required to manage this disease on citrus. Frost damage, wounds, rain, fog, high humidity, and cool temperatures are factors that predispose citrus to infection by *S. sclerotiorum*.

Other Trunk and Branch Diseases

Branch Wilt

Branch wilt, or sooty canker, of citrus is caused by the heat-tolerant fungus *Neoscytalidium dimidiatum* (Penz.) Crous & Slippers (formerly known as *Nattrassia toruloidea* (Nattrass) Dyko & B. Sutton and *Hendersonula toruloidea* Nattrass). The pathogen invades cold injuries, sunburn, or mechanical wounds on the trunks and branches of citrus. This fungus has a wide host range and causes this type of disease in many plants unrelated to citrus.

Symptoms and ecology. Under the bark of the canker, the most conspicuous symptom of the disease is the sooty, brownish black fungal growth that mainly consists of masses of powdery conidia (asexual spores) of the pathogen. These spores are disseminated by wind. Invaded sapwood is light grayish olive to brownish olive. Along the margins of the decayed wood, a thin orange to pinkish line, about 0.04 to 0.12 inch wide, develops. Following the invasion of sapwood, leaf wilt and diebacks develop on branches with cankers. Cankers form on nonshaded, sun-exposed portions of branches or trunks. In southern California and Arizona, trees with sunburn damage are usually highly susceptible to infection.

Economics. In California, the disease is considered of minor importance.

Control. Prevention of injuries and wounds, as well as maintaining healthy and vigorously growing trees, are the main management strategies. Whitewashing sun-exposed branches and trunks, frost protection procedures (e.g., air movement with wind machines, irrigation, orchard heating), and pruning wound treatments, such as zinc-copper-lime or asphalt paints, have been used successfully in preventing branch wilt. Branches should be pruned well below the cankers, removed from the orchard, and destroyed. Pruning equipment should be disinfested with sanitizing chemicals. Proper irrigation schedules to prevent heat injury should also be followed, especially in high-temperature periods of the year.

Rio Grande Gummosis

Rio Grande gummosis is a serious disease in Texas that has also been reported in Florida and California, where it is sometimes called ferment gum. Historically, the disease has been attributed to numerous organisms, including an Actinobacterium, a *Diplodia* species, an unidentified Basidiomycota fungus (a wood rotter), and *Nattrassia toruloidea* (Nattrass) Dyko & B. Sutton (formerly known as *Hendersonula toruloidea* Nattrass). The latter pathogen is also known to cause branch wilt, or sooty canker, of citrus. Currently, the disease is considered to be mainly caused by *Botryosphaeria rhodina* (Berk. & M.A. Curtis) Arx [anamorph: *Diplodia natalensis* Pole-Evans, synonyms: *Botryodiplodia theobromae* Pat. and *Lasiodiplodia theobromae* (Pat.) Griffon & Maubl.]. Symptoms of Rio Grande gummosis range from profuse exudation of clear to frothy amber-brown gum (Botryosphaeria gummosis) to symptoms of sooty cankers (branch wilt) that develop under the bark of affected branches or trunks in later stages of the disease. Symptoms also include psorosis-like bark scaling. When the disease occurs with the unidentified wood rotter, the wood sometimes is discolored vinaceous buff to buff brown. This disease is associated with cold injuries, sunburn, or mechanical wounds on the trunks and branches of citrus (see the sections "Branch Wilt" and "Wood Decays" in this chapter) and is also increased by excessive application of chloride-based fertilizers.

Hyphoderma Gummosis

Hyphoderma gummosis is caused by the Basidiomycota fungus *Hyphodontia sambuci* (Pers.) J. Erikss. (syn. *Hyphoderma sambuci* (Pers.) Jülich, hence the disease name). The fungus has a wide host range, including many tree species, but on citrus it has been reported only from lemon. The disease occurs in the Central Valley and coastal lemon production areas of California.

Symptoms and ecology. The wood decay fungus causes cankers on branches and trunks of lemon trees and is capable of invading the cambial tissue. Symptoms include extensive gummosis along longitudinal cracks that develop around infected tissue. As cankers expand and the pathogen colonizes branches or the trunk, severe wilting and dieback of the tree canopy result. Eventually, tree death occurs. Crusty layers of pink to cream-colored fruiting bodies of the pathogen may form on the surface of diseased branches and the tree trunk under favorable environmental conditions. Basidiospores are wind disseminated and colonize exposed wood such as wounds from pruning or severe environmental conditions (e.g., frost damage). Like other wood decay fungi (see below), *H. sambuci* cannot invade the tree through intact bark.

Economics. In California, the disease can be devastating to orchards in growing areas prone to environmental stress, but its occurrence is localized.

Control. Prevention of injuries and large wood-exposing wounds, as well as maintaining healthy and vigorously growing trees, are the main management strategies. Proper pruning practices include avoiding stub and bench cuts, which prevent rapid wound healing. Pruning should be delayed as long as possible in the spring when no more rainfall is expected, but it should not be delayed until summer if there is a possibility of sunburn injury. Furthermore, wounds should not be exposed to wetness from irrigation. Wood-exposing wounds can be treated with the biocontrol agent *Trichoderma harzianum* (e.g., PlantShield) immediately after pruning. Sanitation measures include the removal of infected plant material from the orchard and burning. There are no curative treatments for this disease.

Wood Decays

Wood decay fungi are organisms that invade exposed wood and cause decay in large roots, trunks, and scaffold branches. Two types of decay fungi have been described in citrus trees: white rots and brown rots. With white rot, all the major components of wood (cellulose, hemicellulose, and lignin) are degraded. Examples of wood decay fungi that cause white rots of citrus are species of *Ganoderma* (e.g., *G. applanatum* (Pers.) Pat.; *G. brownii* (Murrill) Gilb.; and *G. lucidum* (Curtis) P. Karst.); *Hyphodontia sambuci* (Pers.) J. Erikss. (syn. *Hyphoderma sambuci* (Pers.) Jülich); *Oxyporus* species; and *Phellinus gilvus* (Schwein.) Pat. Other wood decay fungi cause brown rots, where only the cellulose and hemicellulose components of wood are degraded, and the brown lignin is left mostly unchanged. Examples of brown rot fungi reported from citrus include *Antrodia sinuosa* (Fr.) P. Karst. and *Coniophora* species. The oak root fungus *Armillaria mellea* is also a wood decay fungus that causes a white rot, and it could be included with this group of organisms; but because this fungus is a serious pathogen that does not need injuries for infection and can invade healthy tissues, it is discussed above under "Armillaria Root Rot."

Symptoms. Trees with wood decay are generally not vigorous and show signs of stress. Leaves are often chlorotic, and with progressive wood decay trees begin to weaken and may eventually have branches that die back, or the entire tree may die. Under the bark or in the decayed wood of infected roots, branches, or trunks, white to brown mycelial growth of the pathogen may be evident. The two kinds of wood decay mentioned above have a distinctive appearance. With white rots, the wood turns spongy and soft and has a whitish appearance. With brown rots, the wood turns into dry, brown cubicles resulting from cracks that develop across the grain. Fruiting bodies (basidiomes) of the fungus may eventually develop on branches or the trunk. With root decays, fruiting bodies are produced near the base of the trunk or on the soil surface. *Ganoderma* species produce large (2 to 4 in wide by 4 to 12 in tall) bracket- or shelf-like fruiting bodies that have fine pores on the underside. *Oxyporus* species have thin, white fruiting bodies with pores on the upper side. Fruiting structures of other wood decay fungi such as *H. sambuci* and *A. sinuosa* are more inconspicuous. The basidiomes of *H. sambuci* consist of a smooth hymenium (fertile layer) forming a cream-colored layer tightly appressed to the bark that becomes crusty with age. The basidiomes of *A. sinuosa* have a poroid hymenium that is initially whitish and eventually tan to light brown and flat to bracket-like on the surface of decayed wood. Fruiting bodies of *Coniophora* species have never been observed on citrus, but they do occur on other woody host plants. The basidiomes have a flat hymenium, with a white margin and a wrinkled brownish central area where the basidiospores are formed.

Wood decay caused by a *Coniophora* species (probably not *C. eremophila*) and *A. sinuosa* has resulted in extensive damage to citrus trees, particularly lemons in California and Arizona, whereas *H. sambuci* has caused extensive damage to lemons in California. All three species incite branch and trunk wood decays of living trees. Brown-rotted wood and limb dieback caused by *Coniophora* species and *A. sinuosa* may lead to tree decline and reduced fruit production. Ultimately, cracking of limbs and breaking of branches occur with heavy crop loads or by wind and rain storms. Extensive branch collapse may lead to eventual tree death. As described above, *H. sambuci* causes a white rot wood decay of citrus. The fungus, however, can invade the sapwood and kill the cambium of lemon tree scaffold branches and trunks. The host responds to the infection by producing gum; the disease is therefore described as Hyphoderma gummosis.

Economics. Wood decay fungi can cause extensive losses in some orchards, but they are generally localized, with only a few trees involved. Decays caused by *Coniophora* species and *A. sinuosa* have caused extensive limb breakage and destruction of orchards when the problem was not managed. Similarly, Hyphoderma gummosis has caused extensive damage due to diebacks of scaffold branches and death of entire trees in lemon orchards in Ventura and Tulare Counties.

Disease cycle and epidemiology. Wood decay fungi may become established on dead stumps, roots, or logs in newly planted citrus orchards, or they may be found on other vegetation nearby. Spores are produced in fruiting bodies and are readily dispersed by wind. Tree injuries, including pruning wounds, cold injuries, gopher injuries, or infection sites caused by *Phytophthora* species, may provide entry points for these fungi. Weakened trees are especially vulnerable to infection. From the entry points, the fungi are able to slowly invade the entire tree. Infected wood loses its strength, and trees are prone to damage by wind or other environmental factors, as well as limb breakage from heavy fruit loads.

Control. Trees should be kept healthy and vigorous. Avoid large pruning cuts or other wood-exposing injuries, especially during wet periods of the year. If disease is spreading via roots (e.g., forming an infection center), remove trees or stumps surrounding the infection area. Before planting new trees, remove stumps and large roots from the orchard and destroy the plant material (e.g., by burning or chipping). Chemical fumigation of the soil may reduce inoculum in wood debris such as infested small roots. Infected branches should be pruned as soon as possible once decay or fruiting bodies (i.e., basidiomes) are detected. Branches should be pruned well beyond the discolored wood to remove all of the infected tissue. Proper guidelines include pruning branches above the branch node to allow for healing, avoiding bench and stub cuts, and pruning during periods of extended dry weather (e.g., 30 days with low rainfall) to allow for wound healing and to prevent the accumulation of water on cut surfaces. As for other pruning practices for managing plant diseases, sanitizing pruning tools with oxidizing agents such as sodium hypochlorite (household bleach) prevents the dissemination of pathogens on orchard equipment.

Foliar and Fruit Diseases

Anthracnose

Anthracnose of citrus in California is caused by *Glomerella cingulata* (Stoneman) Spauld. & Schrenk [anamorph: *Colletotrichum gloeosporioides* (Penz.) Sacc.] The fungus is a facultative pathogen (opportunist) that is commonly associated with citrus as a saprophyte (colonizing dead tissue), but it may cause disease and economic losses in high-rainfall years. Serious preharvest epidemics may lead to postharvest decay. Postbloom fruit drop caused by *Colletotrichum acutatum* J.H. Simmonds is an important disease in Florida, but it has not been reported from California.

Symptoms. Symptoms of anthracnose on leaves are called weather tip and are first evident as chlorotic leaf tips that eventually die and turn brown. Severe leaf infections may lead to leaf drop. On fruit, the first symptom is a very fine, red-brown to black speckling of the rind. These speckles often appear where water has dripped onto the fruit surface and thus appear as tearstains. Sometimes these speckles coalesce to cause large, dark, superficial splotches on the fruit surface. This superficial stain is caused by *Colletotrichum* spores that germinate on the fruit surface and produce microscopic dark-colored appressoria (tip swellings of spore germ tubes during early stages of infection) on the fruit surface. Later, usually in storage, as the fruit ages, these appressoria germinate and penetrate the rind, causing large, dark brown to black sunken lesions that eventually rot the fruit. Infections of twigs may result in dieback.

Economics. The fungus is normally a symptomless invader of fruit that in most years causes little damage unless fruit are injured or overripe. Severe infections, however, lead to rind discoloration by the appressoria of the pathogen and make fruit unsuitable for the fresh-fruit market. Anthracnose has little effect on fruit used for juice, unless the fruit are stored for long periods.

Disease cycle and epidemiology. Anthracnose is a mild-temperature, high-wetness disease. Wetness may be provided by rainfall or extended fog periods. Under these conditions, spores of *C. gloeosporioides* are produced abundantly on dead twigs and leaves and are spread over relatively short distances by rain. Thus, trees with abundant dead twigs provide more inoculum for the disease. Spores germinate on leaves, twigs, and fruit and form appressoria that infect the plant. These infections may remain quiescent until host tissues senesce before causing visible symptoms. Appressoria may also remain dormant and germinate later to infect the host. From these infections, the fungus colonizes the citrus tissue, and under wet conditions new spores will be produced in acervuli. Leaf and twig infections rarely cause severe damage, but occasionally the disease can be economically important on fruit, causing tearstain and postharvest decay.

Control. Elimination of dead twigs in the canopy may be useful. Multisite compounds (e.g., copper) or fungicides with a single-site mode of action (e.g., demethylation inhibitors and quinone outside inhibitors) applied prior to the rainy period usually manage this disease. In most years, this disease is not significant enough to warrant control.

Botrytis Gray Mold

Botrytis cinerea Pers., the cause of gray mold, is a widely distributed fungus with a large host range. On citrus it is a minor pathogen, except where prolonged periods of wetness and cool temperatures occur. Gray mold is a preharvest disease on blossoms, twigs, leaves, bark, and fruit, but it is also a postharvest rot. Lemons are more commonly affected than other citrus species.

Symptoms. A fuzzy, gray fungal growth on the petals is the best sign for identification of blossom infections. The fungus may attack small twigs and branches, causing twig blight, where a yellowish buff lesion quickly enlarges, encircles the limb, and girdles it. Leaves above the lesion quickly wilt, turn brown, and drop. If kept moist, gray conidia form on conidiophores as the mycelium develops on the lesion. *Botrytis cinerea* may also cause gummosis on larger branches that is similar in appearance to gummosis caused by *Phytophthora* species. Gummosis of large scaffold branches is usually associated with frost, injuries, or shell bark. Fruit that was infected in the button stage may have scarred, raised ridges that lower the grade for fresh market sale.

Economics. Botrytis gray mold is a minor disease that is rarely important economically. In some years it may reduce the yield of lemons. Scarred fruit resulting from infection during early fruit development are not suitable for the fresh market. If the disease occurs on young trees in a nursery, losses can be heavy.

Disease cycle and epidemiology. The fungus lives saprophytically on decaying organic matter such as leaves, petals, twigs, and fruit. Spores may be carried to flowers, fruit, or young shoots by wind, water, or insects. Flower petals are most easily infected. Young fruit and twig infections often originate from close contact with infected petals. Fruit may then become scarred or may abscise. Healthy fruit and twigs without injuries are usually not infected. However, if fruit or twigs are predisposed by injuries from frost or sunburn, the pathogen can readily invade the tissues. Continuous wet conditions from fog or rain and cool temperatures around 64°F are optimal for disease outbreaks.

Control. Damage by gray mold is usually minimal, and chemical control is not economical. If conditions are optimal for the fungus, control is difficult because frequent fungicide applications are required. Fungicide applications will reduce damage if applied before rain or fog is expected.

Brown Rot

Brown rot of citrus fruit can be caused by some of the same species of *Phytophthora* that also cause Phytophthora root rot, foot rot, and gummosis, including *P. citrophthora, P. syringae, P. parasitica,* and *P. hibernalis*. Damage from the disease occurs mostly in the winter season in many citrus-growing areas of California. Trees may show symptoms of brown rot alone or in combination with symptoms of Phytophthora root rot, foot rot, or gummosis.

Symptoms. Brown rot develops mainly on fruit in the lower tree canopy, especially those growing near the ground. Infections by *Phytophthora* species cause an olive-brown discoloration of the rind, and the fruit have a distinctive pungent odor. Fruit remain firm and leathery, unless they are invaded by secondary decay organisms. Infected fruit usually fall to the ground. At a very high humidity, fruit become covered by a delicate white growth of the pathogen.

Economics. Brown rot is an important disease of citrus fruit. Losses due to brown rot are very sporadic and are associated with periods of heavy rainfall. Although losses may occur in the orchard, the most serious aspect of the disease is that fruit infected before harvest may not show symptoms. If infected fruit get mixed with healthy fruit, the disease may spread quickly from fruit to fruit in storage and during transit, and many fruit may become affected. In addition, fruit infected with brown rot are readily colonized by wound pathogens, such as *Penicillium* and *Geotrichum* species. Some packinghouses will not accept fruit if the incidence of brown rot in the citrus orchard exceeds 5%.

Disease cycle and epidemiology. *Phytophthora* species that cause brown rot can survive in the soil as persistent chlamydospores, oospores, or as mycelium in decaying roots or fallen diseased fruit. Free water is required for dissemination and infection of the host. At high soil moisture contents, sporangia develop within 18 hours from chlamydospores, oospores, or mycelium inthe soil. Swimming zoospores that are formed in the sporangia are released and may be splashed up onto low-hanging fruit. At temperatures from 57° to 73°F, a continuous wetness period of 3 hours is necessary for fruit infection. New generations of sporangia will release zoospores if wet conditions persist, and serious brown rot epidemics may occur. The length of the continuous rainy period is the most important predictor of brown rot epidemics.

Control. Historically, the most effective control for brown rot is a Bordeaux mixture spray, composed of copper sulfate ($CuSO_4{\cdot}5H_2O$), lime ($Ca(OH)_2$), and water, prior to the rainy period. Many commercial neutral coppers are also very effective. The copper sprays should contain 9.5 to 12.8 ounces of metallic copper per 100 gallons of water in a high-volume sprayer. In order to prevent copper damage, neutral coppers should be "safened" by adding 4.9 to 16 ounces of hydrated lime per pound of metallic copper. Depending on the amount of rainfall, more than one application of copper may be necessary. Preharvest treatments of the canopy with systemic fungicides such as phosphonates (e.g., fosetyl-Al) and phenylamides (e.g., mefenoxam) also provide highly effective control of brown rot before and after harvest when applied up to 6 weeks prior to infection. Postharvest fungicides have historically been ineffective against brown rot. Products with activity against brown rot are being developed. Currently, potassium and calcium phosphite fungicides are registered for postharvest use on citrus, and these products are exempt from tolerances in the United States and many other countries throughout the world.

Clear Rot

Penicillium decays caused by *Penicillium digitatum* (Pers.) Sacc. and *P. italicum* Wehmer are major postharvest rots. If decays are already developing in the field, they are commonly referred to as clear rot. This rarely occurs, but in years with high winds and rainfall, fruit become injured, and clear rot of fruit may develop on the tree.

Symptoms. Clear rot is characterized by large, water-soaked, softened areas of the fruit. At high humidity or in rainy weather, the entire fruit will eventually be covered by abundant green *(P. digitatum)* or bluish *(P. italicum)* spores. Lesions are surrounded by white margins of the advancing mycelial growth of the fungi.

Economics. Clear rot generally is a minor disease in California. Major secondary losses from Penicillium decays, however, are caused when healthy fruit are contaminated by the abundant inoculum from diseased fruit. Fungal spores gain entry into healthy fruit through wounds that occur at harvest, resulting in postharvest decay.

Disease cycle. Species of *Penicillium* are common soil inhabitants and are also found on senescent and decaying plant material in the orchard. The fungus cannot infect the intact fruit surface, but any wound caused by biotic or abiotic factors (e.g., high winds and blowing dust) allows the pathogen to invade the healthy tissues. Because injuries also commonly occur during harvest and handling operations, *Penicillium* species can cause major postharvest losses (see the section "Postharvest Decays," below).

Control. Treatments with protective and systemic preharvest fungicides may help to reduce the amount of inoculum on fruit surfaces and infections that occur at harvest.

Septoria Spot

Septoria spot on leaves and fruit has been reported from most citrus-growing regions in the world. Oranges, lemons, and grapefruit are most frequently damaged, but all citrus varieties are susceptible. In the past, several species of *Septoria* have been associated with Septoria spot. However, this disease is now considered to be caused by a single species, *Septoria citri* Pass.

Symptoms. Under high-rainfall conditions, infections on fruit may begin when the fruit is green and become more conspicuous as the fruit ripens. In California, the fungus is opportunistic and mostly colonizes cold-injured mature fruit. Lesions consist of small depressions or pits 0.04 to 0.08 inch in diameter that extend no deeper than the flavedo. The pits are light tan or buff with narrow greenish margins that become reddish brown. These symptoms sometimes are in the form of tearstains. Small, black spore-producing structures of the fungus, the pycnidia, may be found on fruit lesions. Lesions may enlarge and coalesce into conspicuous brown to black sunken blotches that may extend into the fruit segments. Lesions may continue to develop after fruit are harvested. Septoria spot may sometimes be confused with other fungal diseases (e.g., anthracnose or fruit infections caused by *Alternaria* spp.), as well as copper injury or other abiotic disorders.

Leaf infections result in raised, blisterlike black spots 0.04 to 0.16 inch in diameter and are surrounded by a yellow halo. Later, the center of the spots becomes necrotic and pale brown. Pycnidia may develop in the lesions, especially after leaf drop, when the lesions become brown with a dark margin. Severe leaf drop of the lower portion of the tree may occur when conditions are highly conducive for the disease.

Economics. Septoria spot is generally of minor significance. Damage to fruit rind, however, is of concern when the appearance of fresh fruit is important.

Disease cycle and epidemiology. *Septoria citri* can grow saprophytically on dead citrus twigs and leaves, where pycnidia are often formed abundantly. Conidia from these pycnidia are dispersed by rain splash onto healthy leaves and fruit, where they germinate and cause infections. These infections may not develop into visible symptoms for several months. Infections most commonly occur in low-temperature environments when cold injuries occur to the exposed fruit rind. Low or rapidly fluctuating temperatures are thought to predispose the rind flavedo to injuries and subsequent fungal colonization of damaged tissue. The disease is most severe in years when rainfall is higher than normal. Leaf symptoms often appear in close proximity to irrigation emitters, especially where misters or minisprinklers are placed under the canopy. A risk model based on the accumulation of temperatures below 30°F and total precipitation is available for forecasting the disease in winter months (see sidebar).

Fungicide Applications for Managing Septoria Spot of Oranges

A model has been developed to assist growers in determining the timing of fungicide applications for the management of Septoria spot of oranges. The model is based on two environmental conditions: the accumulation of hours below 30°F and the total amount of precipitation after fruit color break (approximately late October to early November) in a harvest season. Increasing risk factors are assigned to ranges of accumulated parameters.

The values in the following table are risk factors for disease for each of the ranges of accumulated parameters. Note that increasing values for either parameter result in higher risk factors. A risk factor of zero is equal to no risk for disease; a risk factor of 1 (green) is low, 2 (yellow) is moderate, 3 (orange) is high, and 4 (gray) is extremely high.

Using this model, a risk alert exists when the risk factor reaches 2, and a high-risk alert with recommendation for a fungicide application when the risk factor reaches 3. Guidelines include fungicide applications within 30 days after the risk factor of 3 is reached. At risk factor 2, weather forecasts can be used to predict whether increase of disease risk is imminent.

The model is reset after a fungicide application to reaccumulate low-temperature periods and precipitation for the remainder of the season. Regional forecasts can also be done by restarting the model after the region reaches a risk factor of 3.

Numerical risk model for forecasting Septoria spot

Hours with temps less than −1°C (30°F)	Precipitation, mm (in)				
	31–60 (1.2–2.3)	61–90 (2.4–3.5)	91–120 (3.6–4.7)	121–150 (4.8–5.9)	151–180 (6.0–7.1)
< 10	0	1	2	3	4
10–20	1	2	3	4	4
21–30	2	3	4	4	4
> 30	3	4	4	4	4

Control. Preventative copper treatments can be applied in late fall or early winter before rainfall. Additional synthetic fungicides are registered as preharvest foliar treatments, as well as postharvest treatments in the packinghouse.

Other Foliar and Fruit Diseases

Sooty mold consists of dark fungal growth that can be found on fruit, twigs, and leaves, especially when trees are heavily infested by honeydew-excreting insects such as aphids, mealybugs, and whiteflies. On citrus, sooty mold is most commonly caused by *Capnodium citri* Berk. & Desm. Black fungal growth is superficial and can be removed from the plant surface by rubbing. When trees are severely affected, photosynthesis may be reduced and fruit coloring may be delayed. Insect control is the primary management method when sooty mold is serious enough to affect fruit quality and yield.

Powdery mildew caused by *Acrosporium tingitaninum* (J.C. Carter) Subram. (synonym: *Oidium tingitaninum* Carter) has been reported on *Citrus* species in several citrus-producing areas of the world, but it is only found occasionally on mandarins in California. The disease typically forms white circular to elliptical colonies on the upper surface of leaves. The underlying tissue turns dark green and is water soaked at first but becomes yellow. Leaves may wither, dry, and remain attached to the stem. Fungicides such as sulfur or newer compounds that have a single-site mode of action may be used to manage the disease.

Greasy spot caused by *Mycosphaerella citri* Whiteside is found in climates that have prolonged periods of high temperatures and high humidity. The disease is not known to occur in California, although *Mycosphaerella* species have been reported on citrus in the state. A different disease, melanose, is caused by *Diaporthe citri* F.A. Wolf (anamorph *Phomopsis citri*), the pathogen of Diaporthe gummosis, which is present in most citrus-growing countries. Greasy spot and melanose rarely cause damage to fruit in California and are important only where inoculum is abundant and rainfall occurs early in fruit development. Melanose affects mainly fruit. Infections penetrate up to six cells deep into the fruit tissue, and the dead cells become reddish brown. Because it is a superficial disease, it is not economically important if the crop is processed. Both greasy spot and melanose symptoms are often confused with Septoria spot. Stem-end rots, in addition to other fungi, can also be caused by some of the same pathogens that cause Diaporthe and Botryosphaeria gummosis and canker diseases (see the next section, "Postharvest Decays").

Postharvest Decays

The relative importance of postharvest decays of citrus depends on the climate of the production area, host species, cultivar, and time of harvest. Certain fruit diseases are initiated preharvest or during growth and development of the fruit, whereas other diseases originate from injuries inflicted during harvest and handling. Table 21.4 shows the initiation period, site, and type of infection of selected fruit decays in relation to seasonal occurrence.

Penicillium Fruit Rots

Penicillium rots are the most important decays of citrus fruit, especially in areas with little summer rainfall. In more humid climates, Diplodia and Phomopsis stem-end rots may be more economically important than Penicillium decays. Penicillium decays are mainly a postharvest problem, although infections by *Penicillium* species may also occur on the tree during the later stages of fruit development (see the section on clear rot, above). Three decays, caused by three species of *Penicillium*, can be found. Green mold, caused by *P. digitatum* (Pers.) Sacc., is the most common and serious postharvest decay of citrus in many countries. Blue mold, caused by *P. italicum* Wehmer, is also widespread but is generally less prevalent. The third Penicillium decay of citrus fruit, whisker mold, caused by *P. ulaiense* H.M. Hsieh, H.J. Su & Tzean, is much less frequent and rarely of major concern.

Symptoms. Early symptoms of the three Penicillium decays are indistinguishable. Large areas of the fruit appear water soaked, and fruit begin to soften. Aerial mycelium begins to form in the center of the lesions within 2 to 3 days at 68° to 82°F and is soon followed by sporulation of the fungus. The sporulating area for green mold is green to olive-green and is surrounded by a broad band of nonsporulating mycelium. Sporulating lesions of blue mold are bluish and are surrounded by a narrow band of nonsporulating growth. Sporulating lesions of whisker mold are similar in color to blue mold. Conidiophores (simple or branched hyphae bearing fertile cells that produce conidia or mitospores), however, grow in bundles (i.e., synnemata or coremia) that may be arranged in circular patches or concentric circles. Lesion development of whisker mold proceeds at a much slower rate than with green mold and blue mold. The three Penicillium decays may occur on the same fruit.

Economics. Postharvest fruit losses from Penicillium decays are a constant threat to the citrus industry worldwide, and losses are encountered

Table 21.4. Time, site, type of infection, and seasonal conditions that favor the occurrence of major postharvest decays of citrus

Time of infection	Disease	Pathogen	Infection site	Type of infection*	Seasonal conditions
preharvest (infection mainly during growth and development of fruit)	anthracnose	*Glomerella cingulata* (Stoneman) Spauld. & H. Schrenk 1903 *(Colletotrichum gloeosporioides)* (Penz.) Penz. & Sacc. 1884	fruit surface	active or quiescent	warm, wet
	brown rot	*Phytophthora* spp.	fruit surface	active or quiescent	cool, wet
	Botrytis rot	*Botrytis cinerea* Pers. 1794	flower, developing fruit	quiescent	cool, wet
	Diplodia stem-end rot	*Botryosphaeria rhodina* (Berk. & M. A. Curtis) Arx 1970 (*Lasiodiplodia theobromae* 1909; *Diplodia natalensis* Pole-Evans 1911)	flower, developing fruit (button only)	quiescent	warm, wet (spring and fall)
	Phomopsis stem-end rot	*Diaporthe citri* F. A. Wolf 1926 (*Phomopsis citri* H.S. Fawc. 1912)	flower, developing fruit	quiescent	cool, wet (winter)
	Dothiorella stem-end rot	*Botryosphaeria* spp. (*Dothiorella* and other anamorphic spp.)	flower, developing fruit	quiescent	warm, wet
	Alternaria stem-end rot	*Alternaria citri* Ellis & N. Pierce 1902	flower, developing fruit, navel	quiescent	cool, wet
postharvest (infection mainly through wounds inflicted during harvest and handling of fruit)	blue mold	*Penicillium italicum* Wehmer 1894	fruit injuries	active	warm
	green mold	*Penicillium digitatum* (Pers.) Sacc. 1881	fruit injuries	active	warm
	sour rot	*Galactomyces citri-aurantii* E.E. Butler1988 *(Geotrichum citri-aurantii* (Ferraris) E. E. Butler 1988)	fruit injuries	active	warm, wet
	Trichoderma rot	*Trichoderma viride* Pers. 1794	fruit injuries	active	warm

Note: *Some fungi may cause infections that remain quiescent and function as biotrophs in developing and harvested fruit but become active necrotrophic fungi, inciting decay symptoms, during fruit ripening and senescence. Other fungi may cause active decay and function as necrotrophs at the onset of the infection. Nomenclatural rules for fungi changed as of January 2013. Currently, only one fungal name is allowed, and the earliest name published has priority.

annually, especially in arid climates. Secondary damage occurs from spoilage of fruit that are in contact with heavily infected fruit.

Disease cycle and epidemiology. In the orchard, spores of *Penicillium* species from soil, decaying plant material, and fruit on the ground are disseminated by wind to healthy fruit on the tree. Infections leading to postharvest decay are initiated mostly at harvest when minor wounds occur on the fruit and remain undetected during fruit sorting. Decay develops during marketing of the fruit or while in storage. A high incidence of decay may develop during long-term storage of fruit, a common practice for lemons in California. Abundant spore masses are produced on the diseased fruit, and additional healthy fruit are contaminated with spores, leading to infection and potential decay of a large percentage of the fruit lot. Growth of the pathogens is much delayed at temperatures below 41°F. Citrus fruit, however, cannot be stored for long periods at temperatures below 50°F, so temperature management cannot control these postharvest decays.

Control. Sanitation and fungicide treatments are key strategies for decay control. Sanitation measures at harvest should ensure that fruit are not picked up from the orchard floor, and any injury of fruit should be avoided because pathogens rapidly infect the fruit from wounds. Preharvest fungicide applications reduce inoculum levels in the orchard and thus reduce postharvest decays, but this is rarely practiced in California. In the packinghouse, sanitation practices include high-volume chlorine washes of fruit and regular disinfection of equipment. Postharvest treatments with heated bicarbonate solutions are effective in reducing or delaying decay but may cause dehydration and staining of fruit. The use of postharvest fungicides in an integrated system can reduce Penicillium decays to very low levels. Highly active postharvest fungicides are available, but resistant populations of *P. digitatum* have developed against the older materials sodium ortho-phenylphenate (SOPP), imazalil, and thiabendazole. The incidence of populations resistant to imazalil and thiabendazole generally increases and reaches high levels late in the packing season. New reduced-risk fungicides, including fludioxonil, pyrimethanil, and azoxystrobin, have been registered for postharvest use, and registrations of other compounds are planned.

Sour Rot

Sour rot, caused by *Galactomyces citri-aurantii* E.E. Butler (anamorph: *Geotrichum citri-aurantii* (Ferraris) E.E. Butler, formerly described as *G. candidum* Link), is the second-most-important postharvest decay in California. It has been reported from most areas where citrus is grown, and all citrus fruit can be affected. Like Penicillium decays, sour rot is often initiated from wounds on the fruit that occur during harvest and handling. Ripe or overmature fruit are more susceptible to sour rot than are green or immature fruit, especially when fruit are stored for extended periods at high humidity or are shipped without adequate refrigeration.

Symptoms. Sour rot is first visible as a small water-soaked area on the rind; at this stage, it is not easily distinguished from decays caused by *Penicillium* species. The lesion enlarges and becomes very soft, unlike Penicillium decays. There is a distinct border between the very soft lesion and the firm, unaffected rind. The lesion is creamy yellow and at high humidity is covered by a thin layer of whitish fungal growth that may be wrinkled. Rotted fruit have a characteristic yeasty, vinegary odor. At an advanced stage of infection, the fruit is completely macerated by enzymatic activity of the pathogen. The fruit then collapses in a watery mass that drips onto underlying fruit, causing a nest of decay.

Economics. Sour rot is second in importance in citrus to decays caused by *Penicillium* species. All citrus fruit are susceptible, and the disease is most severe in years of high rainfall or on fruit that are stored for long periods, such as lemons. The extent of postharvest fruit losses from sour rot varies from year to year but can be very high because of secondary spread in the storage room.

Disease cycle and epidemiology. The causal organism, *G. citri-aurantii,* is a common soil inhabitant. Spores are disseminated by wind and rain to fruit on the tree, but infections can occur only through injuries. Symptoms quickly develop on very mature fruit when the rind has high moisture content and the fruit is stored at high humidity. Thus, the disease is most serious during and after prolonged wet seasons.

Control. Harvesting and handling practices that minimize fruit injuries reduce the incidence of sour rot. Sanitation practices that keep soil from fruit in harvesting containers or that remove soil particles from fruit are also effective strategies. Washing fruit with chlorinated water or other oxidizing agents and disinfesting handling equipment with sanitizers (e.g., hot water, chlorinated water, quaternary ammonium compounds, formaldehyde) also reduce the incidence of disease. Postharvest wash (spray application) or soak tank treatments with sodium ortho-phenylphenate (SOPP) reduce sour rot but may be phytotoxic. Propiconazole is a new postharvest fungicide in the United States that is very effective against sour rot. In other citrus-producing countries, guazatine and iminoctadine have been used successfully.

Trichoderma Rot

Trichoderma rot is caused by *Trichoderma viride* Pers., a common soil saprophyte that can be a pathogen of citrus fruit. The disease is associated with fruit that are contaminated with soil particles. The decay is generally a minor problem, but it can cause severe losses when fruit are stored for long periods or transported under poor sanitation and shipping conditions. Losses have occurred when fruit are stored under warm conditions and ventilated using ambient air.

Symptoms. Diseased fruit become cocoa-brown and develop a coconutlike odor. This latter characteristic distinguishes Trichoderma decay from other postharvest decays of citrus. The rind remains tough and leathery. Under humid conditions, the fungus develops extensive amounts of mycelium that completely cover the fruit. The fungus sporulates with masses of yellow to emerald-green spores.

Economics. The disease has a sporadic occurrence but may cause extensive losses. Lemons are the most susceptible to decay, followed by oranges and grapefruit.

Disease cycle and epidemiology. The fungus *T. viride* is disseminated by airborne conidia or with soil particles. Fruit in contact with soil particles or soiled harvest bins may become infected. The fungus can grow in wood, and wooden harvest bins can be a source of inoculum for fruit infection. A relatively deep wound associated with damaged oil glands on fruit is required for infection. Infections can start anywhere on the fruit but commonly initiate on the stem or blossom end. Decay of infected fruit often spreads to adjacent fruit (commonly called nesting), especially under humid conditions. Nesting of infected fruit often occurs when juice of infected fruit damages the rind of healthy fruit. Decay develops rapidly on lemons stored at 57°F or higher for several months and on oranges stored at 50°F or higher under ventilation only during shipment. Decay does not develop below 41°F. Lemons stored at 40° to 41°F (a temperature range below the optimum of 50° to 55°F) or oranges stored at an optimal 37° to 48°F, however, have a storage life of only about 4 to 6 and 3 to 8 weeks, respectively.

Control. The disease is managed by integration of proper sanitation, harvesting, and handling procedures, including temperature management. Rapid cooling and storage of cleaned fruit is very effective. The pathogen does not spread from fruit to fruit at 50°F, and decay does not develop when infected fruit are stored at 39°F. Sanitation includes washing fruit with chlorinated water or other oxidizing agents and disinfesting handling equipment with sanitizers (e.g., hot water, chlorinated water, quaternary ammonium compounds, formaldehyde). Lining wooden storage boxes with polyethylene prevents infection of fruit in contact with infested wood. Effective postharvest treatments against Trichoderma rot include warm dip treatments of borax, sodium carbonate, or sodium ortho-phenylphenate, followed by a postharvest fungicide.

Alternaria Rot

Worldwide, *Alternaria* species can cause four distinct diseases of citrus: Alternaria brown spot of mandarins, Alternaria leaf spot of Rough Lemon, *mancha foliar de los citricos,* and Alternaria rot (sometimes called black rot). The pathogen *Alternaria citri* Ellis & N. Pierce can be divided into strains that produce host-specific toxins and those that do not. The strain that causes Alternaria rot is a non-toxin-producing strain. Toxin-producing strains that incite brown spot and leaf spot, as well as other *Alternaria* species (e.g., *A. limicola* Simmons and Palm), have not been reported to cause diseases of citrus in California. Strains on mandarin that cause brown spot have also been referred to as *A. alternata* (Fr.) Keissl. pv. *citri* Solel. Alternaria rot mainly affects navel oranges, where it is also called black rot, and it also affects lemons. Alternaria rot is most important as a stem-end rot on fruit that is stored for a long time. Sometimes the disease develops at the blossom end or at wound sites (cracks at the navel or on the rind) of fruit, causing premature color change and fruit drop in the orchard. In fruit processing, a small amount of Alternaria rot in a fruit lot may give juice a bitter flavor, and small black fragments of rotted fruit may spoil its appearance.

Symptoms. Diseased fruit in the orchard color prematurely and may develop a firm lesion on the rind at the blossom end or at wound sites that is light to dark brown or black and extends as a rot into the core of the fruit. On stored fruit, decay commonly starts at the stem end as a water-soaked softened area that gradually enlarges and turns gray to brown. At high humidity, fuzzy gray mycelium develops on fruit lesions. Some fruit, however, show no external evidence of infection. At storage temperatures of 55° to 59°F, the decay usually appears after 4 to 6 weeks (tree-ripe yellow fruit) or 12 to 16 weeks (green fruit) of storage.

Economics. In California, Alternaria rot can be a major problem on lemons that are stored for extended periods.

Disease cycle and epidemiology. *Alternaria citri* is a facultative pathogen that commonly grows as a saprophyte on dead citrus tissues in the tree canopy and on the orchard floor. Spores are dispersed by wind and can establish infections on fruit with injuries such as split navels of oranges or fruit cracks caused by sunburn. The disease can also appear on fruit that have been stressed by drought or frost. Quiescent infections that occur on the buttons start actively growing once the buttons senesce. Quiescent infections at the stylar end of the fruit may start invading the fruit through growth cracks.

Control. Healthy, high-quality fruit are more resistant to Alternaria rot than are stressed or damaged fruit because *A. citri* is a weak pathogen. Preventing stress can reduce the incidence of splitting and subsequent Alternaria rot. Stylar-end infections generally occur on cultivars with poorly formed navels. Preharvest fungicide treatments are usually ineffective. Delaying harvest until infected fruit have fallen has been used as a strategy to prevent inadvertent inclusion of infected fruit in the harvested crop. Unaffected fruit, however, should be harvested at optimal maturity. The pathogen can also cause a postharvest stem-end rot (see below), but under California conditions this decay is usually minimal if fruit are harvested at optimal maturity. Postharvest treatments with imazalil, 2,4-D, or both have provided some control. The growth regulator 2,4-D delays senescence and thereby restricts colonization of the host.

Other Postharvest Fungal Decays

Several additional postharvest decays may occur on citrus fruit in California, but they are considered to be of minor importance. Diseases caused by *Botrytis cinerea* Pers. mainly develop preharvest on shoots, flowers (especially senescent blossom tissue), and developing fruit. This has been discussed under "Foliar and Fruit Diseases," above. Additionally, lemon fruit may develop raised ridges from preharvest *B. cinerea* infections that result in fruit off-grades. Spores of *B. cinerea* that contaminate developing fruit may cause quiescent infections on the stem end of fruit. After harvest, these infections may result in postharvest fruit decays. Other preharvest diseases that may cause postharvest decays of fruit include anthracnose, Septoria spot, and brown rot caused by *Phytophthora* species (see respective sections above). Additionally, stem-end rots have been reported to be caused by gummosis fungi, including *Botryosphaeria rhodina* (anamorph: *Diplodia natalensis, Lasiodiplodia theobromae*), *Diaporthe citri* (anamorph: *Phomopsis citri*), and other *Botryosphaeria* species (including *Dothiorella* and other anamorphic species), as well as *Alternaria citri*. Stem-end rots are found occasionally in desert areas during monsoon weather but generally do not occur in California. Management of stem-end rots includes orchard practices such as pruning of dead wood, as well as postharvest fungicide treatments and storage temperatures at less than 50°F. Historically, Sclerotinia (cottony) rot has caused severe postharvest decays, but it is rarely seen today because green manures and cover crops that are hosts of the pathogen *Sclerotinia sclerotiorum* (Lib.) de Bary are no longer used in commercial citrus production in California.

References

Adesmoye, A., A. Eskalen, B. Faber, and N. O'Connell. 2011. Multiple *Botryosphaeria* species causing 'Dothiorella' gummosis in citrus. Citrograph 2:2: 32–34.

Bender, G. S., J. A. Menge, H. D. Ohr, and R. M. Burns. 1982. Dry root rot of citrus: Its meaning for the grower. Citrograph 67:249–254.

Bonde, M. R., G. L. Peterson, R. W. Emmett, and J. A. Menge. 1991. Isoenzyme comparison of *Septoria* isolates associated with citrus in Australia and the United States. Phytopathology 81:517–521.

Dreistadt, S. H. 2012. Integrated pest management for citrus. 3rd ed. Oakland: University of California Agriculture and Natural Resources Publication 3303.

Eckert, J. W., and I. L. Eaks. 1989. Postharvest disorders and diseases of citrus fruit. In W. Reuther, E. C. Calavan, and G. E. Carman, eds., The citrus industry. Vol. 5. Oakland: University of California Division of Agriculture and Natural Resources. 179–260.

Feld, S. J., J. A. Menge, and J. E. Pehrson. 1979. Brown rot of citrus: A review of the disease. Citrograph 64:101–106.

Ferguson, A. J., and S. N. Jeffers. 1999. Detecting multiple species of *Phytophthora* in container mixes from ornamental crop nurseries. Plant Disease 83:1129–1136.

Graham, J. H., and J. A. Menge. 2000. Branch and twig diebacks. In L. W. Timmer, S. M. Garnsey, and J. H. Graham, eds., Compendium of citrus diseases. 2nd ed. St. Paul, MN: APS Press. 70–71.

Klotz, L. J. 1978. Fungal, bacterial, and nonparasitic diseases and injuries originating in the seedbed, nursery, and orchard. In W. Reuther, E. C. Calavan, and G. E. Carman, eds., The citrus industry. Vol. 4. Oakland: University of California Division of Agricultural Sciences. 1–66.

Kobbe, B. 1984. Integrated pest management for citrus. Oakland: University of California Division of Agriculture and Natural Resources Publication 3303.

Lutz, A., and J. A. Menge. 1986. Phytophthora root rot. Citrograph 72:33–39.

Munnecke, D. E., M. J. Kolbezen, W. D. Wilbur, and H. D. Ohr. 1981. Interactions involved in controlling *Armillaria mellea*. Plant Disease 65:384–389.

Streets, R. B., and H. E. Bloss. 1973. Phymatotrichum root rot. Monograph 8. St. Paul, MN: APS Press.

Timmer, L. W., S. M. Garnsey, and J. H. Graham, eds. 2000. Compendium of citrus diseases. St. Paul, MN: APS Press.

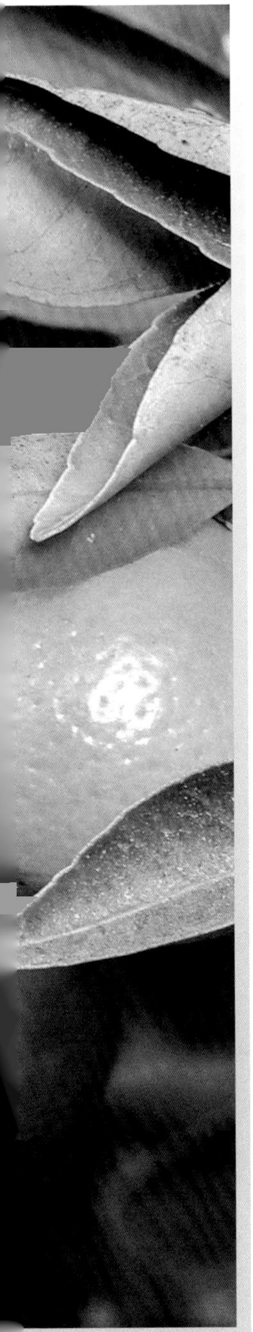

22 Bacterial Diseases

Edwin L. Civerolo and Donald A. Cooksey

This chapter provides basic information about prokaryote-caused diseases of citrus in California to aid growers in understanding how these diseases affect citrus production. Bacterial disease symptoms may be overlooked or confused with other problems. Rapid, accurate, reliable diagnoses are critical for their effective management. Diagnoses should be based on field observations and symptoms, along with confirmatory analysis by greenhouse or laboratory tests. A list of the bacterial diseases of citrus worldwide is presented in table 22.1.

Citrus Blast and Black Pit

Most bacterial plant diseases are economically important in humid regions. Citrus blast and black pit occur in several citrus-growing regions around the world, primarily under wet, cool, and windy conditions. They generally do not occur in hot, dry areas or areas of low relative humidity. Accordingly, these diseases are usually severe and economically important in citrus-growing areas of northern California, where environmental conditions, including long periods of low temperatures, heavy rains, and strong winds in the winter and spring, are conducive to infection, disease development, and spread of the pathogen. Infection is also facilitated by physical injuries to the host caused by wind or hail. However, bacterial diseases are relatively minor diseases of citrus in California. When they do occur, they are usually most severe on the side of the affected tree exposed to wind-driven rain. In lemons, the fruit are generally more susceptible to infection than are leaves and twigs, especially following a puncture injury by thorns while the fruit is wet.

Causal organism. Phytopathogenic pseudomonads are a very diverse group of bacteria with respect to their ecology and plant diseases that they cause. *Pseudomonas syringae* pv. *syringae* (Pss) is readily isolated from young lesions on various artificial nutrient agar-containing media but not from old, dry lesions. Pss can be readily identified by standard bacteriological methods, serology, and biochemical tests. Strains that cause blast and black pit also can be identified by inoculation of young, actively growing susceptible citrus species in the greenhouse.

Symptoms. The symptoms of blast appear on leaves and twigs, and those of black pit appear on the fruit (especially on lemons) (fig. 22.1). Blast lesions usually develop first on the leaf petiole, or wing, of susceptible hosts as small water-soaked or dark spots. These expand rapidly in both directions, upward toward the leaf midvein and downward into the axil and twig. The petiole and twig tissues become severely damaged by girdling, then collapse. The leaves wither, curl, dry, become brown, and eventually drop. Necrosis in the twig is generally limited and usually progresses only a short distance. The lesions of infected twig tissues, beginning at the margins, become reddish brown to chestnut colored and may resemble scabs or calluses. Occasionally, severe infection results in entire twigs becoming girdled and killed. The withered, dry leaves and dead twigs scattered throughout a tree canopy create the "blasted" appearance. Blast symptoms may be confused with frost damage. Black pit lesions initially develop on young fruit as small, light brown spots in the rind. These become reddish brown and finally black. They appear either as small spots or specks, or as larger (0.2 to 0.8 in diameter) sunken spots, depending on the age of the fruit infected.

Epidemiology. Pss strains are common in nature as part of the endemic microbial flora associated with a diversity of plant species, including citrus. The life cycle of these strains includes an epiphytic phase as well as a pathogenic phase. Blast and black pit strains of Pss do not move systemically in citrus hosts

Table 22.1. Prokaryote diseases of citrus worldwide

Disease	Causal organism	Hosts
Australian citrus dieback	phytoplasma?	**Highly susceptible:** grapefruit *(C. paradisi)* **Susceptible:** sweet oranges *(C. sinensis)*; citron *(C. medica)*; mandarins *(C. reticulata)* and mandarin hybrids; bergamot *(C. bergamia)*; sour orange *(C. aurantium)*; yuzu *(C. ichangensis* x *C. reticulata* (formerly *C. junos*)); trifoliate orange *(Poncirus trifoliata)*[1]
bacterial canker (Asiatic type)	*Xanthomonas citri* subspecies *citri*[2]	**Highly susceptible:** grapefruit *(C. paradisi)*; Mexican/Key lime *(C. aurantifolia)*; lemon *(C. limon)*; pointed leaf hystrix *(C. hystrix)* **Susceptible:** limes – Persian/Tahiti lime *(C. latifolia)*, Palestine sweet lime *(C. limettioides)*; trifoliate orange *(Poncirus trifoliata)* and hybrids; citranges (*C. sinensis* x *P. trifoliata*); citrumelos (*P. trifoliata* x *C. paradisi*); tangerines *(C. reticulata)*; tangors (*C. sinensis* x *C. reticulata*); tangelos (*C. reticulata* x *C. paradisi*); sweet oranges *(C. sinensis)*; sour oranges *(C. aurantium)* **Resistant:** mandarins *(C. reticulata)*; citron *(C. medica)* **Highly resistant:** calamondin *(C. mitus* or *Calamansi* x *Citrofortunella microcarpa)*; kumquat (*Fortunella* spp.)[3]
bacterial canker (cancrosis B)	*Xanthomonas fuscans* ssp. *aurantifolii*[2]	Lemons *(C. limon)*; Mexican/Key lime *(C. aurantifolia)*; sour oranges *(C. aurantium)*; pummelo *(C. grandis)*[3]
bacterial canker (cancrosis C)	*Xanthomonas fuscans* ssp. *aurantifolii*[2]	Primarily Mexican/Key lime *(C. aurantifolia)*; however, sour oranges *(C. aurantium)* can also be affected[3]
bacterial spot	*Xanthomonas alfalfae* ssp. *citrumelonis*[2]	Swingle citrumelo (*P. trifoliata* x *C. paradisi*); trifoliate orange *(Poncirus trifoliata)*; grapefruit *(C. paradisi)*[4]
blast and black pit	*Pseudomonas syringae*	**Leaves and twigs:** oranges *(C. sinensis)*; grapefruit *(C. paradisi)* most susceptible to blast[5] **Fruit:** lemons *(C. limon)* most susceptible to black pit[5]
citrus variegated chlorosis	*Xylella fastidiosa*	**Susceptible:** nearly all sweet orange *(C. sinensis)* cultivars **Less susceptible:** lemons *(C. limon)*; limes *(C. aurantifolia)*; mandarins *(C. reticulata)* and mandarin hybrids (e.g., Murcott, Sunburst); kumquat (*Fortunella* spp.); trifoliate orange *(Poncirus trifoliata)*; grapefruit *(C. paradisi)* **Tolerant:** Rangpur lime *(C. limonia)*; citron *(C. medica)*; pummelo *(C. grandis)*[6]
huanglongbing (greening)	"*Candidatus* Liberibacter" spp.	Nearly all commercial citrus species, cultivars, and hybrids, as well as some citrus relatives, are susceptible. **Highly susceptible:** sweet oranges *(C. sinensis)*; mandarins *(C. reticulata)* and mandarin hybrids (tangelo) **Less susceptible:** lemons *(C. limon)*; grapefruit *(C. paradisi)*; pummelos *(C. grandis)*; sour orange *(C. aurantium)*; Rangpur lime *(C. limonia)*; calamondins **Tolerant:** Mexican/Key lime *(C. aurantifolia)*; trifoliate orange *(Poncirus trifoliata)* and some trifoliate orange hybrids[7]
stubborn (Arizona, California, Egypt, Israel, Turkey)	*Spiroplasma citri*	Oranges *(C. sinensis)*, Washington navels, Valencias; grapefruit *(C. paradisi)*; mandarins *(C. reticulata)* and mandarin hybrids[8]
witches'-broom (Iran, Oman, United Arab Emirates)	"*Candidatus* Phytoplasma aurantifolia"	**Natural infection:** Mexican/Key lime *(C. aurantifolia)*; sweet lime *(C. limettioides)*; citron *(C. medica)* **Experimental infection by grafting:** Meyer lemon *(C. myeri)*; rough lemon *(C. jambhiri)*; Rangpur lime *(C. limonia)*; *P. trifoliata*; Troyer citrange (*P. trifoliata* x *C. sinensis*); Eureka lemon *(C. limon)*; Alemow *(C. macrophylla)*; *C. excelsa*; *C. ichangensis*; *C. hystrix*; *C. karna*[9]

[1] Adapted from Broadbent 2000.

[2] Several forms of citrus bacterial canker disease are caused by different species, subspecies, and variants of *Xanthomonas*. The Asiatic form of bacterial citrus canker (canker A) is the most widespread and serious form of the disease and is caused by a number of pathogenic variants of *X. citri* ssp. *citri*. Cancrosis B (canker B) and cancrosis C (canker C) are known to occur only in South America and are caused by pathogenic variants of *X. fuscans* ssp. *aurantifolii*. Citrus bacterial spot disease, known only to occur in Florida, is caused by *X. alfalfae* ssp. *citrumelonis*. For more information, see Schaad et al. 2006.

[3] Adapted from Gottwald et al. 2002; Polek et al. 2007.

[4] Adapted from Gottwald and Graham 2000.

[5] Adapted from Menge 2000.

[6] Adapted from Chung and Brlansky 2005.

[7] Huanglongbing is associated with at least three phloem-limited species of "*Candidatus* Liberibacter": "*Ca.* L. asiaticus," "*Ca.* L. africanus," and "*Ca.* L. americanus." Adapted from Bové 2006; Chung and Brlansky 2006; Garnier and Bové 2000; Polek, Vidalakis, and Godfrey 2007.

[8] Adapted from UC IPM Citrus Pest Management Guidelines; Bové and Garnier 2000a.

[9] Adapted from Bové and Garnier 2000a, 2000b; Chung, Kahn, and Brlansky 2006; OEPPO/EPPO 2006.

beyond the affected tissues. However, limited survival of Pss may occur in apparently healthy twig tissue immediately below blast-affected tissue. Nevertheless, the significance of this as a source of inoculum for perpetuating the disease cycle is unknown. Population levels of Pss strains associated with blast and black pit increase on the surface of citrus leaves and twigs during prolonged periods of wetness (e.g., due to rain or fog) and low temperature. Pss infects susceptible citrus tissue through wounds in young, succulent petioles, shoots, or twigs, as well as on wounds in fruit caused by thorn punctures, wind abrasions, hail, or insect feeding. The most susceptible tissues are young, succulent petioles, shoots or twigs, and fruit. The pathogen is spread, and infection facilitated, by wind-driven rain. There is no evidence that blast and black pit result from Pss transmitted via seed, insects, contaminated equipment, or tools, or through normal citriculture practices. The citrus blast and black pit disease cycle is shown in figure 22.2.

Disease cycle. *Pseudomonas syringae* pv. *syringae* is a normal inhabitant of citrus leaves that becomes increasingly more abundant on leaf and twig surfaces during prolonged periods of wetness due to rain or fog and relatively low temperatures. Infection occurs when bacteria enter injuries in leaves, shoots, or fruit caused by wind, heavy rain, insect feeding, hail, or thorns. Young, succulent shoots and leaves are more susceptible to infection than are mature tissues. The disease rarely progresses when the temperature is below 70°F.

Disease management. The most effective management of citrus blast and black pit is prevention of infection. Accordingly, the timely application of protective copper-containing sprays (e.g., Bordeaux mixture, fixed copper compounds), removal of inoculum sources, and cultural practices can minimize disease development. Protective copper-containing sprays to prevent infection used alone may not be effective. Such sprays during the fall and early winter before the first rains help protect leaves and shoots from Pss infections in late winter and spring. An additional protective spray may be needed in mid to late winter if heavy or extended periods of rain occur. Pss inoculum sources can be reduced by pruning dead or diseased twigs and shoots in the spring after the rainy period. Certain cultural practices also can reduce the incidence and severity of blast and black pit by minimizing infection and pathogen spread. These include planting windbreaks to reduce wind-caused injury and the spread of the pathogen, using bushy cultivars with relatively few thorns to reduce infection sites caused by wounds, and scheduling fertilization and pruning to prevent excessive, particularly susceptible new growth in the fall.

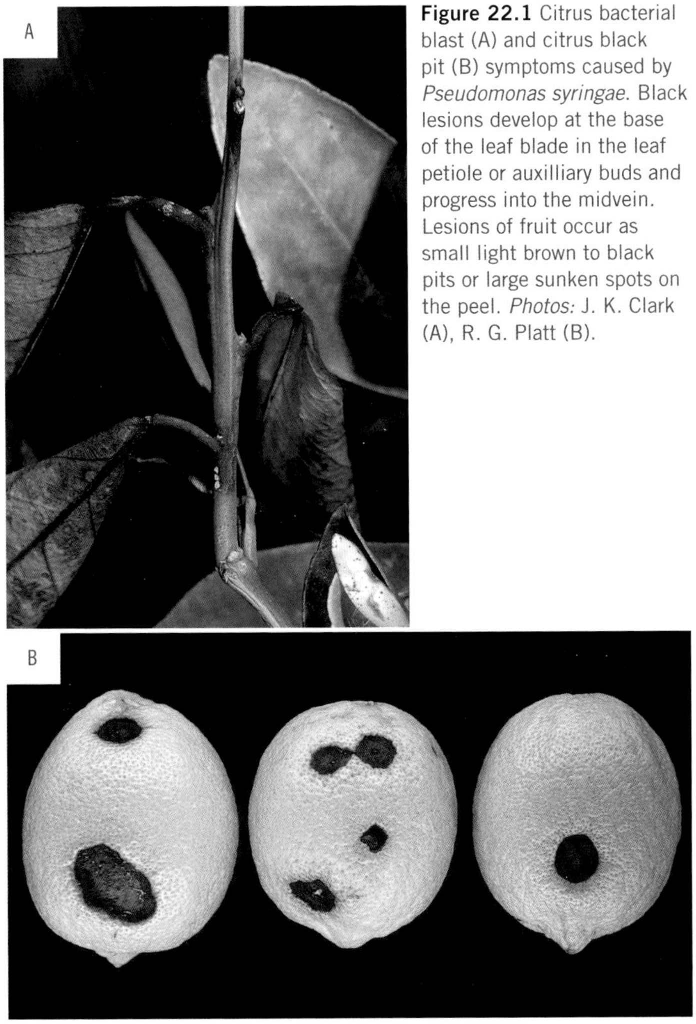

Figure 22.1 Citrus bacterial blast (A) and citrus black pit (B) symptoms caused by *Pseudomonas syringae*. Black lesions develop at the base of the leaf blade in the leaf petiole or auxilliary buds and progress into the midvein. Lesions of fruit occur as small light brown to black pits or large sunken spots on the peel. *Photos:* J. K. Clark (A), R. G. Platt (B).

Stubborn

Stubborn is a systemic disease first recognized in California. It was once considered to be caused by an infectious viruslike agent transmitted to citrus by insects, budding, and grafting. However, the pathogen was subsequently identified as a new, culturable prokaryote designated *Spiroplasma citri.* Stubborn occurs in hot, arid citrus-growing areas, such as the southwestern United States, most North African countries bordering the Mediterranean Basin, and parts of the Middle East (e.g., Israel). Stubborn is more of a problem in young trees than in mature trees. Although stubborn is generally not a lethal disease, trees remain stunted and unproductive. Young trees infected by the pathogen may remain severely stunted, with bushy, upright growth and branches with shortened internodes. Symptoms are usually less obvious after infection of older, mature trees; however, fruit yield can be significantly reduced, and the fruit is of poorer quality than normal. Most citrus species and cultivars can be affected by stubborn.

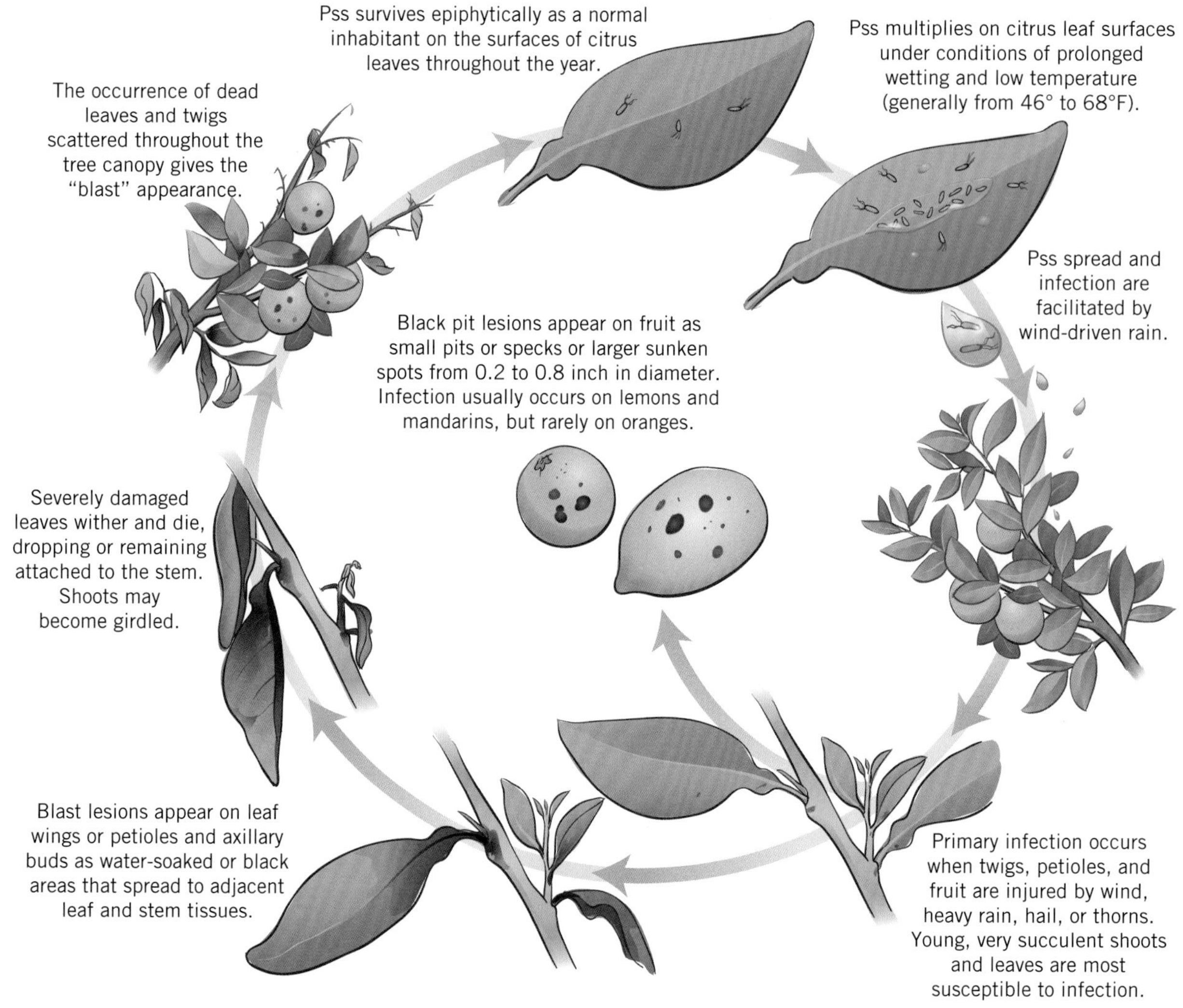

Figure 22.2 Disease cycle and epidemiology of citrus blast and black pit caused by *Pseudomonas syringae* pv. *syringae* (Pss).

Sweet oranges, grapefruit, mandarins, and mandarin hybrids are usually the most severely affected.

Causal organism. On solid artificial nutrient media, *Spiroplasma citri* forms characteristic fried-egg-shaped colonies. It occurs in the phloem tissues, where it disrupts normal plant functions, and is spread by several species of leafhoppers. The most important leafhopper vectors in California are *Scaphytopius nitridus* and *Circulifer tenellus.* In California, *S. citri* is endemic in several noncitrus weed hosts, including periwinkle, many *Brassica* species, wild radish, and *Sisymbrium irio* (London rocket). Leafhoppers become infective when feeding on these species and can transmit *S. citri* to citrus from these inoculum sources.

Symptoms. Stubborn-affected trees are characterized by stunted canopy growth due primarily to shortened internodes of affected shoots (fig. 22.3). Generally, the leaves are small, dense, and abnormally upright, growing close to the stems. Also, leaves may be cupped, thicker than normal, and may have variably chlorotic patterns resembling nutrient deficiency symptoms. Affected shoots may also have rosettes of cupped leaves. Fruit produced on stubborn-affected trees are generally fewer and smaller than those on nonaffected trees, are frequently lopsided or acorn-shaped, and may have an insipid or bitter flavor. Fruit of various ages (crop heterogeneity) may occur on the same tree due to off-season flowering. Fruit may not color at the stylar end during ripening. Also, the albedo of severely affected grapefruit and sweet orange may develop a blue color. Seed in infected fruit may be dark colored and are frequently aborted.

Epidemiology. In California, the leafhoppers *Scaphytopius nitridus* and *Circulifer tenellus* acquire the pathogen from infected weed hosts and transmit it to citrus during feeding. Rates of infection vary from orchard to orchard, depending on leafhopper populations, inoculum sources, vector movement, citrus species and cultivar, and tree age. Leafhopper vector migration from other hosts onto citrus is

Figure 22.3 Stunted tree affected with citrus stubborn disease (foreground), with shortened internodes and abnormal foliage, and unaffected trees (background) (A). Misshapen fruit (B, left panel) from stubborn-affected tree, and normal-appearing fruit (B, right panel) from unaffected tree. *Photos:* R. Yokomi

seasonal (primarily in dry summer months) and is generally related to the availability of natural hosts. Secondary tree-to-tree spread of the pathogen is usually economically significant in young orchards, but it is likely only when there a large number of trees are infected. Occurrence of a few stubborn-affected trees in a mature orchard generally presents less of a threat to nearby trees than do young, diseased trees.

Spiroplasma citri also can be transmitted by budding or grafting from an infected tree. The frequency of this mode of transmission depends on the type of tissue used, the source host species, and environmental factors. Although *S. citri* moves systemically within the tree after infection, it is usually unevenly distributed within infected trees. Thus, some bud progeny from infected trees will be free of *S. citri.* Also, propagations from some species (e.g., navel oranges and some tangelos) result in higher percentages of plants infected with *S. citri* than do propagations from other sources (e.g., Valencia orange and lemons).

Diagnosis. Stubborn can be diagnosed by graft inoculation of sensitive citrus species, such as orange, grapefruit, or tangelo. However, the symptoms, even in sensitive indicator plants, are not highly specific or diagnostic, and the indicator plants must be grown under warm conditions (about 85° to 95°F) with considerable care in order for clear symptom development. Definitive diagnosis is based on isolating *S. citri* by culturing on specific artificial nutrient media, followed by confirmation of its helical structure by microscopy. Additionally, *S. citri* can be identified by molecular methods such as polymerase chain reaction (PCR).

Disease management. The most effective management of stubborn is based on prevention or avoidance through quarantine, budwood certification, and eradication. This may be difficult in areas where the disease and leafhopper vectors are endemic and *S. citri* inoculum sources in noncitrus hosts are present. In areas where stubborn or *S. citri* vectors are not endemic, pathogen-free propagating material (e.g., budwood source mother trees free of *S. citri*) should be used to propagate nursery trees. Propagated nursery trees should be maintained in insect-proof enclosures or in areas free of *S. citri.* Budwood sources can be freed of *S. citri* by shoot-tip grafting. New citrus plantings should not be established in areas in which the potential for endemic infection is high. Diseased trees should be removed and replaced with trees free of *S. citri.*

Huanglongbing

Huanglongbing (HLB) is an economically important disease of citrus. The disease causes catastrophic losses (e.g., reduced fruit quality and yield, dieback, tree decline, and death) and is a major threat to citrus production worldwide (Bové 2006; Gottwald et al. 2007). HLB is a complex pathosystem involving diverse plant hosts, strains, or pathotypes of disease-associated liberibacters (and possibly other disease-causing agents), insect vectors, and environmental conditions. The three forms of the disease, Asian, African, and American, are associated with different species of liberibacters that are transmitted by citrus psyllid insect vectors. Nearly all commercial citrus species, cultivars, and hybrids, as well as some citrus relatives, are susceptible. The disease is also known as mottle leaf in the Philippines, likubin in Taiwan, phloem necrosis and phloem degeneration in Indonesia, dieback in India, and yellow branch and greening in South Africa (Bové 2006; Gottwald et al. 2007). The name "huanglongbing" was adopted by the International Organization of Citrus Virologists for these diseases with symptoms similar or identical to HLB (Moreno et al. 1996 as cited by Gottwald et al. 2007). HLB occurs throughout Asia, including Bangladesh, China (Guangdong and Guangxi), Hong Kong, India, Indonesia (Java and Sumatra), Japan (Ryukyu Islands), Malaysia (Peninsular), Nepal, Pakistan, Philippines, Taiwan, Thailand, and Vietnam; in the Arabian peninsula (Saudi Arabia and Yemen); in Africa (Burundi, Cameroon, Central African Republic, Comoros, Ethiopia, Kenya, Madagascar, Malawi, Mauritius, Rèunion, Rwandi, Somalia, South Africa, Swaziland, Tanzania, and Zimbabwe); in the Caribbean (Cuba, Dominican Republic, and Jamaica); in South America (Argentina and Brazil); in Central America (Belize, Costa Rica, Guatemala, Honduras, Nicaragua, and Mexico); and in North America (Florida and Texas). HLB has also been detected in Alabama, Georgia, Louisiana, Mississippi, and South Carolina (Jepson 2012). A single HLB-affected tree was detected in a dooryard in southern California in 2012; the tree was removed. Subsequent testing of Asian citrus psyllids and plant tissues for HLB-associated liberibacters in California has been negative to date.

Causal organism. The causal agent(s) of HLB have not been conclusively determined. However, three species of "*Candidatus* Liberibacter" ("*Ca.* L. asiaticus," "*Ca.* L. africanus," and "*Ca.* L. americanus") are primarily associated with HLB (Bové 2006; Gottwald et al. 2007). However, at least two phytoplasmas closely related to the witches'-broom phytoplasma (Teixeira et al. 2008) and the aster yellows phytoplasma (Chen et al. 2009) have been associated with HLB in Brazil and China, respectively. Only "*Ca.* L. asiaticus" was identified in the single HLB-affected tree and in Asian citrus psyllids collected from that tree in southern California in 2012. Currently, HLB is not known to occur in California. The HLB-associated liberibacters (as well as the phytoplasmas that have been associated with HLB) are phloem limited.

Figure 22.4 Yellow shoot symptoms of huanglongbing disease. *Photo:* E. E. Grafton-Cardwell.

Symptoms. HLB symptoms are variable and can resemble other disorders of citrus (see Bové 2006; Gottwald et al. 2007). Symptoms considered to be characteristic of HLB include one or more yellow shoots (fig. 22.4) with pale green or yellowish flushes, while other parts of affected trees remain healthy; leaves with "blotchy mottle" (i.e., various shades of yellow and green blending together without distinct limits) that is not symmetrical on either side of the leaf midrib; thickened, leathery leaves; enlarged, swollen, and corky midribs and lateral veins; and leaves with zinc deficiency symptoms and that are upright, making a small angle with the shoot. Eventually, defoliation and twig and shoot dieback occur. Young infected trees may die. Excessive fruit drop occurs from HLB-affected trees. Symptomatic fruit are small, asymmetrically shaped, or lopsided, and have an asymmetric or curved fruit axis. HLB-affected fruit also exhibit a color inversion from yellow-orange to green at the peduncle end while the stylar end is still green. The circular scar left on HLB-affected fruit after removal of the peduncle may be orange rather than pale green as on normal fruit. Seed in HLB-affected fruit are small, brownish, and aborted. The vascular bundles at the peduncular end in HLB-affected fruit are brown. The albedo of HLB-affected fruit may be thickened relative to that of normal fruit. However, aborted seed and thickened albedo are also associated with stubborn-affected fruit.

Epidemiology. There are three forms of HLB: Asian, African, and American (Bové 2006; Gottwald 2010; Gottwald et al. 2007). The Asian form is the most extensive and severe form of HLB. "*Ca.* L. asiaticus" is associated with the Asian form, which is heat tolerant and can develop at temperatures above about 86°F. "*Ca.* L. africanus" is associated with the African form, which is suppressed at temperatures above about 86°F. The American form of HLB is similar to the Asian form with respect to symptom expression and disease severity. However, the American form of HLB, with which "*Ca.* L. americanus" is associated, is heat intolerant, similar to the African form (Gottwald 2007).

The lag time between infection by the HLB pathogen(s) and symptom expression can depend on several factors, including (but not necessarily limited to) specific host cultivar, tree age and condition, efficiency of HLB-associated pathogen transmission, efficiency of citrus psyllids, cultural practices, and

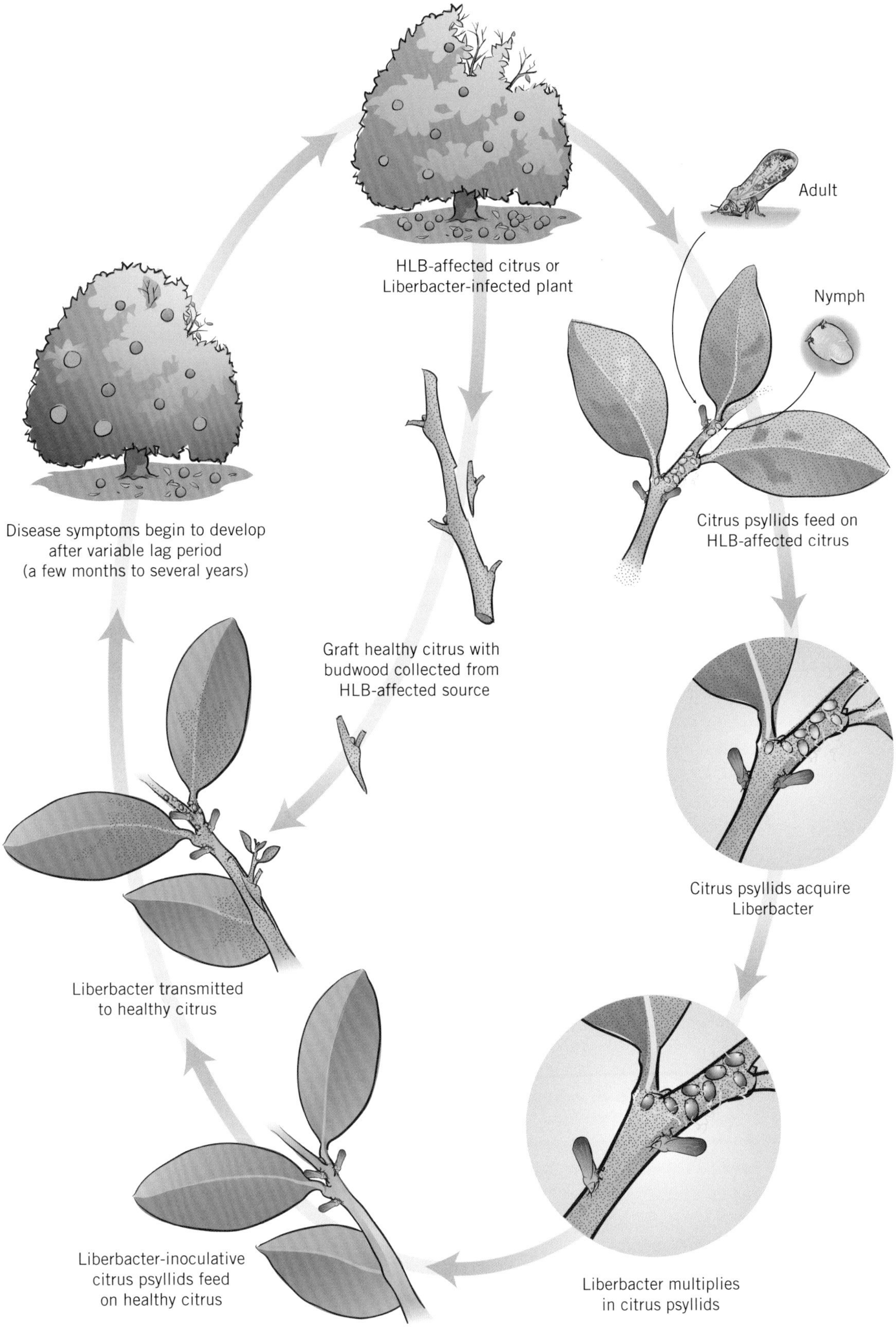

Figure 22.5 Huanglongbing disease cycle.

environmental conditions. HLB symptoms are often nondistinct and may be restricted to different parts or sectors of affected trees, which suggests nonuniform distribution, limited systemic infection, or variable HLB-associated liberibacter titers in different parts of affected trees (Gottwald 2007).

HLB-associated liberibacters are transmitted by the citrus psyllids *Diaphorina citri* ("*Ca.* L. asiaticus" and "*Ca.* L. americanus") and *Trioza erytreae* ("*Ca.* L. africanus"). The effectiveness of natural dissemination of HLB-associated liberibacters may be related to high insect vector populations and extensiveness of inoculum reservoirs (Gottwald 2007). Thus, natural spread of HLB-associated liberibacters is likely greatest when new flush is available and citrus psyllid populations are high. HLB-associated liberibacters can also be transmitted by grafting. Accordingly, HLB can become established by producing nursery trees using budwood from HLB-affected sources. Seed transmission of the HLB-associated liberibacters has not been conclusively established.

Several noncitrus hosts have been identified as alternate hosts for HLB-associated liberibacters and citrus psyllids (Brown et al. 2011; Deng et al. 2007; Halbert and Manjunath 2004; Hung and Su 2000; Manjunath et al. 2008; Phahladira 2012; Walter et al. 2012). However, the role of these alternate hosts of HLB-associated liberibacters and citrus psyllids in the epidemiology of HLB is not clearly or completely understood.

Disease cycle. HLB-associated liberibacters are transmitted from diseased sources by citrus psyllids *(Diaphorina citri* and *Trioza erytreae)* and by grafting. Both of these routes of transmission can establish primary and secondary infections in nurseries and in the field, and are important in dissemination of HLB-associated liberibacters. A generalized disease cycle is presented in figure 22.5.

Disease management. International, national, regional, and local quarantines are critically important for preventing or minimizing long-distance dissemination and establishment of HLB-associated liberibacters (Gottwald 2010). In general, integrated disease management strategies include early detection of liberibacter and HLB; chemical, biological, and area-wide control of citrus psyllid vectors; removal or rogueing of HLB-affected trees; and phytosanitary regulatory measures such as geographical isolation and protection of budwood source trees for propagation, geographical isolation of nursery tree production, citrus nursery tree production in secure and insect-proof structures, and development and use of resistant (genetic or transgenic) commercial citrus cultivars (Bové 2006; Gottwald 2010).

References

Bové, J. M. 2006. Huanglongbing. A destructive, newly emerging, century-old disease of citrus. Journal of Plant Pathology 88:7–37.

Bové, J. M., and M. Garnier. 2000a. Stubborn. In L. W. Timmer, S. M. Garnsey, and J. H. Graham, eds., Compendium of citrus diseases. St Paul, MN: APS Press.

———. 2000b. Witches' broom. In L. W. Timmer, S. M. Garnsey, and J. H. Graham, eds., Compendium of citrus diseases. St Paul, MN: APS Press.

Broadbent, P. 2000. Australian citrus dieback. In L. W. Timmer, S. M. Garnsey, and J. H. Graham, eds., Compendium of citrus diseases. St Paul, MN: APS Press.

Brown, S. E., A. P. Oberheim, A. Barrett, and W. A. McLaughlin. 2011. First report of "*Candidatus* Liberibacter asiaticus" associated with huanglongbing in the weeds *Cleome rutidosperma, Pisonia aculeata* and *Trichostima octandrum* in Jamaica. New Disease Reports 24:25, doi:10.5197/j.2044-0588.2011.024.025.

Chen, J., X. Pu, X. Deng, S. Liu, H. Li, and E. Civerolo. 2009. A phytoplasma related to "*Candidatus* phytoplasma asteri" detected in citrus showing Huanglongbing (yellow shoot disease) symptoms in Guangdong, P. R. China. Phytopathology 99:236–242.

Chung, K.-R., and R. H. Brlansky. 2005. Citrus diseases exotic to Florida: Citrus variegated chlorosis (CVC). Gainesville: University of Florida Cooperative Extension IFAS Fact Sheet PP-223. IFAS website, http://edis.ifas.ufl.edu/pdffiles/PP/PP13700.pdf.

Chung, K.-R., I. A. Kahn, and R. H. Brlansky. 2006. Citrus diseases exotic to Florida: Witches' broom disease of lime (WBDL). Gainesville: University of Florida Cooperative Extension IFAS Fact Sheet PP-228. IFAS website, http://edis.ifas.ufl.edu/pdffiles/PP/PP15000.pdf.

Deng, X., G. Zhou, H. Li, J. Chen, and E. L. Civerolo. 2007. Detection of "*Candidatus* Liberibacter asiaticus" from wampee (*Clausena lansium* Skeels) by nested PCR. Plant Health Progress, doi:10.1094/PHP-2007-0419-01-BR.

Dreistadt, S. H. 2012. Integrated pest management for citrus. 3rd ed. Oakland: University of California Agriculture and Natural Resources Publication 3303.

Garnier, M., and J. M. Bové. 2000. Huanglongbing. In L. W. Timmer, S. M. Garnsey, and J. H. Graham, eds., Compendium of citrus diseases. St Paul, MN: APS Press.

Gottwald, T. R. 2010. Current epidemiological understanding of citrus huanglongbing. Annual Review of Phytopathology 48:119–139.

Gottwald, T. R., and J. H. Graham. 2000. Bacterial spot. In L. W. Timmer, S. M. Garnsey, and J. H. Graham, eds., Compendium of citrus diseases. St Paul, MN: APS Press.

Gottwald, T. R., J. V. da Graça, and R. B. Bassanezi. 2007. Citrus Huanglongbing: The pathogen and its impact. Plant Health Progress, doi:10.1094/PHP-2007-0906-01-RV.

Gottwald, T. R., J. H. Graham, and T. S. Schubert. 2002. Citrus canker: The pathogen and its impact. Plant Health Progress, doi:10.1094/PHP-2002-0812-01-RV.

Halbert, S. E., and K. L. Manjunath. 2004. Asian citrus psyllids (Sternorrhyncha: Psyllidae) and greening disease of citrus: A literature review and assessment of risk in Florida. Florida Entomologist 87:330–353.

Hung, T. H., M. L. Wu, and H. J. Su. 2000. Identification of alternative hosts of the fastidious bacterium causing citrus greening disease. Journal of Phytopathology 148:321–326.

Jepson, S. B. 2012. Citrus greening diseases (Huanglongbing). Oregon State University Extension Service, http://www.science.oregonstate.edu/bpp/Plant_Clinic/Disease_sheets/Citrus%20Greening%20Disease.pdf.

Manjunath, K. L., S. E. Halbert, C. Ramadgu, S. Webb, and R. F. Lee. 2008. Detection of "*Candidatus* Liberibacter asiaticus" in *Diaphorina citri* and its importance in the management of citrus huanglongbing in Florida. Phytopathology 98:387–396.

Menge, J. A. 2000. Blast and black pit. In L. W. Timmer, S. M. Garnsey, and J. H. Graham, eds., Compendium of citrus diseases. St Paul, MN: APS Press.

Moreno, P., J. V. da Graça, and R. K. Yokomi. 1996. Preface. In J. V. da Graça and R. K. Yokomi, eds., Proceedings of the 13th Conference of the International Organization of Citrus Virologists. Riverside, CA. v–vi.

OEPPO/EPPO. 2006. Candidatus *Phytoplasma aurantifoliae.* Bulletin OEPP/EPPO 36:117-119.

Phahladira, M. N. B. 2012. Widespread occurrence of "*Candidatus* liberibacter africanus subspecies capensis" in *Calodendrum capense* in South Africa. European Journal of Plant Pathology 134:39–47.

Polek, M., G. Vidalakis, and K. Godfrey. 2007. Citrus bacterial canker disease and huanglongbing (citrus greening). Oakland: University of California Agriculture and Natural Resources Publication 8218. UC ANR website, http://anrcatalog.ucdavis.edu/Citrus/8218.aspx.

Schaad, N. W., J. B. Jones, and W. Chun, eds. 2001. Laboratory guide for identification of plant pathogenic bacteria. 3rd ed. St. Paul, MN: APS Press.

Schaad, N. W., E. Postinikova, G. Lacy, A. Sechler, I. Agarkova, P. E. Stomberg, V. K. Stromberg, and A. K. Vidaver. 2006. Emended classification of xanthomonad pathogens on citrus. Systematic and Applied Microbiology 29:690–695.

Teixeira, D. C., N. A. Wulff, E. C. Martins, E. W. S. Kitajima, R. Bassanezi, A. J. Ayres, S. Eveillard, C. Saillard, and J. M. Bové. 2008. A phytoplasma closely related to pigeon pea witches'-broom phytoplasma (16Sr IX) is associated with citrus huanglongbing in the Ste. of São Paulo, Brazil. Phytopathology 98:977–984.

Timmer, L. W., and L. W. Duncan, eds. 1999. Citrus health management. St. Paul, MN: APS Press.

Timmer, L. W., S. M. Garnsey, and J. H. Graham, eds. 2000. Compendium of citrus diseases. St. Paul, MN: APS Press.

University of California Statewide Integrated Pest Management Program. UC IPM citrus pest management guidelines. UC IPM website, http://www.ipm.ucdavis.edu/PMG.

USDA-APHIS. 2012. New pest response guidelines: Citrus greening disease. APHIS website, http://www.aphis.usda.gov/plant_health/plant_pest_info/citrus_greening/downloads/pdf_files/cg-nprg.pdf.

Walter, A. J., Y. P. Duan, and D. G. Hall. 2012. Titers of "*Ca.* Liberibacter asiaticus" in *Murraya paniculata* and *Murraya*-reared *Diaphorina citri* are much lower than in *Citrus* and *Citrus*-reared psyllids. HortScience 47:1449–1452.

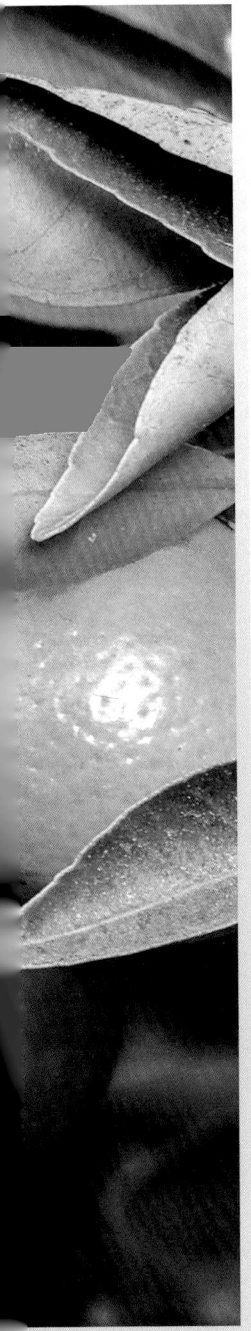

23 Virus and Viroid Diseases

David J. Gumpf, MaryLou Polek, and Georgios Vidalakis

Diseases caused by viruses and viroids have undoubtedly been plaguing humankind directly and indirectly since the dawn of civilization. In today's world, diseases caused by viruses and viroids are some of the most serious maladies of plants, including citrus. A virus is a molecule consisting of a piece of genetic material, either deoxyribonucleic acid (DNA) or ribonucleic acid (RNA), that is surrounded by a protective protein coat or shell. Since a virus is not a cell, it does not reproduce by dividing the way other pathogens like bacteria and fungi do. A virus must enter a living cell, where it interrupts the normal cell activity and directs the cell to start producing more virus nucleic acid and virus protein. The newly formed virus particles can then move from cell to cell and into the vascular system, where they can travel throughout the entire plant. Viruses cause disease by usurping the host cell's reproductive machinery, taking up space, and disrupting cellular processes. As a result, cellular metabolism is upset and leads to the development of abnormal substances and conditions that affect the normal functions of the whole organism. These disruptions result in what we see as symptoms in an affected plant.

A viroid is an even simpler molecule than a virus since it consists of only a piece of RNA without any protein coat (free RNA). In addition, the size of the viroid RNA is 10 to 10,000 times smaller than the nucleic acid found in a virus. It likewise can replicate only when it is within a living cell. Only about 20 plant diseases are known to be caused by viroids, and 7 viroids are found in citrus, causing a variety of diseases and growth abnormalities. To date, viroids have never been shown to be the causal agent for any disease of animals, including humans.

Because of the nature of viruses and viroids and their intimate association with the cells of their host, the diseases they cause are difficult to control. These pathogens are spread primarily by people when producing new trees or topworking existing field trees using infected propagative material and by insufficient sanitation of propagation tools. In addition, insect vectors can naturally spread some of these pathogens. No spray or chemical can be used to cure an infected citrus tree. Instead, virus diseases in tree crops must be managed in such a way as to limit the effect of the pathogen on fruit quality and production. Therefore, the principal method of virus and viroid management involves the propagation and planting of trees that are virtually free of all known diseases or tree varieties that are tolerant of or resistant to the diseases under field conditions. In this way, even if the insect vector is present, the pathogen is not and the disease cannot be transmitted. Programs for the production of such materials were discussed in chapter 7, "The California Citrus Clonal Protection Program." The descriptions that follow are an attempt to inform the reader about the most common virus and viroid diseases that are likely to be encountered in the orchards of California. Exotic pests and the threat they present to California citrus are the subject of chapter 24.

Virus Diseases

Psorosis (*Citrus psorosis virus*, CPsV)

Psorosis is a debilitating bark scaling disease primarily on sweet oranges that results in the slow decline of affected trees. While it generally does not result in tree death, it eventually renders the tree unproductive. The disease is found predominantly in old-budline trees in virtually every citrus-growing area of the world. Described in the 1930s, psorosis was the first citrus disease proved to be caused by a virus. The Psorosis-free Program, which originated in California, was the progenitor of all citrus registration and certification programs.

Disease symptoms and hosts. Classic symptoms of psorosis are scaling and flaking of the bark on the trunk and limbs of sweet orange, grapefruit, and mandarin (fig. 23.1). In some cases, the disease can move very rapidly, and large strips of bark will slough off. These bark lesions seldom develop before 10 years of age on trees propagated with infected buds. As the trunk and limb lesions expand, the xylem or wood beneath becomes impregnated with gum and becomes discolored. Ultimately, most of the xylem becomes plugged with gum, and the tree declines because it lacks a functioning xylem system for water uptake. This wood staining is readily observed in a cross section of the trunk or affected limb. Symptoms on young leaves include chlorotic flecking or clearing over the entire leaf or portions of the leaf (fig. 23.2). Some of the clearing is quite distinct and may occur adjacent to or directly over veins, while in other cases they may coalesce to form larger blotchy areas. These symptoms are best observed on young leaves nearing full expansion after spring and fall growth flushes because the symptoms fade and disappear as the leaves mature.

Figure 23.1 Scaling bark of a CPsV-infected tree. *Photo:* G. Vidalakis.

Figure 23.2 Chlorotic flecking in a CPsV-infected leaf. *Photo:* G. Vidalakis.

Disease agent. The disease agent of psorosis has been difficult to characterize, even though this was one of the first maladies described for citrus. Psorosis is caused by citrus psorosis virus (CPsV), the type species of the genus *Ophiovirus*, with virions characterized as spiral filaments. The nucleic acid of the virus is single-stranded negative-sense RNA, and it is encapsulated with protein to form short and long particles. Three species of negative ssRNA have been detected in CpsV, each of which is contained in a virion characterized as spiral filament. The virus seems to be in low concentration and unevenly distributed within the tree. There also appears to be numerous strains of this virus.

Disease spread. The disease is spread most readily by grafting. The use of buds from an infected source to produce new trees has been the primary method of spread for decades. Some strains of the virus have been transmitted experimentally by mechanical inoculations, as well as using contaminated tools and equipment. In some countries, the disease is spreading naturally on soil, and aphid transmission has been suspected.

Disease diagnosis. The characteristic bark scaling symptoms on the trunk and limbs aid in a tentative field diagnosis of the disease. Positive identification of CPsV is made by graft inoculation of indicator seedlings and the observation of leaf symptoms in the new growth. Seedlings of sweet orange and Dweet tangor are most often used as the disease indicators for psorosis. Laboratory-based detection methods including reverse transcription (RT) polymerase chain reaction (PCR) and serology have been developed, but because of low virus concentration and uneven distribution in the tree, are frequently complemented with bioindexing.

Disease control. The use of propagation budwood from nucellar sources or from sources that have indexed negative for psorosis has eliminated this disease in countries where there is no natural transmission. The virus can be eliminated from valuable clones by thermal therapy or by in vitro shoot-tip micrografting and verified as clean by indexing before further propagation of these clones. Using pruning and budding tools that have been disinfested in a 1% sodium hypochlorite solution (bleach) eliminates any possible transmission of the disease by mechanical contamination.

Tristeza (*Citrus tristeza virus*, CTV)

Tristeza, meaning "sadness" in Spanish and Portuguese, is found worldwide wherever citrus is grown. It is probably the most widely known disease of citrus because of the impacts large-scale epidemics have had in certain areas of the world. Initially known as quick decline, it was reported in southern California in 1939, when trees of sweet orange scions grafted onto Sour Orange rootstock began to collapse. By the mid-1940s, most trees in that area of the state were infected, and the industry moved north to the San Joaquin Valley, where most plantings were made on tolerant rootstock varieties such as Troyer, Trifoliate, and Carrizo.

Disease symptoms and hosts. This virus produces a wide variety of disease symptoms depending on rootstock and scion variety combinations and the virulence of the particular virus strain. The most well known malady of the tristeza disease complex is quick decline (fig. 23.3). Certain isolates of CTV affect sweet orange varieties grafted on Sour Orange rootstock by causing a necrosis of the phloem sieve tubes just below the bud union. This prevents the transport of starch from the scion downward to the root system and debilitates the formation of feeder roots. During times of stress, such as hot temperatures and drought, the tree cannot keep up with water and nutrient needs and rapidly declines. Severe strains can cause tree death within 1 to 2 weeks after the stress event occurs. Shriveled, dehydrated fruit may remain hanging on the tree. This susceptibility is unfortunate because Sour Orange is one of the best rootstock varieties because it provides resistance to root-rotting fungi and is productive even in poor soil conditions.

Stem pitting symptoms can occur in most citrus varieties regardless of the rootstock (fig. 23.4). The vascular tissue in trunks and branches either does not produce new cells or develops abnormally, leaving elongated sunken areas, or pits. In mild cases, the pits cannot be seen unless the bark has been peeled away. Often, the pits are stained light brown. Severe stem pitting isolates of the virus causes "cheesy bark," when the bark cannot be separated from the wood due to the proliferation of phloem cells or the thickening of the walls of ray parenchyma cells. Fruit production of trees affected with CTV stem pitting is reduced, with fruit size reduction and malformation.

Seedling yellows is not a disease symptom observed on field-grown trees, but it does indicate a severe strain of the virus and is a symptom used in greenhouse experiments to determine the severity of a particular virus isolate. Newly formed leaves will be yellow or white (as if dipped in bleach) and much reduced in size. Seedlings infected with this strain of the virus are typically severely stunted.

Disease agent. The causal agent, CTV, is a long, filamentous closterovirus. It is the largest of plant viruses, making the study of its biology very difficult. Adding to the complexity is the existence of virus strains. In general, a strain could be described as a population of CTV that consistently produces a particular symptom or a set of symptoms in a given variety or scion-rootstock combination. An example would be quick decline in sweet orange scions grafted onto Sour Orange rootstock. It has proven difficult, however, to clearly define the CTV strains.

Disease spread. CTV can be disseminated by either graft transmission or by insect vectors. Several species of aphids can transmit CTV, including *Aphis gossypii, A. spiraecola, Toxoptera citricida,* and *T. aurantii.* The rate of spread depends on the aphid species and the particular virus strain. The most efficient vector is *Toxoptera citricida,* the brown citrus aphid. Presently, it does not occur in California but is widespread in Florida since its introduction in 1995. *A. gossypii* is the most efficient vector in California.

Disease diagnosis. Visual diagnosis of CTV in the field can be difficult and usually inconclusive as other diseases exhibit similar symptoms, or the tree may be symptomless. Advanced cases of quick decline may be self-evident; field diagnoses can sometimes be made by peeling back the bark and examining the bud union for honeycombing, thickening in the bark below the bud union, or a band

Figure 23.3 Tristeza quick decline. *Photo:* A. Eskalen and G. Vidalakis.

Figure 23.4 Stem pitting due to CTV infection. *Photo:* G. Vidalakis.

of discoloration (fig. 23.5). In addition, the application of a few drops of iodine to the wood at the graft union causes the starch that has accumulated there to turn purple. Stem pitting strains can be distinguished by peeling back the bark on the trunk or branches and looking for the occurrence of pits.

Historically, detection of CTV was done by inoculation into Mexican lime seedlings. The range of symptoms indicated the severity of the particular strain of the virus. Symptoms that are observed include leaf cupping, vein flecking or clearing, vein corking, stunting, and stem pitting. Further biological characterization of virus strains is done by inoculation into standard host indicators, including lemon, grapefruit, sweet orange, sour orange, and sweet orange grafted onto Sour Orange rootstock under greenhouse conditions.

Although biological assays are quite sensitive and are an accepted diagnostic test, they intensively use space and labor and require about 6 months to 1 year to complete, depending on the variety. These factors are limiting for large-scale testing. Instead, various laboratory assays are used for the diagnosis of CTV. ELISA is the most commonly used in the United States, whereas tissue-blotting is popular in Spain. Molecular methods include hybridization probes and reverse transcription (RT) polymerase chain reaction (PCR) and double-stranded RNA (ds-RNA). Several strains of this virus have been sequenced, but scientists have yet to correlate genomic sequences with a particular symptom.

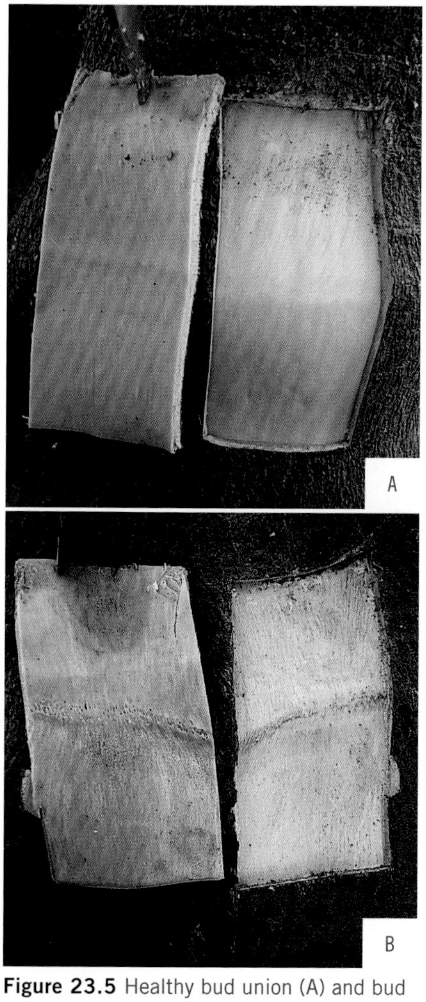

Figure 23.5 Healthy bud union (A) and bud union affected by CTV quick decline (B). *Photos:* A. Eskalen and G. Vidalakis.

Disease control. Currently, the use of CTV-free propagation material and the removal of infected trees are the best and most reliable means of controlling the spread of this virus. All trees used for propagation purposes must be registered with the CDFA registration program and tested annually for CTV. Because CTV is insect transmitted, it is advisable to use tolerant rootstock varieties, although these may be less suited horticulturally or may be susceptible to other endemic diseases. In South America, where the industry had shifted to tolerant varieties, stem pitting isolates have caused losses in these newer plantings. Fortunately, the majority of California CTV isolates characterized to date have been relatively mild.

Where CTV is endemic, such as in South Africa and Brazil, cross-protection is practiced. This last-resort technique uses a mild form of the virus to protect young trees from more virulent strains. Historically, the most successful cross-protecting isolates were selected from surviving trees in areas with a high incidence of stem pitting strains; their efficacy is limited to that specific location of occurrence. More recently it has been determined that the protecting mild strain must be genetically related to the severe strain you wish to protect against. Cross-protection is not a permanent solution, as protection often breaks down and the protecting strain does not confer protection against all severe strains. In Florida, where this method was practiced on a small scale, cross-protection offered no protection after the brown citrus aphid became established. Ultimately, resistant varieties will be developed through breeding or transformation methods.

Citrus variegation virus (CVV), *Citrus crinkly leaf virus* (CCLV), and *Citrus leaf rugose virus* (CLRV)

These related diseases are found in a number of citrus-growing areas of the world but are considered unimportant economically. In years past they were considered part of the psorosis complex because they cause similar symptoms in young leaves.

Disease symptoms and hosts. The most conspicuous symptoms of these diseases are leaf distortion or puckering with or without areas of chlorosis or chlorotic flecking (fig. 23.6). These symptoms are found primarily on lemons, sour orange, citron, grapefruit, and Mexican lime. Generally, the symptoms of all three are very mild in sweet orange and mandarins. CVV can produce some symptoms on fruit, but there are no documented effects on fruit yield.

Disease agent. CVV is caused by the citrus variegation virus; CCLV is caused by the citrus crinkly leaf virus; and CLRV is caused by the citrus leaf rugose virus. These three ilarviruses are serologically related.

Disease spread. These ilarviruses are easily transmitted by grafting and the use of contaminated propagation tools. On occasion, natural infections

Figure 23.6 Puckering of leaves (left) in sour orange due to CVV, compared with healthy leaves (right). *Photo:* G. Vidalakis.

have been observed, but no insect vector has been described. Related ilarviruses have been found in apple, asparagus, and grapes, and it is thought that these crops serve as the primary reservoir, with only occasional spread into citrus.

Disease diagnosis. All three viruses can be detected by graft inoculation into lemon seedlings. Sweet orange, sour orange, and some mandarins may also serve as indicators. CVV, CCLV, and CLRV can be distinguished from one another by inoculations into herbaceous hosts and by ELISA.

Concave Gum and Blind Pocket

Disease symptoms and hosts. Concave gum and blind pocket are graft-transmissible diseases with very similar symptoms, differing mostly in the sharpness of the concavities they cause in the trunks of affected trees (fig. 23.7). Both are common in old-line citrus. Severe deformities can occur in the trunks and main branches of mandarin, sweet orange, and grapefruit varieties, giving them a gnarled or knotted appearance. Gum is deposited in concentric rings in the wood of affected trees and may be seen at the base of the concavity when the bark is removed. In the spring, young leaves may develop chlorotic flecking or "oak-leaf" patterns.

Disease symptoms of blind pocket and concave gum can sometimes be confused with symptoms of disease caused by other pathogens. Cachexia, cristacortis, and tristeza also cause stem pitting symptoms, but blind pocket and concave gum do not cause the bark to thicken abnormally over depressions in the wood. The production of gum is also induced by species of *Phytophthora*. Leaf flecking is symptomatic of cristacortis, impietratura, and psorosis. Old-line citrus is often infected by several of these pathogens, so diagnosis of any one specific pathogen should not be based on leaf symptoms.

Disease agent. The causal agents for concave gum and blind pocket have not been characterized, but they are presumed to be viruses.

Disease spread. Both of these diseases have been widely spread by propagating with infected budwood. There is a report of seed transmission in Trifoliate Orange and Troyer and Carrizo Citranges.

Disease diagnosis. Concave gum and blind pocket can be identified by symptoms in field trees and by the development of oak-leaf patterns in sweet orange, mandarin, or Dweet tangor indicators.

Disease control. Disease control can be accomplished by using disease-free buds when propagating and using virus-free trees for seed production.

Tatter Leaf–Citrange Stunt

Citrus tatter leaf virus (CTLV) was first described in the United States in 1962 in Meyer lemon, a cultivar introduced from China. Initially, tatter leaf and citrange stunt were thought to be caused by two distinct viruses. It is now generally accepted that they are two different isolates of the same virus or virus complex.

Disease symptoms and hosts. CTLV can infect many citrus species, cultivars, and hybrids. Most commercial scion varieties are symptomless hosts,

Figure 23.7 Concavities due to concave gum disease. *Photo:* G. Vidalakis.

but when infected scions are grafted onto rootstocks of Trifoliate Orange or its hybrids, a pronounced bud union incompatibility occurs, causing trees to be stunted and have a chlorotic appearance. Not all isolates of CTLV produce the bud union incompatibility reaction, however.

Disease agent. CTLV is a rod-shaped capillovirus that is closely related to the Apple stem grooving virus (ASGV) and a Lily symptomless virus. ASGV is distributed worldwide in fruit trees such as apple, pear, apricot, and cherry.

Disease spread. CTLV is readily graft transmitted and can also be mechanically transmitted by leaf abrasion and stem slashing. There is no indication of vector spread within citrus, but spread from other noncitrus species may occur.

Disease diagnosis. Most commonly, propagated citrus cultivars infected with CTLV are symptomless unless propagated on Trifoliate Orange or its hybrids. Bud union creasing has several causes, including physiological factors, but creasing induced by CTLV tends to occur in young trees on sensitive rootstocks. Rusk citrange is the most reliable biological indicator for this virus, though *Citrus excelsa* Wester can be used. Indicator plants develop chlorotic leaf blotches and deformed leaves, including a curving of the blade and tattered edges (fig. 23.8). Mechanical inoculation of herbaceous hosts such as cowpea (*Vigna unguiculata* (L.) Walp.) or quinoa (*Chenopodium quinoa* Willd.) is also useful. Although mono- and polyclonal antibody ELISA tests have been used by scientists for research purposes, these methods are not approved for diagnosis of CTLV in California at this time. Molecular methods such as RT-PCR are also available for CTLV detection.

Vein Enation–Woody Gall

Citrus vein enation was first described in California by Wallace and Drake in 1953; woody gall was first described in Australia in 1958. Wallace and Drake demonstrated that these two distinct symptoms were caused by the same agent in 1962. This disease exists worldwide including in the United States (California), Africa (Kenya, Libya, South Africa, and Tanzania), Nepal, Australia, China, Fiji, India, Indonesia, Iran, Japan, New Zealand, Papua New Guinea, Philippines, Spain, South America (Argentina, Brazil, and Peru), and Turkey. Once common in southern California, the use of disease-free propagative material has mostly eliminated this disease.

Disease symptoms and hosts. Symptoms of this disease include enations (overgrowth or thickening) on the underside veins of Mexican lime, sour orange, and Rough Lemon (fig. 23.9), and galls (swellings) on the trunks and branches of Mexican lime, Rough Lemon, and Volkamer Lemon (fig. 23.10). Galls typically form near thorns or in association with wounds. This disease does not have an economic impact on fruit production, except where severe galling occurs on young trees budded to Rough Lemon seedlings. Homeowners may find the galls unsightly. Older, established trees are not affected by gall formations.

Galls develop in Rough Lemon and Volkamer Lemon and in Rangpur Lime and Mexican lime. Vein enation can be seen in the field in lime and sour orange and under cool greenhouse conditions in sweet orange, lemon, mandarin, and Rough Lemon.

Disease agent. A spherical virus has been observed by electron microscopy in phloem cells in enations and in the salivary glands of infective

Figure 23.8 Deformed leaves and chlorotic leaf blotches due to CTLV. *Photo:* D. Gumpf.

Figure 23.9 Vein enation of Mexican lime. *Photo:* G. Vidalakis.

Figure 23.10 Woody gall causes swellings of the trunk. *Photo:* G. Vidalakis.

aphids. Recently a new Enamovirus was identified in association with citrus vein enation disease by deep sequencing of small RNAs. The virus showed similarity with luteovirus sequences and it was tentatively named *Citrus vein enation virus* (CVEV).

Disease spread. This virus is transmitted in a persistent manner by aphids *(Toxoptera citricida, T. aurantii, Myzus persicae,* and *Aphis gossypii).* It is also graft transmissible but is not known to be transmitted by seed.

Disease diagnosis. In addition to the observation of gall formation and vein enation in field trees and bioindicators, a dot-blot hybridization and RT-PCR tests were developed to detect the recently identified CVEV.

Disease control. This disease can best be controlled by avoiding susceptible rootstock varieties and using budwood from disease-free sources.

Viroid Diseases

Viroids are newly identified pathogens. In 1971, spindle tuber disease of potato became the first disease attributed to a viroid, followed by citrus exocortis disease. Currently, seven distinct viroid species are found in citrus. These can be distinguished based on symptom expression in Etrog citron. As viroids were examined by gel electrophoresis technologies, they were placed in groups (CEVd, CVd-I, CVd-II, CVd-III, and CVd-IV) based on their electrophoretic mobility. Today, viroid species are well characterized based on their RNA sequences and biological properties. Citrus viroid species represent four genera of the Pospiviroidae family. These are *Citrus exocortis viroid* (CEVd, genus *Pospiviroid*), *Hop stunt viroid* (HSVd, genus *Hostuviroid*), *Citrus bark cracking viroid* (CBCVd, genus *Cocadviroid*) and *Citrus bent leaf viroid* (CBLVd), *Citrus dwarfing viroid* (CDVd), *Citrus viroid V* (CVd-V) and *Citrus viroid VI* (CVd-VI) of the genus *Apscaviroid.*

Variants are common within each species, and mixed infections of several species are frequently encountered in the field. Sometimes the effect is synergistic, producing disease symptoms more severe than when the viroids exist singly. Mixed infections must be sorted out by sequential polyacrylamide gel electrophoresis (sPAGE), hybridization with specific probes, or RT-PCR analyses followed by nucleotide sequencing.

Cachexia

Cachexia is caused by specific variants of the citrus type of HSVd. It can infect most citrus species and cultivars, but many remain symptomless carriers. Previously, xyloporosis in Palestine sweet lime was considered a separate disease from cachexia, but it was demonstrated to be induced by the same viroid variants (namely CVd-IIb and -IIc). Hence, cachexia and xyloporosis describe reactions on specific host indicator plants and not different diseases.

Disease symptoms and hosts. Mandarins and some mandarin hybrids (tangelo and tangor) and Alemow (*Citrus macrophylla* Wester) are highly susceptible to cachexia. Severely affected trees are stunted and chlorotic. They may decline and, infrequently, die. Discoloration and gum impregnation of the bark can be observed by scraping or cutting away the outer bark. The inner bark surface has rounded bumps, as opposed to the sharp pits caused by CTV and cristacortis.

Disease agent. Specific sequence variants of the citrus types of the HSVd are responsible for this disease. CVd-IIa does not cause cachexia and differ in RNA sequence from the cachexia variants CVd-IIb and CVd-IIc.

Disease spread. The cachexia-inducing viroids are readily transmitted by grafting and by plant sap on contaminated tools. There is no evidence of an insect vector or seed transmission.

Disease diagnosis. On susceptible hosts, cachexia can readily be identified by the symptoms described above. Symptomless hosts can be tested by graft inoculation onto indicator varieties such as Parsons Special mandarin grafted on a vigorous rootstock such as Rough Lemon. Much time is required for symptom development; for example, gum deposits develop near the base of the scion within 6 to 12 months after inoculation under warm conditions. Laboratory-based tests give results more rapidly. These include sequential polyacrylamide gel electrophoresis (sPAGE) nucleic acid hybridization, and reverse transcription (RT) polymerase chain reaction (PCR).

Disease control. Cachexia can be controlled by using viroid-free budwood. Viroids are resistant to heat therapy, so shoot-tip grafting techniques must be used to eliminate cachexia from infected budwood. Pruning and cutting tools must be surface-sterilized in a dilute (1%) sodium hypochlorite solution (bleach). Topworking with susceptible cultivars should be avoided on trees infected with cachexia.

Exocortis

Exocortis, caused by the Citrus exocortis viroid (CEVd), is a bark scaling and stunting disease of trees grafted onto susceptible rootstocks. Fruit quality is typically not affected, but extensive yield losses can occur when infected budwood is propagated on sensitive rootstocks.

Disease symptoms and hosts. CEVd infects most citrus species and cultivars, some citrus relatives, and even some noncitrus hosts. Trifoliate orange, Rangpur Lime, and some citrons and lemons are most sensitive to this viroid; symptoms include stem blotching or bark splitting. Citrons often show leaf epinasty and vein necrosis. Sweet oranges, grapefruit, and mandarins are symptomless when infected with CEVd unless budded onto a sensitive rootstock, in which case bark scaling occurs on the rootstock (fig. 23.11), and the tree may be stunted. Bark scaling appears in trees 4 to 8 years old.

Disease agent. CEVd is the causal agent of this disease.

Disease spread. Because CEVd is readily graft transmitted, this viroid is principally spread by the use of infected propagation material. It is also transmitted by contaminated cutting and pruning tools.

Disease diagnosis. Symptom expression in sensitive varieties is used for disease diagnosis. Indexing for CEVd is typically done by graft inoculation into Etrog citron selection Arizona 861-S-1 and maintained under warm temperatures (90°F). Nucleic acid analysis procedures such as sequential polyacrylamide gel electrophoresis (sPAGE), and molecular hybridization still require inoculation into citron to produce adequate viroid titer. RT-PCR methods are also available for CEVd detection.

Disease control. Only CEVd-free sources should be used, in combination with sensitive rootstocks such as Trifoliate Orange and Rangpur Lime. The spread of CEVd within nursery operations can be avoided by proper sanitation methods, including the treatment of budding tools with a dilute solution of bleach (1%). In orchards, hedging and harvesting equipment should also be cleaned with a bleach solution.

TsnRNA-Ia, -IIa and -IIIb

In certain instances, infection with a particular viroid RNA species may actually have a beneficial effect on a commercial setting. These RNA species have been described as transmissible small nuclear

Figure 23.11 Bark scaling on trifoliate rootstock. *Photo:* E. Calavan.

RNAs, or TsnRNAs. No disease symptoms other than reduction in tree size have been observed with TsnRNA-Ia, -IIa, and -IIIb infections. In some cases, trees inoculated with the TsnRNAs actually produce more fruit per canopy volume than trees without the TsnRNAs. CDFA has approved their commercial use, and budwood containing TsnRNA can be purchased from the CCPP. So far, the only rootstock varieties known to react to TsnRNAs are Trifoliate Orange and Carrizo citrange (fig. 23.12). Various other rootstock and scion combinations are being evaluated. In the absence of true dwarfing scion and rootstock citrus species, the use of TsnRNAs is a unique way to dwarf tree size without negatively impacting fruit production and quality. Orchards can be planted at a greater density, further expanding fruit yield per unit of land surface. Shorter trees may result in a cost savings for harvest and other cultivation practices.

References

French, A. M. 1989. California plant disease host index. Sacramento: California Department of Food and Agriculture.

Garnsey, S. M. 1999. Systemic diseases. In L. W. Timmer and L. W. Duncan, eds., Citrus health management. St. Paul, MN: APS Press. 95–106.

International Organization of Citrus Virologists (IOCV) website, http://www.ivia.es/iocv/.

Karasev, A. V., and M. E. Hilf, eds. 2010. *Citrus tristeza virus* complex and tristeza diseases. St. Paul, MN: APS Press.

Timmer, L. W., S. M. Garnsey, and J. H. Graham, eds. 2000. Compendium of citrus diseases. 2nd ed. St. Paul, MN: APS Press.

Figure 23.12 Clementine on Carrizo rootstock, untreated (left) and treated with TsnRNA Ia+IIa+IIIb (right). *Photos:* G. Vidalakis

24 Invasive Pests: Exotic Plant Pathogens

MaryLou Polek and Georgios Vidalakis

Compared with the rest of the world, the California citrus industry is relatively free of diseases that can impact grower profits. The problems that could result from the introduction of exotic pathogens cannot be overemphasized; these include production losses, fruit quality deterioration, increased production costs, trade barriers, and imposed quarantines that would limit plant and fruit movement. The potential for the introduction of exotic pathogens into California has increased significantly with the expansion of tourism and the immigration of people whose homelands are reservoirs of these pathogens. Further, e-trade has become prolific, and it is difficult to enforce quarantine restrictions on Internet sales. Additionally, actions by the federal government to eliminate trade restrictions, lower the entry standard for pests and pathogens, and preempt a state's authority to ban certain commodities are certain to adversely impact the citrus industry's ability to grow citrus and distribute disease-free trees.

This chapter provides an understanding of exotic citrus pathogens, how their entry has been previously prevented, and how their entry can be prevented in the future.

What Is an Exotic Pathogen?

Exotic pathogens are organisms, including fungi, bacteria, viruses, and viroids, that exist in other countries but are not present in the United States or that exist in other states or regions of the United States but are not present in California. The landmass of the United States contains several widely separated citrus-growing regions whose climatic conditions support the growth requirements of pathogens unique to the region. For example, Asiatic citrus canker is a serious problem in Florida, where it rains most of the year. However, in California, where the climate is drier, the California Department of Food and Agriculture (CDFA) has never detected citrus canker during its annual surveys. Conversely, citrus stubborn, which occurs in areas of high temperatures and low humidity, is present in California but is considered exotic in Florida. Specific pathogen strains or biotypes may also be considered to be exotic. Citrus tristeza virus (CTV) is widespread in Florida and parts of California and therefore cannot be considered an exotic pathogen on the whole; yet certain strains of CTV are specific to Florida and are considered exotic in California, and vice versa. Quarantine regulations prohibit even the sharing of CTV isolates between research scientists in California and Florida.

Unfortunately, exotic plant pathogens may become well established before they are detected. This is primarily because initial symptoms can mimic other endemic diseases, mineral deficiencies and toxicities, and some organisms have a long latent period. The battle against the threat of exotic pathogens is not hopeless, however. Quarantine regulations discussed in this chapter are in place to protect the industry from pathogens. These regulations are intended for the protection of the industry, not to limit the pursuit of new varieties or curtail one's freedom to farm. In addition, CDFA and the citrus industry have become proactive and have implemented trapping, survey, and inspection programs for the early detection of invasive pests. Now, more than ever, individuals must understand the importance of following the proper procedures when obtaining foreign varieties of citrus and the ramifications of bringing untested material into California.

Levels of Quarantine Enforcement

There are three levels of quarantine regulations attributed to the movement of citrus materials, and each level is enforced by a different agency:

- **Federal:** The Code of Federal Regulations regulates the movement of plant materials into the United States from foreign countries and between states. Enforcement is by the United States Department of Agriculture (USDA) Animal and Plant Health Protection Service (APHIS) Plant Protection and Quarantine (PPQ).
- **State:** The California Department of Food and Agriculture (CDFA) regulates the movement of plant materials into and within the state of California.
- **Local:** County agricultural commissioners enforce the regulations governing the movement of plants and plant parts into and within individual counties.

How to Obtain New Varieties

To legally bring citrus plant material, including seed, into the United States, one must obtain a permit from the USDA. For seed, one can obtain a permit online directly from the USDA (see below). Only two individuals within California have the specialized permit to legally import budwood: the director of the Citrus Clonal Protection Program and the curator of the USDA ARS National Germplasm Repository for Citrus and Dates. Anyone interested in importing a new variety should contact one of these agencies. Material will be held in quarantine until it has been tested for all known citrus pathogens. Under no circumstances should plant material in soil be brought into the United States. Soil can contain many bacteria, fungi, and nematodes that are readily transmissible and difficult to control. It is the responsibility of the Office of Homeland Security, USDA-APHIS-PPQ inspectors to monitor customs at international shipping ports and airports. Pertinent contact information is:

USDA APHIS PPQ
Permits for Plants, Plant Products, Soil, Protected Plants and Transit Shipments
Tel: 301-734-0841
Toll-free automated system: 877-770-5990
Fax: 301-734-5786
E-mail: Permits@aphis.usda.gov
http://www.aphis.usda.gov/permits/learn_epermits.shtml

Import Requirements for Plants and Plant Products
Tel: 301-734-0841
Fax: 301-734-3225
http://www.aphis.usda.gov

Citrus Clonal Protection Program
Department of Plant Pathology and Microbiology
University of California
Riverside, CA 92521
Tel: 951-684-8580
http://www.ccpp.ucr.edu

USDA ARS National Germplasm Repository for Citrus and Dates
1060 Martin Luther King Blvd.
Riverside, CA 92507
Tel: 951-827-4399

California Department of Food and Agriculture
Division of Plant Health and Pest Prevention Services
1220 N Street
Sacramento, CA 95814
Tel: 916-654-0317
http://www.cdfa.ca.gov/phpps

Local County Agricultural Commissioners

Your local county agricultural commissioner is an excellent source to help you determine current quarantine regulations. This office is responsible for

- inspection of postal centers, airports, and packinghouses
- verification and issuance of moving permits
- issuance of phytosanitary certificates
- annual surveys of nurseries and nursery stock
- inspection of shipments and subsequent release or rejection
- enforcement of regulations

To locate your county agricultural commissioner's office, visit the CDFA website, http://www.cdfa.ca.gov/exec/county/countymap/.

Trade Issues

Never before have trade negotiations and politics played such a significant role in the erosion of existing safeguards against the introduction of exotic pathogens. In the past, if it was known that a particular country or region harbored pests and pathogens exotic to the United States, the foreign entity could not export that commodity if it was also produced in the United States. The country wishing to export had to prove that the exports did not pose a risk to the

U.S. industry. Now this barrier is being negotiated away by allowing shipments into the United States to nonproduction areas. The burden to prove risk has shifted from the country of export to the U.S. industry; the production areas or states that would be affected seemingly have to prove that the exported product does pose a risk. Sometimes even when solid scientific evidence is presented, achieving a successful trade deal takes precedence.

North American Plant Protection Organization (NAPPO)

The North American Plant Protection Organization (NAPPO) was created in 1976 as a regional organization of the International Plant Protection Convention (IPPC) of the United Nations Food and Agriculture Organization (FAO). It is recognized by the North American Free Trade Agreement (NAFTA). NAPPO's mission is to "provide a forum for public and private sectors in Canada, the United States and Mexico to collaborate in the development of science-based standards intended to protect agricultural, forest and other plant resources against regulated plant pests while facilitating trade" (NAPPO 2013).

The objectives of NAPPO include

- encouraging cooperative efforts among the member countries to prevent the entry, establishment, and spread of quarantine pests
- limiting the economic impact of regulated non-quarantine pests while facilitating international trade in plants, plant products, and other regulated articles
- encouraging and participating in similar hemispheric and global cooperative efforts

NAPPO provides the means for government, industry, and nongovernment organizations in Canada, the United States, and Mexico to address problems caused by emerging pests. This organization develops phytosanitary standards to prevent the introduction and spread of serious quarantine pests, thereby decreasing pest risks and facilitating trade. It serves to harmonize pest management programs such as diagnostic methods and treatment protocols to ensure that the latest technology is used and to coordinate pest survey activities. NAPPO also provides a forum to discuss trade disputes.

Programs in Place

California, with its several control programs, is unique among other states in its management of citrus diseases and pests. These programs were implemented and remain successful through the interaction and cooperation of growers, nursery operators, regulatory personnel, and scientists:

- The Citrus Clonal Protection Program (CCPP) provides disease-tested budwood for all nurseries, growers, and scientists and is a point of entry for all propagative material for the commercial industry brought in from foreign countries. This program is described in detail in chapter 7.
- All propagation material source (mother) trees are registered with the CDFA Nursery Services. It is mandatory that all trees used for propagation purposes be tested for graft-transmissible pathogens. This ensures that diseases are not spread via tree propagation.
- California Citrus Pest District Control Law (Food and Agriculture Code, Division 4, Part 5) provides procedures for the organization, operation, government, and dissolution of districts for the more effective control and eradication of citrus pests. Historically, pest control districts have dealt with California red scale and CTV.
- The Central California Tristeza Eradication Agency (CCTEA) is the union of pest control districts under the legal authority of a joint powers agreement. This agency surveys, detects, and removes any commercial citrus tree that is infected with severe strains of CTV within the participating districts. In the future, the CCTEA is likely to monitor for other important citrus pests and diseases.
- The Citrus Pest and Disease Prevention Program (CPDPP) is a grower-funded program enacted through Assembly Bill 281 in 2009. Its purpose is to fund an operational program against the invasive pests and diseases, such as the Asian citrus psyllid and huanglongbing disease.

Exotic Diseases

Huanglongbing

Causal agent. The agent associated with huanglongbing (HLB), also known as citrus greening disease and yellow shoot disease, is a fastidious (difficult to culture in the absence of host cells) phloem-limited bacterium of the genus *Liberibacter*. It is thought to have originated in Southeast Asia and severely impacts production in the Philippines, Taiwan, Indonesia, China, India, Latin and Central America, and Africa. It was first detected in the United States in 2005 in Florida, then spread to Georgia, Alabama,

Louisiana, Mississippi, and South Carolina. In 2011 it was detected in two commercial orchards in Texas. At this writing, there has been one detection on a residential property in Los Angeles. It has not been detected in Arizona.

Symptoms. Early symptoms include a yellowing of only one limb or sector of the tree canopy. Leaves may develop vein chlorosis or a mottling of all or part of the leaf blade (fig. 24.1). Chronically infected trees are sparsely foliated, display extensive twig and limb dieback (fig. 24.2), and have small leaves that frequently show zinc deficiency symptoms (green veins and chlorotic interveinal areas that are asymmetrical across the midvein). The malady may resemble stubborn and tristeza. Although mandarins are tolerant to certain severe strains of CTV, they are susceptible to HLB. HLB may be distinguished from citrus blight (a disease endemic in Florida and Brazil) in that it does not induce xylem dysfunction or wilting. Fruit tend to fall off the tree prematurely and are frequently small, underdeveloped, and poorly colored (tend to remain green, at least in part). The juice is low in soluble solids, high in acid, and abnormally bitter.

Figure 24.1 Blotchy mottle of leaves due to HLB. *Photo:* E. Grafton-Cardwell.

Identification. Preliminary field identification of HLB can be made by the foliar and fruit symptoms. Further identification requires laboratory analysis using polymerase chain reaction (PCR) or indexing on indicator hosts.

Host range. Three species have been described: *Candidatus* Liberibacter africanus, which is heat sensitive; *Candidatus* Liberibacter asiaticus, characterized by symptom expression under low humidity at a temperature of 68° to 90°F; and *Candidatus* Liberibacter americanus, limited to South American countries, primarily Brazil. The HLB agent can infect all citrus species, cultivars, and hybrids and many other members of the family Rutaceae. Unfortunately, no evidence of resistance has been identified at this time.

Transmission. The HLB agent can be graft transmitted, but rates are highly variable due to the irregular distribution of bacteria within the host plant. It is also vectored by two species of psyllids: *Trioza erytreae,* found primarily in Africa, and *Diaphorina citri,* widespread in Asia, India, South America, Mexico, and the United States. Both insects have host ranges limited to the family Rutaceae, and their preferred hosts include *Murraya paniculata* (mock orange) and Mexican lime. Although nymphs readily acquire the bacteria, adult insects typically transmit the bacterium. It is believed that the bacteria can propagate within the insect, and the psyllid retains the bacteria for life (psyllids can live for several months). The evidence for seed transmission is inconclusive; anecdotal accounts have been reported in China and Indonesia. Bacteria have been isolated from the seed coat but not from the embryo.

Figure 24.2 Healthy tree (left) and HLB-infected tree (right). *Photo:* G. Vidalakis.

Control. Regulation of the importation of propagating materials prevents the introduction of HLB. Where the disease is endemic, it may be controlled by the reduction of inoculum sources (eradication of infected trees), insect control by pesticide application, and planting with only trees certified as HLB-free. Biological control of the psyllid by *Tamarixia radiata,* a parasitic wasp, has had limited success. Tree injection with antibiotics, such as tetracyclines, has been shown to effectively suppress HLB symptoms, but the treatments are expensive, labor intensive, and a regulatory nightmare, while the results are temporary. In other words, the tree is not cured, and after the antibiotic treatments are stopped the disease reemerges. If continuously used, antibiotic residues can be detected in fruit. Budwood sources can be freed of the HLB organism by heat therapy and shoot-tip grafting.

Most areas of the southwestern United States could support the bacteria and its psyllid vector *(Diaphorina citri),* but the vector seems to be sensitive to elevations greater than 2,000 feet.

Citrus Variegated Chlorosis

Causal agent. *Xylella fastidiosa* ssp. *pauca,* a xylem-limited bacterium, causes citrus variegated chlorosis (CVC). This bacterium is the same genus and species as the bacterium that causes Pierce's disease *(Xylella fastidiosa* ssp. *fastidiosa)* of grape, but it is less than 60% genetically similar, and thus they have different subspecies names, *Xf* ssp. *pauca*. It was first identified in Brazil in 1987. Its present distribution includes all citrus-growing areas of Brazil, Argentina, Uruguay, and Paraguay.

Symptoms. Initial symptoms resemble those caused by zinc, boron, and potassium deficiencies. Leaves become chlorotic; on the lower side of the leaf, the chlorotic spots may have a gummy brown center surrounded by a chlorotic halo, typical of bacterial infections (fig. 24.3). The spots may coalesce to form large chlorotic areas. Frequently, certain sectors of the tree canopy are chlorotic, while others remain green. Fruit produced on infected trees may be three times smaller and hard (see fig. 24.3). On severely infected trees, leaf drop may occur.

Figure 24.3 Brown lesions on a leaf and reduced fruit size due to CVC. *Photo:* G. Vidalakis.

Identification. Visual detection of CVC in field trees may be confused with mineral deficiencies. Methods such as culturing the organism, dot immunobinding assays, ELISA, and microscopy should be used to confirm diagnosis. PCR assays are used to distinguish the various subspecies and strains of the bacterium.

Host range. All orange cultivars are susceptible. CVC has not been detected in lemon, Tahiti acid lime, Marsh grapefruit, sour orange, Dancy tangerine, or Ponkan mandarin. Some mandarin and pummelo varieties are symptomless but may harbor the bacterium. Although many rootstock varieties are resistant or tolerant, the disease will still occur on susceptible scions grafted onto these rootstocks.

Transmission. Grafting transmits the bacterium that causes CVC. The disease is also vectored by various species of sharpshooter leafhoppers (Cicadellidae). Seed transmission has also been reported.

Control. Control measures for CVC include removing symptomatic young trees and pruning affected limbs. Weed control of sharpshooter alternate hosts is recommended. Systemic insecticides are used for control of the sharpshooters.

CVC could cause severe yield losses if introduced in the United States. *Oncometopia nigricans* sharpshooters from Florida have been shown to vector the CVC bacterium. In California, the glassy-winged sharpshooter *(Homolodisca vitripennis)* efficiently transmits the bacterium that causes Pierce's disease in grapes and oleander leaf scorch, and work is in progress to determine whether this species can transmit the CVC bacterium.

Citrus Bacterial Canker Disease

Causal agent. Different forms of citrus bacterial canker disease (CBCD) are caused by two species of bacteria belonging to the genus *Xanthomonas: X. citri* ssp. *citri* and *X. fuscans* ssp. *aurantifolii*. Previously these forms of the disease were attributed to pathotypes, *Xanthomonas axonopodis* pv. *citri* (Xac) (formerly *Xanthomonas campestris*) and pv. *citri* (Xcc). Strains of *X. citri* ssp. *citri* are associated with diseases known as Asiatic citrus canker, canker A, and cancrosis A. These are the most economically important and widespread forms of the disease globally. Strains of *X. fuscans* ssp. *aurantifolii* cause canker B, canker C, cancrosis B, and false canker. The pathogenic variants associated with all of these disease forms are considered quarantine pests for many citrus-growing countries and are subject to the same international phytosanitary regulations.

Symptoms. All aboveground tissues as well as roots are susceptible to infection under natural conditions. On leaves, lesions initially appear as tiny, slightly raised, blisterlike eruptions, usually appearing first on the lower surface. As the lesions age, they become tan or brown with water-soaked margins surrounded by a chlorotic halo. Eventually, the lesions become corky or spongy, and the centers may become craterlike (fig. 24.4). Severe infection of susceptible cultivars may result in defoliation. In addition to unsightly blemishes, severe fruit infection may result in premature fruit drop.

Figure 24.4 Fruit lesions due to CBCD. *Photo:* G. Vidalakis.

Identification. A wide variety of tests are used for the identification and differentiation of *Xanthomonas* species. Attached or detached leaves of susceptible hosts such as grapefruit and Mexican lime can be inoculated and observed for the development of lesions. Serological techniques such as ELISA, protein profiles as determined by SDS-PAGE, and DNA analyses such as PCR and RFLP are commonly used.

Host range. Distinct forms of CBCD are caused by different species of the bacterium and affect citrus species and cultivars differently. In general, grapefruit, Mexican lime, and trifoliate orange are highly susceptible, whereas sour orange, lemons, and sweet orange are moderately susceptible. Mandarins are moderately resistant. CBCD-A is the most severe form of the disease, and it affects most citrus varieties. CBCD-B (cancrosis B) severely affects lemons in Argentina, Paraguay, and Uruguay. Mexican lime, sour orange, and pummelo are also susceptible. CBCD-C is associated with Mexican limes in Brazil.

Canker can be a serious disease where rainfall and warm temperatures prevail during periods of shoot emergence and development of young fruit. In recent years, CBCD has also been found in Southwest Asia (Iran, Iraq, Oman, Saudi Arabia, United Arab Emirates, and Yemen).

Transmission. Infection occurs primarily through stomates, other natural openings, and wounds. The presence of leafminers increases disease incidence by creating numerous openings for the entry of bacteria and a place for inoculum to build up. A combination of wind and rain increases the potential for disease spread over short distances; tropical storms, hurricanes, and tornadoes can spread the bacteria up to several miles. All tissues become more resistant to natural infection as they mature. The most critical period for fruit rind infection is during the first 90 days after petal fall.

Long-distance spread of the disease is more apt to occur with movement of infected propagation material and budded trees. Commercial shipment of diseased fruit and deposition of infected culled fruit near an orchard are potential means of spread. There is no record of seed transmission. Nursery workers can carry the bacterium to new locations unless hands, clothes, tools, and equipment are disinfested. Contaminated harvest, pruning, and spray equipment and wooden harvesting boxes can move the pathogen between plantings.

Control. The first line of defense against canker is exclusion. It is most wise to adhere to the rigid restrictions on the importation of propagating material and fruit from areas with canker. Once the disease has been found in a new area, elimination of inoculum by removal and destruction of infected and exposed trees is the most accepted form of eradication. Copper sprays can reduce infection, especially when applied during crucial periods of growth and fruit development.

CDFA conducts an annual survey of commercial citrus for citrus canker. To date, this disease has not been found within California. There is some question about whether this bacterium could become established in the California climate; however, the finds in the Persian Gulf region suggest that it could. Early detection and immediate response to this disease, if introduced, would help protect the California citrus industry.

Citrus Leprosis

Causal agent. Leprosis can be caused by two distinct viruses with a similar morphology (rod-shaped particles) and vector (tenuipalid mites of the *Brevipalpus* genus). *Citrus leprosis virus* cytoplasmic type (CiLV-C) is the prevalent form It is found in the cytoplasm of symptomatic host cells and organelles. *Citrus leprosis virus* nuclear type (CiLV-N) is rare, with reports only in limited areas of Brazil and Panama. In this case, rod-shaped particles are observed only in the nuclei of symptomatic leaf tissue cells. Neither of these virus types moves systemically within the host plant; lesions are localized to areas where virus particles accumulate, typically where the mite has fed. Leprosis is one of the the most economically debilitating virus disease to the Brazilian citrus industry.

Symptoms. Citrus leprosis causes symptoms on leaves, twigs, and fruit. Lesions produced on leaves begin as a chlorotic spot that becomes brown. The center of the lesion may become necrotic (fig. 24.5). Lesions on twigs may be flat or raised necrotic areas (fig. 24.6), whereas lesions on fruit are flat and depressed. Fruit lesions may show concentric patterns (fig. 24.7), are impregnated with gum, and are typically surrounded by a yellow zone. In severe infections, the disease can cause premature fruit and leaf drop, poor canopy development, and dieback. Symptoms on larger limbs may resemble the bark scaling symptoms of psorosis.

Identification. The disease can be identified in the field by the observation of symptoms in trees. Mechanical inoculation using sap from symptomatic tissue onto noncitrus herbaceous hosts, such as species of *Chenopodium,* induces local lesions. Verification of the virus can be done by transmission of the virus to healthy plants by mites or by observing virus particles using electron microscopy. Real-time PCR methods are commonly used.

Host range. There is significant variation in symptom development within members of the Rutaceae, age of the host plant, isolates of the virus, and geographic location. Sweet orange varieties are most susceptible, but sour orange and mandarins can also be infected. Other citrus cultivars do not develop conspicuous symptoms. *Swinglea glutinosa,* a common ornamental in Colombia, was recently identified as the first noncitrus natural host. Under controlled conditions, leprosis was successfully transmitted to non-Rutaceaous hosts by mites. The disease was found in Florida but is thought to have been eradicated. Recently, concern over leprosis has rekindled due to the extensive immigrant population in the Miami area. It is a serious problem in Brazil and other South American countries.

Transmission. Tenuipalid mites of the *Brevipalpus* genus naturally transmit leprosis. Graft transmission is difficult due to the nonsystemic nature of the virus. It can also be mechanically transmitted by sap from citrus to citrus and to herbaceous plants, mainly those of the family Chenopodiaceae.

Control. Successful control of leprosis can be achieved by the use of acaricides to reduce mite populations. Removal of infected tree branches with pruning is an effective way to reduce the inoculum since the virus is not systemic but localized.

Citrus Sudden Death

Causal agent. Two viruses are associated with citrus sudden death in Brazil: CTV and a new marafivirus, tentatively called *citrus sudden death–associated virus* (CSDaV). Since nearly every citrus tree in Brazil is infected with CTV, it is difficult to determine exactly what role CTV plays in this disease.

Symptoms. Citrus sudden death is very similar to CTV quick decline. Symptoms of citrus sudden death begin with a pale discoloration and decreased size of the leaves throughout the canopy. The tree may partially defoliate, produce fewer new shoots, or lack internal shoots. Roots may rot, and large portions of the root system may die. These symptoms become more severe as the disease progresses and

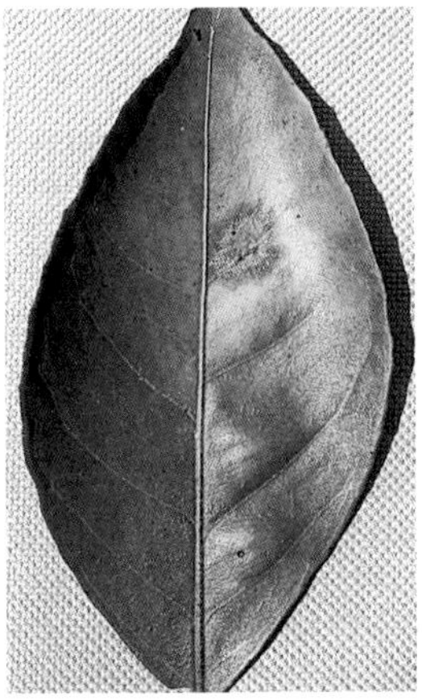

Figure 24.5 Brown necrotic lesions on leaves caused by CiLV. *Photo:* G. Vidalakis.

Figure 24.6 CiLV lesions on twigs. *Photo:* G. Vidalakis.

Figure 24.7 Concentric rings of CiLV lesions. *Photo:* C. Childers.

may eventually result in tree death (fig. 24.8A). The time lapse between the first visible symptoms in the canopy and tree death ranges from 1 to 12 months, depending on the time of year (more rapid in the spring) and cultivar (more rapid in late-maturing cultivars). Tree decline may occur slowly, taking several years with alternating periods of recovery. Symptom expression occurs rapidly when the tree's demand for water is high, as for shoot development and fruit maturation. Under high temperature and long periods of water deficit, the sudden collapse of a diseased tree may occur within a few days.

Identification. Specific to citrus sudden death is the development of a conspicuous yellow-stained layer in the phloem immediately below the bud union of all the sensitive scion-rootstock varieties. In addition, Rangpur Lime rootstock exhibits dramatic changes, including reduction of functional phloem, and the occlusion, collapse, and necrosis of sieve elements, which result in the thickening of the yellow-stained bark at the bud union (fig. 24.8B). PCR methods can confirm the presence of both the citrus sudden death–associated marafivirus and CTV.

Figure 24.8 CSDaV infection: Late-stage infection (A); yellow stained bark layer below the bud union (B) *Photos:* G. Vidalakis.

Host range. Initially, citrus sudden death occurred only in sweet orange varieties such as Westin, Hamlin, Natal, Valencia, Pera, and Rubi grafted onto Rangpur Lime rootstock; however, similar symptoms have been reported when Volkamer Lemon and Rough Lemon rootstocks are used. Scion cultivars differ in susceptibility; Valencia and Westin are more susceptible than Pera and Natal. Other species, varieties, and hybrids are affected, such as Ponkan, Cravo mandarin, Murcott tangor, sweet lime, and Tahiti lime. Symptoms of citrus sudden death are noticeable in trees 22 months or older. This disease occurs only in Brazil at this time and only north of the cities of Sao Paulo and Rio de Janeiro (21° south latitude).

Transmission. In the area where the disease was first observed, the number of symptomatic trees increased from 500 in 1999 to over 300,000 in February 2002. It is graft transmissible. Epidemiological and cage studies have implicated an aerial vector as being responsible for natural spread of the disease, but none have been identified.

Control. The only means of controlling citrus sudden death is to use a tolerant rootstock such as Swingle or Trifoliate. Currently, millions of trees are being inarched in Brazil with tolerant varieties. The predominant use of only one rootstock variety throughout an entire industry should be avoided. This disease may not be a threat to the California industry because Rangpur Lime is not a typical rootstock, and Volkamer is used infrequently as well. Although the continued use of several rootstock varieties may help prevent a similar situation from occurring in California, Brazil's plight should serve as a warning that the CTV can interact with other pathogens, change, or mutate into forms that could potentially have devastating effects on the state's industry. The industry should not become complacent regarding the existing CTV control programs.

Satsuma Dwarf Virus Group

Causal agent. *Satsuma dwarf virus* (SDV) was first described in 1952 in Japan and is now widely distributed throughout many countries. Its spread is attributed to the movement of infected budwood. Three other viruses related to SDV occur in Japan. These include citrus mosaic virus (CiMV), natsudaidai dwarf virus (NDV), and navel infectious mottling virus (NIMV). Little is known about this virus group.

Symptoms. SDV can infect a wide range of citrus species and cultivars, but mandarin varieties are most severely affected. In greenhouse studies, this virus has been successfully transmitted to noncitrus hosts. Infected trees are stunted and yield is reduced. The viruses of this group cause leaves of Satsuma mandarin to become narrow, boat shaped, or spoon shaped; however, SDV affects only the spring flush. Additionally, CiMV causes spotting and blotching of the fruit rind, especially in Satsuma mandarin. NDV produces vein clearing, mottling, and curling of new leaves of natsudaidai and Rough Lemon. Sweet orange is predominantly affected by NIMV, with chlorotic spotting that persists in mature leaves.

Identification. The viruses in this group are distinguished through serology and RNA analyses.

Transmission. SDV is readily graft transmitted between citrus plants and can be mechanically transmitted to both citrus and noncitrus hosts. New infections originate from the movement of infected nursery stock and topworking with infected budwood. A soilborne vector is thought to be involved in the spread of all the viruses within this group, but none have yet been identified.

Control. Shoot-tip grafting can easily free budwood from these viruses. Using disease-free material for propagation, including topworking, can successfully control diseases caused by this group of viruses.

Citrus Yellow Mosaic Virus

Causal agent. *Citrus yellow mosaic virus* (CYMV) is classified as a badnavirus and is found in the cytoplasm of cells from infected plants. The virus was first reported in 1975 in India. This disease is distinct from Citrus mosaic virus (CiMV) previously described. Yield reductions up to 77% have been reported in 10-year-old trees. This virus impacts fruit quality.

Symptoms. CYMV induces bright yellow mosaic symptoms on leaves, yellow flecking along the veins, and vein banding, and it reduces leaf size. Infected trees are stunted. Fruit display depressed yellow patches and elevated green areas.

Identification. CYMV can be diagnosed by graft inoculation into Mosambi sweet orange or pummelo seedlings. ELISA and immunospecific electron microscopy can be used to detect this virus in plant tissue.

Host range. Most citrus species, cultivars, and relatives are susceptible to CYMV. Symptoms are most severe in sweet orange, lemons, mandarins, sour orange, grapefruit, and pummelo. *Citrus decumana* displays mild mosaic symptoms, whereas Mexican lime is symptomless.

Transmission. CYMV is readily transmitted by bud, bark, and leaf patch grafting. It can also be mechanically transmitted by applying leaf sap from infected plants to healthy leaf surfaces. This virus has also been transmitted experimentally by the citrus mealybug *(Planococcus citri)*.

Control. The only recommended means for the control of CYMV is the use of disease-free propagation material and the removal of trees known to be infected.

Citrus Black Spot

Causal agent. Citrus black spot is caused by the fungus *Guignardia citricarpa*. First reported in Australia in 1895, this disease occurs in subtropical regions of the world with summer rainfall including Argentina, Australia, Brazil, China, Indonesia, Japan, Kenya, Nigeria, Philippines, South Africa, Taiwan, Uruguay, Venezuela, and Zimbabwe. It was detected in Florida in April 2010.

Symptoms. Leaves support the growth of this fungus but seldom show symptoms. When present, small necrotic spots with a gray center surrounded by a dark brown ring and yellow halo are characteristic. Lemons most frequently display leaf symptoms. The greatest economic loss is due to symptoms on the fruit. Spots on the rind make the fruit unmarketable, although the internal quality of the fruit remains unaffected. The light-exposed side of the fruit seems to be the most prone to symptom development. Fruit spotting has four phases: hard spot or shot hole, freckle spot, spreading or virulent spot, and false melanose or speckled spot (fig. 24.9). Hard spot is the most typical symptom, occurring mostly on green fruit.

Figure 24.9 Spots caused by the fungus of citrus black spot. *Photo:* G. Vidalakis.

Identification. Identification of this disease is limited to visual symptoms and signs of the fungus on fruit and on living and dead leaves. Diagnostic confirmation is done using real-time PCR methods.

Host range. All commercially grown species of citrus except sour orange are susceptible, but losses are greatest in lemons, Valencia oranges, and grapefruit.

Transmission. Ascospores (sexual) from dead leaves in the orchard provide the main source of inoculum although pycnidial spores (asexual) can provide inoculum to late-hanging fruit such as Valencia oranges. High rainfall, warm temperatures, consistent dew formation, and certain irrigation practices favor fungal infection. The disease is slow to develop when spores are introduced. However, once epidemic proportions are reached, the disease remains serious and is difficult to eradicate.

Control. Preventive fungicide applications such as copper must be applied several times during the growing season. Treatments with systemic fungicides have been used as postinfection applications. Fungicide resistance is a concern. Exclusion, proper cultural practices (especially irrigation), and sanitation are the best means for the control of this disease.

Sweet Orange Scab

Causal agent. Fungal species are responsible for scab diseases of citrus. Sweet orange scab is caused by *Elsinoë australis.* This disease is important on fruit grown for the fresh-fruit market because it produces unsightly blemishes on fruit, rendering it unmarketable. Sweet orange and tangerine varieties are most affected, but other citrus varieties, including grapefruit, can develop symptoms as well. Infected fruit are more likely to drop prematurely. In addition, the disease may stunt young citrus seedlings. On July 23, 2010, the U.S. Department of Agriculture (USDA) Animal and Plant Health Inspection Service (APHIS) confirmed the detection of sweet orange scab on lemon and tangerine trees on a single residential property in Spring, Texas, near Houston. Once a survey was initiated, the disease was also detected in Florida, Louisiana, and Arizona. Intensive surveys were conducted for several years in California, with no detections of the disease until October 2013.

Figure 24.10 Sweet orange scab on lemon. *Photo:* G. Vidalakis.

Figure 24.11 Salmon-pink, discolored wood of a Mal secco fungal infection. *Photo:* A. Eskalen.

Symptoms. Unsightly, scablike lesions develop on fruit rinds and, less often, on twigs and leaves. The damage produced is superficial and does not affect internal fruit quality. Scab pustules, which consist of a mixture of fungal and host tissue, are slightly raised and pink to light brown. In time, the pustule becomes warty and cracked and turns yellowish brown, eventually turning dark gray (fig. 24.10). Microscopic spores are produced within the lesion and are the means of fungus spread. Trees are more susceptible to infection when there is new shoot growth and the petals begin to fall. As the growing tissue matures, it becomes less susceptible.

Identification. The genus *Elsinoë* can be identified by culturing the fungus in the laboratory and noting colony characteristics and spore morphology. However, molecular techniques are necessary to distinguish the species.

Host range. Sweet orange scab occurs in the humid citrus-growing areas of southern South America, including Brazil, Argentina, Paraguay, and Uruguay. While its detection in Florida and Texas is understandable, it is unclear how the fungus became established in Arizona. It affects fruit of sweet oranges, lemons, and many mandarins.

Transmission. Spores are produced within the lesions on leaves and fruit and are dispersed by splashing water onto young, susceptible tissue. Infection occurs on very young leaves, but fruit remain susceptible for up to 2 months after bloom. Spores can spread the disease to susceptible plants if there is a sufficient level of moisture in the environment. The fungus can live through the winter in the tree canopy on limbs and on fruit that were infected during the previous season. Symptoms of the disease can be detected visually at any time of the year. The disease produces symptoms within a few days to 1 week.

Control. A copper fungicide should be applied in the spring to protect developing fruit. Nursery plants must be treated with approved fungicides prior to movement. For current regulations and images, see the APHIS website, http://www.aphis.usda.gov/plant_health/plant_pest_info/citrus/index.shtml.

Mal Secco

Causal agent. Mal secco, Italian for "dry disease," was first observed in 1894 on the Greek islands of Chios and Poros, but the causal organism was not determined until 1929. The fungus *Phoma tracheiphila* is responsible. This disease is confined to citrus-producing countries in Asia Minor, the Mediterranean Basin, and around the Black Sea, with the exception of Spain, Portugal, Morocco, and some areas of the Arabian Peninsula. It primarily affects lemons.

Symptoms. Mal secco attacks trees of any age, but it is more severe on young ones. Infection sites may be detected by vein clearing near a wound on a leaf. Leaves eventually wilt, dry up, and fall. Dieback occurs. The fungus proceeds slowly downward from young shoots to branches and main limbs until the trunk and roots finally become infected. The disease may develop rapidly causing the leaves to dry up and the canopy to collapse. When the bark layer is peeled away, the wood is often discolored, appearing salmon-pink or orange-red due to gum production in the xylem (fig. 24.11). Yield is significantly decreased, and infected trees may die.

Identification. *Phoma tracheiphila* is identified mainly by growth characteristics on artificial agar media and spore morphology. In the absence of sporulation, molecular methods (PCR) and mycelia protein analysis are used.

Host range. Mal secco is most prevalent and severe on lemons (the Italian variety Femminello is highly susceptible) and citron, and some mandarins, tangelos, and tangors are quite vulnerable. This disease has also been reported on *Poncirus, Severinia,* and *Fortunella* as well as *Citrus*. Grapefruit and sweet orange are rarely affected. Susceptible rootstocks include Rough Lemon, Alemow, Sour Orange, and Troyer and Carrizo Citranges.

Transmission. Spores are released by raindrops falling on the inoculum source and are transported short distances by wind-blown rain. Wounds from hailstorms, frost, high winds, pruning, and harvesting predispose trees to infection. Contaminated pruning tools may also be responsible for moving the disease. The fungus can survive in leaf and twig litter on the orchard floor.

Control. Use only disease-free material for propagation. Diseased shoots and branches should be pruned out and burned immediately. Suckers must be removed from trunks and roots to avoid infection of the lower parts of the tree. Stumps of infected trees should be removed and burned. Copper fungicides sprayed on the canopy act as a protectant.

Since this disease occurs in Mediterranean areas with winter rainfall and dry summers, mal secco could easily become established in California. The lemon industry would be most severely impacted, and extreme caution must be taken to use only certified budwood for propagation.

Brown Spot of Mandarin

Causal agent. Alternaria brown spot of mandarins is caused by isolates of *Alternaria alternata* pv. *citri* that produce a host-specific toxin. It was first described on Emperor mandarin in Australia in 1903. It appeared in Florida in 1974, and later in South Africa, Israel, Turkey, and Colombia.

Symptoms. This disease causes serious defoliation, fruit drop, and fruit blemishes on susceptible cultivars. Lesions first appear on young leaves as small brown to black spots, becoming surrounded by yellow halos. Lesions expand and can sometimes constitute large areas of the leaf. The fungus produces a toxin that is responsible for necrosis and can be translocated in the vascular system. Symptoms of this disease are sometimes confused with anthracnose, since that organism readily colonizes the necrotic areas.

Young shoots also can develop lesions, and abscission of young leaves may produce dieback of twigs. Fruitlets can be infected soon after petal fall, and even a small lesion can cause immediate abscission. On mature fruit, symptoms range from small, dark specks to large, black lesions (fig. 24.12). Corky eruptions are often formed, leaving unsightly pockmarks on the fruit surface.

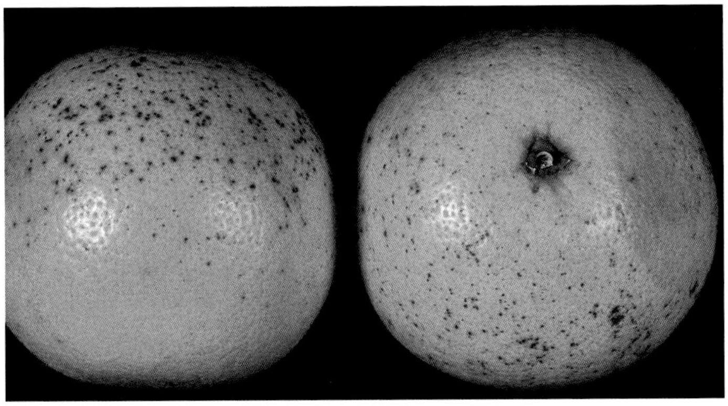

Figure 24.12 Black specks of Alternaria brown spot. *Photo:* D. Gumpf.

Identification. Identification of this disease is limited to visual symptoms and signs of the fungus on fruit and leaves.

Host range. Dancy mandarins and many of its hybrids: Minneola and Orlando tangelos and Sunburst, Emperor, and Murcott mandarins are susceptible. Infrequently, grapefruit can be affected.

Transmission. This disease is spread by spores produced on leaves and twigs but not on fruit. The spores are thick walled and resistant to desiccation. Spore release is triggered by rain and high humidity. Spores become airborne and are carried by winds.

Control. Nursery stock is the main dispersal mechanism for this disease. Growing susceptible cultivars in greenhouses and avoiding wet foliage are the best means of control. Increased spacing between trees and pruning of tree skirts also minimize disease development.

Once the disease is well established, fungicides are usually needed for adequate commercial control. With the increased acreage planted to mandarin varieties in California, this disease, if introduced, may be of increased concern.

Root Rot and Gummosis

Causal agent. *Phytophthora palmivora.* Diseases caused by species of *Phytophthora* are widespread, with nearly all orchards in California harboring one or more species of the Oomycete (water molds) (fig. 24.13). However, *P. palmivora* is not known to occur in California at this time, but it was recently identified in Florida.

Symptoms. *P. palmivora* causes more severe root rot symptoms than do endemic species of *Phytophthora.*

Identification. Laboratory diagnosis by morphological and physiological characteristics and DNA analyses is recommended for species confirmation.

Host range. *P. palmivora* is a devastating pathogen in Brazil on avocado and causes black pod and canker of cacao. Its temperature requirements are intermediate between *P. parasitica* and *P. citrophthora,* so it could survive very well in the California climate.

Figure 24.13 Gummosis. *Photo:* D. Gumpf.

Transmission. Same as endemic species of *Phytophthora.*

Control. This species is considered to be a threat to the California citrus industry because it outcompetes endemic species such as *P. parasitica* in both short- and long-term experiments. It does so by producing a greater number of sporangia at a more rapid rate, and therefore has a high potential to cause an epidemic.

Citrus Gummy Bark

Causal agent. Citrus gummy bark is a disease of uncertain cause that is prominent in sweet orange grown in the Cukurova district of Turkey. Similar disease symptoms, believed to be induced by a viroid, have been reported from other parts of the Mediterranean region. The presence of viroids in infected plant material was confirmed by bioamplification in citron followed by sPAGE. Although samples that tested positive contained a complex of viroids, neither sPAGE nor RT-PCR analyses identified any new or unusual viroid RNA species. Because of these results, symptoms of citrus gummy bark may be the result of a combined reaction of several viroids or with another pathogen.

Symptoms. Discoloration and gum impregnation of the bark occur and are quite similar to the symptoms caused by the cachexia-inducing HSVd variants. When infection is severe, gum pockets are widespread on the trunk and may extend into the scaffold limbs. Only a few gum pockets are observed near the bud union in mild cases. Symptoms do not appear until trees are 8 to 10 years old. The gum impregnations are within the bark and are visible after scraping the top layer of the bark and not by removing a flap of bark (i.e., cutting a window) and exposing the wood tissue as is typically done when examining citrus trunks and branches (fig. 24.14).

Figure 24.14 Gummy bark. *Photo:* G. Vidalakis.

Identification. Detection can be made by scraping into the bark to expose the gum-stained areas. Citrus gummy bark can be distinguished from cachexia by the different type of gum deposits and the lack of any wood pitting (see chapter 23, "Virus and Viroid Diseases"). In addition, cachexia is a mandarin and lime disease, while gummy bark has been reported exclusively in sweet orange.

Host range. Sweet oranges are mostly affected. This disease has been reported in the eastern Mediterranean, North African, and Near Eastern countries, including Egypt, Greece, Iran, Iraq, Libya, Oman, Saudi Arabia, Sudan, and Turkey.

Transmission. This disease is graft transmissible. No vectors are known.

Control. The only recommended means for the control of gummy bark is the use of disease-free propagation material.

Citrus Viroids

Causal agent. A few viroid species and variants have so far not been discovered in California: CVd-I-LSS (low sequence similarity), a variant of *Citrus bent leaf viroid* (CBLVd); CVd-VI (syn. CVd-OS-original sample), originally identified in Japan; and CVd-V, originally identified in Spain. Since their original identification in the 2000s, these viroids have been reported in other countries, such as China, Pakistan, and Iran.

Symptoms. Symptoms induced by these viroids in the bio-indicator host Etrog citron are similar to, and in many cases more severe than, especially when in mix infections, those caused by the Citrus exocortis viroid (CEVd) alone. For example, Etrog citron inoculated with CVd-I-LSS exhibits typical leaf epinasty (leaf drooping or bending downward); however, co-inoculation of CVd-V with Citrus bent leaf viroid results in synergistic interactions that manifest in enhanced leaf epinasty and very pronounced dwarfing.

Identification. sPAGE, RT-PCR, and sequence analysis are the methods used to identify these viroids.

Host range. Most likely, these viroids have a very narrow host range, limited to citrus and citrus relatives typical to the genus *Apscaviroid*. CVd-I-LSS and CVd-VI entered California in Fresno County by the illegal importation of budwood. All trees potentially budded with infected material have been destroyed, and in the absence of any natural vectors for viroid transmission the probability of spread was extremely low.

Transmission. Graft transmission and sap transmission via budding tools can spread these viroids.

Control. Adherence to quarantine regulations regarding the importation of citrus varieties and the use of viroid-tested budding and topworking plant material can best control the spread of these pathogens.

Citrus Blight

Causal agent. Despite that citrus blight was first described in 1874, the cause of this disease remains unknown. In addition to attempts to isolate a pathogenic organism, various soil, nutritional, and cultural factors have been investigated. Citrus blight has been transmitted via root grafts. This suggests the presence of a transmissible agent in the roots and dispels abiotic factors as the cause for this disease. Its distribution includes Florida, Texas, Louisiana, Hawaii, Cuba, Belize, Brazil, Argentina, Colombia, South Africa, and Australia. Citrus blight has never been reported in California.

Symptoms. Typical blight symptoms become apparent in trees 6 to 10 years in age and include wilt and decline of the tree canopy, small and off-color leaves, leaf drop, and twig dieback. New growth flush and flowering are delayed, and fruit set is diminished. Fruit from blight-infected trees are highly acidic and have high soluble solids. As trees decline, fibrous roots die and eventually even scaffold roots may be affected. The wilting of blight-infected trees is attributed to the inability of xylem vessels to conduct water. Amorphous plugs composed of lignin can be found in xylem cells of the roots, trunk, and scaffold limbs. Blighted trees rarely die, but never recover. Once trees become economically unproductive, growers remove infected trees and replant.

Identification. Several diagnostic methods have been developed over the years to detect citrus blight; many of these are physiological in nature. Measurements of zinc accumulation in the bark and wood, the rate of water uptake in the trunk, and the presence of amorphous plugs are some of the best methods. A serological assay detects 12-kD and 35-kD proteins in xylem and leaf extracts that are unique to blight-infected trees.

Host range. All common citrus cultivars and rootstocks can be affected by citrus blight. Trees on Rough Lemon, Rangpur Lime, Trifoliate Orange, Carrizo, and Citrange rootstocks are especially susceptible. Trees on Sweet Orange and Sour Orange are the most tolerant. Florida growers basically choose the fate of their orchards; if they use Sour Orange rootstock to prevent blight infection, they risk infection by CTV.

Transmission. The only known means of transmission of citrus blight is by root grafting. Both tree-to-tree and root piece graft inoculations have successfully resulted in disease transmission. Natural root grafts can transmit the disease in field trees. No transmission has occurred using branch grafts, buds, or soil infestation.

Control. Blight-infected trees should be removed as soon as the diagnosis is made. Replants should be on less-susceptible rootstocks. The potential hazard for blight occurring in California may be low, as this disease has only been reported in areas with a humid climate.

Citrus Chlorotic Dwarf

Causal agent. The causal agent of citrus chlorotic dwarf (CCD) has not been well characterized. It is probably of a viral nature, phloem limited, and exists in very low concentrations within the plant. Recent high-throughput sequencing analysis identified an apparently new geminivirus-like agent associated with symptomatic plants, but the role of this new virus in the etiology of CCD has not been yet determined. It was first recognized in Turkey in the mid-1980s, several years after the introduction of the bayberry whitefly. Since, it has spread rapidly within that area of the Mediterranean. It is believed that its introduction into citrus was via another crop.

Figure 24.15 Crinkled young leaves due to CCD (right); healthy leaves on left. *Photo:* G. Vidalakis.

Symptoms. CCD induces chlorotic patterns and distortion, such as crinkling in young leaves (fig. 24.15). Leaf size is decreased. Trees are severely stunted.

Identification. The only detection method currently available for CCD is by grafting onto a sensitive cultivar such as Sour Orange, Alemow, or Rough Lemon.

Host range. Nearly all citrus cultivars are susceptible to CCD, with lemons, grapefruit, some mandarins and tangelos being especially sensitive. Sweet orange is less affected than other species.

Transmission. The bayberry whitefly *(Parabemesia myricae)* is the primary vector for this pathogen. CCD can also be graft transmitted.

Control. Spread of CCD may be prevented by using disease-free budwood. Control of the bayberry whitefly may slow down the rate of spread but will not curtail long-term movement. Emphasis should be placed on determining other reservoirs for this organism so additional methods for control can be devised. Because of the association with a Mediterranean climate, this disease is a potential threat to the California citrus industry.

References

Bastianel, M., V. M. Novelli, E. W. Kitajima, K. S. Kubo, R. B. Bassanezi, M. A. Machado, and J. F.-Astua. 2010. Citrus leprosis: Centennial of an unusual mite-virus pathosystem. Plant Disease 94:284–292.

Migheli, Q., A. Pane, S. O. Cacciola, D. Erza, V. Balmas, and G. M. di San Lio. 2009. Mal secco disease caused by *Phoma tracheiphila:* A potential threat to lemon production worldwide. Plant Disease 93:852–867.

NAPPO (North American Plant Protection Organization). 2013. NAPPO website, www.nappo.org.

Timmer, L. W., S. M. Garnsey, and J. H. Graham, eds. 2000. Compendium of citrus diseases. 2nd ed. St Paul, MN: APS Press.

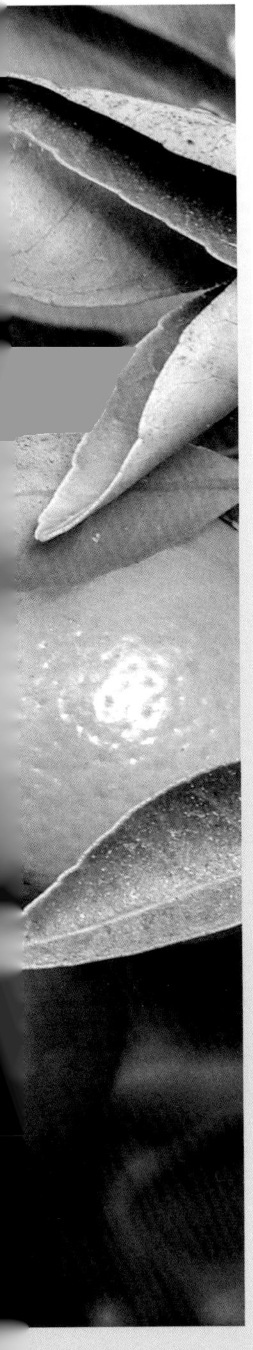

25 Environmental, Physiological, and Cultural Injuries and Genetic Disorders

Robert R. Krueger

Certain disorders of citrus are not currently known to be caused by a pathogenic agent but appear to be inherited, physiologically based, or caused by environmental conditions. The most important of these disorders are rootstock-scion incompatibilities, postharvest fruit disorders, freeze or cold damage, and salt or nutrient deficiencies or excesses, which have been discussed in their respective chapters. This chapter presents information on other disorders of this nature.

Environmental Injuries

Heat Injury and Sunburn

High temperatures have detrimental effects on citrus. In California, heat extremes are most common in the low desert growing areas, but they may also occur in the San Joaquin Valley. High temperatures are accompanied by intense sunlight, and it is not possible to completely separate the effects of heat and light in the field.

Heat and sun damage may cause misshapen fruit, abnormal coloring of the skin (dark yellow or brown spots), burning of the albedo, drying and damage to the pulp (granulation), and seed abortion. The damage varies somewhat with the type of fruit. Mandarins appear to be the most sensitive and may show all three symptoms, whereas lemons may have their pulps desiccated with little external damage, and grapefruit may show some external discoloring with no internal damage.

Peel damage occurs when peel temperature exceeds 120°F, with the extent of damage depending on the duration of high temperatures. Peel temperature is higher than the ambient temperature in most instances. Peel temperatures are lower when there is air movement than when the air is still, and they will also be lower in varieties where the fruit is borne more within the canopy, such as grapefruit, than in varieties where the fruit is more exposed, such as mandarins.

Heat and sun also damage the vegetative growth of citrus trees. Extended periods of heat lead to smaller leaves, yellowing or brown leaves, and the development of lesions. When leaves become reoriented and their lower surfaces are exposed to the sun, they may sunburn and develop irregular, gum-impregnated areas on the blades. Young leaves may be small and misshapen.

Bark of young shoots can separate from the wood, and both bark and wood can be damaged in extreme cases. This leads to reduced translocation of water and nutrients and a consequent reduction in fruit yield and quality. Older bark can be damaged or killed by heat and sun. This is especially true when trees are pruned severely or lack sufficient water. Secondary pathogens can enter through the damaged bark, causing additional problems.

Young trees produced in greenhouses or under shade structures benefit from a gradual acclimation or pretreatment before being planted in very hot climates. Young trees and trees that have been cut back to the scaffolding limbs to allow topworking should be whitewashed to prevent sunburn.

Wind Injury

Wind is often involved in cases of water stress (see below), but it can cause problems independent of the water status of the tree. These problems are due to mechanical damage, such as rubbing of the fruit against the woody structure of the tree (sandblasting). These types of damage are sometimes confused with damage from thrips, scab, and chemicals.

High winds during a flush can cause tattering and distortion of leaves by mechanical damage.

Leaves may also become blasted and hang limply from twigs. These problems may be exacerbated by water stress. More important, fruit may suffer mechanical damage in several ways. For example, slight rubbing of young green fruit against leaves, thorns, twigs, or branches may result in a thin scurf resembling thrips damage. This scurf is mostly superficial, but in lemons, ridges may develop from the production of extra tissue beneath the scurf. More extensive rubbing can cause larger, thicker, scablike areas. These crusts are susceptible to the growth of fungi such as *Alternaria* and *Cladosporium*.

Wind injury of ripe fruit appears as depressed areas on the rind. When the rind skin is broken, oils are released from the vesicles. These sunken areas are usually drab or chestnut brown. Beyond the actual mechanical damage, spreading of the oil can produce characteristic oil spots (see the section on oleocellosis in chapter 26, "Postharvest Handling").

Smog

Smog, particularly ozone, has been shown to have adverse effects on citrus physiological functioning, particularly water relations, photosynthetic assimilation, and leaf senescence and abscission. These can result in reduced growth and yield. Linking small, poor-yielding trees to smog damage without controlled experiments or additional tests is difficult. High levels of ozone can produce a bleached stippling on the upper surface of citrus leaves. The levels of ozone necessary to produce foliar symptoms are higher than those resulting in physiological impairment. Citrus is relatively more tolerant of ozone than some other crops, and a longer period of exposure is necessary to produce these symptoms than for more sensitive crops. Recent research is lacking in this area, and more information is needed.

Flooding

Symptoms of flooding injury are subtle and not always readily apparent. They can also be difficult to differentiate from other types of decline symptoms. Trees that undergo chronic flooding are smaller than normal, with thin, slightly yellowish canopies. Root systems are generally small and shallow, and a drop in the water table after chronic flooding may result in a drought response with leaf wilting, drop, and dieback.

Flooding injury can occur in orchards planted in areas that have high water tables or poor drainage, or it can occur after heavy rains under certain conditions. Roots will survive for several weeks under an oxygen deficit. However, only a few days of truly anaerobic conditions can cause problems. Flooding injury is delayed by cool temperatures because the anaerobic reduction process is slowed. Anaerobic conditions lead to the production of hydrogen sulfide, a toxic product. Disagreeable odors released by digging into the root zone indicate hydrogen sulfide production, which can kill large as well as fibrous roots. In some cases the entire tree can be killed. Flooding also increases the risks of *Phytophthora*-induced fungal rots.

Flooding injury of citrus is not common in California, where the citrus-growing areas have relatively low rainfall that occurs during the cool winter months. Most citrus is planted away from valley floors that are subject to flooding, and most orchards are designed to facilitate drainage (see chapter 9, "Establishing the Citrus Orchard"). Low-volume irrigation may be appropriate in poorly drained areas. If floods do occur, the water should be removed by pumping or other means as soon as possible to prevent tree damage.

Hail Damage

Hail damage to citrus is uncommon in California. When it does occur, it can decrease fruit quality. Hail causes surface damage to the fruit by its impacting the rind. If this damage occurs when the fruit is in active growth, the pockmarks expand as the fruit enlarges. This is more common in Florida, with its summer rains. In California, winter is the rainy season, and most types of citrus are near their full size, although hail damage to new fruit near petal fall is not uncommon in some areas of the San Joaquin Valley.

The damage produced by hail somewhat resembles fruit damage by snails, but it can be distinguished by the larger size of hail damage and the leaf damage associated with snail feeding. In addition, hail damage may provide entrance wounds for secondary pathogens such as *Penicillium*, *Septoria*, and *Colletotrichum* species. More-severe hailstorms can damage leaves and shoots as well as fruit, giving the tree a thin, ragged appearance. In areas where severe hail occurs, the damage from large hailstones can be more extensive, including breaking of small branches, bark shredding, and fruit splitting or dropping. This is extremely rare in California.

Lightning

Damage from lightning is extremely rare in California. Lightning damage, when it does occur, is usually irregular in distribution throughout an orchard, with only a few trees affected. The most common form of lightning damage is for the bark to be killed in a narrow strip extending down the trunk. Injuries are usually confined to the surface of the bark, so the tree can be renewed from below the damaged area. Sometimes bark is killed completely around the trunk base, girdling and subsequently killing the tree. Secondary pathogens such as *Phomopsis, Botrytis, Diplodia, Colletotrichum,* and *Alternaria* species may enter the wounds, causing further damage. On younger tissue, glazed greenish yellow to yellow blotches eventually become elevated, turn brown, and peel away, leaving a rough brown scar. Sometimes islands of living green tissue remain at the bases of petioles on young green shoots before the shoot collapses. Badly injured trees should be removed. Injured twigs should be pruned back and disinfected. Large breaks in the bark should be removed and covered with tree seal.

Physiological Injuries: Water Stress

In simple terms, water stress occurs when a tree's demand for water exceeds its supply. Trees respond in certain ways to cope with this situation, but their response leads to certain physiological consequences for tree and fruit growth. Refer to chapter 11, "Irrigation," for further details on water stress management.

Symptoms of Water Stress

Citrus in California orchards is irrigated, so sustained severe water stress is uncommon; however, short-term water stress can occur when evaporative demand is high. Obvious symptoms of water stress are usually apparent only in extreme cases. Midday wilting of young flush leaves can occur under an extreme lack of soil moisture. This can be exacerbated by wind and low humidity. In more severe cases, older leaves can curl, wilt, and drop. Water stress also can reduce fruit quality by dehydration and can kill cells due to water loss from the rind.

Mesophyll Collapse

Leaves are comprised of several different internal tissues, one of which is mesophyll. Mesophyll can collapse when leaves cannot obtain enough moisture. Factors influencing this collapse include soil moisture supply, temperature, wind, low humidity, saline soil conditions, and small root systems. Red mite infestations appear to exacerbate the problem. External symptoms on leaves are discoloring and chlorosis, drying, and eventually graying and collapse. If invaded by secondary pathogens such as *Colletotrichum, Alternaria,* or *Cladosporium,* leaves may turn dark brown or black.

Twig Dieback

Two types of dieback have been identified. In the first, seemingly healthy twigs die and dry up in place. This dieback is associated with gum in the wood. In the other, more serious, type of dieback, there is no gumming but leaves drop before the twigs or leaves are completely dried out. This second type may affect large branches or large portions of the tree.

Twig dieback is associated with a number of biotic and abiotic factors. Probably the greatest contributing factor is rapid water loss from young tissues. Rapid water loss is usually caused by high temperature, wind, and low humidity. Root problems (including those associated with cold or waterlogged soils), nutritional deficiencies, insects, and diseases can also contribute. Although apparently not the primary cause of twig dieback, some pathogens can exacerbate the damage. Pathogens such as *Colletotrichum, Alternaria, Stemphylium, Coniothecium, Fusarium, Hormodendrum, Diplodia, Phomopsis, Chaetomella,* and bacteria have not been shown to cause dieback; application of fungicides does not prevent dieback caused by water stress.

Management

Water stress is best dealt with by maintaining proper irrigation and fertilization regimes and healthy root systems. Management may include preplant treatments of the soil and an avoidance or reduction of tillage (see part 3, "Citrus Orchard Establishment," and part 4, "Citrus Orchard Management"). Additional stresses, such as foliar applications of agricultural chemicals during hot, windy conditions, should be avoided. It is not possible to entirely avoid problems associated with water stress, but the chances of them occurring can be reduced.

Cultural Injuries: Phytotoxicity

Commercial citrus orchards are treated with various chemicals, including herbicides, insecticides, fungicides, fumigants, and growth regulators. Some of these chemicals can injure citrus if the liquid mixture or volatile vapors come in contact with leaves or fruit. This type of injury is referred to as phytotoxicity. Improper or careless mixing or application of chemicals and chemical drift from adjacent orchards or other crops can cause phytotoxicity. Environmental conditions, particularly high temperatures and wind, may increase the risk of phytotoxicity. Following label instructions on all agricultural chemical products is mandatory; further, proper maintenance and calibration of chemical application equipment also are required to prevent phytotoxicity (see chapter 17, "Citrus IPM").

Herbicides

Currently, the most common system of orchard floor management for citrus is chemical nontillage, where all weeds and plants other than citrus are controlled with herbicides. This reduces or eliminates the chances of root pruning (see below), but injury from herbicides may occur. Injury may be caused by the type of herbicide applied as well as by its concentration in the spray mix, the soil type, rooting depth, rootstock and scion, and environmental conditions. Most commonly, herbicides cause foliar symptoms, but in more extreme cases defoliation and dieback may occur. In rare cases, an entire tree may die; this is most common with small, young trees. Some symptoms of herbicide damage resemble those caused by nutritional or environmental problems, and this may make diagnosis of odd-appearing foliage difficult. Most cases of herbicide injury result from drift rather than from direct contact of the spray mixture, and when the drift is from adjacent crops, it may be even more difficult to diagnose herbicide injury.

Bromacil causes injury by root uptake or foliar contact. Symptoms appear as veinal chlorosis, or in more severe cases, as a general bronzing of the leaves.

Diuron causes injury by root uptake or foliar contact. In the former case, a yellowish white vein clearing appears and can become more pronounced as the amount of diuron taken up into the plant increases. Direct contact of foliage by diuron causes a general chlorosis of leaves that increases with the addition of surfactants. This type of damage also occurs with tank mixes of bromacil and diuron.

Glyphosate injury develops more slowly than do most other forms of herbicide damage due to the nature of its translocation and mode of action. Glyphosate causes narrow, distorted leaves on newly emerging growth or, in more severe cases, multiple bud growth and rosetting. Flower abscission also occurs. Higher concentrations may cause defoliation and dieback. Glyphosate is a systemic herbicide, but it is not active in soil so roots do not take it up. However, it can be transmitted by root grafting, causing symptoms similar to direct contact. Fruit on which glyphosate is sprayed can develop brown or black spots and may abscise.

Norflurazon has a relatively high safety threshold for citrus and rarely causes damage by root uptake. When damage does occur, it appears as a mild chlorosis of the midrib. Contact with norflurazon by tender young foliar growth causes pale yellow spots and distorted leaves.

Oryzalin is a soil-active herbicide that does not cause foliar damage. However, high rates applied as a drench to young trees have been reported to reduce growth.

Paraquat is not soil active, and injury results only with foliar contact. Leaves, fruit, and other green tissues develop chlorotic or necrotic spots or patterns when contacted with paraquat. High rates can result in defoliation or dieback. Since paraquat is not translocated, damage is localized, usually near the base of the tree skirt. Fruit sometimes abscises when it comes in contact with paraquat.

Simazine causes interveinal chlorosis when taken up in high concentrations, and the damage increases with higher concentrations. Problems caused by foliar contact have not been reported.

Trifluralin belongs to the same chemical family as oryzalin and has similar effects.

2,4-D is used as a growth regulator in citrus as well as an herbicide for other crops. It can cause leaf curling and distortion. Contact with concentrated sprays of the amine formulation can result in trunk girdling and root proliferation from the girdled area.

Insecticides and Fungicides

Although pesticides other than herbicides are not formulated to kill or otherwise affect plant growth, they can be phytotoxic. Many agricultural chemicals, including foliar fertilizers, can cause burn or necrosis of leaves or fruit surfaces under certain conditions. Phytotoxic effects are increased by high concentrations, overspraying, extremely cold or hot weather, physiological stress to the plant, and combinations of particular chemicals. In extreme cases, defoliation can occur. The following pesticides are well known for their phytotoxic effects.

Spray oils can enter stomata and lenticels of trees, causing blockage. This can lead to reduced physiological functioning of trees. Foliage may become chlorotic or necrotic, and some oil formulations may blacken bark. In some cases, bark injury may be apparent only when the bark is peeled back. The tree may suffer defoliation, dieback, and fruit drop. Oil sprays have been shown to make navel oranges more susceptible to water spot (see chapter 26, "Postharvest Handling"). These effects were more common with older, heavier types of oils. Narrow-range oils, more commonly used today, cause little damage to citrus if properly used. Tolerance by species or variety to oil sprays is, from most tolerant to least, lemon, grapefruit, Valencia orange, navel orange, tangerine, and lime.

Copper compounds used as fungicides can burn or pit fruit surfaces or intensify other blemishes on fruit. Leaves show necrotic spots on their lower surface that with time may develop through the leaf to the upper surface. Defoliation and dieback may occur. Air pollution containing sulfur and nitrous oxides causes solubilization of copper compounds by acidification, leading to increased sensitivity of citrus to copper fungicides. The possibility of copper burn is increased by warm weather and tank mixing with spray oils or acidic pesticides. Decreasing the concentration of the fungicide or using a neutralizing buffer or lime in the tank mix can reduce the possibility of copper injury.

Sulfur compounds may sometimes cause serious damage to citrus. Sprays of lime sulfur or wettable sulfur can burn leaves and distort subsequent growth. There are fewer effects on smaller, more tender leaves; fruit surfaces may also be scarred.

Fumigants have in the past been applied to nursery soils and in replant situations, including methyl bromide, chloropicrin, metam sodium, methyl isothiocyanate, and dichloropropene. These compounds, particularly methyl bromide, are currently being studied intensely due to their adverse effects on environmental quality. Consequently, their use is expected to decrease or cease in the future. It is possible that damage to citrus from these compounds may be observed in the short term. Fumigants kill the beneficial vesicular-arbuscular mycorrhizal fungi that are associated with citrus roots and contribute to normal growth of citrus. This results in stunting and poor growth of seedling citrus. Roots of the seedlings appear normal but do not have the normal complement of beneficial fungi. Leaves appear small and chlorotic and abscise prematurely. The entire seedling remains alive but in a stunted and unthrifty condition. Proper soil management (improved tilth and drainage) can help avoid this problem.

Cultural Injuries: Equipment

Proper training and supervision of equipment operators is the best way to avoid equipment-related problems. Equipment should be well maintained, appropriately used, and properly operated. Use of herbicides for vegetation control may reduce the number of trips through an orchard and thereby decrease the possibility of damage. However, herbicides themselves can injure citrus and other crops if not used properly. In addition, there are differences in soil properties associated with till versus no-till situations, and the use of chemicals may determine a crop's marketability as organic or conventional. All of these factors should be considered by a grower when selecting an orchard floor management system.

Root pruning by disks or other tillage equipment, the most common type of equipment-related injury in citrus, can reduce root volume or function. This, in turn, can lead to some of the problems related to water stress described above and can provide potential points of entry for secondary pathogens.

Tractor blight is damage inflicted on trees by tractors or other agricultural equipment. The most common damage is breakage of limbs and leaf removal on the periphery of the canopy because of the rounded growth form of citrus trees. This type of injury is of minor importance and is rare in citrus due to typical farming practices in California. It is more common in crops such as grapes, in which the manner of working the crop may result in serious injury to the trunk.

Genetic Disorders

Certain disorders of citrus are not known to be caused by pathogenic agents or abiotic conditions. Some of these disorders were originally thought to be genetic but were later shown to be caused by pathogens; in the future, other genetic disorders may be similarly reclassified. Most of the apparently inherited disorders of citrus involve bud union incompatibilities, as discussed in chapter 3, "Physiology and Phenology." The other inherited disorders are mostly minor, and even the most important of these, wood pocket, is not widespread in California.

Wood pocket occurs in the semi-dense type of Lisbon lemon and in large-fruited acid limes (Persian, Tahitian, and Bearss). Wood pocket in lemon has been reported only from California and has not been seen for many years. More commonly, wood pocket is a disease of the large-fruited acid limes, although these are not commonly planted in California.

The first outward symptoms of wood pocket usually are small longitudinal defects or breaks in bark. A discoloration in the wood is found beneath these lesions and develops before the bark symptoms. In some cases, the bark breaks may be preceded by the deterioration of certain branches or by mature leaf symptoms. As the disease progresses, areas of dead, fissured wood form on one side of affected branches or trunks. These range in length from 1 inch to several feet. The wood beneath these fissures is discolored and filled with gum, appearing irregularly dotted in longitudinal view. Affected branches lose their leaves and gradually die back, or branches die and the leaves suddenly wilt. In extreme cases, most likely with a small tree, tree death is possible. Leaf symptoms occur in some leaves of diseased trees. These symptoms are a type of variegation, ranging from sharply outlined to barely visible. The colors range from green to yellow or white. These leaf symptoms usually occur before the large branch lesions become apparent. Fruit, especially limes, also show a sectored striping. The first symptoms on diseased trees may develop at any age up to 20 years.

In the 1950s, wood pocket in the semi-dense Lisbon lemon was traced back to one original tree. Avoidance of propagation from this source essentially eliminated wood pocket from lemons in California. Currently, wood pocket is found in California in the few large-fruited acid limes. All bud sources of the large-fruited acid limes commercially available in California have had wood pocket. In the 1950s, an extensive search for wood pocket–free trees was made in Florida. After examining over 100,000 trees, a wood pocket–free source was selected. This bud source has recently been introduced into California by the CCPP as VI 708 and has recently been released from quarantine. It is currently being evaluated and observed under California conditions.

Minor inherited disorders: A few apparently inherited disorders are even more uncommon than wood pocket: crinkle scurf of Valencia oranges, a seed-perpetrated disorder of Trifoliate Orange resembling exocortis, and bark rot of sour orange. It is unlikely that a California grower will ever encounter these disorders.

References

Darley, E. F., C. W. Nichols, and J. T. Middleton. 1966. Identification of air pollution damage to agricultural crops. California Department of Agriculture Monthly Bulletin 55(2): 11–19.

Davidson, N. A., J. E. Dibble, M. L. Flint, P. J. Marer, and A. Guye. 1991. Managing insects and mites with spray oil. Berkeley: University of California Division of Agriculture and Natural Resources Publication 3347.

Dreistadt, S. H. 2012. Integrated pest management for Citrus. 3rd ed. Oakland: University of California Agriculture and Natural Resources Publication 3303.

Ketchie, D. O., and J. R. Furr. 1968. Sunburn and heat injury of citrus. California Citrograph 53(7): 252, 270–271.

Klotz, L. J. 1973. Color handbook of citrus diseases. Rev. ed. Berkeley: University of California Division of Agricultural Sciences.

———. 1978. Fungal, bacterial, and nonparasitic diseases and injuries originating in the seedbed, nursery, and orchard. In W. Reuther, E. C. Calavan, and G. E. Carman, eds., The citrus industry. Vol. 4. Berkeley: University of California Division of Agricultural Sciences. 1–66.

Roistacher, C. N. 2000. Wood pocket: A genetic disorder. Citrograph 85(3): 4–6.

Timmer, L. W., S. M. Garnsey, and J. H. Graham, eds. 2000. Compendium of citrus diseases. 2nd ed. St. Paul, MN: APS Press.

Part 6 Citrus Postharvest Handling

26 Postharvest Handling

Mary Lu Arpaia, James E. Adaskaveg, Joseph Smilanick, and Robert Elliott

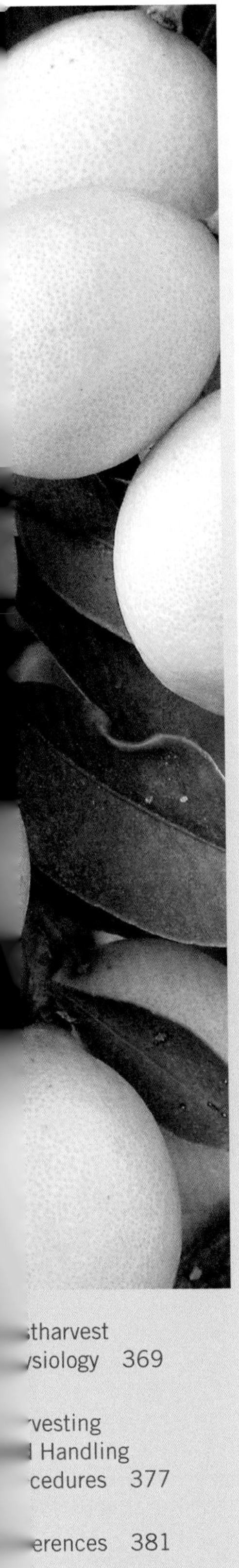

Specific postharvest handling practices have been developed for citrus and other subtropical fruit because these plant species are taxonomically diverse, as reflected in their morphology and compositional characteristics as well as in their responses to handling and environments during harvest, storage, shipping, and display in the marketplace. This chapter discusses these practices as they have evolved in California citrus production.

Subtropical fruit can be grouped into three subgroups according to their relative perishability, as listed in table 26.1. Citrus is considered one of the least perishable commodities from a physiological perspective, but it is very sensitive to decay losses when handled incorrectly. The marketability of subtropical fruit depends on a number of factors, including their relative sensitivity to temperature extremes, physiological disorders, and fungal decay. One or more of the following may influence the commercial postharvest life of citrus fruit:

- respiration rate of fruit and chilling injury as a result of cold storage
- fungal decay
- physiological disorders in general
- water loss (weight loss) and associated shrinkage
- softening

Table 26.1. Relative perishability ranking of selected subtropical horticultural commodities

Highly perishable	Moderately perishable	Less perishable
fresh fig loquat lychee	avocado cherimoya olive persimmon	citrus carob (dry) jujube dried fig pomegranate

Table 26.2 includes quality and maturity factors for citrus in the U.S. Standards for Grades and the California Food and Agricultural Code. Ceponis et al. (1986) discussed the relative importance of physiological disorders and decay in citrus shipments to New York. Rind breakdown was the major physiological disorder reported for oranges, grapefruit, and lemons. Blue and green molds caused by *Penicillium* species were the most important postharvest decays, although sour rot caused by *Geotrichum citri-aurantii* was also a major factor of lemon crop loss. These problems remain predominant to the present day.

Postharvest Physiology

The two essential characteristics of citrus fruit relative to their postharvest physiology are that the fruit are nonclimacteric (static or decreasing respiration rates after harvest) and chilling sensitive. All postharvest procedures should take into account these two characteristics. Citrus fruit have a relatively low rate of respiration, as shown for oranges in table 26.3. Respiration rates for other citrus crops are similar. The fruit also have a low rate of ethylene production that generally is less than 0.1 $\mu L \cdot kg^{-1} \cdot h^{-1}$ at 68°F (Ritenour 2004).

Since citrus fruit are nonclimacteric, they do not undergo the dramatic changes following harvest that are associated with climacteric fruit (e.g., avocado) ripening after harvest. Color changes in citrus occur both on and off the tree. Cold nights followed by warm days are necessary for the green color (chlorophyll) loss and the yellow or orange color (carotenoids) development in citrus fruit. This is the reason citrus fruit remain green after attaining full maturity and still have good eating quality when grown in tropical areas. In contrast, fruit grown

Table 26.2. Quality factors for citrus

	Standard (date issued or revised)	Quality factor
grapefruit	US (1950)	CA and AZ; maturity, color, firmness, shape, skin thickness, smoothness, freedom from defects and decay, size.
	US (1980)	FL; maturity, color (color charts), firmness, smoothness, shape, freedom from discoloration, defects and decay.
	US (1969)	TX and other states; maturity, color, firmness, size, shape, smoothness, freedom from discoloration, defects, and decay.
	CA (1983)	Maturity, as indicated by minimum solids to acid ratio of 5.5 or 6 (desert areas) and > two-thirds of fruit surface showing yellow color (Munsell color no. 0.9GY, 6.40/5.7); freedom from decay, freezing damage, scars, pitting, rind staining, and insect damage.
lemons	US (1964)	Maturity (28 or 30% minimum juice content by volume, depending on grade), firmness, shape, color, size, smoothness, and freedom from discoloration, defects, and decay.
	CA (1983)	Maturity (30% or more juice by volume), freedom from decay, freezing damage, drying, mechanical damage, rind stains, red blotch, shriveling, and other defects.
	CA (1993)	Elimination of maturity standard (30% or more juice by volume).
limes	US (1958)	Color, shape, firmness, smoothness, freedom from stylar-end breakdown.
	CA (1983)	Maturity, freedom from defects and decay.
oranges	US (1957)	CA and AZ; maturity, color, firmness, smoothness, size, freedom from defects and decay.
	US (1980)	FL; maturity, color (color charts), firmness, shape, freedom from discoloration, defects, and decay (used also for tangelos).
	US (1969)	TX and other states; maturity, color, firmness, shape, size, freedom from discoloration, defects, and decay.
	CA (1975)	Maturity (soluble solids to acid ratio of 8 or higher and orange color on 25% of fruit (Munsell color no. 7.5 Y6/6) or soluble solids to acid ratio of 10 or higher and orange color on 25% of the fruit (Munsell color no. 2.5GY 5.6); freedom from defects and decay.
	CA (2003)	Elimination of maturity standard or soluble solids to acid ratio of 10 or higher and orange color on 25% of the fruit (Munsell color no. 2.5GY 5.6).
	CA (2012)	Establishment of the California Standard for navel oranges. California Standard Scale score equal to or in excess of 90, and 90% or more of the navel oranges in any lot, by count, before picking have attained orange color on at least one-fourth of the fruit surface.
tangerines and mandarins	US (1948)	States other than FL; maturity, firmness, color, size, freedom from defects and decay.
	US (1980)	FL; maturity, color (color charts), firmness, size, shape, freedom from defects and decay.
	CA (1983)	Maturity (yellow, orange, or red color on 75% of fruit surface and soluble solids to acid ratio of 6.5 or higher); freedom from defects and decay.

Source: Adapted from Kader 1992.

Table 26.3. Respiration rate of oranges

Temperature		Respiration rate (mg CO_2 $kg^{-1}h^{-1}$)
°C	°F	
5	41	5–8
10	50	6–10
15	59	11–22
20	68	20–31

Source: Adapted from Ritenour 2004.

under California climatic conditions may be fully colored (yellow or orange) while not having attained legal minimum maturity as determined by the soluble sugar to acids ratio or the California Standard. Since chlorophyll levels in the peel are strongly influenced by environmental conditions, it is also possible for citrus to regreen after they have reached full color. This is often observed with Valencia oranges left on the tree until late summer due to the conversion of the chromoplasts back into chloroplasts.

In order to remove the green color from the citrus peel, fruit may be treated with ethylene after harvest because this hastens chlorophyll degradation in the flavedo (the pigmented portion of the peel), but it has little effect on the carotenoid pigment concentration of the peel. This process is known as degreening. Degreening does not generally influence internal fruit composition. The current degreening recommendations for California citrus are 1 to 5 ppm ethylene at 68° to 70°F and a relative humidity of 90 to 95%. Sufficient ventilation in the degreening room to maintain the carbon dioxide concentration below 1% is recommended. The duration of treatment depends on fruit condition, but it ranges from approximately 24 to 72 hours. Decay losses during degreening can be high in some years. To manage these losses, fruit have been historically drenched with imazalil or thiabendazole, but they now may also be drenched with the newly registered postharvest fungicides fludioxonil, azoxystrobin, or pyrimethanil either in bins on field trucks (truck drenches) or after unloading (bin drenches). Degreening does not begin until the treated fruit have dried. Previously waxed fruit should not be degreened, since the coating interferes with the degreening process.

The recommended storage temperature for citrus varies with citrus type and growing region. The current range in storage temperatures recommended for California citrus is given in table 26.4. There has been a recent trend in California, especially with export shipments, to ship oranges at lower temperatures of approximately 37° to 41°F.

Table 26.4. Recommended storage temperatures for California citrus

		Temperature (°F)
grapefruit		58-61
lemon	green	55.5-61
	yellow	44.5-50
lime		48-50
mandarin	(including Minneola tangelo)	39-46.5
	(Satsuma)	34-35.5
orange		37.5-48

Sources: Adapted from Eckert and Eaks 1989; Gross et al. 2004.

Physiological Disorders

Many physiological disorders are attributed to improper postharvest handling of citrus. One is chilling injury, which is associated with low-temperature (nonfreezing) storage. This disorder can occur in the temperature range of 32° to 55°F. The severity of chilling injury is related to temperature and the duration of exposure. Chilling injury symptoms in citrus vary with pre- and postharvest conditions, species, and cultivar. Grapefruit, lemon, and pummelo are considered more susceptible to chilling injury than are orange, kumquat, and mandarin. Table 26.5 summarizes chilling injury symptoms of selected citrus fruit. Other physiological disorders related to fruit maturity, harvesting procedures, and postharvest handling practices include the following.

Table 26.5. Chilling injury symptoms for selected citrus fruit

	Chilling injury symptoms
orange, mandarin, and mandarin hybrids	pitting, brown staining, increased decay
grapefruit	pitting, scalding, watery breakdown, increased decay
lemon	pitting, membrane staining, red blotching, increased decay

Postharvest pitting. This disorder has been reported to occur internationally but was first described in depth in Florida by Petracek and Dou (1998). It was first identified as a peel disorder in the spring of 1994 in white grapefruit. It is microscopically characterized by collapsed oil glands that are scattered throughout the peel and that darken over time. Petracek and Dou demonstrated that this problem is associated with high storage temperature of fruit treated with a fruit coating; their research in Florida indicates that fruit should be cooled as fast as possible after the fruit coating is applied and that delays in cooling by as little as 24 hours can increase the incidence of this disorder.

Rind staining of navel oranges. Rind staining occurs when the peel of the navel orange is physiologically overmature. Symptoms include brownish discoloration of the rind surface that develops 12 to 24 hours after fruit are processed through the packinghouse. Fruit susceptibility varies from season to season, between orchards, within orchards, and even between sectors of a single tree. Rootstock and cultural practices including nitrogen fertilization (high nitrogen promotes the disorder) are known to influence this disorder. Preharvest use of gibberellic acid at color break reduces susceptibility.

Granulation. In granulation, the amount of extractable juice is greatly decreased. This condition is not due to desiccation but rather to gel formation within individual juice sacs, or vesicles. Valencia oranges, tangerines, and tangerine hybrids typically granulate only at the stem end. In this form, granulation is thought to be associated with advanced fruit maturity (i.e., overripe fruit). In navel oranges, granulation tends to occur in the center of the fruit and at the navel end. In the United States, granulation is considered to be a preharvest disorder, whereas in other citrus-growing regions, it has been shown to also develop during storage.

Oleocellosis. This disorder is caused by the phytotoxic action of oil released from the oil glands on the surface cells of the peel. Oleocellosis mostly occurs during cold, wet periods during the harvesting season. As with rind staining of navel oranges, symptoms may not develop until after the fruit have been packed. Management of this disorder can be attained by refraining from harvesting cold, turgid fruit. A similar disorder, oil spotting of lemon, has also been shown to be correlated with fruit turgidity during harvest and can be a serious problem for desert lemons harvested following irrigation.

Stem-end breakdown. This disorder affects the rind tissue surrounding the button of the fruit. It is found primarily on fruit grown in humid, high-rainfall areas. A similar-looking breakdown of citrus, which can be confused with stem-end breakdown, is associated with prolonged low-temperature storage of fruit from arid regions such as California. Breakdown is associated primarily with small oranges. Factors that may influence development of the disorder include low relative humidity after harvest, fertilization and irrigation practices (no clear relationship has been established), and improper management of degreening rooms.

Stylar-end breakdown of limes. This disorder occurs near the nipple at the stylar end of seedless limes. It is initiated by rough handling and involves the leakage of juice vesicle contents into adjacent cells.

Peteca. This is a physiological disorder of lemons that usually expresses itself as dark, sunken pits in the peel and collapse of the albedo. The cause is uncertain. Peteca develops in storage, but susceptibility of fruit appears to be influenced by undetermined growing and climatic conditions. Control is currently limited to holding fruit prior to marketing to allow the disorder to express itself, then eliminating affected fruit by additional sorting.

Table 26.6. Major pathogens of citrus and site of infection

Disease	Pathogen	Infection site
	Preharvest infection	
stem-end rot, black rot	*Alternaria* spp.	flower, young fruit, navel
brown rot	*Phytophthora* spp.	fruit surface
Botrytis rot	*Botrytis cinerea*	flower, young fruit
anthracnose	*Colletotrichum gloeosporioides*	fruit surface (quiescent and active infections)
clear rot	*Penicillium* spp.	fruit rind injuries
	Postharvest infection	
green mold	*P. digitatum*	fruit rind injuries
blue mold	*P. italicum*	fruit rind injuries
Trichoderma decay	*Trichoderma spp.*	fruit rind injuries
sour rot	*Geotrichum citri-aurantii*	fruit rind injuries

Sources: Based on information from Eckert and Eaks 1989 and Adaskaveg et al. 2002.

Postharvest Fruit Decay

The major challenges facing the California citrus industry are maintaining fruit quality and minimizing fruit decay. Postharvest decay is one of the main limiting factors of successful postharvest handling of citrus fruit. All harvesting and subsequent handling steps must take the potential for decay into account. The major postharvest pathogens of citrus in California include fungal species in the genera *Alternaria, Botrytis, Colletotrichum, Geotrichum (Galactomyces), Penicillium, Phytophthora,* and *Trichodema* (table 26.6). Additional minor decays include stem-end rots caused by *Diplodia* and *Phomopsis* species and Septoria spot caused by *Septoria citri.* Infections by the latter pathogens generally occur during wet environments on blossoms, on immature fruit (e.g., stem-end rots), or on maturing fruit exposed to cold temperatures (e.g., Septoria spot). Decay caused by *Penicillium ulaiense* can also be of minor importance in postharvest storage.

Three types of infection can lead to postharvest fruit decays of citrus: preharvest infections that develop during the season as an ongoing disease problem; preharvest quiescent infections that are initiated during the growing season and become active with visible symptoms only after harvest; and infections that occur due to wounding of the fruit during harvest and postharvest handling. The relative importance of these types depends largely on the growing region and microclimatic conditions during fruit development. Quiescent infections (primarily stem-end rots and infections caused by *Colletotrichum* spp.) are more important in humid, tropical growing regions such as Florida. The most serious postharvest diseases arising from preharvest infection after rainfall events in California are brown rot caused by *Phytophthora* species and Septoria spot. In the brown rot disease cycle, the spores of fungus-like organisms (Oomycota) splash onto the fruit, especially those on the lower skirt of the tree. For management of both diseases, foliar fixed copper (or zinc-copper Bordeaux) or synthetic fungicide treatments are commonly sprayed in the fall and winter months to minimize infection (see chapter 24, "Invasive Pests: Exotic Plant Pathogens").

The major postharvest pathogens attacking California citrus fruit are the wound pathogens, especially *Penicillium digitatum.* This fungus needs wounds for infection and does not spread from diseased fruit to nonwounded, healthy fruit. Sporulation of decayed fruit, however, may cause soilage of adjacent healthy fruit, and air contamination by spores can be a significant problem that leads to additional fruit decay from wound infections. Decay due to sour rot can also be very important, particularly in stored lemons in some growing seasons. This disease, once initiated, can spread rapidly among other fruit by contact infections or insects such as fruit flies.

Minimizing decay

Citrus fruit are generally quite resistant to decay. The five major strategies for minimizing postharvest decay of citrus are maintaining fruit resistance to infection by delaying senescence and maintaining the physical integrity of the fruit; reducing fruit injuries during harvest and handling; reducing the pathogen population in the orchard and packinghouse environment; reducing inoculum on the fruit surface using sanitizers; and changing the wound site environment with the use of salts, fungicides, or biological control treatments of fruit. Methods for maintaining host (i.e., fruit) resistance to infection include the following:

- Harvesting mature, noninjured fruit that are not senescent (i.e., avoiding overmature fruit).
- Minimizing mechanical injuries during harvesting and handling.
- Pregrading fruit prior to storage to remove any decayed or weak fruit. In the case of fruit that are degreened in field bins where pregrading is not possible, consider drenching the bins with thiabendazole, imazalil, or another fungicide before ethylene degreening. Be certain the fruit are dry before degreening starts.
- Maintaining proper temperature and relative humidity throughout postharvest handling.
- Delaying rind senescence by a pre- or postharvest application of gibberellic acid (when appropriate) and avoiding unnecessary exposure to ethylene.
- Using a postharvest 2,4-D treatment on lemons to maintain the integrity of the button tissue, which reduces the development of stem-end rots (e.g., Alternaria) in storage, or applying gibberellic acid to slow peel senescence, which reduces sour rot development.

Strategies that reduce pathogen populations in the orchard and packinghouse environment include

- effective preharvest disease control programs, such as removal of fallen fruit (rarely practiced due to labor costs), pruning of dead wood in the canopy, and other orchard sanitation practices
- careful handling of the fruit during harvest, including ensuring that stems are properly clipped, picking bags are free of dirt and debris, and harvest bins are sanitized and have no rough edges that can result in fruit wounding
- regular disinfestation of field containers, packinghouse equipment, and storage facilities with sanitizers such as quaternary ammonium or oxidation treatments such as sodium hypochlorite or ozone (see below)
- use of filters and dedicated areas for handling decayed fruit, either from the field or after storage, to prevent the air in the clean portion of a packing operation to be contaminated with spores of *Penicillium* species from decayed fruit

Treatments

Postharvest treatments for decay control of fruit include sanitizers, salts, fungicides, and biological treatments. Sanitizing treatments can reduce inoculum on fruit. Salts, fungicides, and biological treatments can inhibit growth and reproduction of plant-pathogenic organisms by prevention (e.g., protection), suppression, or therapy of infections. Prevention and suppression are the inhibition of spore germination or growth and development, respectively, whereas therapy is the eradication of established infections. Some treatments have more than one mode of activity and thus fit into more than one category. For greatest effectiveness and safety, these treatments must be used properly.

Sanitizers and salts

Some of the first treatments used to control postharvest decays of citrus fruit were alkaline solutions of borax, sodium carbonate, and sodium bicarbonate (table 26.7). The effectiveness of these treatments is due to accumulation of alkali in potential infection sites on the surface of the citrus fruit. In laboratory studies by Eckert and Sommer (1967), recommended concentrations of borax were lethal to conidia of *Penicillium* species after a 5-minute exposure at 110°F but were only weakly fungicidal at 100°F. Treatments of 6 to 8% borax that are either heated to 110°F or not rinsed after treatment are effective for controlling Penicillium decays and stem-end rots caused by *Diplodia* and *Phomopsis* species. Because visible residues of borax on fruit are not acceptable, commercial treatments are always rinsed with water. Borax is relatively insoluble in water, and this leads to application problems. Thus, commercial treatments usually use 4% borax and 2% boric acid at 110°F. Heated solutions of sodium carbonate or sodium

Table 26.7. Selected commercially developed chemicals registered as postharvest treatments of citrus to prevent crop decays caused by filamentous fungi or yeasts

Material*	Chemical (trade name)/class†	Year introduced	Decay/organisms	Methods of application	Residue tolerance (ppm)‡
RF	azoxystrobin (Diploma) Strobilurin (QoI)	1996	*Penicillium* spp.	drench (aqueous) or spray (aqueous or in fruit coating)	5.0
RF	fludioxonil (Graduate) phenylpyrrole	2005	*Penicillium* spp.	drench (aqueous) or spray (aqueous or in fruit coating)	5.0
RF	imazalil (e.g., Deccocil, Freshgard, Fungaflor) imidazole	1974	*Penicillium* spp.	drench (aqueous) or spray (aqueous or in fruit coating)	10.0
RF	phenylphenols: o-phenylphenol (OPP), sodium o-phenylphenate (tetrahydrate) (SOPP) phenol derivative	1936	*Geotrichum, Penicillium* spp. stem-end rots *Trichoderma, Phytophthora* spp.	wash, sprays, dips, and foams	10.0
RF	propiconazole (Mentor) triazole	1979	*Penicillium* spp.	drench (aqueous) or spray (aqueous or in fruit coating)	5.0
RF	pyrimethanil (Penbotec) anilinopyrimidine	1990	*Penicillium* spp.	drench (aqueous) or spray (aqueous or in fruit coating)	5.0
RF	2-(4-thiazolyl)benzimidazole (thiabendazole, or TBZ) methyl benzimidazole carbamate (MBC)	1968	*Penicillium* spp. stem-end rot	drench (aqueous) or spray (aqueous or in fruit coating)	10 (35 in pulp)
RS	sodium borate (Borax) (sodium tetraborate) inorganic salt	1938	*Penicillium* spp.	dip or drench; rinse with fresh water	exempt
RS	sodium carbonate (soda ash); sodium bicarbonate (baking soda) inorganic salt	1930s	*Penicillium* spp.	dip or drench; rinse with fresh water	exempt
SA	oxidizers (e.g. sodium hypochlorite, chlorine gas, ozone gas) liquids and gases	over many years	bacteria (contamination) and fungi	dips and rinses; rinse with fresh water	GRAS; worker exposure limits

Notes:

*Residual fungicides (RF) are directly toxic to the pathogen; sanitizing agents (SA) are general biocides; residual salts (RS) generally change the pH of the wound, making it a less favorable environment for fungal growth.

†Some fungicides may be premixed and sold separately.

‡Exempt = EPA classification: exemption from tolerance; GRAS = FDA classification: generally regarded as safe.

bicarbonate are slightly more toxic; however, these treatments are less toxic to spores of *Penicillium* species than is borax. Still, these treatments have been shown to be effective under commercial conditions (Smilanick et al. 1997, 1999).

Sanitation treatments reduce populations of microorganisms on equipment, in storage facilities, on the fruit, and in the wash water used to clean the fruit. To clean fruit, sanitizers may be added to water dumps and spray or dip washes. Water alone removes nutrients from fruit surfaces that allow microorganisms to grow and also removes inoculum of postharvest pathogens. Without the use of sanitizers in the water, however, the potential for reinoculation of fruit is high. Thus, sanitation treatments inactivate spores brought into solution from fruit or soil and prevent the secondary spread of inoculum in water. Examples of sanitizing washes for fruit or equipment include halogenated compounds (e.g., hypochlorous acid from sodium hypochlorite or chlorine gas and chlorine dioxide), peroxide solution such as peroxyacetic acid, and ozonated water. Sanitation using hypochlorous acid is one of the most effective, inexpensive, nonresidual ways to reduce microbial contamination from wash water, noninjured fruit surfaces, and equipment (White 1992), and it is being used extensively in postharvest handling of citrus fruit. Sanitation treatments used exclusively on equipment and storage facilities include quaternary ammonium washes and fumigations with liquid or gases toxic to fungal pathogens, such as formaldehyde and ethylene oxide.

Sanitation solutions obtained from sodium hypochlorite (NaOCl) or chlorine gas (Cl_2) produce the microbial biocide hypochlorous acid (HOCl), or active chlorine. Hypochlorous acid rapidly and nonspecifically oxidizes carbonaceous materials in aqueous solutions, resulting in fungicidal and bactericidal activity. Solutions of hypochlorous acid, however, are relatively ineffective in reducing decay if the inoculum is inside the wounds of the fruit because the active compound is rapidly reduced in fruit injuries before it can inactivate the pathogen. Three factors control availability and activity of hypochlorous acid: pH, temperature, and the presence of contaminating organic and inorganic materials.

At higher pH, the amount of active chlorine is dramatically reduced. A low pH (e.g., pH 3 to 6), however, results in the volatilization of chlorine as chloramines. At a very low pH (e.g., pH 2), chlorine gas (Cl_2) will be formed that is highly toxic or lethal to humans. Ideally, pH levels between 6.5 and 7.5 should be maintained, because in this range equilibrium exists between hypochlorous acid and the hypochlorite ion in solution. Under packinghouse conditions, a pH range of 7 to 9 is commonly maintained and acceptable. Also, this pH range reduces the formation of volatile chloramines, which are eye irritants to packinghouse workers.

Temperature also affects hypochlorous acid concentration and contact times: shorter exposures are required at warmer temperatures, but greater volatilization of hypochlorous acid occurs. Furthermore, organic and inorganic materials suspended in water may interfere with the oxidation of microbial inoculum. With high levels of these materials, longer contact times or higher concentrations of hypochlorous acid are required for disinfestation. Nitrogenous compounds (e.g., amines, ammonia, and amino acids) in the wash water decrease the amount of active chlorine and may result in the formation of undesirable combined chlorine.

Microbes are not very sensitive to combined chlorine; however, they react instantly or within seconds to active chlorine in clean water. High temperatures (> 86°F) and organic matter in the chlorinated wash water result in the greatest loss of active chlorine. When chlorination is done using a nonrecirculated wash (e.g., spray bar) on prewashed fruit, the formation of combined chlorine is negligible.

Additional factors affecting the activity of chlorine solutions include the type of microorganism and contact time. In general, higher concentrations of active chlorine result in inactivation of microbial populations using shorter contact times. Although minute concentrations of 1 ppm HOCl (1 ppm of active chlorine) are sufficient to kill decay-causing organisms in clean water, concentrations of 25 ppm HOCl from generic NaOCl or 50 to 200 ppm from labeled NaOCl are commonly used for most commodities to offset changes in the concentration of active chlorine resulting from the amount of crop treated and organic matter accumulated in the wash water during treatment.

The concentration of hypochlorous acid must be monitored periodically. Colorimetric test kits are commonly used and are the most accurate method for routine evaluations. Oxidation-reduction potential (ORP, or REDOX), measured by electrical conductivity across a pair of electrodes, is an automatic method commonly used to determine chlorine concentration. Postharvest use of ORP, however, can result in inaccurate measurements due to other salts dissolved in the water from inorganic and organic soil particles washed off the commodity. Thus, ORP is often used as a guide to detect sudden changes in conductivity that may influence active chlorine concentration.

Cationic detergents such as quaternary ammonium compounds, isopropyl alcohol, live steam, or hot water are currently used for equipment disinfestation. Quaternary ammonium compounds are not effective in reducing decays of fruit, but they are widely used in food processing plants because they are microbial biocides with high water solubility and detergent properties. They also have low mammalian toxicity and are generally noncorrosive at recommended concentrations. The efficacy of live steam depends on obtaining proper temperatures; hot water treatments that are not under pressure may hold little lethal heat energy during short exposure times. Extended exposures to heat treatments may injure the commodity. Isopropyl alcohol is less commonly used.

Fungicides

Fungicide treatments that directly affect fungal pathogens are shown in tables 26.7 and 26.8. Sodium ortho-phenylphenate (SOPP) has residual activity in preventing fruit decay. After treatment, residues accumulate in potential infection sites on the fruit and prevent the development of decay from subsequent inoculation of pathogens. New wounds, however, are not protected. SOPP is quite soluble in water. When used at high concentrations, SOPP treatments are followed by a potable water wash to prevent phytotoxicity. In citrus, a 0.5% solution at pH 11.5 to 11.8 and 109.5°F is commonly used as a dip or spray treatment for management of sour rot caused by *Geotrichum citri-aurantii* and the green and blue molds caused by *Penicillium digitatum* and *P. italicum*, respectively.

During recent years, the older registered postharvest fungicides SOPP, thiabendazole (TBZ), and imazalil were either reregistered or critically reviewed to maintain their registration. With new fungicides being developed by agrochemical companies, fludioxonil (2005), pyrimethanil (2005), azoxystrobin (2006), and propiconazole (2013) have been registered (see table 26.7). In time, these will obtain wide acceptance when maximum residue limits (MRLs) are established in international markets. These new materials should be used in combinations or in alternating regimes with the older fungicides. Unfortunately, there is resistance in *P. digitatum* to the three major older fungicides imazalil, SOPP, and thiabendazole. Still, a potential for resistance development within pathogen populations exists with the new fungicides as well. Their principal value will be the management of decays caused by *Penicillium* isolates resistant to the older fungicides. Good field and packinghouse sanitation and judicious use of fungicides are critical for managing the resistance problem (see chapter 18, "Managing Pesticide Resistance in Insects, Mites, Weeds, and Fungi").

Table 26.8. Selected commercially developed biological control materials and plant growth regulators (PGRs) registered as postharvest treatments of citrus

Category	Organism (product)	Year introduced	Decay organisms or function	Methods of application	Residue tolerance (ppm)*
biocontrol	*Pseudomonas syringae* (Bio-Save)†	1995	*Penicillium digitatum, P. italicum, Geotrichum citri-aurantii*	dip or spray (commonly in fruit coating)	exempt
biocontrol	*Candida oleophila* (Aspire)†	1995	decay pathogens	any type of application	exempt
PGR	Gibberellic acid (Pro Gibb)	1955	delays senescence of rind in oranges and lemons and delays lemon color change in storage (delays onset of decay)	storage fruit coating	exempt
PGR	2,4-D (Citrus Fix)	1942	delays senescence of calyx (button) and reduces calyx abscission (delays onset of decay)	storage fruit coating	5

Notes: *Exempt = EPA classification: exemption from tolerance.

†Bio-Save and Aspire are no longer commercially available.

Biological treatments

Concern about the safety of chemical treatments has been the primary motivation for developing biological control methods using antagonistic organisms such as bacteria and yeasts (see table 26.8). Mechanisms described for biocontrol agents include competition, antibiosis, parasitism, and induction of host resistance (Larkin et al. 1998). These mechanisms may reduce the amount of pathogen inoculum, protect the infection site, limit disease development after pathogen infection, or induce resistance in the host. For postharvest use on citrus, two biological control treatments were registered in the mid-1990s (see table 26.8). Bio-Save is a preparation of the antagonistic bacterium *Pseudomonas syringae,* whereas Aspire is a preparation of the yeast *Candida oleophila.* The commercial efficacy of Bio-Save and Aspire has been inconsistent, and these materials are no longer in use. Like other biological controls, these treatments never completely prevented fruit decay. In general, biological controls provide only a partial level of control, they often have inconsistent results, and they provide no curative activity. New biocontrols should be regarded as complementary tools for the management of postharvest decays and used together with other strategies as part of an integrated pest management program.

Synthetically produced plant growth regulators that act as plant hormones are also being used for managing postharvest decays of citrus (see table 26.8). By changing the plant's physiology, plant growth regulators have an indirect effect on the fruit's susceptibility to postharvest decay caused by opportunistic (weak) pathogens. Thus, any treatment that delays plant senescence not only delays ripening but may also reduce susceptibility to pathogens that favor senescent tissues for infection. For example, postharvest treatment of lemons with gibberellic acid reduces ethylene production, delays ripening, and consequently delays the onset of sour rot caused by *G. citri-aurantii* (Coggins et al. 1965). For control of stem decays of citrus, 2,4-D treatments delay the senescence of lemon fruit buttons and thus delay the development of Alternaria stem-end rot. Diplodia and Phomopsis stem-end rots of oranges and Penicillium, Alternaria, and Colletotrichum decays of mandarins have been commercially controlled with postharvest fruit treatments of 2,4-D.

Harvesting and Handling Procedures

Citrus fruit, with the exception of lemons, are normally picked in one harvest. Lemons can be picked three or four times during the commercial harvest season, based on color and size. Most citrus fruit in California are clip-harvested. Some Florida citrus fruit are snap-picked (twist-and-pull method). Table 26.9 summarizes the major steps in postharvest handling of California navel and Valencia oranges.

Minimum maturity standards for California fruit are listed in table 26.2. Except for lemons and navel oranges, minimum maturity is based on a combination of the soluble solids to titratable acidity (sugar to acid) ratio and peel color. Minimum maturity for navel oranges is based on a combination of peel color and the sugar and acid content of the fruit. As of 2012, the California Standard provides a different way to measure the relationship of the soluble sugars and titratable acidity. The fruit sample must have a minimum score of 90 to pass (Arpaia and Obenland 2011; Obenland et al. 2009). The calculation for the California Standard (CS) is

$$CS = [\text{soluble solids} - (4 \times \text{titratable acidity})] \times 16.5.$$

Minimum maturity for lemons is based on juice content. Following harvest, lemons are usually sorted into four color classes: dark green, light green, silver (greenish yellow), and yellow. Lemons grown in coastal areas may be stored for prolonged periods prior to marketing. Flow diagrams of packinghouse operations for citrus are shown in figures 26.1 to 26.3.

Preharvest Factors That Affect Postharvest Quality

Postharvest quality depends not only on how fruit are handled after harvest but also on how citrus trees are grown. Losses from postharvest decay or physiological breakdown are ultimately more costly than preharvest losses because it is necessary to account for the added economic inputs of harvesting, transportation, packing, and distribution to market. There is also potential loss of markets when customers lose confidence in the ability of a marketer to deliver quality citrus and may take their business elsewhere. The following cultural practices have been shown to affect postharvest quality of citrus:

- **Nitrogen fertilization.** High levels of nitrogen, as determined through leaf analysis, has been correlated with increased susceptibility to rind staining of navel oranges, as well as softer peels.
- **Irrigation.** Irrigation within several days of harvest has been shown to increase susceptibility of lemons to sour rot and oleocellosis. High fruit turgidity at the time of harvest may also enhance peel disorders.
- **Selection of rootstock.** In addition to several fruit quality factors, such as smoothness of the

Table 26.9. The major steps in the postharvest handling of California navel, Valencia, and miscellaneous other oranges

Step	Comments
Harvest and initial fruit-handling steps	
assessment of minimum maturity	Maturity based on sugar to acid ratio (8:1 minimum maturity) and peel color for Valencia and other oranges and California Standard and peel color for navel oranges.
harvesting*	Fruit clipped and placed in 40–60 lb picking bag. Cotton gloves used to minimize damage during harvesting process. No fruit taken from the ground; some growers skirt-prune trees to keep fruit off the ground. Fruit transported to packinghouse in bins of about 1,000 lb on day of harvest. Sanitary facilities for workers provided in field.
degreening	Early-season navels and late-season Valencia oranges. 1–5 ppm ethylene at 68° to 70°F, 90–95% RH. Carbon dioxide maintained below 1%. Duration of treatment varies.
bin dump and initial packline handling	May have exhaust fan at dump to pull off any spores on fruit especially following degreening. Chlorine spray (100–200 ppm) common after bin dump. Trash and small fruit eliminator.
pregrading and large/small fruit elimination†	Graders wear gloves to minimize damage to fruit. Good lighting at all grading steps essential to maximize detection of defects and to differentiate between culls, processed products, and fresh market grade fruit. Initial separation of fruit into culls (destined for landfill or feed), processed products, and fresh market grade fruit. Culls often dropped into flume system and carried to accumulation site. Fruit removed for processed products taken to accumulation bins designated for this purpose.
Fruit cleaning and sanitation	
Option 1: Tank treatment, duration generally 1.5 to 2 minutes, 4 minutes maximum; solutions vary but are listed in order of importance	Treatment choices: Sodium carbonate (soda ash): 3% wt/vol at 105°F, pH 11.5. Sodium bicarbonate: 3% wt/vol with chlorine (200 ppm) at 68° to 82°F, pH 8.3. Sodium ortho-phenylphenate (SOPP): 0.5% wt/vol, 90° to 115°F, pH 11.7–12.0. Borax/boric acid: (4%/2% wt/vol) at 105°F, pH 10–11. Lime sulfur: 3% wt/vol at 105°F, pH 10.0. Tanks often heated to 140°F overnight; tank mixtures changed routinely, approximately every 2 weeks. About 30% of orange houses in CA use tank treatments.‡
Option 2. Fruit sanitation with ortho-phenylphenate (OPP)	If no tank or high-pressure washing, fruit may pass over series of brush beds; OPP applied as sanitizing agent, followed by rinsing with chlorinated water.
Option 3. High-pressure washer (HPW)	Introduced into CA in 1990, HPW units are now used in most orange houses. Originally introduced as a way to remove scale and sooty mold from fruit surface. 80–300 psi in chlorinated water (100–200 ppm). Trend is toward adding sodium bicarbonate in wash water. Recirculating water system with filtration system to remove particulate matter. Water replenished continuously; completely replaced every 24 hours. Washing followed by chlorinated water rinse (100–200 ppm).
water elimination	Sponge rolls to remove excess water.
Fruit grading	
manual grading	Manual grading to remove rots and other culls and further differentiate fruit into processed products and fresh market grade. Use of UV light to highlight surface defects.
electronic grading (may be used in addition to manual grading)	About 25% of orange houses have some sort of electronic grading equipment. Allows sorting of fruit by defect, color, weight, and freeze damage. Reduces manual handling and the potential for damage to the fruit.
fruit waxing and drying	For good wax application, fruit should have a minimum pulp temperature of 55°F. pH 8.0–9.0 waxes based on shellac, carnauba, wood rosin, or combination. Fungicides may be applied by drench or water spray prior to waxing or may be added to the wax. Water applications of imazalil may be heated. Dryer temperature from 90° to 140°F. Duration in drying tunnel averages 3 to 5 minutes.
postwax grading	Final grading for fresh market, processed products, or culls.
fruit sizing and stickering	Primarily accomplished with electronic sizers. Fruit sent to bulk accumulation bins by size.
fruit packing	38 lb carton; hand or electronic pattern pack. Workers handling fruit wear cotton gloves to minimize damage to fruit. Fruit (choice grade, small sizes) may be packed in bulk bins or bagged (poly or net bags).
Final processing	
palletization, cooling, and loading for transport	Packed fruit in separate facilities from fruit from field. Truck loading usually done in isolated area away from rest of packinghouse.
general packinghouse sanitation	A variety of materials are available for use to clean harvest bins, packline, and the general facility, such as quaternary ammonium products, isopropyl alcohol, and high-pressure steam. Monitoring for plant pathogen levels and development of fungicide-resistant strains. The common practice is to clean rollers and brush beds during breaks and at the end of the day. Isolation of different tasks and fruit grades to minimize contamination potential. Rodent control practiced. Safety training for workers on hygienic behavior and etiquette.

Notes:

*Care is taken in the field during harvest to minimize damage to fruit, since the consequences of injury are increased decay from wound pathogens, water loss, and the potential for peel breakdown in subsequent handling.

†Fruit that are culled include those with rots, punctures, splits, and similar defects. Fruit diverted to processed products would include defects such as scarring due to wind, insects, or limb rub. Fruit shape, color, and protruding navels may also cause fruit to be diverted to processed products. Fruit destined for the fresh market must be free of splits and punctures and must have a minimum amount of scarring due to insects, wind, or limb rub.

‡C. Orman and R. Elliott, Sunkist Growers, personal communication.

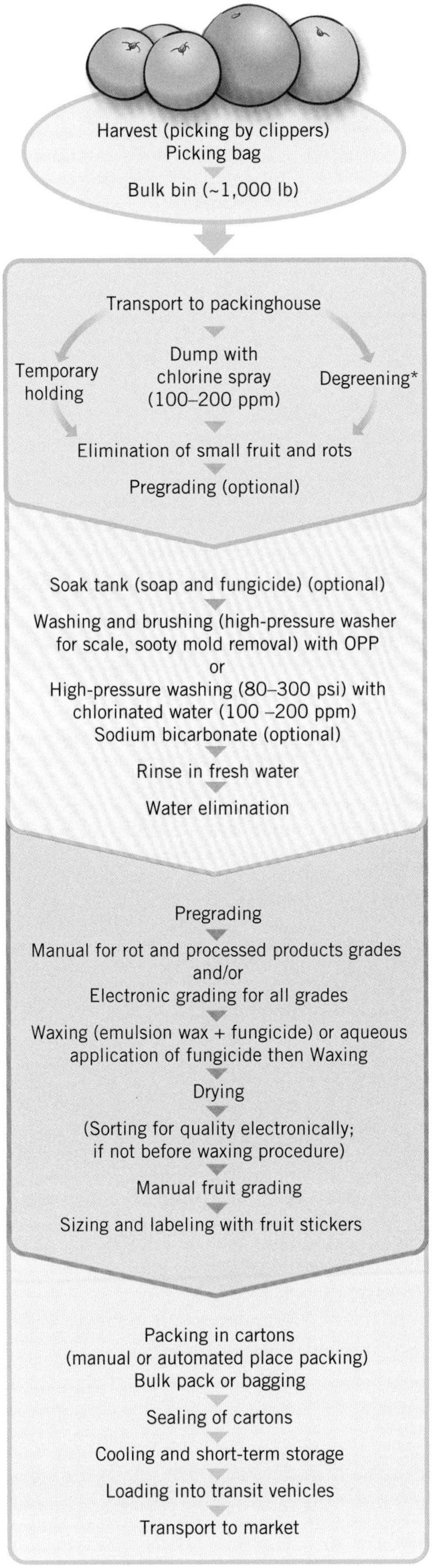

Figure 26.1 Postharvest handling systems for oranges and grapefruit. *Note:* *For oranges, the fruit may be washed and then color sorted before degreening.

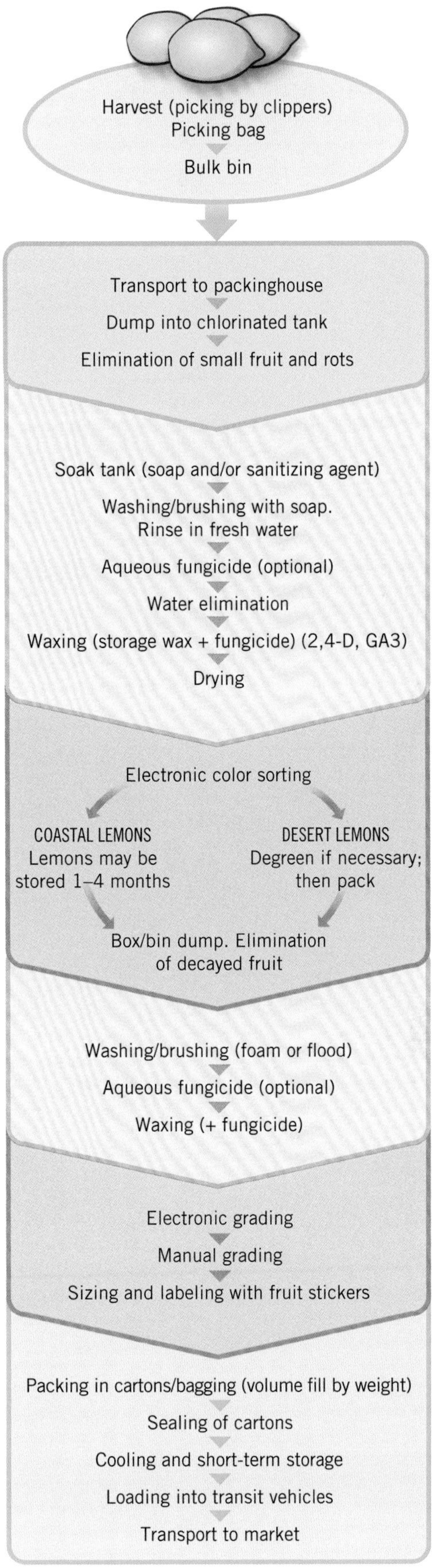

Figure 26.2 Postharvest handling systems for lemons.

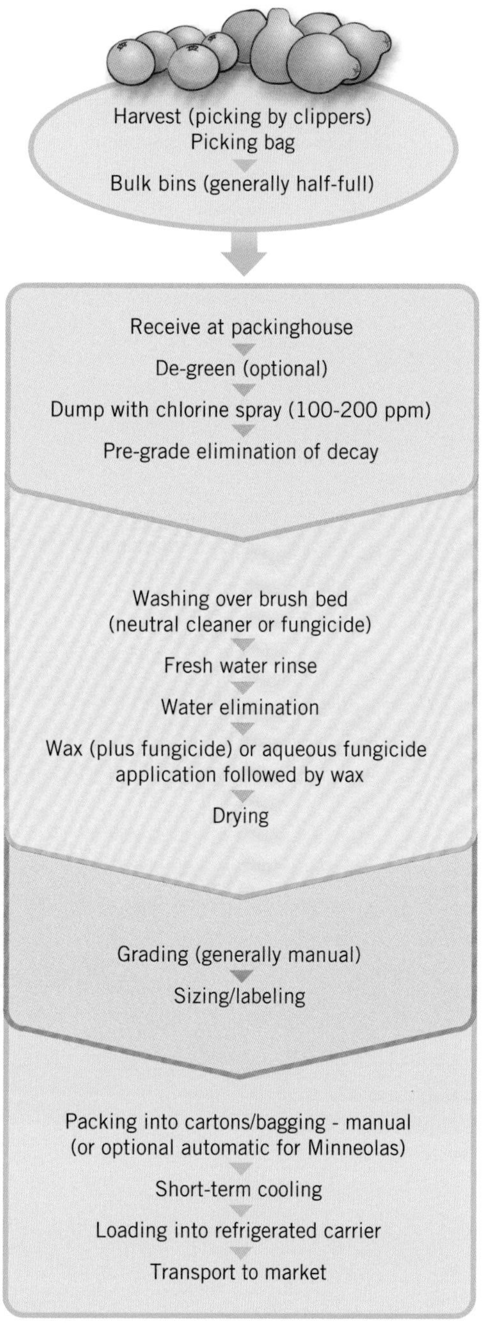

Figure 26.3 Postharvest handling systems for mandarins (including Minneola tangelos).

peel and the solids to acid ratio, rootstock can also affect susceptibility to rind staining of navels and granulation.

- **Plant growth regulators.** Preharvest application of gibberellic acid slows maturing of the rind, which delays senescence and reduces the susceptibility of navel oranges and lemons to decay and of navel oranges to rind disorders such as staining.

Harvest factors

Timing of harvest and how citrus fruit are picked can significantly affect postharvest quality of fruit. Table 26.10 provides a list of factors to consider when preparing for harvest. Supervision is essential to ensure that due care is taken during all harvesting and handling procedures. Picking injuries are caused by long stems, clipper cuts, plugged fruit (snap-picked fruit where the peel is torn), and fingernail scratches (if gloves are not used). Of these, long stems, which can result from snapping or improperly clipping the fruit, are perhaps the most critical because they can injure other fruit. A stem as short as ⅛ inch can create injury sufficient to cause decay. Other causes of peel damage include soiled picking bags, leaf material, or twigs that can cause surface abrasion or fruit punctures. Harvest bins should be inspected for general cleanliness before use. Growers should work with packinghouses to ensure that bins are thoroughly washed prior to drop-off. Dropped fruit, beyond fruit safety considerations, should never be picked off the ground, since the likelihood of fruit damage and inoculation with a soilborne pathogen such as *Phytophthora, Penicillium,* or *Geotrichum* species are high.

Growers must analyze their handling procedures and adopt practices that minimize rough handling of the fruit. This includes grading farm roads that will be used to transport fruit out of the field and maintaining field equipment. Harvested fruit should be kept in the shade if at all possible prior to transport to the packinghouse. This is especially important when picking during warm periods or when fruit will not be transported for several hours. Prompt transport to the packinghouse is essential. This helps minimize subsequent fungal decay and the development of physiological disorders.

Harvesting lemons or navels when fruit turgor pressure is high can cause oleocellosis. This may occur following rain, foggy weather, or a recent irrigation. Oleocellosis results when oil cells are ruptured and the oil is released, injuring adjacent peel tissue. Prior to harvest, a tool consisting of a 9 mm piston attached to a pressure gauge, such as a

Table 26.10. Factors to be considered when preparing for harvest

communication	Have a good working relationship with packinghouse and coordinate harvesting with them. Work closely with picking crew foreman or contract crew to ensure proper handling of the fruit. Visit the orchard and observe the picking crew. Periodically watch fruit run over the packline and understand why and how fruit are handled.
preharvest preparation	Make sure that all necessary fungicidal sprays have been properly applied to protect fruit from subsequent decay. Grade roads that will be used to transport fruit out of the orchard to minimize unnecessary jostling of fruit in the bin. Wash and clean all equipment used for harvesting fruit. Make sure that ladders, bins, clippers, and picking bags are in good condition.
weather	Do not pick when there is moisture on the fruit. Delay harvesting until the fruit is dry or until the oil release pressure is greater than 7 lbf. Winter harvesting: Harvest in the warmer part of the day if possible to minimize the development of oleocellosis. If harvesting after a period of rain, monitor peel turgidity with oil release method. Summer harvesting: Maintain good water status in trees and avoid harvesting fruit from trees that are water-deficit stressed (wilted). Monitor fruit oil release pressure if irrigation has recently occurred to minimize problems with subsequent peel disorders.
training the picking crew	Make sure that all workers have received training in the proper use of clippers and ladder placement in the tree. Make sure that all members of the picking crew use gloves and have short fingernails. Make sure that the picking foreman does not allow bags or bins to be overfilled. Do not allow dropped fruit to be picked up off the ground.
after harvest	Protect fruit from direct sun. Move fruit promptly to the packinghouse. Place harvest bins in shade prior to pickup if transport will not occur for several hours.

hand-held pressure gauge, can be used to test the pressure required to release peel oil. A small piece of rind tissue is placed under the piston, which presses the flavedo until the first evidence of oil is observed. That pressure is recorded in pounds per square inch (psi). Citrus fruit harvested when rind oil release pressure is below 7 psi are most prone to oleocellosis.

Good agricultural practices

In 1998 the U.S. Food and Drug Administration (FDA) published *Guidance for Industry: Guide to Minimize Microbial Food Safety Hazards for Fresh Fruit and Vegetables.* This document was developed in response to several outbreaks of foodborne illness associated with consumption of fresh produce. It outlines guidance for growers, harvesters, and packers to follow in regards to food safety issues. While these are guidelines and not regulations, much of the retail grocery and food service trade has adopted them as required practices for their fresh produce suppliers. This trend is growing. It is worthwhile for growers to become familiar with these practices and consider their adoption.

References

Adaskaveg, J. E., H. Förster, and N. F. Sommer. 2002. Principles of postharvest pathology and management of decays of edible horticultural crops. In A. A. Kader, ed., Postharvest technology of horticultural crops. 3rd ed. Oakland: University of California Division of Agriculture and Natural Resources Publication 3311. 163–195.

Arpaia, M. L., S. Collin, K. Fjeld, J. Sievert, and D. Obenland. 2011. The science behind the proposed maturity standard change. Citrograph 2(3): 25–33.

Brown, G. E., and W. R. Miller. 1999. Maintaining fruit health after harvest. In L. W. Timmer and L. W. Duncan, eds., Citrus health management. St. Paul, MN: APS Press. 175–188.

Burns, J. K. 2004a. Grapefruit. In K. C. Gross, C. Y. Wang, and M. Saltveit, eds., The commercial storage of fruit, vegetables, and florist and nursery stocks. Agriculture Handbook 66. USDA ARS BARC website, http://www.ba.ars.usda.gov/hb66/072grapefruit.pdf.

———. 2004b. Lime. In K. C. Gross, C. Y. Wang, and M. Saltveit, eds., The commercial storage of fruit, vegetables, and florist and nursery stocks. Agriculture Handbook 66. USDA ARS BARC website, http://www.ba.ars.usda.gov/hb66/084lime.pdf.

———. 2004c. Mandarin (tangerine). In K. C. Gross, C. Y. Wang, and M. Saltveit, eds., The commercial storage of fruit, vegetables, and florist and nursery stocks. Agriculture Handbook 66. USDA ARS BARC website, http://www.ba.ars.usda.gov/hb66/090mandarin.pdf.

Ceponis, M. J., R. A. Cappellini, and G. W. Lightner. 1986. Disorders in citrus shipments to the New York market, 1972-1984. Plant Disease 70:1162–1165.

Coggins, C. W., H. Z. Hield, I. L. Eaks, L. N. Lewis, and R. M. Burns. 1965. Gibberellin research on citrus. California Citrograph 50:457–466, 468.

Eckert, J. W., and I. L. Eaks. 1989. Postharvest disorders and diseases of citrus fruit. In W. Reuther, E. C. Calavan, and G. E. Carman, eds., The citrus industry. Vol. 5. Oakland: University of California Division of Agriculture and Natural Resources. 179–260.

Eckert, J. W., and N. F. Sommer. 1967. Control of diseases of fruit and vegetables by postharvest treatments. Annual Review of Phytopathology 5:391–432.

Gross, K. C., and J. L. Smilanick. 2004. Lemon. In K. C. Gross, C. Y. Wang, and M. Saltveit, eds., The commercial storage of fruit, vegetables, and florist and nursery stocks. Agriculture Handbook 66. USDA ARS BARC website, http://www.ba.ars.usda.gov/hb66/082lemon.pdf.

Gross, C., Y. Wang, and M. Saltveit, eds. 2004. The commercial storage of fruit, vegetables, and florist and nursery stocks. Agriculture Handbook 66. USDA ARS BARC website, http://www.ba.ars.usda.gov/hb66/090mandarin.pdf.

Guidance for industry: Guide to minimize microbial food safety hazards for fresh fruit and vegetables. Updated regularly. U.S. Food and Drug Administration Guidance Documents website, http://www.fda.gov/Food/GuidanceComplianceRegulatoryInformation/GuidanceDocuments/default.htm.

Kader, A. A. 1992. Standardization and inspection of fresh fruit. In A. A. Kader, ed., Postharvest technology of horticultural crops. 2nd ed. Oakland: University of California Division of Agriculture and Natural Resources Publication 3311. 191–200.

Larkin, R. P., D. P. Roberts, and J. A. Gracia-Garza. 1998. Biological control of fungal diseases. In D. Hutson and J. Miyamoto, eds., Fungicidal activity: Chemical and biological approaches to plant protection. New York: Wiley. 149–191.

Obenland, D., S. Collin, B. Mackey, J. Sievert, K. Fjeld, and M. L. Arpaia. 2009. Determinants of flavor acceptability during the maturation of navel oranges. Postharvest Biology and Technology 52:156–163.

Petracek, P. D., and H. Dou. 1998. Reducing postharvest pitting incidence of citrus fruit. Packinghouse Newsletter 184 [University of Florida].

Ritenour, M. A. 2004. Orange. In K. C. Gross, C. Y. Wang, and M. Saltveit, eds., The commercial storage of fruit, vegetables, and florist and nursery stocks. Agriculture Handbook 66. USDA ARS BARC website, http://www.ba.ars.usda.gov/hb66/100orange.pdf.

Smilanick, J. L., B. E. Mackey, R. Reese, J. Usall, and D. A. Margosan. 1997. Influence of concentration of soda ash, temperature, and immersion period on the control of postharvest green mold of oranges. Plant Disease 81:379–382.

Smilanick, J. L., D. A. Margosan, F. Mlikota, J. Usall, and I. F. Michael. 1999. Control of citrus green mold by carbonate and bicarbonate salts and the influence of commercial postharvest practices on their efficacy. Plant Disease 83:139–145.

Snowdon, A. L. 1990. A color atlas of post-harvest diseases and disorders of fruit and vegetables. Vol. 1: General introduction and fruit. Boca Raton, FL: CRC Press.

Wardowski, W. F., S. Nagy, and W. Grierson, eds. 1986. Fresh citrus fruit. Westport CT: AVI Publishing.

White, G. C. 1992. Handbook of chlorination and alternative disinfectants. 3rd ed. New York: Van Nostrand Reinhold.

Part 7 Developing Issues and Opportunities for Citrus Production

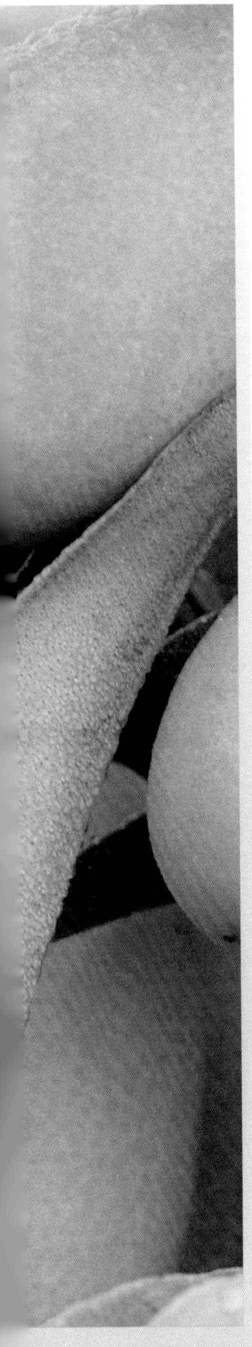

27 Surface and Ground Water Quality

Laosheng Wu

After several decades of effort, technologically based controls through the National Pollution Discharge Elimination System and industry wastewater pretreatment have substantially reduced the amount of pollutants from point sources. On the other hand, according to the National Water Quality Inventory of the 2004 Reporting Cycle (EPA 2010), about 44% of assessed stream miles, 64% of assessed lake acres, and 30% of assessed bay and estuarine square miles were not clean enough to support designated uses such as fishing and swimming (impaired). Agriculture was the no. 1, no. 3, and no. 6 leading sources of impairments, respectively, for the assessed bodies of water.

Applied agricultural chemicals, such as fertilizers, pesticides, and herbicides, can travel to surface water bodies via runoff or to ground water by leaching. The sources of contaminants found in water come from two basic categories: natural and human induced. Agricultural activities can cause large-scale environmental impacts on surface and ground water quality. In regions with intensive agriculture, environmental problems, such as nutrient enrichment (eutrophication), heavy metals, and pesticides and herbicides in surface and ground water, are closely related to runoff and leaching.

Unlike point sources of pollution, which are typically controlled through treatment facilities and authorized by permits, nonpoint sources are diffuse and not readily controlled by conventional wastewater treatment means. Controlling nonpoint source pollution requires the use of preventive plans and practices by those directly involved in such activities. Nonpoint source pollution is also highly dynamic and highly variable; its entry point to receiving waters is usually difficult to identify. The most common types of nonpoint source pollutants associated with surface water quality problems are the runoff of sediment, pesticides, herbicides, and plant nutrients. Nitrate, pesticides, and herbicides that leach to ground water are also related to agricultural activities such as disturbance of the land surface, fertilizer use, and application of pesticides and herbicides.

In California, statewide assessments of water quality show that nonpoint source pollution has the greatest effect on the state's surface and ground water quality. Citrus is one of the largest crops in California. If not properly managed, application of nitrogen, pesticides, and herbicides in citrus production can cause surface and ground water quality problems. It is in the best interest of growers to address water quality issues proactively and adopt best management practices (see Ingels 1994; Seelig 1996; Buttler et al. 1998) that can reduce the potential for surface and ground water contamination.

Water Quality Regulations on Nonpoint Source Pollution

Federal and state surface and ground water quality regulations that may significantly impact California agriculture, including citrus production, are Section 303(d) of the Clean Water Act (CWA) passed by the U.S. Congress in 1972 and ground water protection regulations such as the Ground Water Protection Area (GWPA) adopted by the California Department of Pesticide Regulation (DPR) in 2004. Conditional waivers of discharges from irrigated agriculture may also affect agricultural activities. Several regional water quality control boards (RWQCBs) also have developed regulations on limiting nitrate leaching to ground water.

Environmental lawsuits and court decisions have revitalized requirements of the CWA that were overlooked until recently. The CWA established the regulatory concept of a maximum loading rate of a pollutant that a receiving water body can assimilate without resultant water quality impairments with respect to applicable water quality standards. The CWA also requires states to identify all of its water bodies that fail to meet applicable water quality standards and to establish a total maximum daily load (TMDL) for each of these polluted water bodies. A TMDL defines how much of a pollutant a water body can tolerate (absorb) daily and still comply with applicable water quality standards. It requires that the pollution coming from both point sources (discrete conveyances) and nonpoint sources (everything other than point sources) be reviewed. The latter includes landscape-scale sources, such as storm water and agricultural runoff, as well as dust and air pollution that find their way into water bodies. All pollutant sources in the watershed combined, including point and nonpoint sources, are limited to discharging no more than the TMDL.

Excess sediments, nutrients, pesticides, herbicides, metals, and pathogens are the leading causes of water quality impairments, and many of the pollutants are associated with agricultural activities. According to the interpretation of the California State Water Resources Control Board (SWRCB), the Porter-Cologne Water Quality Control Act (California Water Code Section 13000 et. seq.) requires an implementation plan when a TMDL is incorporated into water quality control plans. The Porter-Cologne Act requires each regional water quality control board to formulate and adopt water quality control plans for all areas within its region. It also requires that an implementation program be developed that describes how water quality standards will be attained. TMDLs can be developed as a component of the implementation program, which requires description of the implementation features, or TMDLs can be established as a water quality standard. When TMDLs are established as a standard, the implementation program must be designed to implement them. Typically, a revision of the implementation program is needed whenever a new standard is adopted. The regional water quality control boards' conditional waivers of discharges from irrigated agriculture can also require citrus growers to protect ground water quality.

To protect ground water quality in California, the DPR has adopted regulations to establish GWPAs and to protect wellheads from runoff water containing pesticide residues. A ground water protection area is defined as "an area of land that has been determined by the director [of DPR] to be vulnerable to the movement of pesticides to ground water. The determination of a ground water protection area is based on factors such as soil type, climate, and depth to ground water that are characteristic of areas where legally applied pesticides or their breakdown products have been detected and verified in ground water" (DPR 2013). The regulations classify ground water protection areas as "leaching ground water protection areas" and "runoff ground water protection areas." In leaching ground water protection areas, pesticide residues move from the soil surface downward through the soil matrix and percolate into the ground water. Soils in these areas are coarse, with relatively rapid infiltration rates. In runoff ground water protection areas, pesticide residues are carried in runoff water to the ground water, or through more direct routes such as dry or drainage wells, poorly sealed production wells, or soil cracks. Soils in runoff ground water protection areas typically have a hardpan layer or low infiltration rates.

These DPR regulations may have a significant impact on California agriculture. If an operation is located within either a runoff or leaching GWPA, and the operator wants to use any of the materials on the list of regulated pesticides, he or she must obtain a permit for possession or use of the material in a ground water protection area and comply with one of the management practices specified in the regulations. Products containing active ingredients that are regulated by DPR may be applied only by a certified applicator. The permit is issued to the property operator, who may or may not be the applicator. Since the pesticide is a restricted material, the applicator must be certified to apply it if any one of the following mitigation measures is met (DPR 2013):

The soil is disturbed, within 7 days before the pesticide is applied, by using a disk, harrow, rotary tiller, or other mechanical method.

- The pesticide is incorporated on at least 90% of the area treated, within 48 hours after the day the pesticide is applied, by mechanical methods or by low-flow irrigation (0.25 to 1 in), including chemigation if allowed by the label, provided runoff does not occur.
- The pesticide is applied as a band treatment immediately adjacent to the crop row so that not more than 33% of the distance between rows is treated.
- The pesticide is applied between April 1 and July 31.
- Control of runoff:
    - ▹ All irrigation runoff and all precipitation on and drainage through the field must be retained onsite for 6 months after application. The retention area on the field should have a percolation rate of no more than 0.2 inch per hour.

- ▹ All irrigation runoff and all precipitation on and drainage through the field must be stored offsite for 6 months after application, provided the channel transporting the runoff and the holding area do not have a percolation rate of more than 0.2 inch per hour.
- ▹ Runoff must flow onto an adjacent unenclosed fallow field for 6 months following application. The receiving field must be a minimum of 300 feet long and not be irrigated for 6 months after application.

Leaching GWPAs contain coarse soils with relatively rapid infiltration rates. Listed materials may be applied by a permitted applicator if any one of the following mitigation measures can be met for 6 months following application of the pesticide (DPR 2013):

- No irrigation takes place.
- No contact with irrigation water. Pesticides must be applied to the planting bed or the berm above the level of irrigation water in the furrow or basin so there is no contact with leaching irrigation water.
- Irrigation management: Irrigation must be managed so that the ratio of the amount of irrigation water applied divided by the net irrigation requirement is 1.33 or less for 6 months following application of the pesticide.

According to the proposed regulations regarding wellhead protection, the following activities are prohibited within 100 feet of an unprotected well (including domestic, municipal, agricultural, dry or drainage, monitoring, or abandoned wells) (DPR 2013):

- Mixing, loading, and storage of pesticides.
- Rinsing of spray equipment or pesticide containers.
- Maintenance of spray equipment that could result in spillage of pesticide residues on the soil.
- Application of preemergent herbicides.

These prohibitions do not apply if the well is protected from irrigation or rainfall runoff by being above the grade of drainage or being surrounded by a berm.

Managing Pollutants in Runoff Water

Pesticides, herbicides, nitrogen, and phosphorus in runoff water, either in dissolved form or attached to sediments, are the most common pollutants of surface water. For example, the phosphorus of particulate-associated form eroded during flow events can account for 91% of the total phosphorus entering fresh water annually (Gillingham and Thorrold 2000; Hart et al. 2004). To reduce nitrogen and phosphorus loss, growers must increase the efficiency of nitrogen and phosphorus use by considering crop nutrient demands, balancing the inputs from fertilizers and from other sources, such as manure and compost application, and reducing the volume of runoff water. Since most of the pesticides and herbicides are adsorbed on soil particles, reducing pesticide application by adopting integrated pest management (IPM) and minimizing soil erosion and sediment runoff are some of the best management practices to protect surface water quality. Mulch and vegetative filters or strips also effectively reduce pollutant loading to surface water.

Sediment is also a pollutant. Movement of sediment in runoff water is influenced by soil slope, rainfall or irrigation intensity, and soil surface characteristics (see O'Geen and Schwankl 2006). Conservation and contour tillage, buffer strips (see Grismer et al. 2006), surface mulching, and cover crops are some of the effective practices used to reduce runoff water. Conservation tillage (no-till and reduced tillage) and cover crops may increase percolation rate and reduce surface runoff. The use of cover crops in citrus production in California is not a very popular practice, but studies have shown that cover crops can prevent surface sealing and thus improve water infiltration (fig. 27.1). Also, more water can be retained on the surface with cover crops. Increased infiltration rate and more water retention on the surface can effectively reduce the runoff (O'Geen et al. 2006).

BMPs for Reducing Sediment, Nitrogen, and Phosphorus Surface Runoff

- » Conservation measures such as cover crops and surface mulching to control soil erosion.
- » Source control of nitrogen and phosphorus by balancing the inputs and crop use, and by avoiding or minimizing soil nitrogen and phosphorus buildup in excess of crop requirement.
- » Incorporating fertilizer into the soil to make it less available for surface runoff.
- » Using buffer strips to reduce storm water runoff.
- » Applying polyacrylamide (PAM).
- » Changing the design and setup of a sprinkler irrigation system.

A multiyear erosion control demonstration was conducted in Ventura County using municipal yard

Figure 27.1 Cover crops can prevent surface sealing and improve infiltration on orchard soils. *Photo:* J. K. Clark.

trimmings as mulch in commercial citrus orchards (Faber 1999). This study indicated that mulch can conserve soil-water, since water evaporation from the soil surface and transpiration by weeds are reduced; improve infiltration and reduce soil erosion, since mulch can reduce surface sealing and increase surface water detention; improve soil fertility due to the decomposition and release of nutrients from mulch into the soil; and reduce the growth of weeds so that less herbicide needs to be applied.

Thick vegetation or buffer strips along waterways that can slow runoff and capture pollutants are another way to reduce nitrogen and phosphorus in surface runoff to water bodies. Shrubs and ground cover can absorb up to 14 times more rainwater than a grass lawn and require no fertilizers. Studies have shown that buffer strips can effectively reduce levels of storm water–suspended sediments (Lee et al. 2003).

Polyacrylamide (PAM) has been rapidly accepted in recent years as a soil-water conservation technology in irrigated agriculture. Application of PAM to irrigation water or soil has proven to be effective and economical in stabilizing soil structure, increasing water infiltration, reducing soil erosion, and improving runoff water quality.

If runoff occurs from a surface-irrigated orchard, the grower should consider changing the irrigation design and setup to minimize surface runoff, retaining the runoff water, and installing tailwater return systems (see Schwankl et al. 2007a, 2007b).

Managing Pollutants to Ground Water

Nitrogen (N), phosphorus (P), and potassium (K) are the nutrients plants require in the highest amounts and are also the most commonly applied fertilizers. Among them, phosphorus and potassium have relatively low mobility in soil and therefore usually have less impact on ground water quality than does nitrogen, especially nitrate-nitrogen (NO_3^--N). The behavior of synthetic pesticides and herbicides in soil is determined by many factors such as pesticide or herbicide property, soil property, climate conditions, and agricultural management practices (Ferruzzi and Gan 2004), while nitrate leaching is mainly controlled by water management. Due to their difference, the management practices for reducing the flow of nitrate and pesticide or herbicide pollution to ground water are discussed separately in the following sections.

Reducing Nitrate Leaching

Ground water contamination with nitrate has been linked to the rate of nitrogen application. Nitrate nitrogen concentration in soil solution extracts was found to correlate with the amount of nitrogen applied to soil: as applied nitrogen increases, the amount of nitrate nitrogen in the soil also increases. The average nitrogen application in citrus, 100 to 200 pounds of nitrogen per acre, makes citrus one of the largest recipients of nitrogen in the state. The adoption of pressurized irrigation practices and the inclusion of nitrogen fertilizers in irrigation water appear to have resulted in an increase in the total amount of nitrogen applied to citrus over the past 10 to 15 years.

BMPs to Reduce Nitrate Leaching

» Account for the amount of available nitrogen in soils to obtain efficient use of applied nitrogen fertilizer.

» Match nitrogen input with crop seasonal requirements; use slow-release fertilizer.

» Avoid leaving excess nitrate in soils with sandy or gravelly textures because of the high potential for leaching.

» Use foliar applications of nitrogen.

» Schedule irrigations according to evapotranspiration demand.

» Increase irrigation application frequency in course-textured soils.

A study conducted in Florida citrus production (Paramasivam et al. 2002) suggests that efficient irrigation management, aimed to replenish only the water deficit within the rooting depth, along with increased application frequency of fertilizers in small quantities, improves the nutrient uptake efficiency while minimizing leaching loss.

Citrus growers can reduce nitrate ground water pollution and increase profits by using foliar urea fertilization. Lovatt (1995) reported that foliar application of nitrogen to avocado trees could supply a portion of the nitrogen to be applied in a given year and thus reduce the amount of nitrogen applied to the soil. The results also demonstrated that a late-spring application of low-biuret urea to the foliage is a cost-effective method to fertilize navel oranges. The increase in yield of larger-sized fruit observed in this study for the late-May foliar application of low-biuret urea resulted in a net increase in return revenue to the grower each year. Since the grower will likely fertilize with nitrogen sometime during the year, foliar application of urea in late spring would seem to afford many benefits over soil-applied nitrogen.

Proper irrigation management is critical to protect water quality, since most California citrus production relies heavily on irrigation. Scheduling irrigation according to soil moisture measurement or potential ET and good maintenance of irrigation systems to ensure high uniformity are the key steps to reduce nitrogen pollution (see chapter 11, "Irrigation"). The most common way to determine when and how much water to apply is to take a direct soil moisture measurement or use the water budget approach: applying only the amount of water to meet the evapotranspiration (ET) requirement and leaching requirement if salinity may become a problem (see chapter 8, "Soil and Water Analysis and Amendment"). Improvement of irrigation system uniformity can also increase crop water use efficiency and reduce nitrate leaching potential.

Reducing Pesticide and Herbicide Leaching

The most important properties that control the persistence and mobility in the soil of a pesticide or herbicide are the organic carbon adsorption coefficient and the biological degradation half-life. The former describes the relative affinity or attraction of the pesticide or herbicide to soil materials, and therefore its mobility in the soil; the latter is a measure of persistence in the soil. Soil properties also affect pesticide and herbicide transport and persistence. Fine-textured soils usually have higher organic matter content, which can retain more applied chemicals. Because soil percolation rate is much lower in clay soils than in sandy soils, a sandy soil is more vulnerable to pesticide and herbicide leaching. Depth of the ground water table is also important in assessing pesticide or herbicide vulnerability: it is easier for the chemicals to reach the ground water if the ground water table is shallow.

BMPs to Reduce Pesticide and Herbicide Leaching

» Adopt IPM (tillage, residue and mulching, biological control, pest-resistant cultivars, etc.).

» Use less-mobile and lesstoxic pesticides and herbicides.

» Proper irrigation scheduling.

» Install backflow prevention valves.

» Avoid pesticide and herbicide application before intensive rainfall or irrigation.

Adopting best management practices can reduce the potential of pesticides and herbicides to contaminate ground water. For example, IPM can usually reduce pesticide application rate and frequency, so the total amount of application can be reduced. The most common IPM practices include the following (see chapter 17, "Citrus IPM"):

- Reduce or eliminate the use of highly mobile or persistent pesticides; this generally reduces the potential for ground water contamination, particularly for areas with coarse-textured soil and a shallow water table.
- Grow tolerant or resistant cultivars to help avoid plant diseases and pests.

- Use tillage to control pests and weeds where it is appropriate.
- Use biological control of pests when it is available and when its effectiveness has been demonstrated.
- Preemptive management may be the most effective means of dealing with certain pest problems. Responsive management options can be quite limited if pests are allowed to reach outbreak levels.
- Implement management strategies that maintain pests below threshold levels to help avoid expensive or less-effective methods of pest control.
- Use pest and crop models to help predict pest problems and guide management decisions.
- Optimize the timing of pesticide applications according to pest life cycles and economic thresholds of damage.
- Rotate pesticides to prevent development of pest resistance (see chapter 18, "Managing Pesticide Resistance in Insects, Mites, Weeds, and Fungi").

Maintaining soil organic matter through soil conservation practices and surface mulching can help adsorb pesticides or prevent the leaching of pesticides through the soil profile. As organic matter decreases, so does the soil's ability to adsorb pesticides that move through it. Organic matter on the soil surface can also improve soil permeability and water infiltration, which may result in greater potential for leaching to ground water. The balance between increased adsorption and infiltration must be weighed for each management recommendation in every case.

Soil factors such as slope and hardpan can affect the surface runoff of pesticides and herbicides. Soil conservation practices that reduce the force of runoff water should be used. Although reduced tillage is beneficial with respect to soil erosion and maintaining organic matter, it may promote movement of pesticides through direct pathways. In cases where preferential flow is significant, conservation tillage should include some method of surface disruption.

Pesticide application should be avoided prior to intense rainfall or irrigation events. The largest losses of pesticide occur during the first runoff event after application. The amount of loss decreases with each additional day between application and intense rainfall.

As in managing nitrogen leaching, irrigation provides opportunities to reduce the potential of pesticide and herbicide leaching. When injecting chemicals into an irrigation system, chemigation equipment and backflow prevention valves should be used.

References

Buttler, T. M., A. G. Hornsby, D. P. Tucker, J. L. Knapp, and J. W. Noling. 1998. Managing pesticides for citrus production and water quality protection. Gainesville: University of Florida Cooperative Extension Service Circular 974.

DPR (California Department of Pesticide Regulation). 2013. DPR regulations website, http://www.cdpr.ca.gov/docs/emon/grndwtr/gwp_regs.htm.

EPA (U.S. Environmental Protection Agency). 2010. Watershed assessment, tracking, and environmental results. EPA website, http://iaspub.epa.gov/waters10/attains_nation_cy.control#wqs.

Faber, B. A. 1999. Erosion control using mulch or cover crops in lemon orchards. In Proceedings of the Cover Crop Workgroup Annual Meeting, Parlier, CA. 45–52.

Ferruzzi, G., and J. Gan. 2004. Pesticide selection to reduce impacts on water quality. Oakland: University of California Division of Agriculture and Natural Resources Publication 8119. ANR CS website, http://anrcatalog.ucdavis.edu/SoilWaterIrrigation/8119.aspx.

Gillingham, A. G., and B. S. Thorrold. 2000. A review of New Zealand research measuring phosphorus in runoff from pasture. Journal of Environmental Quality 29:88–96.

Grismer, M. E., A. T. O'Geen, and D. Lewis. 2006. Vegetative filter strips for nonpoint source pollution control in agriculture. Oakland: University of California Division of Agriculture and Natural Resources Publication 8195. ANR CS website, http://anrcatalog.ucdavis.edu/SoilWaterIrrigation/8195.aspx.

Hart, M. R., B. F. Quin, and M. L. Nguyen. 2004. Phosphorus runoff from agricultural land and direct fertilizer effects: A review. Journal of Environmental Quality 33:1954–1972.

Ingels, C. A. 1994. Protecting groundwater quality in citrus. Oakland: University of California Division of Agriculture and Natural Resources Leaflet 21521.

Lee, K. H., T. M. Isenhart, and R. C. Schultz. 2003. Sediment and nutrient removal in an established multi-species riparian buffer. Journal of Soil and Water Conservation 58:1 (Jan./Feb.): 1–8.

Lovatt, C. J. 1995. Avocado growers can reduce soil nitrate groundwater pollution and increase yield and profit. FREP Contract 95-0523. CDFA Inspection Services Division website, http://www.cdfa.ca.gov/is/docs/Lovatt95.pdf.

O'Geen, A. T., and L. J. Schwankl. 2006. Understanding soil erosion in irrigated agriculture. Oakland: University of California Division of Agriculture and Natural Resources Publication 8196. ANR CS website, http://anrcatalog.ucdavis.edu/SoilWaterIrrigation/8196.aspx.

O'Geen, A. T., T. L. Prichard, R. Elkins, and G. S. Pettygrove. 2006. Orchard floor management practices to reduce erosion and protect water quality. Oakland: University of California Division of Agriculture and Natural Resources Publication 8202. ANR CS website, http://anrcatalog.ucdavis.edu/SoilWaterIrrigation/8202.aspx.

Paramasivam, S., A. K. Alva, A. Fares, and K. S. Sajwan. 2002. Fate of nitrate and bromide in an unsaturated zone of a sandy soil under citrus production. Journal of Environmental Quality 31:671–681.

Schwankl, L. J., B. R. Hanson, and T. L. Prichard. 2007a. Causes and management of runoff from surface irrigation in orchards. Oakland: University of California Division of Agriculture and Natural Resources Publication 8214. ANR CS website, http://anrcatalog.ucdavis.edu/SoilWaterIrrigation/8214.aspx.

———. 2007b. Managing existing sprinkler irrigation systems. Oakland: University of California Division of Agriculture and Natural Resources Publication 8215. ANR CS website, http://anrcatalog.ucdavis.edu/SoilWaterIrrigation/8215.aspx.

Seelig, B. D. 1996. Improved pesticide application BMPs for groundwater protection from pesticides. Fargo: North Dakota State University Extension Circular AE-1113.

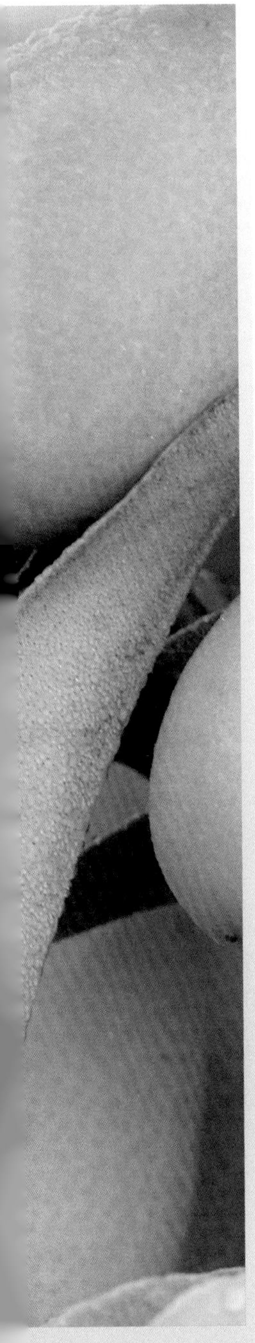

28 Ozone Air Pollution

David A. Grantz

Smog, mostly ozone, has become a reality in citrus production areas around the world. Southern California's former dominance of citrus production in the state has given way to the heavy air pollution and urban pressures of the Los Angeles metropolitan area. As production has moved into the rural San Joaquin Valley, so too have thousands of new residents, with automobiles and industries that provide them jobs. Inevitably, so too has ozone. Because the climatic factors associated with productive citriculture are also associated with air pollution, ozone stress will be a factor in many citrus production areas that share an urban influence on air quality. This includes many rural areas in California and around the world that are some distance downwind from urban growth. There are no current management practices to combat the effects of ozone, but it is important for growers to recognize the foliar damage and the potential for yield suppression by ozone and to be alert for the emergence of genetic and cultural advances in ozone tolerance.

Ozone

Air Quality and Geography

Ozone is an unusual air pollutant. It is a secondary pollutant that is formed in the atmosphere from reactions of primary pollutants such as nitrogen gases and reactive organic vapors. These are emitted directly from numerous natural and human-made sources. Ozone is now a regional pollutant, affecting rural areas where local pollution sources may be minimal. It is only moderately phytotoxic, but it is so widespread and frequent that it is the most damaging air pollutant to natural and agricultural ecosystems.

Worldwide, the climatic factors associated with productive citriculture are also associated with production and accumulation of ozone. The coincidence is striking and potentially troublesome (Grantz and Sanz 2004). This is particularly true in the San Joaquin Valley (SJV) of California (fig. 28.1).

The SJV is an urbanizing rural area, the current center of California's citrus industry, and the location of numerous local and upwind pollution sources. The SJV is a deep air basin, at an elevation of about 300 to 400 feet above sea level and surrounded by mountains up to 14,000 feet tall to the east and 4,000 feet tall to the south and west. The only opening to the sea is through the narrow Carquinez Straits, at the northern end of the SJV. The result is an air mass that has restricted exchange with the bulk atmosphere.

The Fresno Eddy circulates this stagnant air in a counterclockwise direction flowing southward from the relatively rural northern and western SJV along Interstate 5, with its contributions of nitrogen oxides (NO_x) and hydrocarbons, then northeast past the oil fields and refineries of Kern County, with their abundance of hydrocarbons, and circling the urban areas of Bakersfield, Visalia, and Fresno. Nitrogen oxides originating from high-temperature combustion sources, such as cars and industry, along with hydrocarbons from cars, industry, and vegetation, mix and react in the atmosphere, driven by the ultraviolet component of the abundant sunlight. The result is photochemical smog, mostly ozone, concentrated along the eastern margins of the SJV—that is, in the citrus production areas.

The SJV is designated as an extreme nonattainment area for ozone, a distinction shared only with the Los Angeles air basin, the former center of California's citrus industry. Air quality substantially improved from the 1960s to the 1980s as the federal Clean Air Act

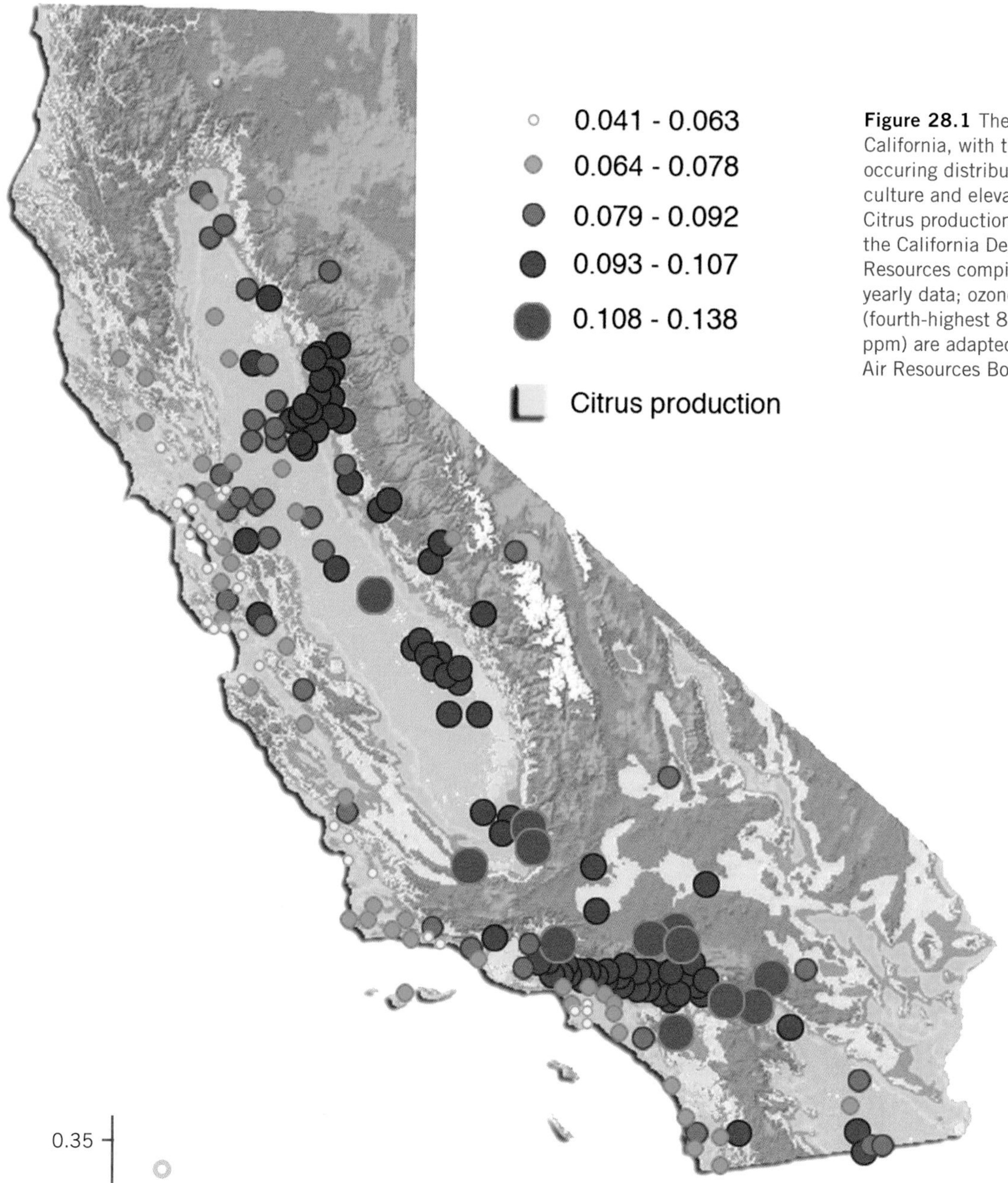

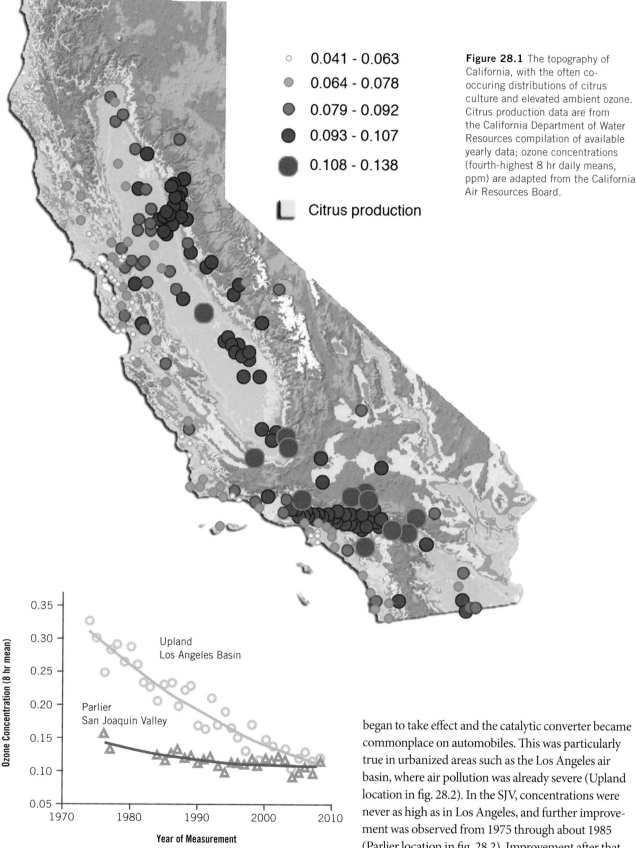

Figure 28.1 The topography of California, with the often co-occuring distributions of citrus culture and elevated ambient ozone. Citrus production data are from the California Department of Water Resources compilation of available yearly data; ozone concentrations (fourth-highest 8 hr daily means, ppm) are adapted from the California Air Resources Board.

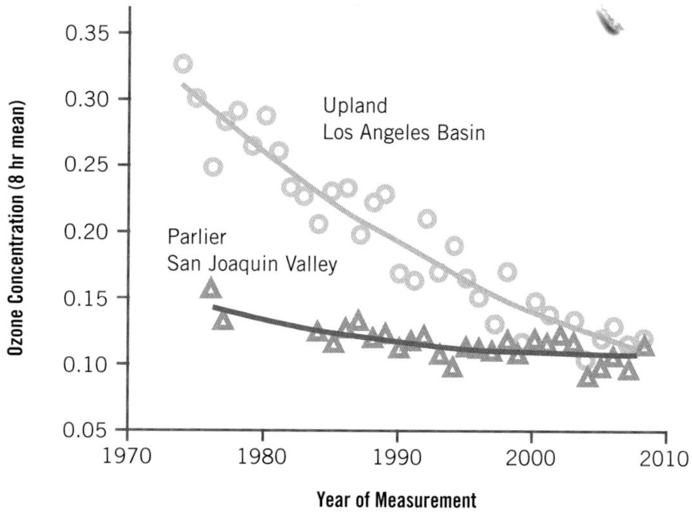

Figure 28.2 Trends of air quality in the citrus production areas of the Los Angeles basin and the San Joaquin Valley. Data are taken from the California Environmental Protection Agency, Air Resources Board. Shown are the highest 8-hour average ozone concentrations observed at Upland and Parlier for each year (ppm).

began to take effect and the catalytic converter became commonplace on automobiles. This was particularly true in urbanized areas such as the Los Angeles air basin, where air pollution was already severe (Upland location in fig. 28.2). In the SJV, concentrations were never as high as in Los Angeles, and further improvement was observed from 1975 through about 1985 (Parlier location in fig. 28.2). Improvement after that has proven to be more difficult in both locations, although by late 2013 the SJV was on track to achieve attainment of the older 1-hour standard for ozone. The current regulatory standard in the United States is based on the maximum daily 8-hour mean (< 0.075 ppm; 3 yr average of the fourth-highest annual value). The 8-hour mean is a metric that plant scientists have long favored over the previous 1-hour standard as an effective measure of the ozone threat to vegetation. For consistency with current regulatory standards, ozone concentrations in the research studies considered below have been converted to 8-hour daylight means using a uniform conversion factor that assumes a sinusoidal diurnal time-course.

Ozone Effects on Crop Production

Ozone is now considered primarily a public health problem in the United States. Interestingly, in other parts of the world, ozone is considered primarily a threat to crops and forests and only secondarily as a direct threat to public health (Heck and Furiness 2001).

Ozone was first discovered as an important air pollutant during the search for the cause of a plant disease in southern California. Agricultural scientists at the University of California, Riverside, began to report injury to several crops in the Los Angeles basin for which no pathogen could be found (Middleton et al. 1950). Ozone was finally identified as the causative agent of this syndrome (Richards et al. 1958; Heggestad and Middleton 1959) and eventually as a serious threat to crop production in general.

During the 1980s the U.S. Environmental Protection Agency (EPA) and the U.S. Department of Agriculture (USDA) instituted a National Crop Loss Assessment Network to identify the impacts of ozone on major crops in the United States. Agricultural crop loss due to ozone was estimated to cost about $1.5 billion nationally and about $300 to 400 million in the San Joaquin Valley (Heck et al. 1988).

Ozone damage to vegetative growth and horticultural yield is not necessarily correlated with visual symptoms. However, ozone often does cause visible injury, typically on the leaves of many plant species (see Costello et al. 2003; Dreistadt et al. 2004). These can be identified for a wide variety of plant species in the color plates appearing in Darley et al. (1966), Jacobson and Hill (1970), and Flagler (1998).

Visible ozone damage in citrus may appear similar to other types of injury (table 28.1). It may involve the classic oxidant stipple, or bronzing, on upper leaf surfaces, or a more generalized chlorotic mottling of both surfaces (fig. 28.3A). High concentrations may lead to nonspecific leaf curling (fig. 28.3B). Growers must remember that no visible warning signs may appear, even when costly ozone damage is already occurring in the orchard.

Figure 28.3 Ozone symptoms (A) on mature leaves of Valencia orange showing adaxial (upper) and abaxial (lower) surfaces with classical chlorotic mottling; and (B) on grapefruit showing a leaf cluster with curled leaf margins and chlorotic mottling. *Photos:* R. Platt.

Ozone Effects on Citrus Yield and Quality

Early Studies

Lemons and oranges were among the first crops reported to be damaged by ozone in Los Angeles air. In experiments conducted during the 1960s, ozone was found to increase fruit drop, decrease the amount of stem wood production, and decrease the longevity of green leaf area on the trees (Thompson and Taylor 1969). In Cucamonga and Upland, lemons (cv. Eureka) lost 50 to 60% of their yield,

Table 28.1. Possible visible symptoms of ozone damage to citrus crops, adapted from generalized symptoms of broadleaf plants

Acute injury	Chronic injury
Bleaching, usually upper surface but can be both surfaces. Can be isolated white or purple spots or involving entire surface. In extreme cases, may involve death of entire leaf, usually excepting the major veins.	Bronzing, progressing to red or brown coloration of leaf surface, more evident on upper surface.
Flecking of small necrotic areas of dead cells. Spots can be white or brown and may have a metallic appearance.	Premature senescence of normally long-lived leaves.
Stippling of very small spots of a few dead underlying cells. May be white, black, red, or purple.	Reduced root growth and proliferation through the soil. Possible subtle changes in plant water status.

Source: After Flagler 1998.

Note: Hardy citrus leaves are relatively resistant to visible symptoms, but this does not mean no ozone damage is occurring.

whether as fruit weight or fruit number, due to ozone. Fruit size was not affected. A related study with mature (16-year-old) navel orange trees exhibited similar yield losses of 50 to 55%; similar changes in fruit weight and number of fruit implied little effect of ozone on the size of individual fruit (Thompson 1969).

Citrus is often subject to high rates of abscission of flowers and young fruit. Control of fruit drop is known to be responsive to plant hormones, but the physiological control remains poorly understood under any conditions. The responses to ozone stress support theories that reduced carbohydrate supply to the developing fruit, whether by inhibition of photosynthesis or of transport of sugars to the developing sinks, causes physiological adjustment of the crop load that compensates for reduced carbohydrate supply.

In these early investigations (Thompson et al. 1967, 1970; Thompson and Taylor 1969), the ambient atmosphere was used as the air pollution stressor. During the period of this study a number of oxidant pollutants, notably peroxyacyl nitrates (PAN) and nitrogen dioxide (NO_2) were abundant in this production area. To achieve the various treatment concentrations, ambient air was diluted with charcoal-filtered air to achieve a range of fractions of the (largely unknown) ambient pollutants.

A clever series of differential air filtrations in these experiments showed that the observed air pollution effects were due in large measure to ozone, though other oxidants in the Los Angeles–area atmosphere also had detrimental effects. Studies performed in plastic greenhouses near Upland with navel oranges (Thompson et al. 1971) imposed either ambient air or charcoal-filtered air to which ambient or twice-ambient levels of NO_2 were added. Leaf drop was greater and fruit yield was less in the ambient air than in the ambient or twice-ambient NO_2 treatment. This eliminated NO_2 as the principal causative agent and implicated other oxidants such as ozone as the most damaging components in the ambient environment.

While the dominance of ozone was demonstrated, the additive effects of other pollutants was shown by effects on summertime fruit abscission in navel oranges (table 28.2). In ambient air, inside or outside the exposure chambers, navel orange trees lost 75 to 80% of fruit that had formed. In filtered air, trees lost only 51% of fruit. However, adding ozone equal to the concentration of oxidants in the ambient air resulted in only a 63% fruit drop. Other chemical species were responsible for about half of the air pollution impact on fruit loss in these experiments. Fortunately, while ozone has remained a problem since that time, many of the other oxidants have been reduced considerably in current ambient air.

The high ozone concentrations imposed in these studies reflected ambient ozone concentrations at the time (1960s) in the Los Angeles basin. For example, exposure in September in Upland exceeded 0.1 ppm for more than 9 hours per day, and daily maximum levels rose to about 0.3 ppm on most days and to

Table 28.2. Impact of ozone on fruit drop in 16-year-old navel orange trees, 1 season exposure

Ozone (O_3) treatment (8 hr mean)	Ozone level in treatment (ppm)	Summer fruit drop (% of fruit set)
charcoal filtered	0.028	50.8
ambient	0.264	76.2
outside	0.264	79.6
charcoal filtered + ambient O_3 only	0.264	62.8

Source: Thompson et al. 1972.

0.5 ppm on some days. These extremely damaging concentrations are no longer observed in the United States due to regulatory actions and improved technology. In the most recent decade in the Los Angeles air basin, no days exceeded 0.3 ppm. However, heavily polluted air basins in some developing countries still approach these levels.

Later Studies

More recent studies with Valencia oranges on Troyer Citrange rootstock (Olszyk et al. 1990, 1991) differ from those mentioned above and may be more indicative of current conditions (fig. 28.4). Levels of oxidants were varied using charcoal-filtered air, undiluted ambient air, and a 50:50 mixture of the two. Vegetative growth, canopy volume, and total mass of abscised leaves of the trees were not influenced by ozone. However, ozone increased the weight of individual leaves, particularly between April and October, and decreased yield, averaged over 2 on-crop years of alternate-bearing trees by 31%. This was associated with reductions in both number and size of individual fruit (Olszyk et al. 1990), despite the lack of impact on vegetative growth that would be visible to a grower.

Ozone reduced vegetative growth in Ruby Red grapefruit (fig. 28.5). Trees were exposed to ozone for 8 months in field exposure chambers. There was a substantial effect of rootstock on ozone sensitivity, which was unexpected since ozone cannot penetrate the soil (Eissenstat et al. 1991). Growth on Sour Orange rootstock led to much-reduced leaf area production (fig. 28.5A) and stem biomass production (fig. 28.5B), relative to the same scion grafted to Volkamer Lemon rootstock. The more physiologically active and more rapidly growing Ruby Red–Volkamer trees exhibited considerably greater sensitivity to ozone than the less active and less responsive Ruby Red–Sour Orange trees. This is consistent with information for other crop species, suggesting that plants that grow more rapidly are more sensitive to ozone. This may reflect greater stomatal opening to support photosynthesis, allowing greater entry of carbon dioxide but also of ozone. Alternatively, it may reflect a lack of physiological hardening in rapidly growing tissues. Whatever the mechanism, on Volkamer rootstock, these grapefruit trees exhibited a decrease in leaf production of about 20% in ambient air and up to about 35% at twice-ambient ozone concentrations (fig. 28.5A). A similar pattern was observed for stem elongation growth (fig. 28.5B). In contrast, on Sour Orange rootstock there were few consistent differences associated with increasing ozone concentration.

Figure 28.4 Ozone symptoms on a mature leaf cluster of Valencia orange showing adaxial (upper) surface bronzing and chlorotic mottling. *Photo:* R. Platt.

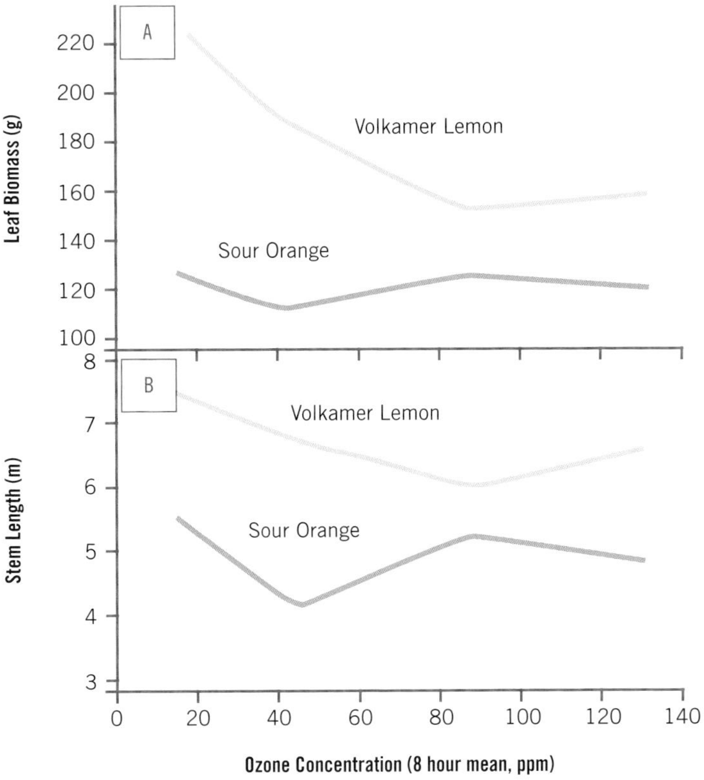

Figure 28.5 Effect of ozone exposure on (A)leaf biomass and (B)stem elongation of Ruby Red grapefruit on two rootstocks. Ozone had a significant effect on leaf biomass and stem elongation in trees grafted to Volkamer Lemon but not on trees on Sour Orange. *Source:* Adapted from Eissenstat et al. 1991, with ozone exposure estimated from data presented.

Fruit Quality

Fruit quality, specifically fruit size and color (Olszyk et al. 1990, 1991), was also affected by ozone. Valencia oranges had a less intense orange color with increasing ozone concentration. On a green to orange scale, with a range of 3 to 13, ozone reduced orange color from 11.5 to 10.8. Neither total soluble solids nor titratable acidity was affected by ozone concentration in these experiments. Individual fruit weight was reduced (Olszyk et al. 1989).

Yield Loss Equations

By the 1980s the importance of exposing plants to a range of ozone concentrations was recognized. This allowed an understanding of the effects of ozone on yield in any one location where experiments were conducted and provided the data needed to develop yield loss equations. These relationships allowed extrapolation (with caution) of yield losses to new locations where experiments could not be, or had not been, conducted. A study of this type was conducted over several years at the University of California Citrus Experiment Station in Riverside (Olszyk et al. 1990, 1991). For data averaged over 1986 and 1987, yield of Valencia oranges was mathematically related to ozone concentration by the following yield loss equation:

$$\text{Yield} = (53.7\ \text{kg/tree}) - (261.1 \times \text{ozone}). \quad \text{Eq. 1}$$

where 53.7 kg per tree is an estimate of yield in unpolluted air and ozone (in ppm) is the daylight seasonal mean ozone concentration in the environment of the trees. Several independent yield loss relationships are shown (fig. 28.6) for Valencia and navel oranges. Though evaluated in different chamber designs and different locations and years, the similarity of the response relationships is striking.

By assuming a clean air ozone concentration of 0.025 ppm, the ozone sensitivity can be expressed in terms of relative yield loss, as

$$\%\ \text{loss} = 10.8 - 434 \times \text{ozone}. \quad \text{Eq. 2}$$

This relationship shows that fruit yield was substantially affected by ozone. Fruit numbers were also reduced by ozone, according to the following loss equation:

$$\text{Number} = (83\ \text{fruit/tree}) - (245 \times \text{ozone}). \quad \text{Eq. 3}$$

Later studies (Leung et al. 1982) evaluated the relative sensitivity to ozone of production in lemons, navel oranges and Valencia oranges. Estimated losses due to ozone in 1975 in Orange County, which by then had not yet become urbanized, were about 1.7% for lemons and 12.7% in Valencia oranges. In the already heavily polluted San Bernardino area, these estimated losses were about 14.4% for lemons and 38.1% for Valencia oranges (Leung et al. 1982).

Using the yield loss equation (Eq. 1) developed by Olszyk et al. (1990) for Valencia oranges, an approximate yield loss for oranges under today's conditions can be calculated for the eastern San Joaquin Valley. Assuming the same background ozone concentrations of about 0.025 ppm in unpolluted air (there is always a little ozone in ambient air) and using the 2008 seasonal 8-hour mean ozone concentration of 0.075 ppm (Parlier; mid-June through mid-September) to represent citrus production areas in the San Joaquin Valley, one can calculate a yield loss of about 28%. Because ozone concentrations vary considerably from day to day, and because it is not fully known which part of the year to consider for evergreen citrus trees, it is difficult to generate accurate predictions of yield loss. However, these numbers suggest that ozone may remain a significant determinant of citrus yield in the San Joaquin Valley under current conditions.

Yield loss functions for citrus and other crops have been developed using a wide variety of exposure chambers, with different chamber designs and in different locations and years. The resulting ozone sensitivities are generally similar (Heck et al. 1988) and similar to more recent results obtained with alternative open-air exposure systems. In the majority of cases, crop sensitivity to ozone is similar inside and outside the chambers, with plants inside the chambers often underestimating the true sensitivity. In the study of Thompson et al. (1970), the outside-grown trees suffered a yield loss of 76%, slightly greater than the 61.8% yield loss in the chambers exposed to the ambient treatment. Similarly, in the studies of Olszyk et al. (1989, 1990, 1992), yield of the trees in the ambient chambers (exposed to unfiltered air) was greater than the yield of similar trees in ambient air outside the chambers. Like all experimental manipulations, use of open-top exposure chambers may alter the experimental system, but they remain an important method of estimating crop loss to ozone. Recent efforts to express ozone exposure in terms of the amount of ozone that is taken up through the stomata rather than in terms of the atmospheric concentration may further strengthen the linkage between results obtained under a variety of exposure conditions and those in commercial field environments.

Ozone Effects on Citrus Management

Alternate Bearing

Ozone did not have any impact on off-crop fruit drop in oranges in the studies performed in 1986–1987 at the Citrus Experiment Station in Riverside (Olszyk et al. 1990, 1991). This is consistent with the hypothesized role of ozone exposure in reducing carbohydrate supply to developing fruit, since the already-reduced fruit set imposes only a modest drain on plant reserves in an off-crop year.

Water Use

Ozone was shown to have a significant impact on the water requirements of a citrus crop. In the studies of Thompson et al. (1967), the rate of water use by lemon trees was reduced with increasing concentrations of ambient air pollutants (fig. 28.7). The trees were exposed in the field environment, enclosed in plastic-covered greenhouses. The reductions in water use were associated with reductions in photosynthesis, suggesting that ozone (most likely) reduced stomatal conductance.

Averaged over the Cucamonga and Upland sites, and over lemons and oranges, the amount of water use by citrus declined with increasing ozone exposure. Water use was estimated as supplemental irrigations required to maintain the soil-water potential above about 50 centibars. Between 1962 and 1966, trees on a 15-day irrigation cycle required only 5 supplemental irrigations in ambient Los Angeles air but 29 supplemental irrigations in ozone-free filtered air. Under severely water-limited conditions, ozone could reduce water use and thereby avoid inhibition of photosynthesis by water stress. In these irrigated experiments, the trees growing in filtered air exhibited greater photosynthesis on a leaf-area basis and were able to fix more carbon dioxide into sugars on a whole-tree basis, and they produced a bigger crop. In addition to inhibiting photosynthesis in individual leaves, ozone also causes premature leaf senescence and abscission, which reduced leaf area duration. For example, in lemon trees, exposure to ambient air caused twice the number of leaves to drop as in filtered air between September and April, reducing the photosynthetically active leaf area and further reducing potential productivity.

In the studies of Thompson et al. (1970), navel orange water use was again related to the concentration of air pollutants, and photosynthesis was

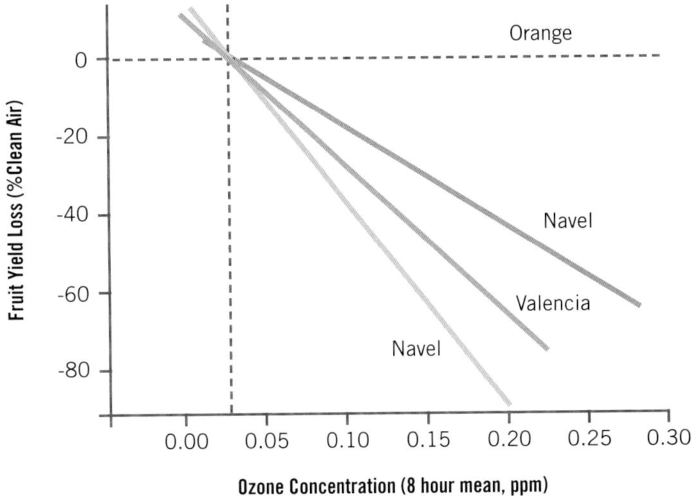

Figure 28.6 Ozone-yield response curve for two types of oranges. Data for the upper navel orange curve are from Thompson et al. 1972; data for the lower navel orange curve are from Kats et al. 1985 and Olszyk et al. 1988; data for the Valencia orange curve are from Olszyk et al. 1990. Note the general similarity despite the differences in genotype, location, and the years of the experiments.

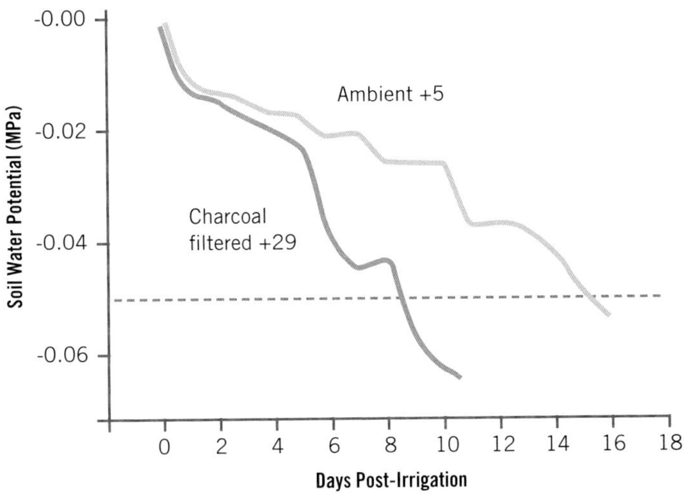

Figure 28.7 Effect of ozone on water use in lemons. Note the more rapid extraction of soil-water by lemon trees grown in clean air than by those grown in polluted ambient air. The red dotted line indicates the soil-water potential at which supplemental irrigations were applied (5 irrigations for ambient trees and 29 for charcoal-filtered trees). *Source:* Adapted from Thompson et al. 1967.

reduced by 34% in ambient air. There were six treatments in this study, including one group of trees that were exposed to (highly artificial) pure charcoal-filtered air. This group of trees used significantly more water than the other treatments.

There are fewer studies evaluating ozone sensitivity of citrus species other than oranges and lemons. The studies of Matsushima et al. (1985) evaluated transpiration and stomatal responses of Satsuma mandarin to short-term exposure to a range of ozone concentrations. In these studies, the relative tolerance of citrus to ozone was attributed to rapid stomatal response to ozone. Equilibrium stomatal conductance (inferred from leaf temperature) was strongly related ($r^2 = 0.72$) to ozone concentration over the range of 0 to 1.2 ppm. Stomatal closing responses to the introduction of ozone in the leaf environment began almost immediately, and equilibrium values were generally attained within approximately 8 minutes. This is more rapid than stomatal kinetics in many other species (e.g., soybean; see Grantz and Zeiger 1986).

It is interesting to note that while ozone reduced stomatal conductance and water use by orange trees (Olszyk et al. 1991), it also caused plant water potential to become more negative, or drier. This usually indicates more, not less, water use. In the case of these orange trees, ozone stress increased the difficulty of acquiring water from the soil. This is consistent with an ozone impact on root development. Inhibition of sugar transport, from the source leaves down to sink tissues such as fine roots, has been demonstrated in cotton (Grantz and Farrar 1999). Ozone was also shown to reduce the hydraulic conductance of these roots (Grantz and Yang 1996). This is an area of current research that in the future may lead to a better understanding of the mechanism of ozone action on citrus productivity. This knowledge will be required as a first step toward developing trees with improved resistance to ozone and to developing management strategies to cope with damaging ambient levels of ozone.

The effects of ozone on water use in citrus are attributable to the reduced rate of leaf production and increased leaf abscission, to the extent of stomatal closure in response to ozone, and to reduced root hydraulic capacity and soil exploration.

Pest Management

Chronic exposure of Valencia orange trees to ambient or charcoal-filtered air for 4 years (Fenn et al. 1989) reduced the total population of leaf surface microorganisms. Indices of diversity were not substantially altered, indicating that all species were reduced to approximately the same extent. However, in studies with other crop plants, differential species sensitivity among microorganisms to ozone led to changes in pathogen-antagonist relationships that could lead to difficulties with biological control (Hibben and Taylor 1970). Short-term exposures to these ambient concentrations of ozone had little effect on the microbiota (Fenn et al. 1989). The range of services provided by these leaf surface organisms is not well understood, and ozone impacts on their abundance and diversity could impact nutrient relationships, sequestration of leaf-deposited materials in the canopy, and resistance to exotic pest species.

Ozone may alter the competitive relationships between crop and weed species, particularly below ground (Grantz and Shrestha 2005). Some effects may be explained by the relative sensitivity of root growth to ozone in the competing species. Other effects are due to changes in seed or tuber production, and therefore in weed spread, which may have serious consequences for management strategies. These interactions remain to be explored specifically in citrus.

Similarly, there have been no studies of ozone impacts on insect pests or root pathogens in citrus. Studies in other crops suggest that there may be significant interactions between ozone and pest species. These are typically mediated by altered movement of sugars to the roots, affecting predation below ground, and altered fecundity of insect pests, altering predation above ground. Further work in citrus on these important potential interactions may become necessary as ozone encroaches upon rural production areas.

Conclusions

Much of the citrus industry in California has moved from the Los Angeles basin up into the San Joaquin Valley. While this has largely been driven by urban sprawl in southern California, it has also had the effect of removing citriculture from an area with the nation's highest ozone air pollution to a region that is considerably less polluted. However, rising population and increased vehicle miles traveled in the San Joaquin Valley have created an ozone problem in this still largely rural area. As urbanization continues and population shows no sign of leveling off, it is clear that ozone concentrations are likely to remain a substantial challenge to the production of high-quality food and fiber in this highly productive agricultural region.

A subject that has received relatively little research attention, and deserves more, is the interaction between ozone and biotic stressors. In citrus these interactions may be very important. It is not known whether plants that are stressed by ozone are more or less susceptible to mites, thrips, or other insect pests. It is not known whether there is increased or decreased susceptibility to viral challenges such as tristeza or to root pathogens such as nematodes or *Phytophthora*. Given ozone impacts on the sugar budgets of affected plants, and particularly on roots, these are areas of potentially serious impact.

Acknowledgments

The author is grateful for the assistance of Kris Lynn of the Geographic Information System facility at the Kearney Agricultural Center in developing the spatial analyses and to Gwen Conville of the Graphic Arts facility at Kearney for assistance with preparation of the figures.

References

Costello, L. R., E. J. Perry, N. P. Matheny, J. M. Henry, and P. M. Geisel. 2003. Abiotic disorders of landscape plants: A diagnostic guide. Oakland: University of California Division of Agriculture and Natural Resources Publication 3420.

Darley, E. F., C. W. Nichols, and J. T. Middleton. 1966. Identification of air pollution damage to agricultural crops. California Department of Agriculture Bulletin 55:11–19.

Dreistadt, S. H., J. K. Clark, and M. L. Flint. 2004. Pests of landscape trees and shrubs. Oakland: University of California Division of Agriculture and Natural Resources Publication 3359.

Eissenstat, D. M., and J. P. Syvertsen. 1991. Interaction of simulated acid rain with ozone on freeze resistance, growth, and mineral nutrition in citrus and avocado. Journal of the American Society for Horticultural Science 116:838–845.

Eissenstat, D. M., J. P. Syvertsen, T. J. Dean, G. Yelenosky, and J. D. Johnson. 1991. Sensitivity of frost resistance and growth in citrus and avocado to chronic ozone exposure. New Phytologist 118:139–146.

Fenn, M. E., P. H. Dunn, and D. M. Durall. 1989. Effects of ozone and sulfur dioxide on phyllosphere fungi from three tree species. Applied and Environmental Microbiology 55:412–418.

Flagler, R. B. 1998. Recognition of air pollution injury to vegetation: A pictorial atlas. 2nd ed. Pittsburgh, PA: Air and Waste Management Association.

Grantz, D. A., and J. F. Farrar. 1999. Acute exposure to ozone inhibits rapid carbon translocation from source leaves of Pima cotton. Journal of Experimental Botany 50:1253–1262.

Grantz, D. A., and M. J. Sanz. 2004. Common co-occurrence of citriculture and ozone air pollution: Potential for yield reductions. Paper No. 97. Proceedings of the 10th Congress of the International Society of Citriculture.

Grantz, D. A., and A. Shrestha. 2005. Ozone reduces crop yields directly and alters crop competition with weeds such as yellow nutsedge. California Agriculture (April-June): 137–143.

Grantz, D. A., and S. Yang. 1996. Effect of O_3 on hydraulic architecture in Pima cotton: Biomass allocation and water transport capacity of roots and shoots. Plant Physiology 112:1649–1657.

Grantz, D. A., and E. Zeiger. 1986. Stomatal responses to light and leaf-air water vapor pressure difference show similar kinetics in sugarcane and soybean. Plant Physiology 81:865–868.

Heagle, A. S., R. B. Philbeck, and W. W. Heck. 1973. An open-top chamber to assess the impact of air pollution on plants. Journal of Environmental Quality 2:365–368.

Heck, W. W., and C. S. Furiness. 2001. The effects of ozone on ecological systems: Time for a full assessment. Environmental Manager (October): 15–24.

Heck, W.W., O. C. Taylor, and D. T. Tingey, eds. 1988. Assessment of crop loss from air pollutants. London: Elsevier.

Heggestad, H. E., and J. T. Middleton. 1959. Ozone in high concentrations as cause of tobacco leaf injury. Science 129:208–210.

Hibben, C. R., and M. P. Taylor. 1970. Ozone and sulphur dioxide effects on the lilac powdery mildew fungus. Environmental Pollution 9:107–114.

Jacobson, J. S., and A. C. Hill. 1979. Recognition of air pollution injury to vegetation: A pictorial atlas. Pittsburgh, PA: Air Pollution Control Association.

Kats, G., D. M. Olszyk, and C. R. Thompson. 1985. Open-top experimental chambers for trees. Journal of the Air Pollution Control Association 35:1298.

Leung, S. K., W. Reed, and S. Geng. 1982. Estimations of ozone damage to selected crops grown in Southern California. Journal of the Air Pollution Control Association 32:160–169.

Matsushima, J., K. Yonemori, and K. Iwao. 1985. Sensitivity of Satsuma mandarin to ozone as related to stomatal function indicated by transpiration rate, change of stem diameter, and leaf temperature. Journal of the American Society of Horticultural Science 110:106–108.

Middleton, J. T., J. B. Kendrick, and H. W. Schwalm. 1950. Injury to herbaceous plants by smog or air pollution. Plant Disease Report 34:245–252.

Olszyk, D. M., H. Cabrera, and C. R. Thompson. 1988. California statewide assessment of the effects of ozone on crop productivity. Journal of the Air Pollution Control Association 38:928–931.

Olszyk, D. M., B. K. Takemoto, and M. Poe. 1991. Leaf photosynthetic and water relations responses for Valencia orange trees exposed to oxidant air pollution. Environmental and Experimental Botany 31:427–436.

Olszyk, D. M., G. Kats, C. L. Morrison, P. J. Dawson, I. Gocka, J. Wolf, and C. R. Thompson. 1990. Valencia orange fruit yield with ambient oxidant or sulfur dioxide exposures. Journal of the American Society for Horticultural Science 115:878–883.

Olszyk, D. M., B. K. Takemoto, G. Kats, P. J. Dawson, C. L. Morrison, J. Wolf Preston, and C. R. Thompson. 1992. Effects of open-top chambers on 'Valencia' orange trees. Journal of Environmental Quality 21:128–134.

Richards, B. L., J. T. Middleton, and W. B. Hewitt. 1958. Air pollution with relation to agronomic crops v. oxidant stipple of grape. Agronomy Journal 50:559–561.

Thompson, C. R. 1969. Effects of air pollutants in the Los Angeles basin on citrus. Proceedings of the 1st International Citrus Symposium 2:705–709.

Thompson, C. R., and O. C. Taylor. 1969. Effects of air pollutants on growth, leaf drop, fruit drop, and yield of citrus trees. Environmental Science and Technology 3:934–940.

Thompson, C. R., G. Kats, and E. Hensel. 1971. Effects of ambient levels of NO_2 on navel oranges. Environmental Science and Technology 5:1017–1019.

———. 1972. Effects of ambient levels of ozone on navel oranges. Environmental Science and Technology 6:1014–1016.

Thompson, C. R., O. C. Taylor, and B. L. Richards. 1970. Effects of photochemical smog on lemons and navel oranges. California Agriculture 24(5): 10–11.

Thompson, C. R., O. C. Taylor, M. D. Thomas, and J. O. Ivie. 1967. Effects of air pollutants on apparent photosynthesis and water use by citrus trees. Environmental Science and Technology 1:644–650.

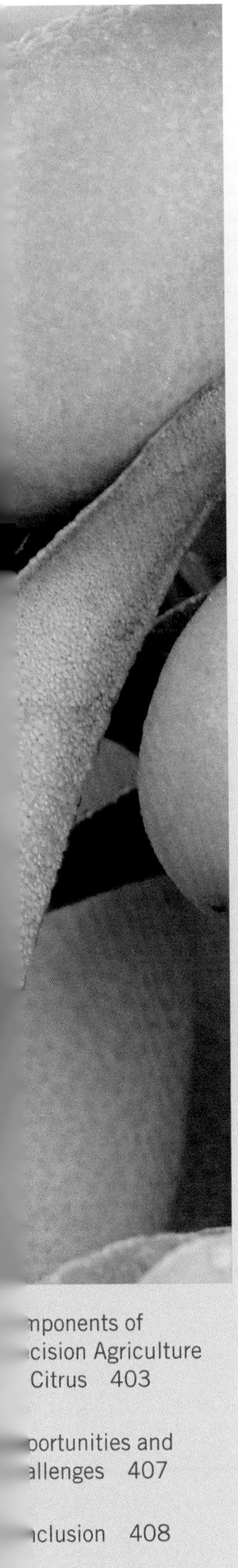

29 Precision Agriculture

Reza Ehsani and Patrick Brown

Precision agriculture is a relatively new farm management concept in which the farm manager uses site-specific information as a basis for management decisions. The goal of precision agriculture is to manage each crop production input on a site-specific basis to reduce waste, increase profits, and maintain the quality of the environment. These goals can be achieved by monitoring local yield, local environmental conditions, and local plant productivity to determine the exact amount and timing of input demands. Precision agriculture uses soil sampling, yield monitoring, remote sensing, and variable-rate applications of herbicide, pesticide, and fertilizer, as well as global positioning systems (GPS) and geographic information systems (GIS), to optimize yield and minimize costs.

Also known as site-specific crop management, precision agriculture was initially developed during the late 1980s in the U.S. Midwest, where it was adopted on large acreages of corn and soybean production. It represents a powerful new advancement in production practices. Perhaps the greatest current constraints to further adoption of precision agriculture are the cost associated with measuring yields and the environmental determinants and designing the means to deliver inputs on a spatially and temporally defined manner. In the broad-acre crops, including soybean and corn, the cost of implementing precision agriculture systems frequently exceeds any benefits to yield or net income, and usage has been limited to fields with the greatest net productivity or those planted to uniquely valuable cultivars (high-oil, high-protein, dedicated end-use products, etc.). Precision technology does not replace conventional farming knowledge and wisdom but complements it by providing better and more accurate information that leads to a better-informed decision.

The high cost of the technology would suggest that adoption of precision agriculture should occur first in the high-value commodities, where the cost of the technology could be justified, or in areas that suffer from significant environmental constraints, including nutrient and pesticide runoff or proximity to residential areas. In this regard, vegetable and orchard crops in California represent a great potential market for this technology.

The lack of a yield monitoring system that allows for individual tree yield determination for many vegetable and orchard crops is a major issue in adoption of precision agriculture for these crops. Since all new management techniques will be evaluated on the basis of their effect on yields, the inability to determine yield on a site-specific basis will entirely constrain the adoption. This constraint is the most important issue limiting adoption of this technology and is perhaps the most important first innovation that will be required if this technology is to be valuable to the industry.

Components of Precision Agriculture for Citrus

Although the precision agriculture concept is the same for all crops, the way it is implemented is crop- and location-specific. Usually, yield monitoring is considered the first step in applying precision agriculture technology. It involves collecting geo-referenced yield data. However, other types of field data are needed in addition to yield data in order to apply precision agriculture. Canopy size and density, pests and weeds, rootstock and tree age, remote sensing images, soil fertility, soil type, soil electrical conductivity, and yield are examples of data that can be used for a citrus precision production system.

Field data can be converted to useful information by conducting a visualization process. For example, raw yield data is just several pages of numbers that in itself provides very little information. By converting the raw data into a map, it becomes a one-page figure that can easily show the areas of high or low yield. The same can be said for data collected for pests and weeds. A large amount of data must be collected in this type of system. Some data may be beneficial, whereas other data may not apply directly to the production operation. The conversion of data to meaningful visual information refines the data into a form that allows the grower to ask so the grower may ask "how" and "why" questions.

The second process is converting the information into knowledge. Knowledge is the important first step for asking the correct "how" and "why" questions before management decisions are made. In this process, layers of information are integrated to add more value. For example, a yield map only shows low or high yield spots in the orchard, but it does not state why the yield is low at a particular spot or what the causes of yield variability are. To answer these questions, more information is needed. For example, having a tree canopy size and density distribution map can be very useful in explaining the cause of yield variability. The process of converting information into knowledge involves integrating different layers of information using technologies such as GIS in combination with mathematical or statistical models. The information resulting from this step can be used by the orchard manager to make sound and informed management decisions.

What makes the above system work successfully is technology. The level of technology accepted is the determining factor for bringing the largest percentage of success to the orchard production. The hand-held computer is one example of such technology. It can help the grower to collect site-specific information such as weed and pest distribution data and also to organize and effectively manage data. Computer software, including spreadsheets, databases, GIS, and other types of applications, are also needed to organize the data. New, low-cost GPS equipment has given growers the ability to locate position information within a few feet. By combining position data with other field data, growers can use the capabilities of GIS to create useful maps.

Precision agriculture is not a single technology but rather a set of many technologies from which growers can select to form a system that meets their unique needs and management style. Not all technologies that were developed for precision agriculture are directly useful for citrus production. A system can be as simple as a variable-rate fertilizer spreader with a couple of canopy sensors that can apply the fertilizer based on the tree size. Such a system can easily improve profitability, increase efficiency, and improve environmental performance, and the growers can recoup the investment costs of the equipment very rapidly.

Yield Monitoring

Collecting yield data is often considered to be the first step in the application of precision agriculture. Growers are usually aware of the total yield for a given block or entire orchard, but yield variability from tree to tree or row to row is unknown. Yield monitors can provide and document the amount of yield variability at a smaller scale and can lead the way to manage the needs of each individual tree rather than treating the entire block of trees uniformly. Yield monitoring is the process of measuring fruit yield for a given location and integrating it with GPS-obtained coordinate information. Currently, commercial yield monitoring systems are available for hand-harvested citrus fruit that can be installed on trucks and can record the locations and weigh the individual tubs and bins. Yield monitoring systems are not yet commercially available for citrus mechanical harvesting machines, but researchers at the UF-IFAS Citrus Research and Education Center are currently developing such a system. Several mapping software packages are commercially available that can read yield data and create a yield map. Most of these software packages can provide some detailed statistical information regarding the collected data. For example, they can provide maximum, minimum, average, and standard deviation of yield. They can also categorize several years of collected data to create different management zones. In addition, the mapping software can overlay the yield variability maps on aerial images of the orchard to provide an enhanced visualization of the data. This is helpful to understand the cause of yield variability. Eventually, the data will be used to create profitability maps that show where orchard profit was maximum or minimum or which areas lost money. It is also possible to create application maps based on different management zones for different crop inputs. Yield maps from several years of data can also be used to monitor yields and find trends of declining yield for a given area of the orchard. This could be an indication of possible diseases or pests in that area of the orchard.

Figure 29.1A shows a yield map for a section of a citrus orchard that is overlaid on an aerial image. The yield map demonstrates the yield variation pattern, with a high production area roughly in the center of the orchard. The amount of variability between high and low yield is about 50%, which is a very significant difference that was not recognized by the grower, who previously harvested the block as a whole and had not observed this trend. To explain the cause of yield variability, other sources of data and information are usually needed. The major causes of yield variability in a citrus orchard are variability in tree size and volume, soil type, elevation, and disease. Figure 29.1B shows a tree volume variability map for the same area. The tree volume variability map shows a similar pattern as the yield map and can explain the yield variability to a large extent. This map, which was developed at the Citrus Research and Education Center of the University of Florida, is generated by a sensor that can measure tree size. Figure 29.1C shows the soil electrical conductivity (EC) variability map for the same area. Soil EC maps usually correlate well with soil type, especially when soil salinity is not an issue. In this case, EC, to some degree, relates to canopy and yield variability.

In addition to this general trend, information can also be obtained on the performance of individual trees; hence, growers can use this information to remove underproductive trees and to select superior trees for propagation. The potential to identify improved cultivars is very great indeed. With information such as this, growers can investigate the cause of the yield loss by observation, soil tests, alteration of irrigation patterns, and other factors. If irrigation changes can correct the problem, the return on investment can also be calculated by estimating average yield loss, number of acres impacted, and cost of remediation. Since citrus can be alternate bearing, care must be taken in interpreting this data; nevertheless, it is clear that very significant differences in yield potential exist throughout this block. Previously, and consistent with most grower practices, this block was harvested as a single unit, and the yield variability within the field went unrecognized.

The great opportunity (and challenge) of precision farming is the ability to determine the cause of this yield variability and to develop management strategies to alleviate it. This could be achieved by combining yield mapping with other technologies, including remote sensing, soil sampling for pH and nutrient concentration and water availability, and disease and pests analysis. Since orchards are high-cost investments, generate significant income, and have a long lifespan, there is considerable value in optimizing management in the orchard, since the dividends should be realized for all subsequent years of production.

Figure 29.1 (A) Contour yield map of citrus orchard, Fort Basinger, FL. Yield expressed as boxes of fruit (90 lb). (B) Contour map of tree volume (in cubic meters) variation map. (C) Soil electrical conductivity map.

Remote Sensing

Remote sensing is one of the main components of precision agriculture technology. It commonly refers to the multiband images that are taken from a remote platform, such as airplanes, helicopters, or satellites. These images contain spectral reflectance information from the tree canopy and soil. Analyzing these remotely collected multiband images can provide cost-effective and timely information about tree health and soil condition. For example, it can help in spotting drainage problems in the field or detecting moisture stress in orchards. Traditional images were limited to a maximum of five bands (three in visible range and two in the near-infrared range). However, recent advances in technology provide images in a series of narrow and contiguous wavelength bands. These images are known as hyperspectral images, and although they are more costly to acquire, they contain more data and can provide more detailed information than multispectral images.

Citrus growers can use remote sensing to assess tree health and soil condition and to implement recommended remedial actions. Large-scale citrus growers can use this technology for counting the trees and measuring the canopy-covered ground area, eventually using that information for forecasting crop quality and yield early in the season. Insurance companies can use similar information to evaluate tree damage for insurance claims. Figure 29.2A shows an aerial image of a citrus orchard. A software program can use the information contained in the image to count the number of trees (fig. 29.2B) and place a red dot where it was able to recognize a tree. This information is useful for maintaining a tree inventory and determining replacement tree needs.

Variable Rate Technology

Variable rate technology (VRT) for inputs is another important component of precision agriculture that provides economic benefits to growers through reduction in application of fertilizer, agrochemicals, and irrigation water, as well as positive environmental impacts. The variable-rate application of materials can be map based or sensor based. In a map-based system, the controller system uses a prescription map that is loaded into the memory of the controller system to control the amount of input at any given location. A prescription map is a database that determines how much input should be applied at any given point. The controller gets the location information from a GPS, then using the prescription map, controls a valve or an actuator to vary the input application rate on the go. In the sensor-based system, the controller reads the input from a sensor to determine the amount of input and changes the amount of input accordingly; there is no need for a GPS or application map in this system. The map-based system is more common for row crops, where determining the exact amount of input is a function of several factors and it is not possible to measure them on the go.

The sensor-based, on-the-go, variable-rate system is more suitable in tree crops like citrus. For example, in citrus, the amount of fertilizer applied for a tree is a function of tree height and width. Therefore, the fertilizer spreader uses sensors that can detect the height and width (canopy volume) and change the fertilizer rate accordingly. Different types

Figure 29.2 Application of aerial imaging for citrus tree inventory: original picture (A); software-counted image (B).

of sensors are commercially available to measure canopy volume or size. Ultrasonic, photoelectric, and laser sensors manufactured by different companies were tested for VRT applications in citrus. The most common applications are dry inputs (granular fertilizer, soil pesticides, or lime), liquid inputs (liquid fertilizer and pesticides), and irrigation water.

The variable-rate spreader can apply dry granular fertilizer, lime, and certain granular pesticides. Spinner spreaders and pneumatic spreaders are the most commonly used applicators. More than one hopper is usually used to hold the desired inputs. Separate drive units and metering devices for each hopper determine the amount of each type of granule input to be applied. The rate of fertilizer can be changed by the speed of a belt, height of a gate, or rotation of an auger, depending on the type of metering device.

Variable-rate liquid applicators are mainly sprayers or liquid fertilizer applicators that vary the application rate according to the size of the tree in the orchard. The application rate is mainly in volume per tree. On-the-go sensors are generally used to detect the height or canopy volume of the tree and open space between trees. The signal from the sensors is fed to an on-board computer and controller to spray a predetermined amount to the tree. To maintain a preset application rate to the particular tree, travel speed and output rate are monitored so that the lag time between the detection of the tree by the sensor and the application of chemicals is matched. When chemicals are applied on the basis of the height of the tree, zones are created according to heights. When canopy volume is used as the basis of variable-rate application, volume sizes are considered to predetermine the application rate.

Very few studies have been made to assess the exact economic and environmental impact of VRT in citrus orchards. A study by the University of Florida found that a VRT spreader would cost approximately $30,000 to $40,000. Even a 15% reduction in fertilizer use would save $21 per acre; 1,380 acres of use would pay for the spreader (Ehsani et al. 2009).

One of the most significant input costs for orchards is the application of water and nutrients. Incorrect application can result in yield loss, runoff, and enhanced disease pressure (waterlogging, drought, etc.). Optimizing irrigation and fertilization strategies requires information on the yield of each individual tree, which allows the application of inputs to be based specifically on the local demand. While the technology for spatially variable delivery of water (irrigation) and water-dissolved chemical fertilizers (fertigation) does not yet exist, it is of great interest and is a subject of ongoing experimentation. Studies have shown that optimized irrigation using spatially variable systems increases yield more than uniform distribution. Spatially variable management can minimize water usage and runoff, as well as mitigate excess fertilizer application, which is wasteful and can leach into ground water.

The development of technologies to deliver tree-specific doses of fertilizers and water is essential to the full implementation of a precision agriculture system for orchard crops. Most of the current developments in spatially variable irrigation have occurred in center pivot and linear move systems for field crops, although intensive research and development is under way at UC Davis. In orchards, significant variation between trees means individual tree management may be worthwhile. A single pistachio tree can produce more than $5,000 in nuts over its lifetime. Growers have significant incentive to optimize productivity and profit by bringing each tree to its maximum yield potential while minimizing waste.

Opportunities and Challenges

Successful growers make management decisions based on the best available information. Currently, most growers use orchard or block yield information to determine management decisions (fertilization, pruning, pest control, etc.), since that is the smallest unit for which they have information. The ability to know what the yield is at a scale smaller than a whole orchard provides growers with more information and allows them to improve management. Further, the capacity to easily determine yield will provide researchers, growers, and extension agents with a greatly improved ability to conduct research and test new management strategies. The ability for growers to easily test new technologies on their own fields is essential for the adoption of best management practices.

To truly optimize orchard performance and minimize the waste of valuable inputs such as water and nutrients, citrus producers must develop increasingly site-specific management practices. Only by precisely matching tree demand with input supply can efficiency be realized. The potential benefits to the adoption of precision agriculture in citrus are perhaps even greater than they are in pistachio, since systems could be established to also monitor citrus quality, (size, damage, color, etc.), which provides information that is more valuable than yield alone. Citrus production is increasingly subject to constraints of price and environmental compliance that threaten profitability. Citrus is also frequently planted on land that varies in soil type, terrain, and environment, all of which increase the benefits of site-specific management.

To date there has been only very limited adoption of precision farming systems in citrus, with most growers citing the poor level of technology development, the lack of a truly effective yield monitoring system, cost, and the difficulty of interpreting the results as key reasons for their slow adoption (Sevier and Lee 2004).

Several technologies suitable for use in precision agriculture for citrus currently exist, including

- sensor-based variable-rate applicators (e.g., Tree-See)
- prescription map–based variable-rate applicators (e.g., Legacy 6000)
- pest scouting and mapping (e.g., EntoNet)
- weed scouting and mapping
- remote sensing (e.g., aerial or satellite imagery)
- GPS receiver (e.g., boundary mapping)
- soil variability mapping
- water table monitoring (e.g., automated irrigation scheduling)
- harvesting logistics (e.g., mapping Brix, acid, and sugar levels to determine peak harvest time)
- yield monitoring (e.g., Goat yield monitoring system)

The greatest current limitation is a viable yield monitor. Development of such a monitor is simplified in machine-harvested crops such as pistachio, but is also possible in hand-harvested crops, although additional technical difficulties exist.

Conclusion

There is very great potential for improvements in citrus production if a viable precision agriculture system could be developed. Indeed, it is quite likely that stricter environmental legislation will demand that the application of all inputs (water, nutrients, and pesticides) be tightly linked to real tree demand, which is primarily determined by yield. Since growers can currently determine only the yield of large areas, the very real potential exists for many trees to be overtreated while others will be undertreated, resulting in both yield loss and environmental damage. The benefits of adoption of precision agriculture extend beyond mere yield optimization; they greatly enhance the ability of growers and researchers to conduct meaningful research, to identify superior individual trees, and to remove poor trees. Ultimately, technology will allow precision agriculture to be conducted in citrus.

Resources

Companies that sell precision agriculture components for citrus include the following (provided for information only and does not imply a recommendation by the author):

Farm Works Information Management
A Division of Trimble
6795 S. State Road 1
PO Box 250
Hamilton, IN 46742-0250
Tel: 1-800-225-2848
http://www.farmworks.com/

Trimble Navigation Limited
935 Stewart Dr.
Sunnyvale, California 94085
Tel: 408-481-8000
http://www.trimble.com/agriculture/

Tree-See Control Systems
Charles F. Roper
120 S. Dillard St.
Winter Garden, FL 34787
Tel: 1-800-872-5032
http://www.treesee.com

Dickey John Corporations
5200 DICKEY-john Rd.
Auburn, IL 62615
Tel: 217-438-3371
www.dickey-john.com/

TeeJet Technology
PO Box 7900
Wheaton, IL 60189-7900
Tel: 630-665-5000
Fax: 630-665-5292
www.mid-tech.com

Durand Wayland
Rick Cordero
PO Box 1404
Lagrange, GA 30241-1404
http://www.durand-wayland.com

References

Ehsani, R., A. Schumann, and M. Salyani. 2009. Variable rate technology for Florida citrus. University of Florida Agricultural and Biological Engineering Department, Florida Cooperative Extension Service Publication AE444. EDIS website, at http://edis.ifas.ufl.edu/pdffiles/AE/AE44400.pdf.

Sevier, B. J., and W. S. Lee. 2004. Precision farming adoption by Florida citrus producers: Probit model analysis. University of Florida IFAS Extension Publication CIR1461. IFAS website, http://edis.ifas.ufl.edu/ae283.

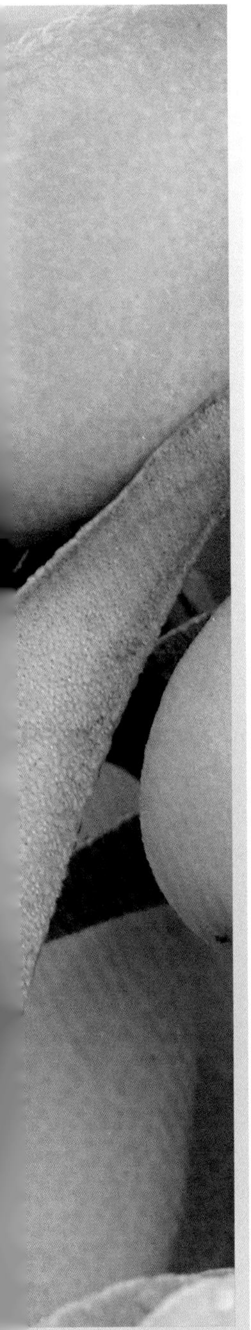

30 Biotechnology

Mikeal L. Roose

Although "biotechnology" refers to the use of organisms to produce useful products, the term is generally used to refer to the development of products or diagnostic methods that employ molecular biology. It is likely that biotechnology will become increasingly important for citrus growers, as it leads to improved varieties, new diagnostic tests for diseases and variety identification, and development of management practices tailored to the physiological state of the tree or fruit. Certain applications of biotechnology are controversial and could affect marketing of citrus fruit. This chapter outlines the major biotechnologies that are being applied in citrus so growers can make better-informed decisions about whether to use this technology. Most research in citrus biotechnology is directed at development of improved varieties, but there are also important potential applications to crop and postharvest management. Progress in plant biotechnology is rapid; see the websites in the sidebar for current, detailed information.

Biotechnology Information on the Web

- USDA-APHIS (regulates transgenic plants): http://www.aphis.usda.gov/publications/biotechnology/index.shtml
- UC Agricultural Biotechnology Information Center (science-based information and resources on agricultural biotechnology): http://ucbiotech.org/index.html
- GMO-Compass (European site on biotechnology issues): http://www.gmo-compass.org/eng/home/
- Free downloads of ANR publications on various aspects of biotechnology: http://ucanr.edu/freepubs/.

Citrus Variety Development

To understand the promise of biotechnology for development of improved citrus varieties, it is essential to understand the history of the citrus varieties we now grow. The citrus varieties currently grown originated from wild species by hybridization and selection of varieties and bud sports with desirable traits (see chapter 4, "Scion Cultivars"). For example, much evidence indicates that the orange originated in China from a hybrid between a mandarin and a pummelo, perhaps followed by crosses to other mandarins. Subsequently, the variety was propagated vegetatively or by nucellar seed, and growers began to identify and select variants with distinctive characteristics such as blood oranges, navel oranges, seedless oranges, and early and late ripening dates. This process explains the many orange varieties that we have today. Similar processes evidently occurred in lemon, grapefruit, and certain mandarin groups such as Satsumas and Clementines. In mandarin and pummelo, many new varieties originated as hybrid seedlings rather than by mutation. Modern breeding by deliberate hybridization and selection has resulted in some successful new varieties, but mainly within the mandarin group. Attempts to improve oranges by hybridization and selection have not been productive because breeders have found it nearly impossible to produce orangelike fruit by hybridizing oranges or by hybridizing among mandarins, oranges, and pummelos. The orange is a unique hybrid. The situation is rather similar to a human couple hoping to produce two successive children who are identical or nearly so. The chance of success in the genetic lottery that determines which particular genes are passed from each parent to a hybrid is too low.

How, then, are we to produce oranges in which various defects are corrected? The traditional approach has been to search for natural mutations that occur as bud sports. However, useful mutations have never been found for many traits, for example, an orange that is resistant to *Phytophthora*, or that peels like a Clementine, or has higher content of certain health-promoting chemicals. Biotechnology should be able to produce varieties with such improvements because it permits scientists to add a specific gene or genes to an existing variety. The genetic lottery is not involved in this process, so the variety will retain its original characteristics, except for those altered by the introduced gene. Another type of application is called marker-assisted selection, in which the breeder uses biotechnology to follow the inheritance of particular genes and thereby improve the odds of success in the genetic lottery.

Genes and Characters

The information that determines the characteristics of a citrus variety is stored in the DNA, a long molecule composed of subunits called nucleotides that come in four different forms, designated by the letters G, C, A, and T. The order in which these various nucleotides occur in DNA determines what particular information is present, just as the order of letters determines what information is in a word or sentence. A typical gene consists of a few hundred to a few thousand of these nucleotides, and each cell of a citrus tree has perhaps 30,000 genes that are assembled into nine different chromosomes. Control sequences, generally located near the ends of the gene, determine when during plant development, in which tissues, or in what environment a gene is expressed. Individual genes can be isolated and changed in the laboratory using the standard techniques of biotechnology. Specific nucleotides in a gene can be changed, and sometimes these changes alter the function of a gene, just as changing letters in a sentence can alter the meaning of the sentence. Various genes and control sequences can be isolated, separated from each other, and put back together in new combinations. For example, a gene normally expressed in roots can be attached to a control sequence that causes it to be expressed in rind tissue. If the gene produces a substance that repels insects, the rind would now contain this substance.

It is clear that these same processes of change (mutation) and rearrangement of DNA occur naturally, but very infrequently. Biotechnology can achieve specific designed changes in the DNA, hence the term "genetic engineering," which is often used to describe this process. Most genes affect traits by specifying a particular protein or enzyme that performs a function inside the cell. Many characters are expressed only when several particular proteins are present in the cell. Changing one gene (and therefore one protein) has the desired effect only if all of the other genes necessary are already present. Thus, genetic engineering of a trait can be complex and is not always successful. As our knowledge of genes and the proteins they specify improves, it is becoming easier to predict the effect of specific changes in DNA on functional characteristics of plants that express the altered gene.

Genetic Transformation

Altering a gene in a test tube in the laboratory does not affect a plant until the gene is placed back into a plant. This is accomplished by the process of genetic transformation, the transfer of an isolated fragment of DNA into an organism where it becomes part of the normal DNA complement. The most common method for transforming plants, and citrus in particular, uses a naturally occurring soil bacterium called *Agrobacterium* to introduce the DNA into plant cells and incorporate the new DNA fragment into the plant DNA (fig. 30.1). If these cells can be induced to regenerate into a whole plant, then all cells of the plant may have the introduced gene. In the laboratory, the scientist can add the desired gene to a specific part of the *Agrobacterium*'s own DNA. Plant cells are then infected with the *Agrobacterium*, and the DNA is transferred into the plant cell. It subsequently becomes part of the normal plant DNA, and as the cells divide, a piece of transformed tissue develops. DNA can also be introduced into plants using the biolistic method, in which the DNA is coated onto tiny beads that are shot into the plant. Occasionally, some of this DNA becomes incorporated into the plant chromosome.

Scientists often link a gene called a selectable marker to their gene of interest in order to obtain a higher frequency of transformed plants. A selectable marker is a gene that gives the transformed cells the ability to survive or grow in the presence of an antibiotic, herbicide, or other stress that kills the normal cells. Only those cells that survive the selection treatment can regenerate into a plant, and these plants will contain both the desired transgene and the selectable marker. As discussed below, the presence of an antibiotic resistance or herbicide resistance gene is likely to complicate or prevent release of new varieties. The process of transformation is somewhat random in that the new gene can be inserted anywhere in the plant's DNA, but the gene may function better in some locations than others. Therefore, it is

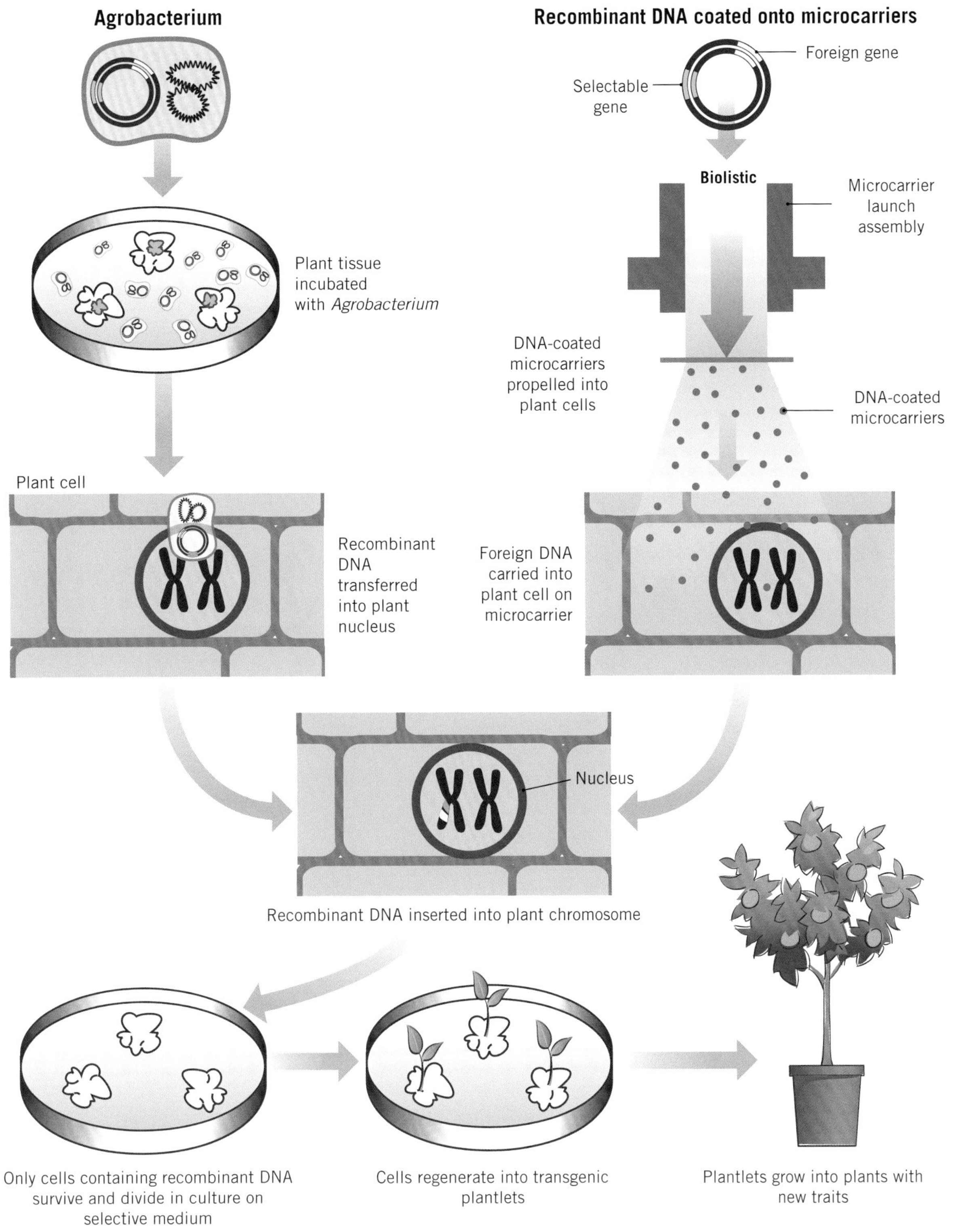

Figure 30.1 Transformation of plant cells and the regeneration of plants from these cells. The diagram illustrates two approaches that can be used to introduce genes into plant cells: *Agrobacterium* transformation (left) and biolistic transformation (right). In both methods, tissue culture is used to select transformed cells and regenerate plants from them.

nearly always necessary to produce many different plants with a given gene and evaluate them to find those that work as expected.

When scientists have produced a transformed (also called transgenic, because it has a transformed gene) plant that may be useful, it must be tested in the field to ensure that it has the expected characteristics and no unexpected ones. As with any citrus field trial, this will take many years. In citrus, a few transgenic varieties are currently in field trials, but none appear close to release. Unless the time to fruiting is reduced, it seems probable that commercialization of transgenic citrus will take at least 10 years. Under current regulations, commercial release of a transgenic variety requires a battery of tests to ensure that the new variety is safe for human health and the environment. This testing can be very costly, so the value of the new variety must be high to justify this expense. The presence of the antibiotic resistance genes added as selectable markers may block release in many countries. New methods are needed to develop transgenic citrus that do not contain such genes.

Cotton, corn, and soybean engineered with genes that cause herbicide and insect resistance have been very successful in the United States and some other countries. Another success story is the development of transgenic papaya that is resistant to papaya ringspot virus. The transgenic variety is generally credited with saving the Hawaiian papaya industry from complete destruction by this virus. Transgenic varieties of many other crops have shown promise in experimental tests, but few have been commercialized, in part due to the regulatory cost involved. Many transgenic citrus trees have been produced and are being tested under containment in laboratories and greenhouses, and a few are now being tested in the field (table 30.1). Transgenic citrus is considered to be one of the most promising methods to maintain a viable industry in the presence of huanglongbing and other exotic pathogens for which strong, natural resistance has not been found in citrus.

Transgenic plants, also called GMOs (genetically modified organisms), are controversial because many consumers, particularly in Europe and some other countries, do not believe that such food is natural or safe. Despite these caveats, many scientists and industry leaders believe that transgenic citrus has great potential to improve our existing varieties. Alternative ways are being sought to alter specific traits. Some of these involve directed mutation, in which a specific gene that already occurs in citrus is changed so that it produces a more desirable property in the organism. These methods are still being researched in experimental organisms, and their application to citrus will be further in the future. They are also more limited in scope than transgenics because they allow alteration only of the genes already in the plant rather than introduction of a completely novel gene.

Experiments are in progress to develop methods of gene silencing using transgenic or virus-infected rootstocks to silence specific genes in a scion. This is possible because genes that exist and are expressed only in the rootstock can produce mobile signal molecules that then alter expression of genes in the scion. This approach would produce fruit that do not carry transgenes, and therefore the cost of testing before release may be much lower than for varieties in which the fruit contains a transgene.

Variety Identification

Biotechnology can also be used for a variety of diagnostic tests, such as identification of varieties. The environment can influence the observable characteristics of a tree or fruit so that it may be difficult to say what variety produced a particular fruit. All cells of an organism have the same DNA, and this should also be true for different trees of the same variety that result from vegetative propagation such as budding. Therefore, if the identity of a tree or fruit

Table 30.1. Transgenic citrus that may reach commercial release by 2020

Trait	Transformed gene	Status in 2013
CTV resistance	parts of viral genome	field trial
CTV resistance	gene from trifoliate orange	lab test
insect resistance	spinach defensin	field trial
insect resistance	snowdrop lily lectin	field trial
HLB resistance	antimicrobial peptides	field trial
HLB resistance	plant defense genes	field trial
canker resistance	antimicrobial peptides	field trial
canker resistance	plant defense genes	field trial
dwarfing	rolABC genes from bacteria	lab test
drought tolerance	proline biosynthesis gene	lab test
salinity tolerance	HAL2 gene from yeast	lab test

is in question, DNA can be isolated from the tree or fruit and compared with DNA from trees or fruit of a panel of possible varieties (fig. 30.2). There are many different ways of comparing two DNA samples. Many tests produce a DNA fingerprint consisting of a number of bands of DNA that looks something like a barcode. These methods test for differences in a very small portion of the total DNA, which includes about 380 million nucleotide letters in citrus. DNA fingerprinting works well for varieties that originate as hybrids, for example, different tangelos, or C35 and Carrizo rootstocks, because such varieties differ for many nucleotide letters (perhaps 1 to 10 differences per 1,000). However, if the variety belongs to a group comprised of different bud sport selections, such as oranges, lemons, most limes, Clementines, or Satsumas, the problem is much more difficult. The DNA of different varieties within these groups is extremely similar. In theory, two different orange varieties might differ by only one of the 380 million letters (the number of letters in about 380 copies of this book). It is thus not easy to find such a difference when we do not know where to look in the DNA book. At present, it is possible to use biotechnology to identify all citrus varieties that have originated by hybridization but very few of those that have originated by selection of mutations. This may change in the future as technology improves. If a genome has already been sequenced once (as is now true for citrus), it may be possible to essentially determine the complete DNA sequence of a new individual for modest ($1,000) cost.

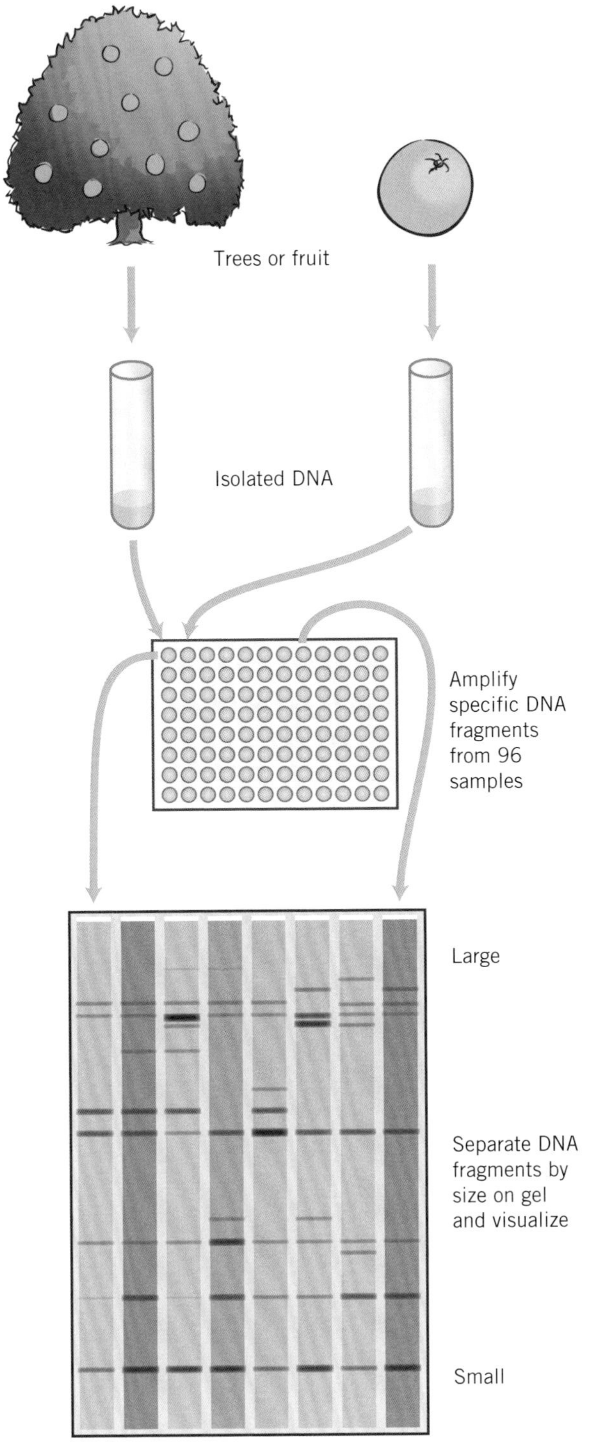

Figure 30.2 Use of DNA fingerprinting to identify citrus varieties. DNA is purified from trees or fruit, then specific DNA fragments are amplified using the polymerase chain reaction (PCR). This produces enough of each specific fragment of DNA to be detected. The fragments are separated by size by loading them onto a porous gel and applying an electric current, which causes the fragments to move through the gel. Smaller fragments move faster because they are less retarded by the pores. DNA specific stains or other methods are then used to visualize the fragments.

Disease Detection

Biotechnology provides excellent tools to determine which, if any, pathogens are present in citrus. This was one of the earliest uses of biotechnology in citrus. For example, detection of citrus tristeza virus (CTV) is routinely done using ELISA, a method of detecting specific proteins produced by the virus. More-sensitive methods have since been developed that detect the genetic material of the virus, and these can be used when necessary. Many other pathogens can also be detected using biotechnology, including exocortis viroid and some exotic diseases not currently present in California.

Tree Management

In the future it may become possible to evaluate the health of citrus trees using biotechnology. Many environmental factors influence tree growth and fruiting. Trees respond to these influences by altering the subset of genes that are expressed—that is, which genes are turned on and to what extent. A tree stressed by lack of water expresses a somewhat different set of genes than one stressed by iron deficiency or salinity. Biotechnology may eventually develop simple, economical tools to monitor expression of genes that indicate these stresses. As with much biotechnology, such developments will come first from research to develop methods to monitor human health. However, because human DNA differs from citrus DNA mainly in the particular sequence of letters present, many of the techniques can be adapted to citrus. The challenge will be for researchers to identify unique signatures for each disease or stress factor and validate these across a range of environments.

Biotechnology has already contributed to better methods to detect certain citrus diseases, and it is being used to develop improved cultivars in citrus and many other crops. It will likely impact growers in the future as it accelerates the development of new varieties and introduces new traits not previously available in citrus. It is likely to lead to improved methods to diagnose plant health, including tests that growers can apply themselves. As with any new technology, there will be risks to growers and industries who adopt it early, but there will also be the potential for a competitive edge.

Measurement Conversion Table

U.S. Customary	Conversion factor for U.S. Customary to Metric	Conversion factor for Metric to U.S. Customary	Metric
		Length	
inch (in)	2.54	0.394	centimeter (cm)
foot (ft)	0.3048	3.28	meter (m)
yard (yd)	0.914	1.09	meter (m)
mile (mi)	1.61	0.62	kilometer (km)
		Area	
acre (ac)	0.4047	2.47	hectare (ha)
square foot (ft^2)	0.0929	10.764	square meter (m^2)
square mile (mi^2)	2.59	0.386	square kilometer (km^2)
		Volume	
fluid ounce (fl oz)	29.57	0.034	milliliter (ml)
quart, liquid (qt)	0.946	1.056	liter (l)
quart, dry (qt)	1.1	0.91	liter (l)
bushel (bu)	0.035	28.37	cubic meter (m^3)
gallon (gal)	3.785	0.26	liter (l)
inch (in)	305	0.00328	millimeter
acre-inch (ac-in)	102.8	0.0097	cubic meter (m^3)
acre-foot (ac-ft)	1,233	0.000811	cubic meter (m^3)
cubic foot (ft^3)	28.317	0.353	liter (l)
cubic yard (yd^3)	0.765	1.307	cubic meter (m^3)
gallon per acre	9.36	0.106	liter per hectare (l/ha)
gallon per ton	4.173	0.239	liter per metric ton
		Mass	
ounce (oz)	28.35	0.035	gram (g)
pound (lb)	0.454	2.205	kilogram (kg)
ton (T)	0.907	1.1	metric ton (t)
pound per acre (lb/ac)	1.12	0.89	kilogram per hectare (kg/ha)
ton per acre (T/ac)	2.24	0.446	metric ton per hectare (t/ha)
		Pressure	
pound per square inch (psi)	6.89	0.145	kilopascal (kPa)
pound per square inch (psi)	0.0689	14.5	bar
		Energy	
BTU	1.055	0.947	kilojoule (kJ)
		Power	
horsepower (hp)	0.745	1.34	kilowatt (kW)
		Temperature	
Fahrenheit (°F)	°C = (°F – 32) ÷ 1.8	°F = (°C x 1.8) + 32	Celsius (°C)

Glossary

abiotic disorder. Disease-like condition caused by nonliving factors in a plant's environment.

anabolism. A series of enzyme-catalyzed chemical reactions that require energy and use small molecules to construct larger, complex organic molecules that comprise the tissues of living organisms.

annual. Plant that normally completes its life cycle in a single year.

anther. The organ of a flower that produces pollen.

Anthophyta. Flowering plants, identified by the their reproductive structure and the presence of a carpel that encloses one to many ovules, which in turn mature into a fruit that encloses one to many seed.

anoxia. Condition in which oxygen is absent.

apothecium. Sexual fruiting body of a fungus in the phylum Ascomycota.

appressorium. Tip swelling of a spore germ tube during early stages of infection.

ascospore. Sexually produced reproductive spore in certain fungi.

autotrophic. Cells or organisms that synthesize energy-rich organic molecules from simple inorganic molecules and energy from sunlight (photoautotrophic) or chemical reactions (chemoautotrophic).

axillary bud. Undeveloped branch or flower tissue (a bud) formed in the narrow angle between a petiole and the stem.

basipetal. Proceeding or developing from the growing tip of a shoot or root toward the main axis of a plant.

biennial. Plant that grows vegetatively in one year and bears fruit and dies in the next year.

biotic disorder. Disease-like condition caused by the living factors in a plant's environment.

broad-spectrum insecticide. Insecticide that kills a large number of unrelated insects.

bud break. Initiation of floral or vegetative growth from a bud.

budding. Propagation by insertion of a desired scion bud into a seedling rootstock.

calyx. Structure in a flower formed by a whirl of sepals.

carpel. Female reproductive structure of the Anthophyta flower, including the stigma, style, and ovary that encloses one to many ovules.

Casparian strip. A band of hydrophobic, waxlike material (suberin) in the cell wall of endodermal cells surrounding the xylem and phloem tissue of a plant root.

catabolism. A series of degradative metabolic reactions catalyzed by enzymes that beak down large, complex molecules into smaller components, typically with the release or transfer of energy.

cellular respiration. The complete oxidation of energy-rich food molecules to carbon dioxide and water using oxygen as the final electron acceptor.

chlamydospore. Thick-walled, single-celled survival spore of certain fungi.

chloroplast. Organelle that contains chlorophyll and other pigments that capture light energy, and the enzymes that catalyze the chemical reactions of photosynthesis.

conidiophore. Simple or branched hypha bearing fertile cells that produce conidia or mitospores.

conidium. Asexual spore of fungi.

cotyledon. The first leaf produced by a seed plant after germination.

cyme. Inflorescence in which each floral shoot terminates in a flower.

determinate. Plant growth in which branches or stems terminate in a floral bud.

dicot. Plant whose embryo has two cotyledons.

DNA. Deoxyribonucleic acid, a nucleic acid found in some viruses, all bacteria, chloroplasts, mitochondria, and animal and plant cell nuclei; the genetic material of all organisms except RNA viruses.

ds-RNA. Double-stranded RNA molecules, which are not present in virus-free plants; their presence can be used as an indicator of virus infection.

ectoparasite. Parasite that lives predominantly on the outside of its host.

ELISA. Enzyme-linked immunosorbent assay, a serological test to determine the presence and amount of a given virus in a sample.

endoparasite. Parasite that lives primarily within its host.

entire. A leaf shape that is smooth, with an undivided outline.

epinasty. Growth in which leaves droop or bend downward, a common symptom of viroid infection.

erumpent. The breaking open of a lesion on a host plant.

exogenous. Introduced from outside an organism.

fastidious. Organisms that cannot be cultured on artificial nutrient media or under laboratory conditions and can survive only in specific environments.

floral intensity. The number of inflorescences per tree.

ground water protection area (GWPA). An area of land that has been determined to be sensitive to the movement of pesticides to ground water.

host race. Nematode populations that are differentiated by their ability or inability to reproduce on selected plant species.

hymenium. Fertile layer in a fungal fruiting structure.

hypha. Growing filament that makes up the mycelium of fungi.

hypoxia. Condition in which the oxygen concentration is lower than normal.

indeterminate. Plant growth in which stems or branches do not terminate in floral buds and in which floral buds are borne laterally.

instar. Life stage of an insect between two molts.

internode. Area of a stem between two nodes.

lanceolate. Leaf shape that is narrow and tapering at both ends.

larva. Immature feeding state of an insect with complete metamorphosis that hatches from an egg and develops into a pupa.

leachate. Portion of rainfall or irrigation water that drains below the crop root zone.

mesophyll. Leaf tissue located between the upper and lower epidermis of a leaf, consisting of palisade mesophyll and spongy mesophyll, in which photosynthesis occurs.

mitochondria. Organelles that contain the enzymes by which chemical energy is produced by the process of cellular respiration.

monocot. Plant whose embryo has a single cotyledon.

monoembryonic. Cultivar in which only one seedling emerges from a seed through strictly sexual seed parentage.

mycelium. Body of a fungus, made up of hyphae.

node. The part of a stem where buds form.

nucellar. Citrus seedling that is genetically identical to its maternal parent and is commonly used for propagation of rootstocks.

nymph. Immature state of an insect with incomplete metamorphosis that hatches from an egg and develops into an adult.

obovoid. Leaf shape resembling an egg, with the narrow end attached to the stem.

organelle. Specialized cell structure resembling an organ.

PAGE and sPAGE. Polyacrylimide gel electrophoresis, a method used to detect viroids.

parasite. Organism that lives on or in the body of another organism (the host) without killing it.

parasitoid. Insect that lays its egg in or on an insect host and whose larva develops by feeding on that host and eventually killing it.

PCR. Polymerase chain reaction, a technique that allows for the detection of genetic material occurring in minute amounts.

perennial. Plant that may live three or more years.

petiole. Stalk that connects a leaf to the stem.

pH. The relative concentration of hydrogen ions in a solution; a measure of the acidity or alkalinity of a solution on a scale of 0 to 14 in which increasing values represent increasing alkalinity.

phenology. The biological stages in the life cycle of a plant or animal.

phialospore. Thin-walled, single-celled asexual dissemination spore produced by fungi.

photosynthesis. The process by which light energy is captured by chlorophyll and other pigments in the chloroplast and converted to chemical energy using water and carbon dioxide to produce sugar.

phytosanitary. Concerning the health of agricultural goods crossing borders, especially their freedom from pests and diseases requiring quarantine.

phytotoxic. Material that damages leaves or fruit of a plant.

polycarpic. Plants that flower and produce fruit and seed many times before dying; perennial plants.

predator. Organism that attacks and consumes its prey.

preemergent herbicide. Herbicide applied before the target plant sprouts from the ground.

postemergent herbicide. Herbicide applied after the target plant sprouts from the ground.

prokaryote. Single-celled organism that lacks a distinct cell nucleus and organelles and whose DNA is not organized into chromosomes.

pupa. The non-feeding, transforming stage of an insect with complete metamorphosis that passes from larva to pupa to adult

pycnidial spores (pycniospores). Asexual spores produced by certain fungal species.

pycnidium. Dark, asexual, spore-producing structure of a fungus.

reactive oxygen species. Chemically reactive oxidizing agents that damage other molecules by removing an electron or proton and interfering with normal metabolism.

resistant. Inherited ability to withstand pesticides or diseases that harm other individuals of the same species.

rhizome. A modified stem that is an extension of the main stem growing underground. It can survive fragmentation such as by tillage operations and can reestablish in new locations.

rhizomorph. Rootlike aggregation of hyphae.

rhizosphere. Soil zone characterized by interactions between living plant roots and microorganisms.

RNA. Ribonucleic acid, a nucleic acid that is the genetic material of most plant viruses and is also involved in protein synthesis.

rosette. A cluster of leaves formed by very short internodes.

scion. The top part of a tree above a graft; a grafted cultivar or variety that bears fruit.

sclerotium. Firm, pigmented mass of hyphae for fungal survival during a dormant stage.

sepal. Leaflike structure that protects a bud.

sessile. Having one or more stages of an insect life cycle fixed in place; settled or unable to move about.

signal transduction. A sequence of events initiated by a signal (typically light, a hormone, or a stress) that ultimately brings about a change in cell metabolism, altering a plant's physiology.

sporangium. Saclike reproductive structure containing zoospores of certain fungal-like organisms.

stele. The central part of a root.

stigma. The uppermost portion of the carpel on which pollen lands and germinates, producing the pollen tube.

stolon. A horizontal modified stem that usually grows along the soil surface and can establish following fragmentation, such as by tillage or mowing.

stomate. A microscopic pore in the epidermis of a leaf through which gas exchange (respiration) is regulated.

stylar. The part of a fruit opposite the stem; point where a flower style joined the ovary.

subpyriform. Somewhat pear shaped.

tolerant. Capacity of an organism to withstand the negative effects of a substance such as a pesticide or conditions such as a harsh environment.

tuber. A modified stem that is the swollen end of a rhizome; usually spherical and produced underground.

turgor pressure (hydrostatic pressure). The force exerted by water in a plant cell.

viroid. A submicroscopic piece of RNA this has no protein coat and can infect plant cells, replicate, and cause disease.

virus. A submicroscopic infective agent containing a protein coat along with DNA and RNA.

zoospore. Motile asexual fungal spore.

zygote. A cell that is the product of fertilization via sexual reproduction and that develops into an embryo.

Index